Lecture Notes in Mechanical Engineering

About this Series

Lecture Notes in Mechanical Engineering (LNME) publishes the latest developments in Mechanical Engineering—quickly, informally and with high quality. Original research reported in proceedings and post-proceedings represents the core of LNME. Also considered for publication are monographs, contributed volumes and lecture notes of exceptionally high quality and interest. Volumes published in LNME embrace all aspects, subfields and new challenges of mechanical engineering. Topics in the series include:

- Engineering Design
- Machinery and Machine Elements
- Mechanical Structures and Stress Analysis
- Automotive Engineering
- Engine Technology
- Aerospace Technology and Astronautics
- Nanotechnology and Microengineering
- Control, Robotics, Mechatronics
- MEMS
- Theoretical and Applied Mechanics
- Dynamical Systems, Control
- Fluid Mechanics
- Engineering Thermodynamics, Heat and Mass Transfer
- Manufacturing
- Precision Engineering, Instrumentation, Measurement
- Materials Engineering
- Tribology and Surface Technology

More information about this series at http://www.springer.com/series/11236

Vidosav Majstorovic · Zivana Jakovljevic
Editors

Proceedings of 5th International Conference on Advanced Manufacturing Engineering and Technologies

NEWTECH 2017

Springer

Editors
Vidosav Majstorovic
University Belgrade
Belgrade
Serbia

Zivana Jakovljevic
University of Belgrade
Belgrade
Serbia

ISSN 2195-4356 ISSN 2195-4364 (electronic)
Lecture Notes in Mechanical Engineering
ISBN 978-3-319-56429-6 ISBN 978-3-319-56430-2 (eBook)
DOI 10.1007/978-3-319-56430-2

Library of Congress Control Number: 2017937923

Printed on acid-free paper

This Springer imprint is published by Springer Nature
The registered company is Springer International Publishing AG
The registered company address is: Gewerbestrasse 11, 6330 Cham, Switzerland

Preface

The 5th International Conference on Advanced Manufacturing Engineering and Technologies (NEWTECH) will be at University of Belgrade, Faculty of Mechanical Engineering, Belgrade, Serbia, from 5 to 9 June 2017. It is organized by the University of Belgrade, Faculty of Mechanical Engineering. Over 100 delegates will be attending NEWTECH 2017—academics, practitioners and scientists from 12 countries—presenting and authoring 36 papers. The Conference Programme includes 11 keynote speeches (opening/closing sessions), breakout sessions and workshop discussions and a closing session on the final day.

The previous conferences on Advanced Manufacturing Engineering and Technologies (NEWTECH) were as follows:

- 1st Conference—2009 at Dunarea de Jos University of Galati, Galati, Romania. Conference Founder and Chair was Prof. Viorel Paunoiu.
- 2nd Conference—2011 at Brno University, Brno, Czech Republic. Conference Chair was Prof. Miroslav Piska.
- 3rd Conference—2013 at KTH Royal Institute of Technology, Stockholm, Sweden. Conference Chair was Prof. Cornel Mihai Nicolescu.
- 4th Conference—2015 at Wroclaw University of Technology, Poland. Conference Chair was Prof. Jerzy Jedrzejewski.

The main goal of the conference is to bring together experts from academia and industry, and it represents an excellent occasion for exchange of knowledge, ideas, experiences, research results and information in the field of manufacturing.

We acknowledge the outstanding contributions of the following keynote and opening/closing sessions' speakers at NEWTECH 2017.

Plenary Session 1: Recent Trends in Manufacturing:

- Steven Y. Liang, "Process and Microstructure in Materials-Affected Manufacturing";
- Viorel Paunoiu, Doina Boazu, "Hydro-Multipoint Forming, a challenge in sheet metal forming";

- Miroslav Piska, Petra Ohnistova, "On the Advanced Milling Technology of Thin-Walled Parts for Aerospace Industry";
- Károly Szipka, Theodoros Laspas, Andreas Archenti, "Mechanistic Approach for the Quasi-Static Capability Evaluation of Machine Tools";
- Katarina Monkova, Peter Monka, Dušan Manduľák, Miroslav Džupon, "Some Aspects Influencing Production of Porous Structures with Complex Shapes of Cells";
- Wit Grzesik, Krzysztof Żak, "High-Precision Machining of Hard Steel Parts Using Special Abrasive Operations".

Plenary Session 2: A Roadmap to Industry 4.0:

- Detlef Kochan, Ronald Miksche, "Advanced Manufacturing and INDUSTRIE 4.0 for SME";
- Zivana Jakovljevic, Vidosav Majstorovic, Slavenko Stojadinovic, Srdjan Zivkovic, Nemanja Gligorijevic, Miroslav Pajic, "Cyber-Physical Manufacturing Systems (CPMS)";
- Slavenko Stojadinovic, Vidosav Majstorovic, Numan Durakbasa, "An Advanced CAI Model for Inspection Planning On CMM";
- Djordje Cica, Branislav Sredanovic, Stevo Borojevic, Davorin Kramar, "An Integration of Bio-Inspired Algorithms and Fuzzy Logic For Tool Wear Estimation In Hard Turning";
- N.A. Fountas, Srđan Živković, R. Benhadj-Djilali, C.I. Stergiou, Vidosav Majstorovic, N.M. Vaxevanidis, "Intelligent CNC Tool-Path Generation Methodology for Sculptured Surfaces by Using a Cyber-Physical Metrology Inspection System".

The conference has the following special, workshop and ordinary sessions:

- Product Design;
- Machining Optimization;
- Product Lifecycle Management;
- Production Planning and Control;
- Cyber-Physical Manufacturing Systems;
- Machine Tools and In-Machine Measurements;
- FabLabs in Science and Education.

NEWTECH 2017 can be regarded as very successful due to its special features: (i) it presents new research directions in the field of advanced manufacturing (bio-inspired manufacturing, Cyber-Physical Manufacturing, Industry 4.0), and (ii) offers practical applications and solutions for various problems in manufacturing.

We express our gratitude to all people involved in conference planning, preparation and realization, and especially to:

- Founder and Honorary Conference Chairs (Prof. Viorel Paunoiu, Prof. Miroslav Piska, Prof. Cornel Mihai Nicolescu and Prof. Jerzy Jedrzejewski).

- All authors, and especially keynote speakers who have contributed to the high scientific and professional level of the conference.
- All members of the International Programme Committee for reviewing the papers and chairing the Conference Sessions.
- Springer and Mr. Pierpaolo Riva for publishing the conference proceedings within the series "Lecture Notes in Mechanical Engineering".
- Ministry of Education, Science and Technological Development of the Republic of Serbia for supporting the conference.

We wish to express a special gratitude to M.Sc. students at the University of Belgrade, Faculty of Mechanical Engineering, Ms. Julija Kostić and Ms. Dragana Jevđović for their effort in preparing and managing the conference in the best way.

The host of the next 6th Conference on Advanced Manufacturing Engineering and Technologies—NEWTECH 2019 will be the Laboratory of Manufacturing Processes and Machine Tools (LMProMaT), School of Pedagogical and Technological Education (ASPETE), Athens, Greece. The Conference Chair will be Prof. Nikolaos M. Vaxevanidis.

Belgrade, Serbia
March 2017

Vidosav Majstorovic
Zivana Jakovljevic

Contents

An Integration of Bio-inspired Algorithms and Fuzzy Logic for Tool Wear Estimation in Hard Turning

Djordje Cica, Branislav Sredanovic, Stevo Borojevic and Davorin Kramar

Abstract One of the most important factors in hard turning is tool wear, since tool condition affects the quality of the product, tool life, and, consequently, the efficiency of the machining process. Modern methods of cooling and lubricating such is high pressure cooling provides possibility to reduce intensive wear of cutting tool due to better penetration of the fluid into the chip-tool and workpiece-tool interfaces. This paper investigates the potential of fuzzy expert system, where the fuzzy system is optimized using two bio-inspired algorithms, namely genetic algorithm (GA) and particle swarm optimization (PSO), for tool wear prediction in hard turning. Experiments have been conducted on a 100Cr6 (AISI 52100) steel workpieces with 62 HRC hardness using inexpensive coated carbide tools under high pressure cooling conditions. The estimated values of tool wear obtained from developed GA and PSO based fuzzy expert systems were compared with the experimental data and very good agreement was observed.

Keywords Fuzzy logic · Bio-inspired algorithms · Tool wear

D. Cica (✉) · B. Sredanovic · S. Borojevic
Faculty of Mechanical Engineering, University of Banja Luka,
Vojvode Stepe Stepanovica 71, 78000 Banja Luka, Bosnia and Herzegovina
e-mail: djordjecica@gmail.com

B. Sredanovic
e-mail: sredanovic@gmail.com

S. Borojevic
e-mail: stevoborojevic@hotmail.com

D. Kramar
Faculty of Mechanical Engineering, University of Ljubljana,
Askerceva 6, 1000 Ljubljana, Slovenia
e-mail: davorin.kramar@fs.uni-lj.si

V. Majstorovic and Z. Jakovljevic (eds.), *Proceedings of 5th International Conference on Advanced Manufacturing Engineering and Technologies*,
Lecture Notes in Mechanical Engineering, DOI 10.1007/978-3-319-56430-2_1

1 Introduction

The hard turning process is defined as the single point turning of materials that are hardened above 45 HRC and up to 65 HRC. This relatively a new approach in the machining area has become possible due to improvements in the performance of the new cutting tool materials. Among available hard turning tool materials, cubic boron nitride (CBN) has been widely used because of its characteristics such is high abrasive wear resistance, high hardness and chemical stability at high temperatures. However, the cost of CBN turning tools is relatively high, so it is necessary to investigate the possibility of using inexpensive cutting tool materials such as coated carbide tools which also can effectively utilize the hard turning process. On other hand, cutting edge of a tool in hard turning is subject to a very high temperatures which causes intensive wear of carbide tools and significant reduction in tool life. Modern methods of cooling and lubricating, where cutting fluids act simultaneously as coolants and lubricants, provides possibility to reduce these intensive wear of carbide tools. High pressure cooling (HPC) allows a better penetration of the fluid into the chip-tool and workpiece-tool interfaces, thus providing a better cooling effect and decrease in tool wear [1–3].

One of the most important factors in hard turning process is tool wear, since tool condition affects the quality of the product, tool life, and, consequently, the efficiency of the machining process. Accurate prediction of remaining useful cutting tool life is critical to achieve to effective condition based maintenance and reducing overall maintenance cost. However, due to very complex and highly nonlinear phenomena involved in wear, predicting or monitoring tool wear is a very difficult task. Nowadays, there have been many successful applications of artificial intelligence (AI) methods such as artificial neural networks (ANN), fuzzy logic (FL), genetic algorithms (GA), particle swarm optimization (PSO), etc. for modelling tool wear in turning process as function of cutting conditions, tool geometry, tool and work materials, etc. A detailed review of papers related to tool wear and tool condition monitoring in turning by means of ANN was presented by Sick [4]. The article compares the methods and methodologies applied in 138 publications used to carry out simulation experiments, to evaluate and to present results, etc. Although prediction of tool wear using AI methods has been extensively studied by many researchers, there are only a few papers dealing with the implementation of these methods for tool wear prediction in hard turning. Predictive models for conventional turning may not be adequate for hard turning, because a conglomeration of phenomena, such as tool wear and chip formation, exhibits unique behaviour not found in conventional turning operations.

Ozel et al. [5] have been investigated tool flank wear and surface finishing in finish turning of AISI D2 steels (60 HRC) using ceramic wiper design inserts. Multiple linear regression models and ANN models were developed for predicting these two parameters. Scheffer et al. [6] were investigated various aspects associated with hard turning in order to develop an accurate system for monitoring tool wear using CBN tools. The novel formulation of the proposed model provide an accurate solution for monitoring crater and flank wear during hard turning. Ozel and Karpat [7] utilizes ANN models to predict tool flank wear and surface roughness for variety of cutting conditions in finish hard turning using CBN tools. The performance of ANN and the regression model have been tested with experimental data, and the ANN model is found to be capable of better predictions for tool wear and surface roughness. Quiza et al. [8] were investigated the influence of cutting speed, feed and time on flank wear in hard turning of D2 AISI steel with ceramic tools using multiple regression analysis and ANN, where parameters of the design and the training process for the ANN model have been optimized using the Taguchi method. Palanisamy and Shanmugasundaram [9] use regression and ANN models for predicting tool wear and surface roughness in hard turning using CBN content inserted tool. The performance of the both models has been tested with experimental data and ANN model has been found to be better predictions for tool wear and surface roughness. Gaitonde et al. [10] employed a multilayer feed-forward back-propagation ANN to analyse the effects of cutting speed, feed, and machining time on specific cutting force, surface roughness, and tool wear during hard turning of AISI D2 cold work tool steel with three different ceramic inserts. Wang et al. [11] designed a novel ANN based estimator for CBN tool wear modelling in hard turning. The estimator is based on a fully forward connected ANN and trained using the extended Kalman filter algorithm in order to speed up the learning convergence. Cica et al. [12] investigate the potential of ANN and adaptive neuro-fuzzy inference system (ANFIS) for tool wear and surface roughness prediction in hard turning operations. The estimation results obtained by both models are compared with experimental results and very good agreement is observed.

As this review of the literature reveals, there has been not much work reported on prediction of tool wear in hard turning using AI methods, compared to conventional turning. Furthermore, all predictive models are based on the use cutting tool materials such as ceramics and CBN. A review of literature cited above also clearly indicates that ANN are most applied AI method for estimation of tool wear. Apart from ANN, fuzzy logic also has great potential for modelling and optimization of machining processes. So the objective of this study is to develop fuzzy expert systems for prediction of tool wear in coated carbide cutting tool during hard turning process under high pressure cooling conditions, where the fuzzy systems are optimized using bio-inspired algorithms.

2 Experimental Design and Set-Up

The machining experiments were conducted in longitudinal turning process on conventional lathe fitted with a high-pressure plunger pump of 250 MPa pressure and 10 l/min capacity. The fluid used was the PRIMOL 3000, a 3% emulsion on the basis of vegetable oil mixed with water. All experiments were performed on 100Cr6 steel with hardness of 62 HRC. Due to high fatigue strength and hardness this material is used mainly for small and medium-sized bearing components. Usually in practice, the 100Cr6 steel is machined with CBN tools. As previously mentioned, CBN tool inserts are much more expensive comparing to the carbide tool inserts. Therefore, it is necessary to investigate the possibility of using inexpensive cutting tool materials such as coated carbide tools in hard turning. The cutting tool inserts used in the experiments were coated carbide cutting tools CNMG 12 04 08 MF5 SECO TH1000 with 0.8 mm nose radius. TH1000 is a TiSiN-TiAlN nanolaminate PVD coated grade.

Machining experiments were conducted according to the following high pressure cooling parameters: pressure 50 MPa and 2.0 l/min jet flow capacity. Because the study was performed at pressure levels much higher than commercially used in practice, smaller nozzle diameters (0.4 mm) were applied. These lead to more precise jet direction which is directed to the cutting edge at the angle of 30° with the rake face at the distance of 30 mm, and smaller momentary lubricant delivery rates for the same jet momentum. Tool wear measurements were acquired with a CCD camera mounted on a Mitutoyo TM510 microscope aided with imaging software.

In this study, five levels of feed (0.08, 0.125, 0.160, 0.180 and 0.200 mm/rev) and four levels of cutting speed (65, 85, 100 and 120 m/min) were used as cutting conditions, while the depth of cut was held constant at a 0.5 mm. Depending on the progress in the amount of wear, the cutting process was interrupted periodically and flank wear depth was measured by using a microscope. In this study, tool life is measured by the machining time taken by the same insert until the flank wear reaches its allowable limit of approximately 0.25 mm or if an interval of time of cutting reaches value of 20 min. The schematic layout of the experimental setup is shown in Fig. 1.

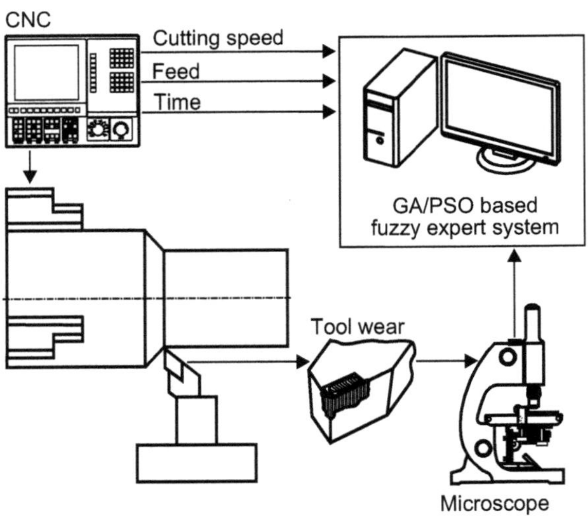

Fig. 1 Schematic layout of experimental setup

3 Development of Fuzzy Expert Systems

It is very well known that fuzzy logic is one of the most successful of today's methods for developing rule-based models of mathematically complex or ill-defined systems. Fuzzy logic provides a formal methodology for representing and implementing a human's heuristic knowledge. Fuzzy logic can model nonlinear functions of arbitrary complexity and therefore it is suitable in defining the relationship between the system inputs and the desired system outputs.

Fuzzy inference system (FIS) is the major unit of a fuzzy logic system that uses if-then rules, and based on these rules the decisions are made. Normally, FIS can be considered to be composed of four components (Fig. 2). They are a fuzzification interface, a knowledge base (fuzzy rule base), an inference system (decision-making logic), and a defuzzification interface.

The fuzzy rule-based system uses rules defined in the form of if-then logical statements, in which the preconditions and consequents involve linguistic variables. Fuzzy rule-based system has been developed to model input-output relationships in hard turning process. There are three inputs, namely, feed, cutting speed and time, and one output, tool wear. A triangular shape of a membership functions (MFs) is employed to describe the fuzzy sets for input and output variables. All variables have been partitioned according to the experiment parameter ranges (Fig. 3).

In the next step, fuzzy rules are formulated to establish the relationship between the input and the output in a fuzzy system. These rules for the knowledge base were usually constructed manually. This task is usually accomplished through an observation of a human expert within a specific domain of knowledge. Thus, the

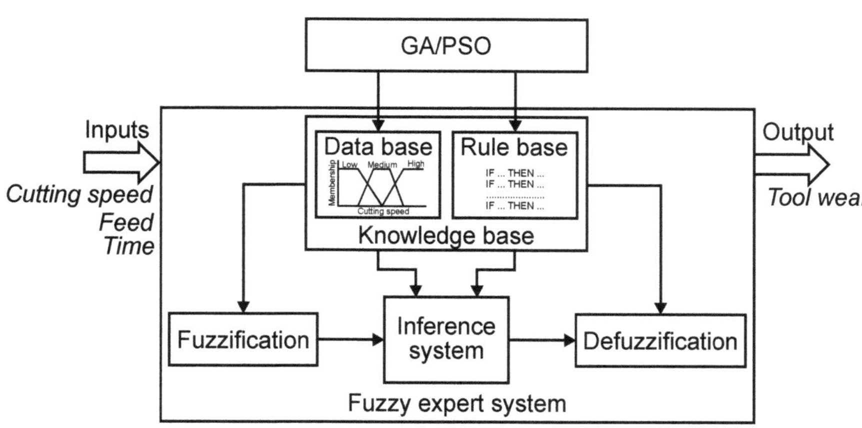

Fig. 2 A schematic diagram showing GA/PSO trained fuzzy expert system

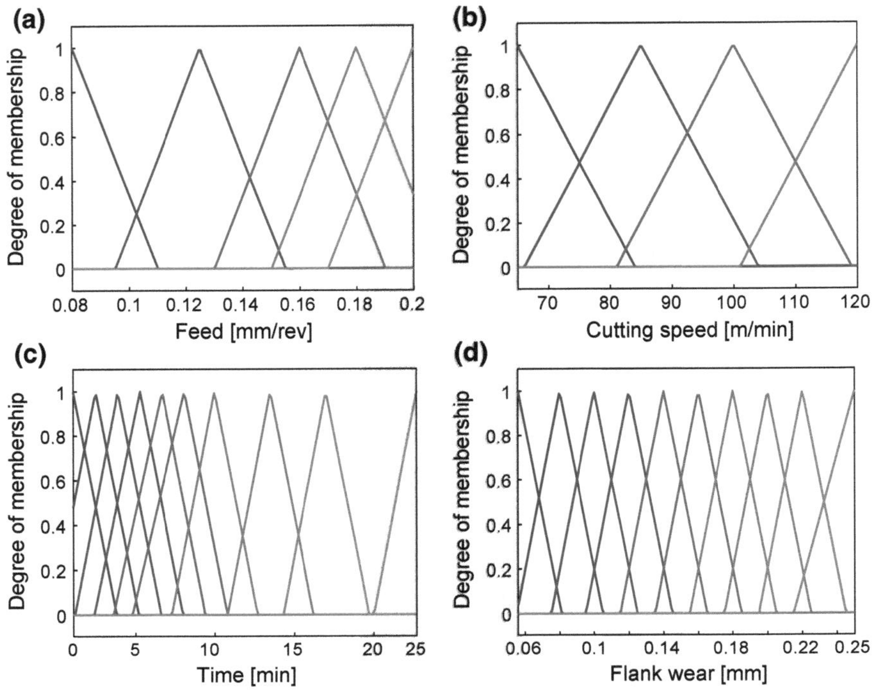

Fig. 3 Membership functions: **a** feed, **b** cutting speed, **c** time, and **d** tool wear

fuzzy rules obtained in these way are strongly depending on the priori knowledge of the human expert. Inadequate fuzzy rules or boundaries of MFs have considerable impact on the final performance of the fuzzy expert system. Finally, the fuzzy rules obtained in these ways are not adaptive. These are main reasons why many researchers working on automatic tuning techniques for fuzzy expert systems.

Recently, bio-inspired algorithms have been shown effective in the derivation and adaptation of the fuzzy rule base. In order to provide an adaptive and self-tuning fuzzy logic expert system, in this paper GA and PSO were used for optimization of an already existing fuzzy expert system through a tuning process. As the Fig. 2 shows, the fuzzy system lies at the core of the hybrid structure. Fuzzy system fuzzifies the input state, performs the inference based on the fuzzy rules and aggregates the result of the inference process into a crisp output, while bio-inspired algorithms are used to improve the performances of the fuzzy system [13].

3.1 GA Based Fuzzy Expert System

Genetic algorithms are optimization algorithms that use principles inspired by natural population genetics to evolve solutions to problems. GA simulates the evolutionary principle of survival of the fittest to produce better approximations to a solution. The GA starts with a randomly generated population of chromosomes as the initial solution and uses the objective function to measure the fitness of each individual in initial population. The GA advances by selecting a group of individuals in the current population to be parents for the next generation according to certain selection criteria and rejecting the remaining solutions. Then new child chromosomes are generated by selection, crossover and mutation and the new population is used in the next iteration of the algorithm. Commonly, the algorithm terminates when it reaches a certain number of generations or a satisfactory fitness level has been reached for the population.

GA have been used successfully in a wide variety of fields, including the optimization of fuzzy rule-based system. There are a lot of parameters in a fuzzy rule-based system that can be tuned in order to improve its performance, but in general, the GA have been used to evolve a fuzzy system by optimizing the data base (DB), rule base (RB), or knowledge base (KB), i.e. both data base and rule base. Data base optimization involves utilization of the genetic tuning process to slightly adjust the shape of the MFs to which the linguistic labels in the fuzzy rules refer, according to a fitness function that specifies the design criteria in a quantitative manner. Since developed fuzzy rule-based system employed triangular MFs, optimization of MFs distributions is reduced to changes of the base widths. Furthermore, fuzzy rule weights were also tuned by GA.

The objective of the search process is to maximize or minimize a fitness function that describes the desired behaviour of the system. The fitness of a GA string is calculated as:

$$f = \frac{1}{n}\sum_{i=1}^{n}\left|\frac{T_i - O_i}{T_i}\right| \tag{1}$$

where T_i and O_i represent the target and predicted values, respectively, and n denotes the number of outputs.

Because performance of a GA depend heavily on its parameters, a very careful study must been carried out to determine the set of optimal parameters, where the value of one parameter is varied at a time, while other parameters have fixed values. The evolutionary parameters were 340 and 510 for population size and number of generations, respectively. The genetic operations reproduction, crossover and mutation were also used. Probability of reproduction, crossover and mutation were 0.12, 0.2 and 0.02, respectively. Figure 4 shows the optimized MFs distributions of the input-output MFs that are obtained after a fine tuning using a GA.

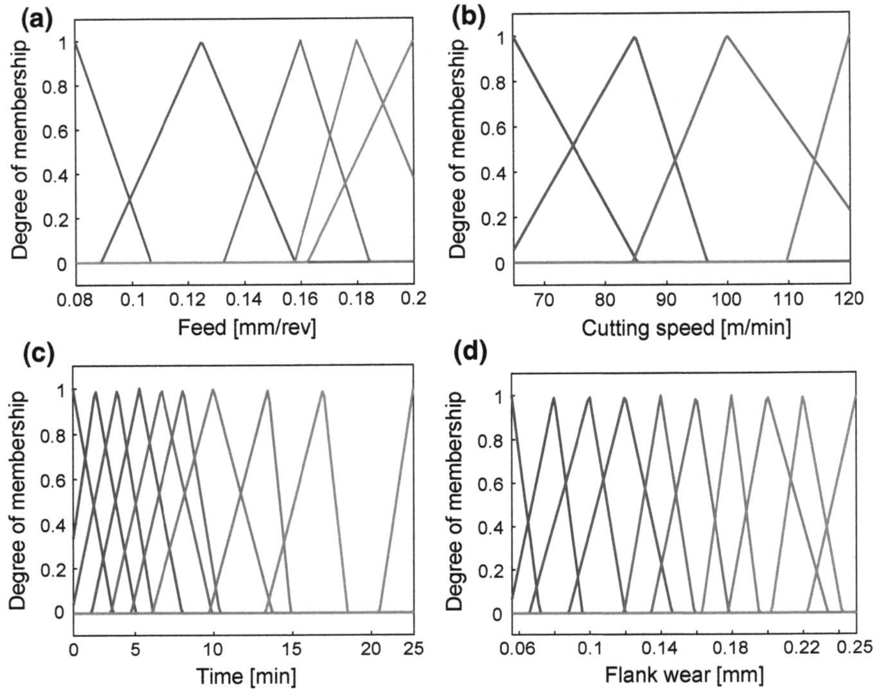

Fig. 4 Genetic algorithm optimized membership function distributions: **a** feed, **b** cutting speed, **c** time, and **d** tool wear

3.2 PSO Based Fuzzy Expert System

Swarm intelligence is an innovative optimization technique that originally took its motivation from the collective behaviour of social insects such as ants, bees as well as animal societies such as flocks of birds or schools of fishes. Particle swarm optimization shares many similarities with genetic algorithms. Like GA, PSO is also start with a population of random solutions and searches for the optimum by updating successive generations. However unlike GA, PSO does not have genetic operators such as crossover and mutation. In standard PSO, the potential solutions (particles) moving interactively through the problem space by following the current optimum particles. Each particle represents a solution to the optimization problem and each particle is influenced on its search direction by cognitive and social information. Cognitive information is related to the best solution or fitness that particle has achieved so far, while social information is associated with the best value obtained so far by any particle in the swarm. The optimization took place by changing the velocity of each particle or accelerating randomly each particle toward its personal best position and global best position of the swarm.

PSO is considered as the one of the most powerful methods for resolving optimization problems and in this paper, PSO is used to optimize data base of the fuzzy expert system. Similar to the previous chapter, a careful study have been carried out to determine the set of optimal parameters. The optimization process takes place with the values of cognitive acceleration, social acceleration, maximum number of generation and population size. The optimal values of cognitive acceleration, social acceleration, number of generations and population size were 0.5, 1.2, 480 and 280, respectively.

4 Results and Discussion

In this chapter the developed GA and PSO based fuzzy expert systems were tested for tool wear prediction in hard turning by comparing the predicted results with the experimental data. In order to estimate the prediction capability of the developed bio-inspired fuzzy expert systems, errors were analyzed using mean absolute percentage error and maximum absolute percentage error. For GA and PSO based fuzzy expert system mean absolute percentage error were 3.7 and 3.6%, respectively. Maximum absolute percentage error for the GA based fuzzy expert system was 22.8%, whereas for the PSO based fuzzy expert system model these value was 17.6%. Therefore it is evident that there is good agreement between estimated and experimental values of tool wear for both models. The author-defined fuzzy expert system, derived from their heuristic knowledge of the hard turning, predicts tool wear with a mean absolute percentage error of 15.4%, while maximum absolute percentage error for this system was 63.8%. Hence, it is evident that the hybridization between fuzzy logic and bio-inspired algorithm can significantly improve the system performance. Figure 5 shows the results of a comparative study between predicted tool wear using proposed GA and PSO based fuzzy expert systems and experimental results. It is obvious that the tool wear depends, in first order, on cutting speed, and also on feeds, but in smaller percentage compared to the influence of cutting speed. This can be explained by intense heat generation that inevitably occurs during hard turning of hard-to-machining materials. The results presented in Fig. 5 indicate that minimal tool wear was obtained when using feed 0.08 mm/rev and cutting speed 85 m/min, while high intensity of tool wear was obtained when using feed 0.125 mm/rev and cutting speed 120 m/min (Fig. 6).

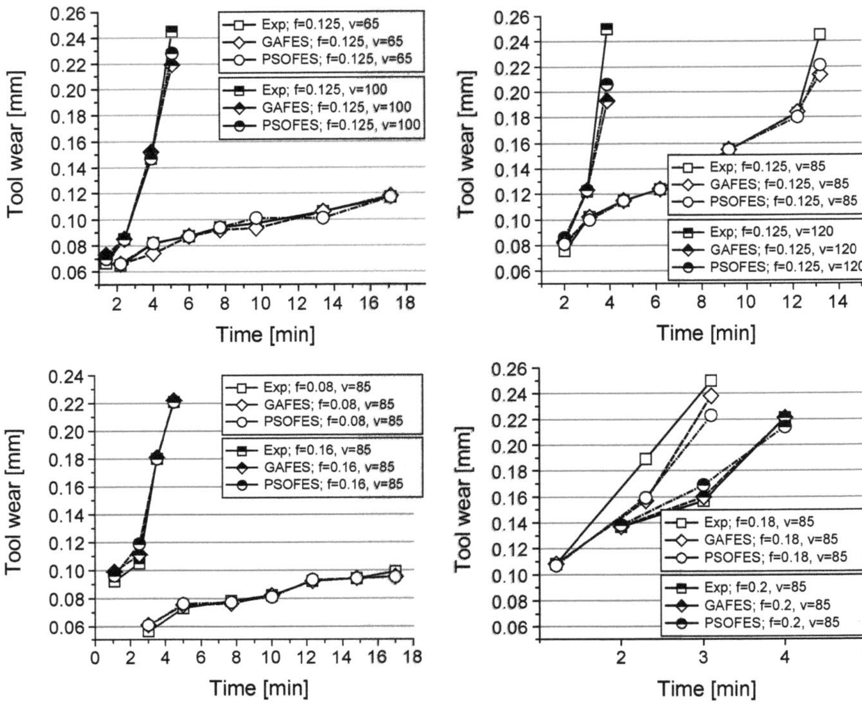

Fig. 5 Comparison of results of GA and PSO based fuzzy expert system for tool wear with experimental results

5 Conclusions

In this paper, two fuzzy expert systems for tool wear prediction in hard turning were developed and discussed. All experiments have been conducted on a 100Cr6 steel workpieces with 62 HRC hardness using inexpensive coated carbide tools under HPC conditions. The fuzzy rules were developed based on actual experimental data that shown the relationship between input parameters represented by cutting speed, feed and machining time with the flank wear as the output response. In order to improve performances of fuzzy expert systems, two bio-inspired algorithms, namely GA and PSO, were utilized for tuning the data base of the author defined fuzzy expert system. The predicted results of tool wear achieved via these two fuzzy expert systems were compared to the experimental dataset. The result demonstrated a very good settlement between the GA and PSO based fuzzy expert systems and experimental results with 96.3 and 96.4% mean accuracy, respectively. Thus, bio-inspired fuzzy expert systems can be used successfully in order to predict tool

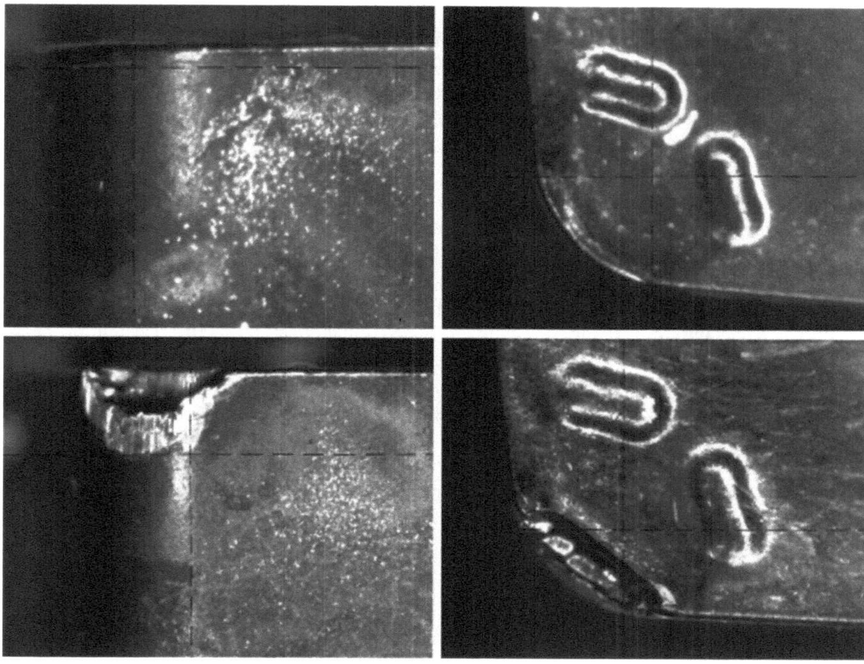

Fig. 6 Flank wear (*left*) and crater wear (*right*) for 0.125 mm/rev feed and 120 m/min cutting speed at the beginning (*above*) and at the end of a cutting process (*below*)

wear in hard turning operations within the range of input parameters under consideration. The results of this study have an interesting advantage for industrial application. It is shown that inexpensive coated carbide tools can be also successfully used for hard turning if appropriate methods of cooling and lubricating, such as high pressure cooling, are applied.

References

1. Ezugwu EO, Bonney J (2004) Effect of high-pressure coolant supply when machining nickel-base, Inconel 718, alloy with coated carbide tools. J Mater Process Technol 153–154:1045–1050
2. Klocke F, Sangermann H, Kramer A, Lung D (2011) Influence of a high pressure lubricoolant supply on thermo-mechanical tool load and tool wear behaviour in the turning of aerospace materials. Proc Inst Mech Eng Part B: J Eng Manuf 225:52–61
3. Kramar D, Krajnik P, Kopac J (2010) Capability of high pressure cooling in the turning of surface hardened piston rods. J Mater Process Technol 210:212–218
4. Sick B (2002) Online and indirect tool wear monitoring in turning with artificial neural networks: a review of more than a decade of research. Mech Syst Signal Process 16:487–546
5. Ozel T, Karpat Y, Figueira L, Davim P (2007) Modelling of surface finish and tool flank wear in turning of AISI D2 steel with ceramic wiper inserts. J Mater Process Technol 189:192–198

6. Scheffer C, Kratz H, Heyns PS, Klocke F (2003) Development of a tool wear-monitoring system for hard turning. Int J Mach Tool Manuf 43:973–985
7. Ozel T, Karpat Y (2005) Predictive modeling of surface roughness and tool wear in hard turning using regression and neural networks. Int J Mach Tool Manuf 45:467–479
8. Quiza R, Figueira L, Davim JP (2008) Comparing statistical models and artificial neural networks on predicting the tool wear in hard machining D2 AISI steel. Int J Adv Manuf Technol 37:641–648
9. Palanisamy P, Shanmugasundaram S (2010) Modelling of tool wear and surface roughness in hard turning using regression and artificial neural network. Int J Mach Mach Mater 4:76–94
10. Gaitonde VN, Karnik SR, Figueira L, Davim P (2011) Performance comparison of conventional and wiper ceramic inserts in hard turning through artificial neural network modeling. Int J Adv Manuf Technol 52:101–114
11. Wang X, Wang W, Huang Y, Nguyen N, Krishnakumar K (2008) Design of neural network-based estimator for tool wear modeling in hard turning. J Intell Manuf 19:383–396
12. Cica D, Sredanovic B, Kramar D (2015) Modelling of tool life and surface roughness in hard turning using soft computing techniques: a comparative study. Int J Mater Prod Technol 50:49–64
13. Kramar D, Cica D, Sredanovic B, Kopac J (2016) Design of fuzzy expert system for predicting of surface roughness in high-pressure jet assisted turning using bioinspired algorithms. Artif Intell Eng Des Anal Manuf 30(1):96–106

Hob Mill Profiling Method for Generation of Timing Belt Pulley

Nicuşor Baroiu, Virgil Teodor, Florin Susac and Nicolae Oancea

Abstract In this paper is proposed an algorithm based on the complementary method of surface en wrapping the method of generating relative trajectories for profiling hob mill designed for machining of timing belt pulleys, with circle frontal profile. A profiling algorithm for hob mill is proposed. The algorithm is developed in analytical form and is compared with a graphical method developed in CATIA design environment for a real example. The algorithm starts with analytical equations determination of profile to be generated. For this profile is determined the family of trajectories described in the relative motion between tool and blank. The family of normal's to these trajectories is determined and is selected the normal which pass through the gearing pole. The point corresponding to this normal belongs to the characteristic curve, so it will belong to the tool's surface too. In this way, the profile of intermediary surface is determined. The intermediary surface is an auxiliary surface which represents the surface of rack gear tool generating the desired profile. Consequently, knowing the form of the generating rack gear, the profile of hob mill is determined. For this, a helical motion is applied to the rack gear's profile. The axial form of hob mill is presented as base for generation of this tool's type.

Keywords Hob mill profiling · CATIA · Timing belt pulley · Machining technology

N. Baroiu (✉) · V. Teodor · F. Susac · N. Oancea
Department of Manufacturing Engineering, "Dunarea de Jos" University of Galati, Domneasca Street 111, 800201 Galati, Romania
e-mail: Nicusor.Baroiu@ugal.ro

V. Teodor
e-mail: virgil.teodor@ugal.ro

F. Susac
e-mail: Florin.Susac@ugal.ro

V. Majstorovic and Z. Jakovljevic (eds.), *Proceedings of 5th International Conference on Advanced Manufacturing Engineering and Technologies*, Lecture Notes in Mechanical Engineering, DOI 10.1007/978-3-319-56430-2_2

1 Introduction

The issue of complex surface generation using profiled tools and, also, CNC machine tools is a hard concern of international research teams.

The machining using the hob mill is a machining process for obtaining ordered curls of profiles as: spur gears; splined shafts; sprocket wheels etc. The process is characterised by an increased productivity, precision corresponding to the IT 7–8 classes and roughness corresponding to R_a = 6.3–12.5 μm.

Regarding others machining processes for teething, the hobbing is less expensive and characterised by a good precision.

The machine-tools are automatic machines and can manufacture a wide dimensional domain, from small parts to pieces with diameter up to 3 m.

The hob mill's teeth successive cut the material and the piece's profile is obtained by rolling.

The primary peripheral surface of tool may be regarded as generated by the surface of a rack-gear with straight teeth (the intermediary surface) which rotate around an axis. This kinematics is on base of the intermediary surface profiling algorithm.

While machining, the tool and the blank have revolving motions around two disjunctive axes. In addition, in order to cut the entire teeth width, the tool and the blank have a translation relative motion, along the piece's axis, simultaneous with the rolling motion.

The hob mill profiling issue for generation of an ordered curl of profiles is an in-plane enwrapping problem, with single point contact. The solving of such problem is done based on the second theorem of Olivier for enwrapping surfaces which depend by two independent parameters.

The second theorem of Olivier for study an enwrapping problem with single-point contact starts from the knowing of a surfaces family which depend by two independents variable parameters, in principle, in form:

$$\left(\Sigma_{\alpha,\beta}\right) : \mathrm{F}(X, Y, Z, \alpha, \beta) = 0, \tag{1}$$

where α and β are independents variable parameters.

The enwrapping of this surfaces' family is obtained according to the second theorem of Olivier by associating, to the surfaces' family, the specific enveloping conditions:

$$\begin{aligned} \dot{F}_{\alpha}(X, Y, Z, \alpha, \beta) &= 0; \\ \dot{F}_{\beta}(X, Y, Z, \alpha, \beta) &= 0, \end{aligned} \tag{2}$$

representing the partial derivatives of the surfaces' family equations regarding the variable parameters α and β. Between α and β cannot be established any dependency.

The equations assembly (1) and (2) eliminating the α and β parameters determine a new surface, let it $S(X, Y, Z) = 0$, which is reciprocally enwrapping with the $\left(\Sigma_{\alpha,\beta}\right)$ family.

The equations assembly:

$$\left.\begin{array}{l} F(X, Y, Z, \alpha, \beta = cst.) \\ \dot{F}_{\alpha} = 0 \end{array}\right\} C_{\Sigma,\alpha} \\ \left.\begin{array}{l} F(X, Y, Z, \alpha = cst., \beta) \\ \dot{F}_{\beta} = 0 \end{array}\right\} C_{\Sigma,\beta} \tag{3}$$

determines onto the Σ surface an intersection point, the characteristic point, representing the contact point between the surface to be generated and the flank of the hob mill conjugated with Σ, the S surface.

The geometric locus of the contact points, in the motion generated by the α and β variables parameters, determines the surface conjugated with the $\left(\Sigma_{\alpha,\beta}\right)$ surfaces family, namely the $S(X, Y, Z) = 0$ surface, the primary peripheral surface of the hob mill.

Obviously, a solution based on the Gohman theorem also exists. This solution uses the intermediary surface's method. The intermediary surface is the surface of the rack-gear which generate the ordered curl of profiles.

The method of intermediary surface leads at changing the single-point contact problem in two successive problems of linear contact of enwrapping surfaces: the rack-gear determination and, next, starting from this, the characteristic curve determination. The characteristic curve is the contact curve between the rack-gear tool and the primary peripheral surface of the hob mill [1].

The performances of surface generation depend to the cutting tools geometry. Li et al. [2] approach the issue of helical tools cutting edges, using the fundamentals of surface enveloping, for machining the back face of the cutting edges of tools with helical flutes. They are proposed more technological solutions for grinding with preformed grinding wheels and it is 3D modelled the sharpening process. The 3D modelling allows a simple and rigorous analysis for actual geometry of sharpened cutting tool.

The issue of surface generation by enveloping has initially analytical approach. This type of study for generating process by enveloping for complex surfaces is often used by researchers.

Complementary theorems were proposed and applied: the "minimum distance" theorem and the "substitutive circles" theorem [1]. These theorems allow the solving of hob mill profiling problem, for generation of an ordered curl of surfaces, [3].

Wu and Hsu [4] deals with a complex problem for generation of compressor rotors using helical tools (helical grinding bodies), and, also, Yang [5] solve the problem using the generalized coordinates theory and the theory of surface enveloping.

Stosic et al. [6] analyse the constructive forms and generating tools for the compressors rotors.

Fig. 1 Timing belt pulley

Teodor et al. [7] create an algorithm for generation of surfaces known in discrete form, by substituting the in-plane generatrix of these surfaces with Bezier polynomials.

It is obviously that the analytical modelling of surface generation is fundamental in this domain, but that it can and should be completed, or, where the methodology allows, replaced with graphical methods, based on graphical design software, easier to use and with minimal errors from technical point of view.

The developing of graphical design environment allows the approaching of surface enveloping issue, for cutting tools design or for modelling the generating errors, using the capabilities of AutoCAD, CATIA or other graphical design software.

Berbinschi et al. [8, 9], Teodor et al. [3, 10], Baroiu [11] have developed applications in graphical design environment, solving problems of tools profiling for: ordered curls of profiles associated with a pair of rolling centrodes; profiling of tools bounded by revolution surfaces, generating cylindrical helical surfaces with constant pitch. All these approaches give solutions for enveloping problems based on the surface enveloping fundamentals, the Olivier or Gohman theorems.

The graphical design environment allow to develop graphical algorithms [8, 9], which lead to intuitive, rigorous and, at the same time, fast solutions for this type of problems.

An original solution was developed based on the complementary theorem of the "in-plane generating trajectories" [11], applied for the study of enwrapping surfaces associated with a pair of rolling centrodes.

The presented method can be a base for profiling hob mills which generate ordered curls of surfaces (spur gears; worms of helical compressors; splined shafts; rotors of helical pumps).

The 3D models of hob mills can be used for machining these mills using CNC equipment, as so as, for automated control of the manufactured tools.

In this paper is proposed the relative generation trajectories method for profiling the hob mill which generates an ordered curl of profiles—timing belt pulley with circular profile (Fig. 1). A specific algorithm is presented.

2 The Profile to Be Generated. Reference Systems. Generating Rack-Gear

The profile of the belt pulley, reference systems and relative motions of the pulley and generating rack-gear are presented in Fig. 2.

The reference systems are defined:

- *xyz* is the global reference system, with z axis overlapped to the rotation axis of the pulley;
- *XYZ* mobile reference system, joined with the centrode with R_r radius;
- $\xi\eta\zeta$ mobile reference system joined with the generating rack-gear, with η axis overlapped to the centrode associated with the rack-gear

The generating kinematics includes the rotation of the pulley's centrode, with φ angle, and the translation of the rack-gear along the η axis, with λ parameter.

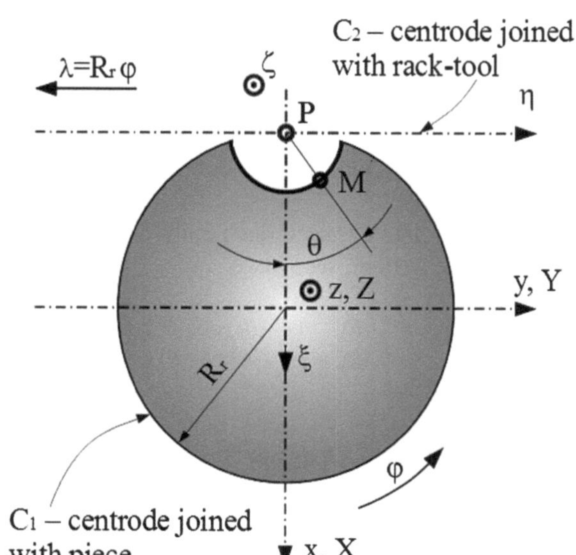

Fig. 2 Frontal profile of the belt pulley; reference systems

The rolling condition of the two centrodes, C_1 and C_2 is:

$$\lambda = R_r \cdot \varphi. \tag{4}$$

The equations of the profile to be generated are defined in the *XYZ* reference system:

$$C_\Sigma \left| \begin{array}{l} X = -R_1 + r \cdot \cos\theta; \\ Y = r \cdot \sin\theta; \\ Z = t, \end{array} \right. \tag{5}$$

With r radius of the circular profile and θ and t variable parameters. Initially, $t = 0$.

From the relative motion:

$$\xi = \omega_3^T(\varphi) \cdot X - a, \tag{6}$$

after developments, the analytical form of the $(C_\Sigma)_\varphi$ surfaces' family results:

$$(C_\Sigma)_\varphi \left| \begin{array}{l} \xi = (-R_1 + r\cos\theta)\cos\varphi - r\sin\theta\sin\varphi + R_r; \\ \eta = (-R_1 + r\cos\theta)\sin\varphi + r\sin\theta\cos\varphi + R_r\varphi; \\ \zeta = t, \end{array} \right. \tag{7}$$

with t variable scalar.

The enveloping of this surfaces' family represents the generating rack-gear.

The rack-gear's form is determined based on the "relative generating trajectories" theorem.

This requires to knows the normal in the M current point of the $C_\sum$ surface, see Fig. 2.

$$\vec{N}_{C_\Sigma} = (X_{C_\Sigma} - \kappa\cos\theta) \cdot \vec{i} + (Y_{C_\Sigma} - \kappa\sin\theta) \cdot \vec{j} + t \cdot \vec{k}. \tag{8}$$

Coordinates X_{C_Σ} and Y_{C_Σ} are defined by Eq. (5) and κ is a scalar variable value along the normal.

Now, the relative trajectories family of the normal (10) is defined regarding the $\xi\eta\zeta$ reference system, see (8).

$$(N_{C_\Sigma})_\varphi \left| \begin{array}{l} \xi = [-R_1 - (r-\kappa)\cos\theta]\cos\varphi - (r-\kappa)\sin\theta\sin\varphi + R_r; \\ \eta = [-R_1 + (r-\kappa)\cos\theta]\sin\varphi + (r-\kappa)\sin\theta\cos\varphi + R_r\varphi; \\ \zeta = t. \end{array} \right. \tag{9}$$

A normal, which belongs to the $(N_{C_\Sigma})_\varphi$ trajectories family, have to pass through the gearing pole, with coordinates:

$$P\left|\begin{array}{l}\xi = 0;\\ \eta = R_r \cdot \varphi;\\ \zeta = t.\end{array}\right. \tag{10}$$

This leads to:

$$\varphi = \arcsin\left(\frac{R_1}{R_r}\sin\,\theta\right) - \theta. \tag{11}$$

The condition (11) represents the enwrapping condition which, associated with the $(C_\Sigma)_\varphi$ family, determines the shape of the generating rack-gear:

$$I\begin{cases}\xi = (-R_1 + r\,\cos\theta)\cos\,\varphi - r\,\sin\,\theta\,\sin\,\varphi + R_r;\\ \eta = (-R_1 + r\,\cos\theta)\sin\,\varphi + r\,\sin\,\theta\,\cos\,\varphi + R_r\varphi;\\ \zeta = t;\\ \varphi = \arcsin\left(\frac{R_1}{R_r}\sin\,\theta\right) - \theta.\end{cases} \tag{12}$$

In this way, the rack-gear surface I is reduced at:

$$I\begin{cases}\xi = \xi(\theta);\\ \eta = \eta(\theta);\\ \zeta = t,\end{cases} \tag{13}$$

see Fig. 3.

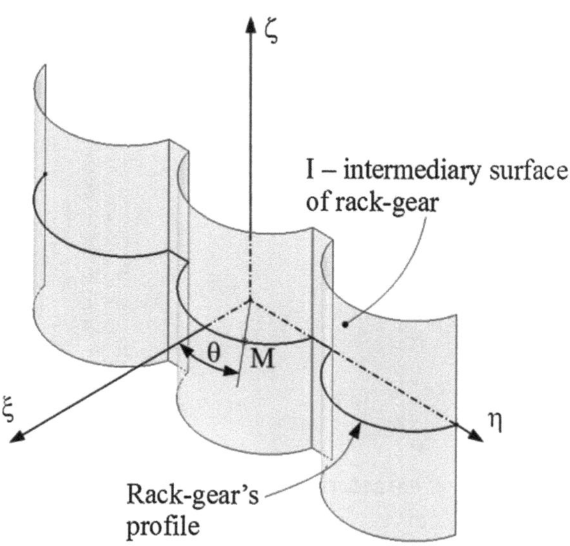

Fig. 3 The shape of rack-gear tool in $\xi\eta\zeta$ space

3 Primary Peripheral Surface of Hob Mill

The reference systems joined with the rack-gear and the future primary peripheral surface of hob mill are presented in Fig. 4.

Reference systems:

- xyz is the global reference system, with z axis overlapped to the axis of the belt pulley;
- $x_0y_0z_0$ auxiliary fixed reference system, joined with the axis of the future hob mill, $\vec{V}$;
- $\xi\eta\zeta$ mobile reference system, joined with the generating rack-gear, with η axis overlapped to the rolling line;
- XYZ mobile reference system joined with the belt pulley;
- $X_1Y_1Z_1$ mobile reference system, joined with the worm which represents the primary peripheral surface of the hob mill, with $\vec{V}$ axis overlapped to the axis Y_1
- $\xi_1\eta_1\zeta_1$ mobile reference system, joined with $\vec{A}$ axis, parallel with axis $\vec{V}$

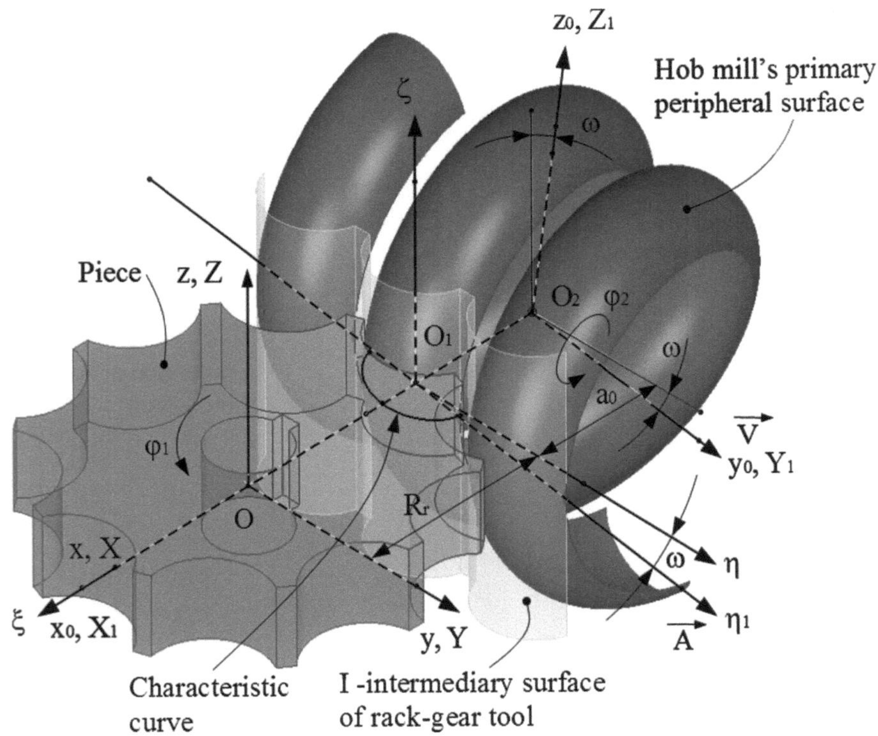

Fig. 4 Reference systems

3.1 Kinematics of the Generating Process

Let decompose the helical motion, with $\vec{V}$ axis and p helical parameter, in an assembly of motions: translation along the $\vec{T}$ generatrix of the cylindrical surface of rack-gear and rotation around the $\vec{A}$ axis. The $\vec{A}$ is parallel with $\vec{V}$ and at a_0 distance from this, see Fig. 4:

$$(\vec{V},p) \sim (\vec{T},v) + \left(\vec{A},\omega_A\right). \tag{14}$$

In Eq. (14), v is the translation motion's parameter and ω_A is the rotations angular speed around the $\vec{A}$ axis,

$$\varphi_1 = 2 \cdot \pi \cdot \omega_A. \tag{15}$$

The distance a_0 has the value $a_0 = p \cdot \tan\left(\frac{\pi}{2} - \omega\right)$, with ω inclination angle of the $\vec{V}$ axis regarding the plane of ξ axis, see Fig. 4.

Note We have to notice that, for the composed motion (14), in the translation component along the $\vec{T}$ axis, the generatrix of the cylindrical surface is self-generated. Consequently, the characteristic curve of the surface (13) in the movements assembly (14) will depend only by the rotation around the $\vec{A}$ axis, see Fig. 4.

The surfaces' family generated by the rack-gear (13), in the $X_1Y_1Z_1$ reference system, has the equations:

$$(I)_{\varphi_2} \left| \begin{aligned} X_1 &= [\xi(\theta) + a_0]\cos\varphi_2 - [\eta(\theta) + p_m\varphi_2\cos\omega]\sin\omega\sin\varphi_2 \\ &\quad - t\cos\omega\sin\varphi_2; \\ Y_1 &= [\eta(\theta) + p_m\varphi_2\cos\omega]\cos\omega - t\sin\omega; \\ Z_1 &= [\xi(\theta) + a_0]\sin\varphi_2 + [\eta(\theta) + p_m\varphi_2\cos\omega]\cos\varphi_2\sin\omega \\ &\quad + t\cos\omega\cos\varphi_2. \end{aligned} \right. \tag{16}$$

The enwrapping of $(I)_{\varphi_2}$ surfaces family (16) represents the helical surface—the primary peripheral surface of the hob mill.

3.2 Specific Enwrapping Condition

According to the generating trajectories family method, the normal to the I surface is determined. The directrix parameters of the normal to the I surface are:

$$\vec{n}_I = \dot{\eta}_\theta \cdot \vec{i} - \dot{\xi}_\theta \cdot \vec{j}, \tag{17}$$

so, the vector of normal to I in the current point of the rack-gear's surface is:

$$\vec{N}_I = (r - \kappa) \cdot \dot{\eta}_\theta \cdot \vec{i} + (r - \kappa) \cdot \dot{\xi}_\theta \cdot \vec{j} + t \cdot \vec{k}, \tag{18}$$

with κ variable scalar parameter.

The $\vec{N}_{I_{\xi_1\eta_1\zeta_1}}$ normal's family results after rotation around the $\vec{A}$ axis:

$$\left(\vec{N}_I\right)_{\varphi_1} \left| \begin{aligned} \xi_1 &= (r+\kappa)\dot{\eta}_\theta \cos\varphi_1 + (r+\kappa)\dot{\xi}_\theta \sin\omega \sin\varphi_1 - t\cos\omega\sin\varphi_1; \\ \eta_1 &= (r+\kappa)\dot{\xi}_\theta \cos\omega - t\sin\omega; \\ \zeta_1 &= (r+\kappa)\dot{\eta}_\theta \sin\varphi_1 - (r+\kappa)\dot{\xi}_\theta \sin\omega\cos\varphi_1 + t\cos\omega\cos\varphi_1. \end{aligned} \right. \tag{19}$$

From condition that the $\left(\vec{N}_I\right)_{\varphi_1}$ normals' family intersect the $\vec{A}$ axis:

$$\vec{A} \left| \begin{aligned} \xi_1 &= 0; \\ \zeta_1 &= 0, \end{aligned} \right. \tag{20}$$

by removing the κ scalar parameter is reached the form

$$q(\theta, t) = 0, \tag{21}$$

equivalent with $t = 0$, the enwrapping condition.

The (16) and (21) assembly of equations represents the helical surface, primary peripheral surface of the future hob mill—the S surface:

$$S \left| \begin{aligned} X_1 &= X_1(\theta, \varphi_2); \\ Y_1 &= Y_1(\theta, \varphi_2); \\ Z_1 &= Z_1(\theta, \varphi_2). \end{aligned} \right. \tag{22}$$

The profile of the S surface, for $\varphi_2 = 0$, is:

$$\begin{aligned} X_1 &= r\cos\theta + a_0; \\ Y_1 &= r\sin\theta\cos\omega; \\ Z_1 &= r\sin\theta\sin\omega, \end{aligned} \tag{23}$$

with $a_0 = p \cdot \tan\left(\frac{\pi}{2} - \omega\right)$.

The a_0 value represents the radius of cylinder of helical surface where take place the rolling with the centrode of the generating rack-gear.

4 Numerical Application

A numerical application is proposed for the case of a timing belt pulley having the dimensions:

- circular pitch, $\frac{2\pi R_r}{z_{teeth}} = 31.416\,\text{mm}$;
- rolling radius $R_r = R_e = 50$ mm;
- axial pitch of hob mill, $p_W = \frac{p_r}{\cos\omega} = 31.741$ mm; p_r—pitch of generating rack-gear.

$$p_r = \frac{2\pi R_r}{z_{teeth}} = 31.416\,\text{mm}. \tag{24}$$

Inclination angle of the worm's helix, ω, with $R_{rt} = 35$ mm—radius of rolling cylinder for worm tool:

$$\omega = \arcsin\left(\frac{p_r}{2\pi R_{rt}}\right) = 8.213^{\circ}. \tag{25}$$

- helical parameter of the hob mill:

$$p_w = \frac{p_W}{2\pi} = 5.052. \tag{26}$$

Rolling radius of the hob mill is chosen from technological reasons:

$$a_0 = \frac{p_W}{2\pi} \cdot \frac{1}{\tan\omega}. \tag{27}$$

From (27) and (25) results $a_0 = R_{rt}$.

The graphical method assumes to determine the profile of the rack-gear tool [5].

After this, in *DMU Kinematics* module of CATIA design environment is generated a gear-type mechanism which is composed from rack-gear and hob mill, positioned according to the generation kinematics, see Fig. 4. The hob mill is regarded as fixed element and the profile of the rack-gear is rotated around the $\vec{A}$ axis.

In the part file of the rack-gear component was draw a line perpendicular to the profile and which pass through an arbitrary point of this profile.

While the rack-gear is rotated around the $\vec{A}$ axis, the distance between the normal and the axis of the hob mill is monitored and, when this distance become 0, it is accomplished the condition that the normal to the profile pass through the worm axis. This means that the point through the normal pass belongs to the characteristic curve.

The point is recorded in the part file of the hob mill and its position is isolated in order to not change when the mechanism is modified.

The normal to the profile of the rack-gear is drawn from other point of the profile and the algorithm is resumed.

In this way, some points from the characteristic curve are determined and the form of characteristic curve can be determined drawing a spline through these points.

The form and coordinates of points belonging to the characteristic curve and axial section of the hob mill are presented in Fig. 5 and Table 1.

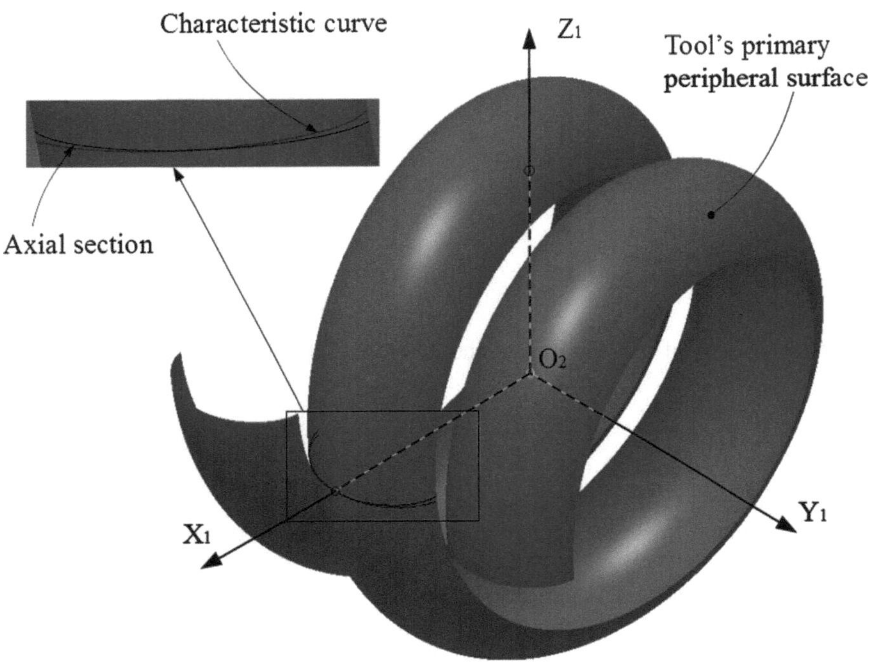

Fig. 5 Hob mill for generating timing belt pulley

Table 1 Coordinates of points from characteristic curve and axial section

Characteristic curve			Axial section		
X_1	Y_1	Z_1	X_1	Y_1	Z_1
35.219	−13.494	−0.893	35.230	−13.622	0.000
38.393	−12.114	−0.493	38.479	−12.127	0.000
41.149	−9.839	−0.329	41.210	−9.814	0.000
43.246	−6.922	−0.205	43.275	−6.892	0.000
44.555	−3.572	−0.100	44.563	−3.552	0.000
45.000	0.000	0.000	45.000	0.000	0.000
44.555	3.572	0.100	44.563	3.553	0.000
43.246	6.922	0.205	43.275	6.892	0.000
41.149	9.839	0.329	41.210	9.814	0.000
38.393	12.114	0.493	38.479	12.127	0.000
35.219	13.494	0.893	35.230	13.622	0.000

5 Conclusions

The "relative generation trajectories" method is a new complementary method [5], which if it is applied in graphical form allows to approach the profiling issue for hob mill designed to generate a ordered curl of surfaces, as example the teeth of a timing belt pulley.

The graphical method determines the generating rack-gear for a timing belt pulley and, after that, determines the primary peripheral surface of a generating hob mill.

The method is simple, easy to apply, and allows to avoid the rough errors which may be overlooked when using an analytical method. It is obviously simplicity of the method if compared to the analytical one, see Eqs. (19) and (21).

It is presented a numerical example for which the results are presented in Table 1 and Fig. 5.

Acknowledgements This work was supported by a grant of the Romanian National Authority for Scientific Research and Innovation, CNCS—UEFISCDI, project number PN-II-RU-TE-2014-4-0031.

References

1. Oancea N (2004) Generareasuprafeţelorprinînfăşurare (Surfacesgeneration bywinding), vol I–III. Galati University Press, Galati, ISBN 973-627-170-4, ISBN 973-627-107-2 (vol I), ISBN 973-627-176-6 (vol II), ISBN 973-627-239-7 (vol III)
2. Li G, Sun J, Li J (2014) Modeling and analysis for clearance machining process of end mills. Int J Adv Manuf Technol 75:67–675. doi:10.1007/s00170-014-6154-3

3. Teodor V, Frumuşanu G, Oancea N, Berbinschi S (2015) Analytical method to profile the hob mill generating an ordinate whirl of surfaces with non-involute profile. Proc Manuf Syst 10(2):71–76
4. Wu Y-R, Hsu W-H (2014) A general mathematical model for continuous generating machining of screw rotors with worm-shaped tools. Appl Math Model 38:28–37
5. Yang J, Abdel-Malek K (2006) Verification on NC machining processes using swept volumes. Int J Adv Manuf Technol 28:82–91. doi:10.1007/s00170-004-2352-8
6. Stosic N et al (2011) Geometry of screw compressor rotors and their tools. J Zhejiang Univ Sci A (Appl Phys Eng) 12(4):310–326
7. Teodor VG, Popa I, Oancea N (2010) The profiling of end mill and planning tools to generate helical surfaces known by sampled points. Int J Adv Manuf Technol 51:439–452. doi:10.1007/s00170-010-2655-x
8. Berbinschi S, Teodor V, Oancea N (2012) 3D graphical method for profiling tools that generate helical surfaces. Int J Adv Manuf Technol 60:505–512. doi:10.1007/s00170-011-3637-3
9. Berbinschi S, Teodor V, Oancea N (2012) 3D graphical method for profiling gear hob tools. Int J Adv Manuf Technol. doi:10.1007/s00170-12-3989-3
10. Teodor V, Baroiu N, Susac F, Oancea N (2016) The rack-gear tool generation modelling. non-analytical method developed in CATIA, using the relative generating trajectories method. IOP Conf Ser Mater Sci Eng 161:012028. doi:10.1088/1757-899X/161/1/012028
11. Baroiu N, Teodor V, Oancea N (2015) A newform of plane trajectories theorem. Generation with rotary cutters. Bulletin of Polytechnic Institute of Iassy, Tome LXI (LXV), Machine Construction, pp 27–36. ISBN 1011-2855

Cyber Physical Production Systems—An IEC 61499 Perspective

Zivana Jakovljevic, Stefan Mitrovic and Miroslav Pajic

Abstract Recent developments in the field of cyber physical systems (CPS) and internet of things (IoT) open up new possibilities in manufacturing. CPS and IoT represent enabling means for facilitating the companies' adaptation to the ever-changing market needs and adoption of individualized manufacturing paradigm. It is foreseen that implementation of CPS and IoT will lead to new industrial revolution known as Industry 4.0. The fourth industrial revolution will bring about the radical changes in manufacturing process control through distribution of control tasks to intelligent devices. On the other hand, over the last decades International Electrotechnical Commission invested significant efforts in generation of the standard IEC 61499 for distributed automation systems. In this paper we provide the outline of the interconnection of IEC 61499 standard and Reference Architecture Model Industrie 4.0 in cyber physical production systems. The findings of the paper are illustrated using a realistic case study—an example of pneumatic manipulator that is made of CPS devices.

Keywords Cyber physical production systems (CPPS) · Industry 4.0 · IEC 61499

Z. Jakovljevic (✉) · S. Mitrovic
Faculty of Mechanical Engineering, University of Belgrade,
Kraljice Marije 16, 11000 Belgrade, Serbia
e-mail: zjakovljevic@mas.bg.ac.rs

S. Mitrovic
e-mail: limvo.da.zimvo@gmail.com

M. Pajic
Department of Electrical and Computer Engineering, Duke University,
100 Science Dr, Hudson Hall Room 206, Durham, NC 27708, USA
e-mail: miroslav.pajic@duke.edu

V. Majstorovic and Z. Jakovljevic (eds.), *Proceedings of 5th International Conference on Advanced Manufacturing Engineering and Technologies*,
Lecture Notes in Mechanical Engineering, DOI 10.1007/978-3-319-56430-2_3

1 Introduction

Integration of physical system and its cyber model through the concept of Cyber Physical Systems (CPS) opens up new possibilities in the production processes control. By interconnecting physical processes and modules for data processing through real-time interaction, CPSs integrate physical process and its virtual/cyber model. Smart devices (sensors and actuators) with integrated data processing and communication modules along with the development of Internet of Things and Services (IoT), facilitate a new level of distribution of control tasks between production resources.

The five level automation hierarchy formalized through Purdue Reference Enterprise Architecture (PERA) [1] and standardized through IEC 62264: Enterprise-control system integration [2] will give the way to the distributed control systems [3, 4] as shown in Fig. 1. All functional elements of the automation hierarchy will remain, but instead of strict pyramidal structure, their functionality will be performed through interoperability of the system elements.

Control system will be distributed to the system's components/resources, and complete functionality of the system will be achieved through components communication and interoperability. Smart devices are inherently modular, and they enable ad hoc interconnection and configuration of production resources and control systems according to the current needs. Modularity of resources and software with the support of cloud services will make it possible to system integrator to, by downloading appropriate applications, configure the system as needed and when it is needed [5]. In this way, full individualization of applications and high level resources consumption can be achieved. Key characteristic that enables this kind of reconfiguration is open system characterized by devices interoperability, system reconfigurability and software portability [6].

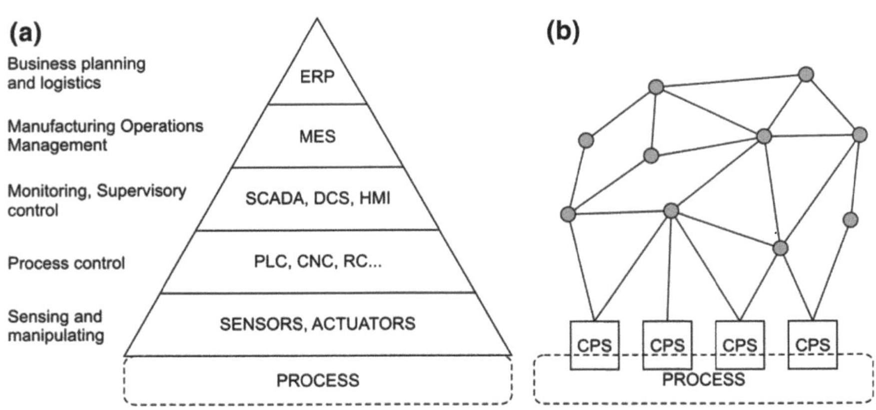

Fig. 1 **a** Automation hierarchy. **b** Distributed (networked) control system

CPS can be implemented at various levels of manufacturing process, starting from the elements of manufacturing resources, their subsystems, individual machines, cells, up to the control of the whole production systems or production networks. Cyber Physical Production Systems (CPPS) [7] represent the highest level of implementation of CPS concept in manufacturing. These systems can be regarded as reconfigurable production systems, based on interoperable and autonomous cyber-physical resources, that are configured according to the current production/market needs. This type of reconfigurability is very important for manufacturing companies to face the challenge of manufacturing individualized products according to the customer demands.

The changes in production systems have the potential will bring about new industrial revolution known as Industrie 4.0 [8]. To meet these challenges, German professional associations and industrial enterprises have established strategic initiative Platform Industrie 4.0, and one of the main results of this initiative is the Reference Architecture Model Industrie 4.0 (RAMI4.0) [9]. To incorporate vertical, horizontal and end-to end integration of manufacturing systems into one model, RAMI4.0 introduces three axes: (1) Hierarchy Levels based on IEC 62264, (2) six Layers to define IT representation of components, and (3) Life Cycle and Value Stream.

On the other hand, the use of distributed control in automation systems leads to increased complexity in control software, especially when an easy reconfiguration is required. Control software based on IEC 61131-3 programming languages is made of monolithic blocks and the changes during reconfiguration require significant efforts and interventions in many parts of the program. Recently, it has been shown that in flexible manufacturing systems (FMS) the software was limiting factor that made the FMS inflexible [10]. To facilitate the modeling, simulation, and reconfiguration of distributed control systems, international standard IEC 61499 [11] is introduced.

In this paper we present the modeling of distributed control systems in CPPS using IEC 61499, as well as an overview of RAMI4.0 Layers from IEC 61499 perspective. The paper is structured as follows. In Sect. 2 we provide a brief overview of RAMI4.0 hierarchy levels and layers. Section 3 presents IEC 61499 and its relation to RAMI4.0 architecture, while in Sect. 4 we further illustrate this relation using a case study. Finally, Sect. 5 provides concluding remarks.

2 RAMI4.0 Hierarchy Levels and Layers—A Brief Overview

Reference Architecture Model Industrie 4.0 (RAMI4.0) [9] keeps all hierarchy levels from IEC 62264 and adds two extra levels: Product and Connected World. The highest level of IEC 62264 architecture is the enterprise. Nevertheless, the globalization and developments in the field of cloud services fuzzify the borders of companies, and enable generation of value adding networks of enterprises. As a result, the highest level of enterprise control system integration within RAMI4.0 becomes the connected world. In addition, Industrie 4.0 product is a smart product that can be tracked along the whole life cycle, and also represents an element of the control hierarchy.

It should be emphasized that RAMI4.0 hierarchy is rather considered as a functional hierarchy, it presents the allocation of functionalities, and it does not follow the pyramid of PERA and IEC 62264 (Fig. 1a). It assumes the distribution of control tasks between different devices.

The interoperability of devices at all functional hierarchy levels is crucial for correct performances of distributed control system, and it is achieved through several interoperability layers. RAMI4.0 layers are derived from Smart Grid Architecture Model (SGAM) [12] that contains Component, Communication, Information, Function and Business layer. RAMI4.0 architecture replaces the component layer with the asset layer. The asset layer refers to the real world, and it includes not only the physical components of the system—devices (cylinders, axes, fixtures, tools, etc.), but also all other resources, such as humans. Since the assets, in a number of cases, have only the possibility of passive communication, RAMI4.0 contains an additional layer—Integration layer, thus having a total of six interoperability layers: Asset, Integration, Communication, Information, Functional, and Business layer (Fig. 2).

Business layer is the highest layer, and it considers business policy, objectives, regulatory rules, and similar aspects. Functional layer defines the functions of the system independently from the actors that will carry out these functions [12]; control system is integrated at this level [9]. Information layer provides the information that is exchanged between the functions of the system. At this level, the physical components obtain their virtual representation. This level provides the logical interface between the assets, and it can be observed as the most important layer. For the exchange of information between functions, canonical data model and common semantics between functioning elements of the system are crucial. On the other hand, there are several standardized industry data formats whose data formats need unification. Automation ML (Markup Language) [13] represents an effort to standardize the exchange of data in production systems in order to facilitate the

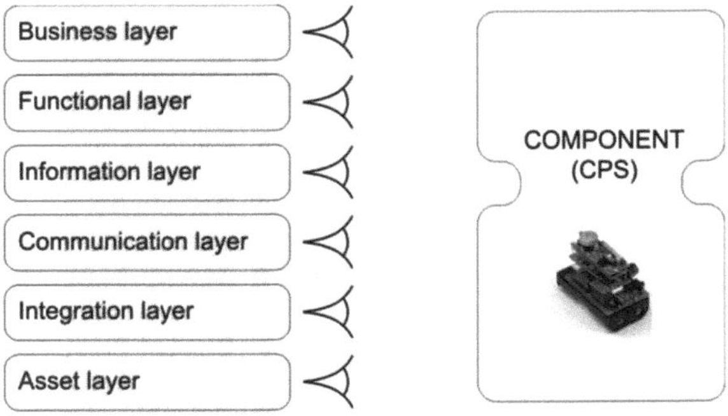

Fig. 2 RAMI4.0 interoperability layers

interconnection of different engineering tools. It is an XML-based data format that integrates existing XML based formats employed in different application areas, such as CAEX, COLLADA and PLCopen XML.

Communication layer enables the exchange of data between system components. Layers 1–4 of the OSI (Open Systems Interconnection) communication model [14] are well standardized, and a number of protocols such as IEEE 802.3, IEEE 802.11, IEEE 802.15.4, as well as IEEE 802.2 are currently employed. However, the standardization of upper layers of OSI model is not at appropriate level. In this context, OPC Foundation has been developing the OPC Unified Architecture [15, 16] as a platform independent (Windows, Linux, UNIX, embedded systems) middleware communication protocol that is foreseen as enabling communication technology for CPPS [7]. The combination of Automation ML semantic description of production plant and OPC UA secure communication opens up new possibilities in horizontal and vertical integration of CPPS [7, 15].

A number of assets in industrial environment, such as products or humans, inherently have only the passive communication capability with respect to the protocols used in communication layer. The role of integration layer is to get the information from assets, to translate it into the form readable by communication and information layers, and to provide information and generate events that will be utilized within higher interoperability levels.

3 Modeling of Distributed Control Systems Using IEC 61499

Reconfigurability of production resources and distribution of control tasks changes the way industrial control systems are developed. Each CPS component has its own software, and, to enable relatively fast integration and re-use of modules during system reconfiguration, it is convenient to encapsulate component software in the form of objects [17]. On the other hand, object oriented software development requires highly skilled software specialists, thus deteriorating one of the highest advantages of IEC 61131-3 [18] based control systems programming languages—the possibility to be developed and used by software non-specialists.

IEC 61131-3 standard defines function blocks (FBs) as program organization units that encapsulate certain algorithms similarly to object class. One of the characteristics of function blocks is that the designers of control systems can easily integrate function block instances into the control software. Function blocks can have different complexity depending on the algorithm they execute. Starting from the properties of function blocks, International Electrotechnical Commission developed international standard IEC 61499: Function blocks [11]. The scope of this standard is the reference architecture for utilization of function blocks for effective system integration in distributed control systems.

In IEC 61499 FB models software or hardware component's functionality and, from the object-oriented software perspective, FB can be regarded as a class that defines the behavior of instances [6, 10]. Each CPS component can have its own FB that can be downloaded from software library, and the system integrator does not need to understand FB's internal structure to employ it. FBs at input and output have events and data associated to these events, and they execute certain algorithms depending on the input events (Fig. 3).

The distributed system as a whole carries out one or more functions. In the context of IEC 61499, a function of the system represents a network of function blocks—application. Function blocks are invoked/activated by input events, and after execution of the corresponding algorithm, FBs generate output events to invoke/activate the following function blocks in the network. Finally, after execution of the algorithm FB goes to idle state.

At the physical level, the distributed system is composed of a number of devices that have interfaces to the process and to communication services. Devices can contain one or more resources—processors within the device or multiple tasks within a single processor [6]. The standard defines service interface function blocks (SIFB) that are used to communicate the data between different devices, as well as between devices and processes. Within the scope of RAMI, devices can be regarded as assets.

In the system integration workflow, the starting point is the design of an application—a network of function blocks that defines the functionality of the system. From the RAMI point of view, the applications define functional layer.

Function blocks are interconnected by events and data exchanged between them. In this way, a common data model and semantics are generated and using events flow logical interface between FBs is designed. Thus, the data and events flow defines the information that is exchanged between functioning elements and represents the information layer of RAMI.

After the application design, its parts (i.e., sub-networks of FBs) are mapped to the resources that will execute them. In this way, some of the connections between FBs are broken and, within the resource FB network, they have to be connected to communication SIFBs. Communication SIFBs enable data and events flow between resources, effectively modeling a communication layer of RAMI. However, it

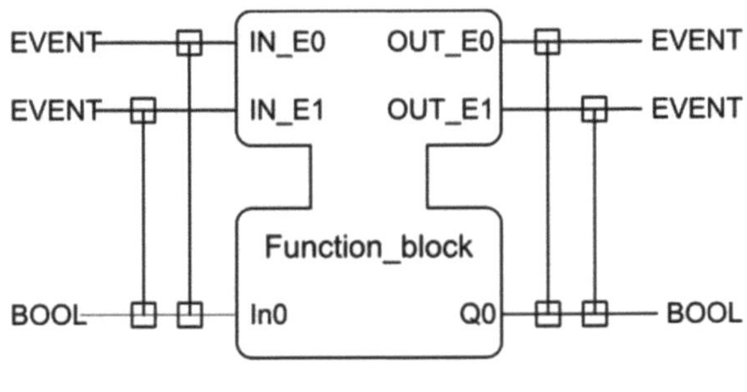

Fig. 3 IEC 61499 function block

should be emphasized that IEC 61499 is a generic standard and it does not define concrete communication protocols [17].

4 Case Study

To illustrate the observations regarding the interconnection of RAMI layers and an IEC 61499 distributed system model, in this section we present a case study that refers to the modeling of distributed control of a pneumatic manipulator. Modeling is performed using software FBDK 2.4 (The Function Block Development Kit) Version 20081003 by Holobloc Inc. [19]. To simplify presentation, the considered pneumatic manipulator (Fig. 4) has 2 DOF and it consists of two intelligent double acting pneumatic cylinders (C1 and C2) and one intelligent gripper (G). Each pneumatic cylinder has integrated 5/2 directional control valve, proximity sensors (one for each limit position) and a control device with communication capability.

Pneumatic cylinder function block type is presented in Fig. 5a. It has two events at input—event CLK that is used for simulation purposes and event MOVE. On receipt of MOVE event, cylinder moves from advanced (end) to retracted (home) position, and vice versa, depending on the position it was in at the event arrival. At the output of function block there are two events: HOME_E and END_E. These events are issued when cylinder reaches home and end position, respectively. In addition, cylinder FB has three outputs: POS, HOME and END. HOME and END represent Boolean variables that correspond to home and end proximity sensor, and they are employed for communication of cylinder status to HMI (Human Machine Interface). POS output carries information on current position of cylinder and it can be used for simulation purposes.

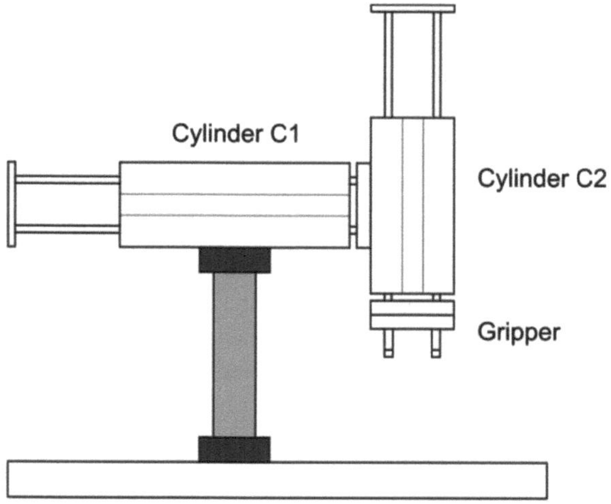

Fig. 4 2 DOF pneumatic manipulator

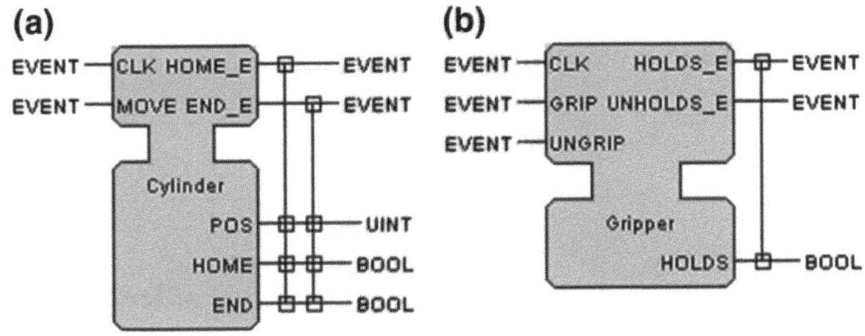

Fig. 5 Function blocks: **a** Cylinder. **b** Gripper

Intelligent gripper includes control valve and control device with communication capabilities. Its FB is presented in Fig. 5b and it has three input and two output events, as well as one output variable. Input events GRIP and UNGRIP are used to initialize gripping and releasing, and input event CLK is employed for simulation purposes. Output events HOLDS_E and UNHOLDS_E are released when holding or releasing finishes, and Boolean variable HOLDS is used to communicate gripper status to HMI. It should be noted that it is not necessary for system integrator to know the internal structure of FBs to be able implement them—they can be observed as black boxes.

The working cycle of pneumatic manipulator is described by the following sequence:

$$C2 + G + C2 - C1 + C2 + G - C2 - C1- \tag{1}$$

that represents the functionality of the system. From (1) it can be observed that:

- Cylinder C1 changes position each time cylinder C2 reaches home position—on C2 HOME_E event.
- Cylinder C2 changes position at the beginning (after system initialization), after gripping and releasing, and when cylinder C1 reaches end position. This corresponds to events: INIT (initialization), G HOLDS_E, G UNHOLDS_E, and C1 END_E.
- Gripper changes between gripping and releasing when cylinder C2 reaches end position. For every first C2 END_E event, gripping in initialized, and for every second C2 END_E event, releasing is initialized.

Starting from given explanation of system functionality, we have designed the application presented in Fig. 6 in FBDK. In this application, besides cylinders and griper FBs instances, three sub-networks of FBs can be observed. Sub-network of FBs E_RESTART, E_DELAY and E_CYCLE generates the sequence of events for simulation purposes. Second sub-network consists of four E_MERGE blocks and it merges the events needed for activation of C2 cylinder FB instance. Finally, sub-network of FBs E_CTU and E_SWITCH insures that for every first C2 END_E event gripping, and for every second C2 END_E event releasing is initialized.

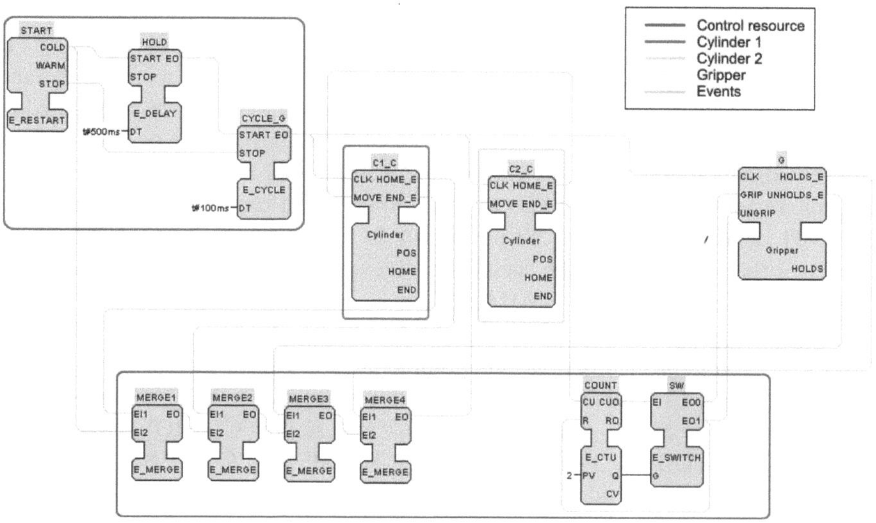

Fig. 6 Application developed in FBDK

The application from Fig. 6 represents the functionality of the system. In addition, the flow of events defines the logical interface between the functions of the system, and it represents canonical data (events in this case) model with common semantics for all functional elements. In this way a model of the information flow is generated.

At physical level the system consists of four devices (assets):

- Cylinder: C1
- Cylinder: C2
- Gripper: G
- Men to machine interface: MMI

as presented in Fig. 7a. In our model, MMI device contains three resources: (1) CYCLE that will generate event cycle for simulation purposes, (2) MMI1 that represents panel, and (3) CONTROL that will implement the control logic of the system.

To model the distribution of control tasks to different devices of the system, parts of application (Fig. 6) are distributed to different resources. Cylinder FBs carry out

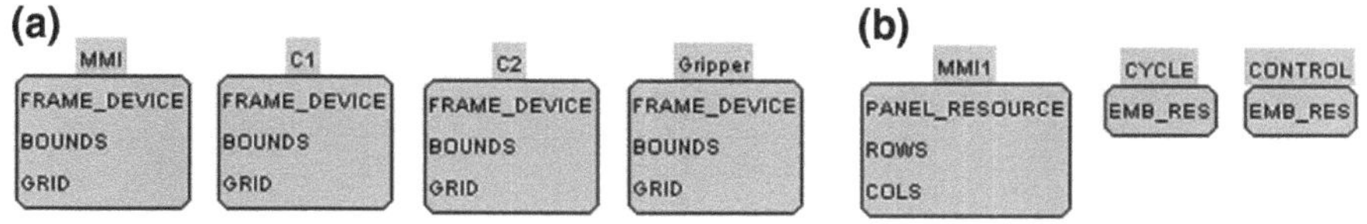

Fig. 7 **a** Devices within the system. **b** Resources within MMI device

the control of the cylinders—reading inputs from proximity sensors, and writing output to directional control valve depending on the cylinder position. Cylinder FB instances C1_C and C2_C are mapped to devices C1 and C2, respectively and Gripper FB instance G to Gripper device. In our control system, we have opted to map FBs involved in control of events flow to CONTROL resource of MMI device. If other control architecture was chosen, these FBs could be mapped to C2 and G devices. In addition, we added FBs referring to output of process Boolean variables to MMI1 resource; these FBs are not shown in Fig. 6 for clarity of presentation. Finally, FBs that generate the sequence of events for simulation purposes are mapped to resource CYLCE of MMI device.

During mapping of C1_C FB to C1, the flow of events in application (Fig. 6) is broken. To re-establish this flow, the communication SIFBs that will transfer the events between C1 resource and other interested resources in the system have to be added. These are SUBL_0 FB instances CCYCLE and C1_MOVE for events receiving, and PUBL_0 FB instances CC1_HOME and CC1_END for events transmitting (Fig. 8). Besides, one PUBL_2 SIFB instance—CC1_HOME_END is introduced to transfer the data regarding end positions of cylinder to MMI1 resource. Similar SIFBs are added to FB networks of C2, G, and CONTROL resource (Figs. 9 and 10). In this way the communication between assets of the system is defined. To check the performances of the developed distributed control system, we have run the simulation, and the results, obtained using MMI1 resource, are presented in Fig. 11. The sequence of reaching cylinders limit positions and gripping from Fig. 11 is in accordance with relation (1).

The developed model of distributed control system for pneumatic manipulator contains first three functional levels of automation pyramid (Fig. 1a): Sensing and

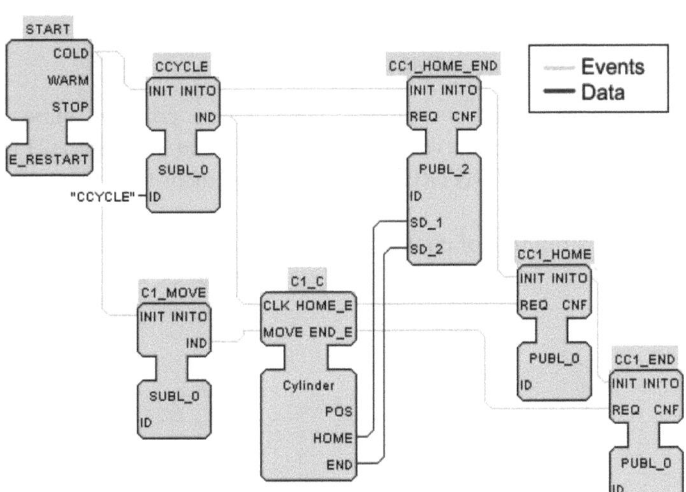

Fig. 8 Network of FBs assigned to C1

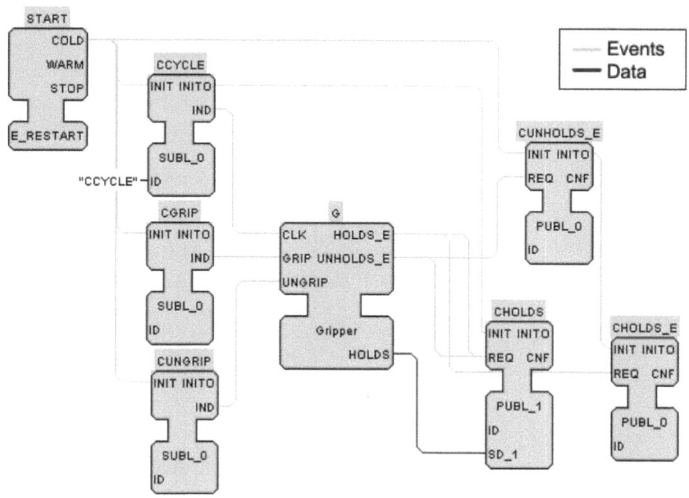

Fig. 9 Network of FBs assigned to G

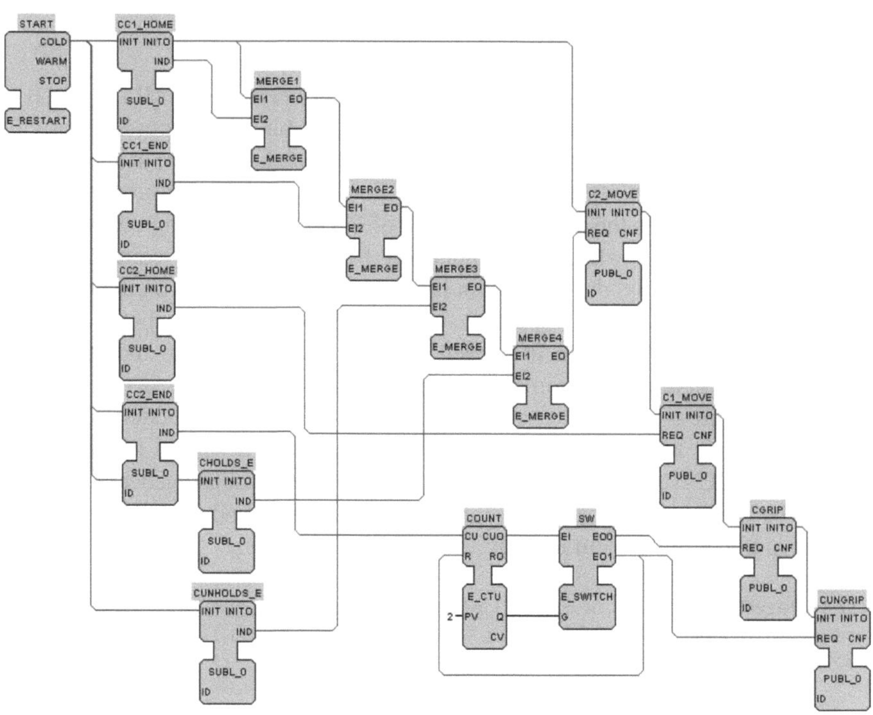

Fig. 10 Network of FBs assigned to CONTROL

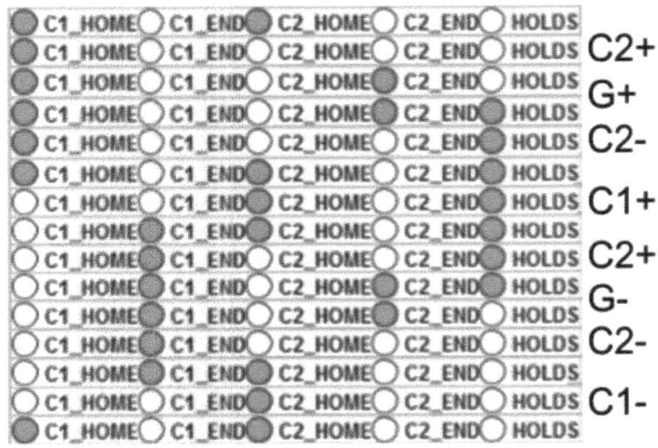

Fig. 11 Results of simulation—MMI1 device

manipulating, Process control, and Monitoring distributed over four intelligent devices (CPS).

5 Conclusion

In this paper, we have explored the interconnection between RAMI4.0 and IEC 61499 standard. It has been shown that IEC 61499 application represents an efficient means for representation of RAMI4.0 functional layer since it defines the functioning of a control system as a whole. Further, the connection of FBs I/O events and data models the information layer of RAMI4.0 providing common data model and semantics, while communication SIFBs define the communication between devices. Finally, devices within IEC 61499 control system model represent assets with communication capability (CPS). These interconnections between RAMI4.0 and IEC 61499 are illustrated using a case study—the model of distributed control system of pneumatic manipulator.

In addition, the considered case study presents how different functionality levels of IEC62264 can be distributed over intelligent devices. It should be noted that according to IEC 61499 the whole manipulator can be encapsulated into one FB that can be further used for the design of manufacturing cell and eventually the whole manufacturing system control.

Acknowledgements This work is a part of research which is supported by the Ministry of Science and Technological Development of Serbia, with grant numbers TR35007, TR35020 and TR35023.

References

1. Williams TJ (1994) The Purdue enterprise reference architecture. Comput Ind 24(2/3): 141–158
2. International Organization for Standardization (2013) IEC 62264-1 Enterprise-control system integration—Part 1: models and terminology
3. VDI/VDE-Gesellschaft Mess- und Automatisierungstechnik (2013) Cyber-physical systems: Chancen und NutzenausSicht der Automation. https://www.vdi.de/uploads/media/Stellungnahme_Cyber-Physical_Systems.pdf Accessed 14 Feb 2017
4. Monostori L (2014) Cyber-physical production systems: roots, expectations and R&D challenges. Procedia CIRP 17:9–13
5. Koren Y, Shpitalni M (2010) Design of reconfigurable manufacturing systems. J Manuf Syst 29(4):130–141
6. Christensen JH, Strasser T, Valentini A, Vyatkin V, Zoitl A (2012) The IEC 61499 function block standard: software tools and runtime platforms. ISA Autom Week. http://www.holobloc.com/papers/iec61499/61499_SWT_RTP.pdf Accessed 14 Feb 2017
7. Monostori L, Kádár B, Bauernhansl T, Kondoh S, Kumara S, Reinhart G, Sauer O, Schuh G, Sihn W, Ueda K (2016) Cyber-physical systems in manufacturing. CIRP Annals—Manufact Technol 65(2):621–641
8. Kagermann H, Wahlster W, Helbig J (2013) Recommendations for implementing the strategic initiative industrie 4.0. Acatech—National Academy of Science and Engineering. http://www.acatech.de/fileadmin/user_upload/Baumstruktur_nach_Website/Acatech/root/de/Material_fuer_Sonderseiten/Industrie_4.0/Final_report__Industrie_4.0_accessible.pdf Accessed 14 Feb 2017
9. VDI/VDE-Gesellschaft Mess- und Automatisierungstechnik (2015) Reference Architecture Model Industrie 4.0 (RAMI4.0). http://www.zvei.org/Downloads/Automation/5305%20Publikation%20GMA%20Status%20Report%20ZVEI%20Reference%20Architecture%20Model.pdf Accessed 14 Feb 2017
10. Vyatkin V (2012) IEC 61499 function blocks for embedded and distributed control systems design. ISA. ISBN 978-1-9360007-93-6
11. International Electrotechnical Commission (2012) IEC 61499-1 function blocks—part 1: architecture
12. CEN-CENELEC-ETSI Smart Grid Coordination Group (2012) Smart grid reference architecture. https://ec.europa.eu/energy/sites/ener/files/documents/xpert_group1_reference_architecture.pdf Accessed 14 Feb 2017
13. AutomationML consortium (2014) AutomationML whitepaper: part 1—architecture and general requirements. https://www.automationml.org/o.red/uploads/dateien/1417686950-AutomationML%20Whitepaper%20Part%201%20-%20AutomationML%20Architecture%20v2_Oct2014.pdf Accessed 14 Feb 2017
14. International Organization for Standardization (1994) ISO/IEC 7498-1 Information technology—open systems interconnection—basic reference model: the basic model
15. AutomationML consortium, OPC Foundation (2016) AutomationML whitepaper: OPC unified architecture information model for automationML
16. Leitner SH, Mahnke W (2006) OPC UA – Service-oriented Architecture for Industrial Applications. GI softwaretechnik-trends 26(4). ISSN 0720-8928
17. Lewis R (2008) Modelling control systems using IEC 61499. The Institution of Engineering and Technology, London
18. International Electrotechnical Commission (2013) IEC 61131-3 programmable controllers—part 3: programming languages
19. HOLOBLOC, Inc. (2016) FBDK 2.5—The function block development kit. http://www.holobloc.com/fbdk2/index.htm Accessed 14 Feb 2017

Virtual Model in Monitoring and Optimization of a Selective Waste Collection Integrated System

Cicerone Laurentiu Popa and Costel Emil Cotet

Abstract This paper presents some results of a research project aiming to increase the usage degree, productivity and storage capacity of a selective waste collection system and the integration of the system in the smart city infrastructure. Companies that have waste collection and recycling as their target activity are looking for solutions in order to increase the collection levels for all types of waste. In our project we plan to modernize a waste collection system that will considerably increase the waste collection level, that will meet the requirements and the needs identified by the economic agent and that will adapt to the new technologies so that it will be integrated in the infrastructure of the smart city. The first objective of the project is to increase the use of the system, improve its productivity and its collected waste storage capacity. The second objective is to integrate the system in the infrastructure of the smart city. For the activities planned in order to meet these objectives, the paper proposes a material flow management optimizing algorithm based on a virtual model of the selecting and processing architecture of the system. The material flow management of this system will be based on its virtual model in order to identify and eliminate material flow concentrators and increase productivity. Simulation will be used to diagnose the initial performance of the system structural elements as well as to validate the optimized system performances after eliminating the bottlenecks.

Keywords Selective waste collection · Integrated system · Material flow management · Virtual model · Modelling · Optimization · Diagnosis

C.L. Popa (✉) · C.E. Cotet
University Politehnica of Bucharest, Bucharest, Romania
e-mail: laur.popa79@gmail.com

C.E. Cotet
e-mail: costelemilcotet@gmail.com

V. Majstorovic and Z. Jakovljevic (eds.), *Proceedings of 5th International Conference on Advanced Manufacturing Engineering and Technologies*,
Lecture Notes in Mechanical Engineering, DOI 10.1007/978-3-319-56430-2_4

1 Introduction

The European Union-wide strategy suggests filling the lack of natural resources required to manufacture products with the use of recycled parts and products that are at end of their lifecycle, resulting thus in a transfer to a closed loop recycling economy, a new concept backed by the European Commission, which will be developed and implemented in the 2014–2020 time period [1].

With a rapid decrease of the available natural resources and due to increased costs of raw material extraction, the European strategy suggests the replacement of the natural resources required to manufacture products with materials from recycled parts and products [2]. In line with the European trend, The New National Strategy for Waste Management (NSWM) presents the Romanian strategic objectives for 2020. It also proposes a framework describing the measures that would ensure the transition from the current development model based on production and consumption to a model based on waste generating prevention and to the use of recovered materials from the industry, thus ensuring the preservation of national natural resources [3].

The promoted ideas by the NSWM include:

- encouraging the prevention of generating waste;
- resources sustainably and resources reuse;
- increasing the quantity of recycled materials in order to reduce the consumption of natural resources;
- increasing the recycling rate and improving the quality of recycled materials;
- encouraging green investments.

At present, Romania has a level of waste collection well below the EU average and runs the risk of not achieving her targets and of failing to make the transfer to a circular economy. According to a Eurostat report, Romania's recycling rate for 2015 was of 3%, while the UE average was 28% (for municipal solid waste). At national level a target of 50% regarding the preparation for reuse and recycling of waste must be achieved by 2020, otherwise Romania risks penalties of 200.000 Euro/day [4, 5].

Studies show that in Romania about 84% of municipal solid waste could be recycled, therefore companies involved in waste collection and recycling are looking for solutions to increase the collection levels for all types of waste.

In this context, we used an optimization algorithm for a waste collection system to increase the waste collection level and meet the requirements and needs identified by the economic agent. As a result of our algorithm, the main optimized performance parameter was productivity. Important results were also obtained in increasing the system usage degree and storage capacity. The first step of this algorithm was to diagnose the existing system performances. The second step was to propose a new modified architecture of the system in order to improve productivity. The third step was to include new technologies in this optimized architecture design in order to assure the integration in the infrastructure of the smart city. We used two virtual prototypes of the system: one for the current architecture (Fig. 1) the second for the modified one (Fig. 5). After modelling the current architecture using Witness Horizon, we performed a simulation and obtained a diagnosis of the current system in order to identify the bottlenecks and the necessary systems modification. The modified system's configuration supposes an optimized version for the waste flow collection. There are two kinds of solutions for eliminating these concentrators by modifying the current model. Functional remodelling potentiates the system performance by modifying the structural elements placement or functional parameters but is not introducing or eliminating structural elements. Technological remodelling allows system changes also when structural elements should be added or removed. Functional and technological remodelling are the optimizing tools used to transform the current model in the modified one. For this optimized model another material flow simulation is performed for quantifying the increasing performance. In the end a profit validation is needed, comparing the investment in optimizing the system versus the gain from the increased system performance. Direct beneficiaries of this optimized system are all waste recyclers (plastic, glass, paper/cardboard, electronics etc.) because the solutions resulting from our project will increase the capacity of waste collection and processing, the productivity and the efficiency of the activities involved.

The paper is organized as follows: Section 2 presents the main research steps necessary to achieve the proposed objectives. Section 3 presents the current architecture of the selective waste collection system modelled in Witness Horizon. Section 4 reports the diagnosis results regarding the key performance indicators associated with the waste flow in the current system. In Sect. 5 the necessary systems modification and its optimized architecture are presented. A new simulation in Witness Horizon is made in order to validate the optimized system performances. Results regarding the usage degree, the productivity and the storage capacity are presented, while Sect. 6 discusses the conclusion of the study and future work.

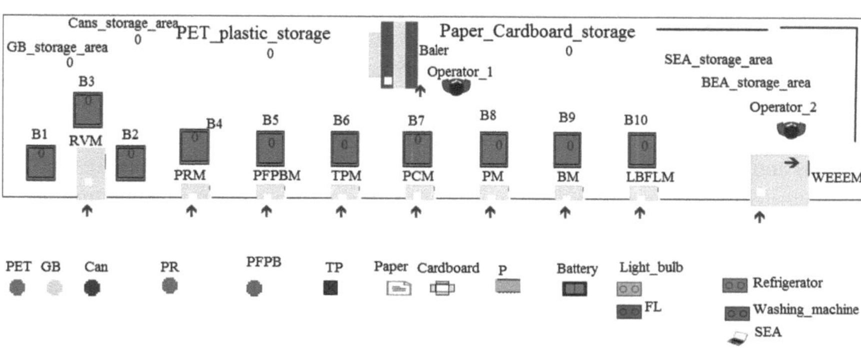

Fig. 1 Virtual model of the current selective waste collection made in Witness Horizon

2 Methodology

The main research steps for achieving the project objectives are the following:

a. Based on data provided by the company (project partner) and on studies conducted to the current system, a detailed description was given for each module of the waste collection system. Based on these data, the waste collection system's preliminary architecture modelling was performed in Horizon Witness. Also, the parameterization of each component was done, the relationships between the structural elements needed for the simulation of waste flow from the current system were established and the virtual model of the current waste collection system was obtained.

The main Material Flow Management (MFM) [6] algorithm steps are starting from this point and are based on all the above information, as follows.

A preliminary waste processing architecture virtual modelling is done with the necessary parameters for the work points, buffers, transport and transfer systems. The trajectories of the mobile entities are established. For this preliminary architecture virtual prototype the material flow is simulated for a performance diagnosis report.

Using this report, the flow concentrators were identified and eliminated by modifying the processing architecture. We use functional remodeling to modify the structural elements placement or functional parameters and technological remodelling to allow system architecture changes when structural elements should be added or removed. After functional and technological remodelling, another material flow simulation is performed for quantifying the increasing performance of the optimized system architecture. The last stage is the profit validation based on the comparison of the investment in optimizing the system architecture and the gain from the increased system performance.

As one can see we refer here at the MFM algorithm in a restricted acceptation, as a set of simulation and optimization rules for duration and costs using a virtual architecture model. This MFM algorithm for optimizing waste processing architectures performances should be adapted to the specific requirements of various kind of waste [7]. We define a processing architecture as a set of structural elements interacting such that to contribute to a product manufacturing. We recognize here four main structural elements (work points, transport systems, transfer systems, buffers) and two auxiliary structural elements (mobile entities and human resources) for a waste processing architecture. Establishing a processing architecture for a product or a family of products involves designing and structuring the relationships between structural elements, processes, procedures and human resources.

We agree to define the role of the structural elements as follows. The work points (reverse vending machines, waste processing equipment, machine tools, assembly points etc.) are the structural elements covering all the technological stages necessary for the part manufacturing. Transport systems (conveyors, Automated Guided Vehicles–AGVs, Rail Guided Vehicles–RGVs etc.) are moving the mobile entities between the long distance situated structural elements. The transfer systems (robots and manipulators) are orienting and fixing the mobile entities, and are making the connection between near structural elements. Buffers are storing the mobile entities for a limited period of time. The mobile entities are the various kind of waste [8, 9].

b. According with the presented MFM steps a diagnosis of the key performance indicators associated with the waste flow in the current system was made. Based on the virtual model, a waste flow simulation using Witness Horizon was performed in order to achieve the system's diagnosis. The simulation was conducted for a working week timeframe (4200 min) and the results show the waste collection rate for each module component. The results should be similar with the data provided by the company. The problems encountered on each system's module are presented.
c. Based on the diagnosis results, the changes needed for the system redesign are described in order to increase the usage degree, the productivity and storage capacity of a selective waste collection system. The proposals concerning the characteristics and configuration of each module are described in order to obtain an optimized system version.
d. Using Witness Horizon, the system configuration has been modified so as to correspond to the new characteristics and to obtain an optimized system version.
e. A new waste flow simulation based on this new system configuration has been made. Relevant reports were interpreted to show if the results confirm that the optimized system version has a significant increase collection capacity for each category of waste, an increase in productivity and an increased waste storage capacity.

3 Virtual Model of the Selective Waste Collection System

Witness Horizon simulation software (version 2016) was used for modelling the current system architecture, for conducting a diagnosis of the key performance indicators associated with waste flow in the current system and to identify necessary changes in order to optimize the system as required by the company (project partner).

The virtual model made with Horizon Witness for the selective waste collection system in the current configuration is presented in Fig. 1. In order to obtain a virtual model, all components of the system have been defined and classified in the software's specific categories: work points, transport systems, transfer systems, storage systems, human resources, mobile entities. Next, using the database available in Witness Horizon or by using the "*Import*" function, the structural elements were positioned in the working area. All the system's components were parametrized in accordance with the current system functioning. The relationships between the system's elements were made in order to obtain an identical waste flow with the one from the real system.

The waste flow modelled in Witness Horizon is as follows:

The mobile entities PET (recipients of maximum 2 litres), cans (maximum 0.5 litres) and GB (glass bottles) are introduced in the Reverse Vending Machine (RVM) according to the defined relationship. RVM will take, identify and sort the waste, and will distribute the waste to the containers B1 (PET), B2 (cans) and B3 (GB). In Fig. 2 the RVM equipment used now in the current system can be seen. When the containers B2 (maximum capacity of 2000 recipients) and B3 (maximum capacity of 600 recipients) are full, the operator will empty the containers and will store the waste to Cans_storage_area (shows the number of full containers) and to GB_storage_area (shows the number of full containers). When the container B1 is full (maximum capacity of 600 recipients) the operator will transport the recipients to the baler and the resulted bale will be stored to PET_plastic_storage (shows the number of bales).

PR (plastic recipients), including PET over 5 litres, are mobile entities that cannot be accepted by the RVM Tomra equipment. PR are introduced in PRM (Plastic recipients module) and stored in the B4 container. When the B4 container is full (maximum capacity of 200 recipients) the operator will transport the recipients to the press and the resulted bale will be stored to PET_plastic_storage (shows the number of bales).

PFPB (Plastic film and plastic bags) mobile entities are introduced in PFPBM (Plastic film and plastic bags module) and stored in the B5 container. When the B5

Fig. 2 Reverse vending machine from TOMRA, T83 Hcp model [10]

container is full the operator will transport the waste to the press and the resulted bale will be stored to PET_plastic_storage (shows the number of bales).

TP (Tetra Pack) mobile entities are introduced into TPM (Tetra Pack module) and stored in the B6 container (maximum capacity of 400 recipients).

Paper and Cardboard mobile entities can have two flow routes depending on the amount brought by the client. Rules have been defined in Witness Horizon so that amounts smaller than 0.5 kg of paper or paperboard will be introduced in PCM and stored in the B7 container. When B7 is full the operator will sort the paper and cardboard and will transport them to the baler. If the amount brought by the customer exceeds 0.5 kg, the operator will weigh the waste and will transport it to the baler thus resulting a bale of about 50 kg which will be deposited in Paper_Cardboard_storage (shows the number of bales).

P (polystyrene) mobile entities are introduced into PM (polystyrene module) and stored in the B8 container.

Battery mobile entities (A, AA and AAA type etc.) are introduced into BM (Battery module) and stored in the B9 container (maximum capacity of 400).

Light_Bulb and FL (fluorescent lamp) mobile entities are introduced into LBFLM (Light_bulb and fluorescent lamp module) and stored in the B10 container (maximum capacity of 200).

Refrigerator, Washing machine, TV and SEA—small electronics and appliances (including phones, laptops, PC drives, printers, video consoles, microwave etc.) mobile entities will enter the WEEEM (waste electrical and electronic equipment module) according to the rules defined at system's modelling [11, 12]. For WEEEM a 5 min average cycle time was defined for loading, weighing, voucher issuance and storage. An output rule for WEEEM has been defined (see Fig. 3) to set the Operator to transport the Refrigerator, Washing machine and TV mobile entities to

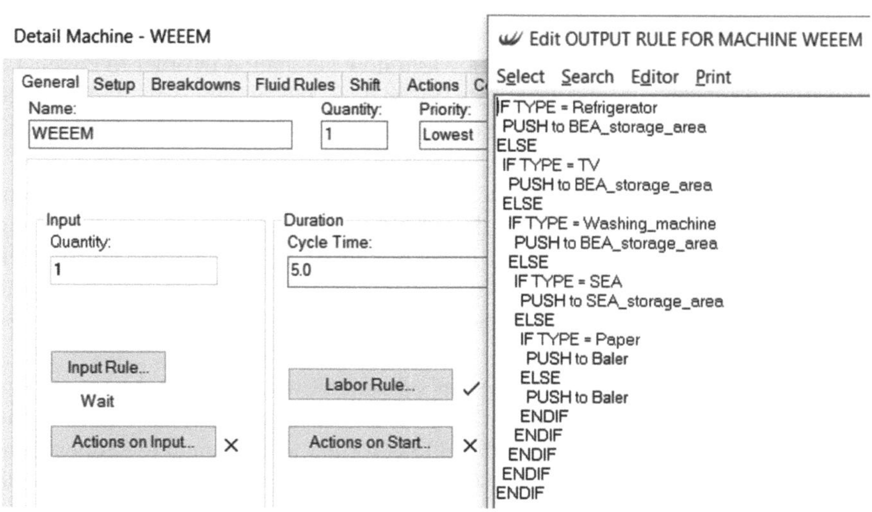

Fig. 3 WEEEM work point parametrization

BEA_storage_area—Big electronics and appliances storage area (shows the number of mobile entities stored) and the SEA mobile entities to SEA_storage_area—Small electronics and appliances storage area (shows the number of mobile entities stored).

The input data for the mobile entities parametrization were provided by the company (project partner) and represent an average of system's inputs. Each mobile entity has been linked with the dedicated working point. For example, PET, GB and Can mobile entities have "Output" rules to RVM, Light_bulbs and FL to LBFLM, electronics and appliances to the WEEEM etc. For parametrization, distribution laws have been use because the amounts of the waste entered into the system may vary. For example, for paper and cardboard we defined a "*IUniform*" distribution law. We defined values between 0.1 and 20 kg. The output rule contains an "*IF*" conditional, so, if the amount is smaller than 0.5 kg, cardboard/paper goes to PCM (Paper and Cardboard Module), and if the quantity brought for recycling is greater than 0.5 kg, cardboard/paper will be weighed on the scale platform from the WEEE module.

In the virtual model two operators (Operator_1 and Operator_2) have been defined. They were set to intervene in the system's waste flow in the same manner with the real system. Thus, the operator will weigh the appliances/electronics large and small, the paper and the cardboard on the WEEE module, then, depending on the waste type will transport it to storage (electronics/appliances) or to Balers (paper and cardboard). Also, the operator will empty the containers B1, B2, B3, B4, B5 and B7. The operator was set so as to use the baler for waste pressing, tying and for bales storage.

4 System's Diagnosis

The virtual model is used to perform a waste flow simulation in order to conduct a diagnosis. The diagnosis is necessary to analyse the current key performance indicators and to determine the input data necessary to optimize the system's waste flow [13, 14].

The simulation was made for a working week timeframe by defining a function: 60 min × 10 h × 7 days to reflect the 10 h/day and 7 days per week (including Saturday and Sunday) according to the actual program. We decided for a week diagnosis because according to the data provided by the company, the waste is taken weekly for transportation to the recycler. In some cases, exceptions to this rule may occur.

Figure 4 shows a report of the system's diagnosis resulted from the waste flow simulation made in Witness Horizon. The "*Name*" field contains the names of the system's component. The "*Total In*" field shows the total of mobile entities from the storage systems (Buffers). The "*Total Out*" field shows the number of mobile entities taken by the recycler. The "*Now In*" field shows the number of mobile entities that are currently in containers or in the storage areas (coinciding with

WITNESS

Buffer Statistics Report by On Shift Time

Name	Total In	Total Out	Now In	Max	Min
Cans_storage_area	1	0	1	1	0
SEA_storage_area	14	0	14	14	0
B5	350	0	350	350	0
B6	700	0	700	700	0
B8	35	0	35	35	0
B9	140	0	140	140	0
B10	88	0	88	88	0
BEA_storage_area	12	0	12	12	0
Paper_Cardboard_storage	12	0	12	12	0
PET_plastic_storage	8	0	8	8	0
GB_storage_area	2	0	2	2	0

Fig. 4 Diagnosis results for containers and storage areas

"*Total*" because the waste will not have yet been taken by the recycler), "*Max*" and "*Min*" shows the maximum and the minimum of mobile entities who were ever stored on that item. Cans_storage_area and GB_storage_area show the number of full containers, Paper_Cardboard_storage and PET_plastic_storage show the number of bales obtained after pressing and binding. BEA_storage_area and SEA_storage_area show the number of electronics and appliances stored. B6, B7, B8, B9 and B10 show the number of mobile entities stored in these containers.

The RVM module dedicated to the collecting of PET, cans and glass bottles was set to take a maximum of 42 entities per minute. It was necessary to use distribution laws because in reality this module is not required continuously. There may be times during the day when no customer comes to bring waste or may be times in a day when small queues of clients are created. Also, taking into account the fact that in reality not all the customers are introducing the waste with the same pace into RVM, results in the need for using distribution laws in the simulation in order to obtain results similar with the real system's inputs and outputs. The results obtained from reports (simulation for one week) shows the following:

- A container has been filled with the cans (Cans_storage_area), also in B2 there are currently 50 cans. Given that B2 have a capacity of 2000 cans, a total of 2050 cans are collected.
- Two containers were filled with glass bottles (GB_storage_area),and in B3 there are currently 200 glass bottles. Since B3 have a capacity of 600, a total of 1400 glass bottles are collected.
- PET_plastic_storage' s report shows that eight bales were stored.

The main problem is that after the RVM equipment identifies and sorts the waste, the remaining operations are carried out by the operator: emptying B1, B2, B3 containers, the operator transports the waste for storage etc.

The Report on the B6 container shows that 700 different packaging or Tetra Pak were collected. The container is not full, its maximum capacity being 750 packages. Given that in the simulation the average values were used, there could be cases when the container could be filled faster, which would jeopardize the collection.

Paper_Cardboard_storage reports show a total of 12 bales obtained after pressing and binding the paper and paperboard. From the 12 bales, 8 are bales with cardboard and 4 are bales with paper. Taking into account that a bale has about 50 kg, it results that 400 kg of cardboard and 200 kg of paper are collected. These are average values and peak areas may exist when the amounts may be higher.

The Report on the B8 container shows that 35 pieces of polystyrene were collected. The container is not emptied by the operator because its maximum capacity is 50 pieces. Report on B9 container shows that 140 batteries were collected. The container is not emptied because its maximum capacity is 400 pieces. If we consider the average values used for the simulation, we can estimate that for filling the B9 container it would take approximately 20 days.

Report on B10 container shows that 88 light bulbs and FL were collected. The container is not emptied since its maximum capacity is 200 pieces. The detailed report shows that 53 light bulbs and 35 FL (fluorescent lamps) were collected. If we consider the average values used for the simulation, we can estimate that for filling the container it would take approximately 15 days. Report on SEA_storage_area shows that were collected 14 pieces of electronics and small appliances. Report on BEA_storage_area shows that 12 pieces were collected. An identified problem in the collection of WEEE relates to the storage capacity, resulting that in the current system configuration the values approaches the maximum storage capacity.

5 System's Optimization

To identify the changes needed to redesign the current system a waste flow optimization will be made based on the virtual model and on the diagnosis made in Witness Horizon.

Increasing the number of operating hours from 10 to 24 h/day is one of the main objectives. Thus, on the simulation system for the optimized version we will consider this requirement, and the reference for the simulation will become 60 min $\times$ 24 h $\times$ 7 days, which means 10,080 min of operation time in a week in comparison with 4200 min (in the current system waste can be collected between 9 a.m. and 7 p.m.) in its current version. Of course, we will take into account that between 11 p.m. and 6 a.m. the amounts of collected waste will be lower. However, between 6 a.m. and 9 a.m. and between 7 p.m. and 11 p.m. significant quantities of waste will be collected. In feedbacks questionnaires, most of the customers said they prefer a system that is available 24/7.

Modules considered a priority for the redesign and automation are: the RVM module (PET, cans, glass bottles), the Paper and cardboard module and the WEEE module. The optimized version of the system configuration made in Witness Horizon is presented in Fig. 5.

The RVM module dedicated to the collection of PET, aluminium cans and glass bottles was modified so that it can contain two similar equipment for loading, identifying and sorting of the waste: RVM and RVM2. In this way the bottlenecks caused by an equipment error will be eliminated and waiting queues during pick hours will be reduced. After sorting the mobile entities are transported by C1 to storage containers. The cans are transported to B2 (with the modified capacity of 3000 from 2000) and to B2_1 (with a capacity of 3000 units) containers. The glass bottles are transported to B3 (with the modified capacity of 2000 from 600) and to B3_1 (with a capacity of 2000 units) containers. PETs are redirected to the C2 conveyor, which will transport them to the Baler. The obtained bale is picked up by the AS/RS (Automated storage and retrieval system) and will be stored in PET_plastic_storage. The new storage configuration allows the vertical storage of bales. The design of this module will be done in 2017–2018.

Plastic recipients module (PRM), Plastic film and plastic bags module (PFPBM) and Tetra Pack module (TPM) were not modified.

The Paper/Cardboard module will be modified so that the collected waste will no longer be weighed on the scale dedicated to the WEEE module. The Paper and Cardboard module will have a dedicated scale. The Paper and Cardboard module (PCM) was repositioned next to the WEEE modules. After pick up and weighing, the paper or the cardboard will be compacted and tied at the B7. Bales of

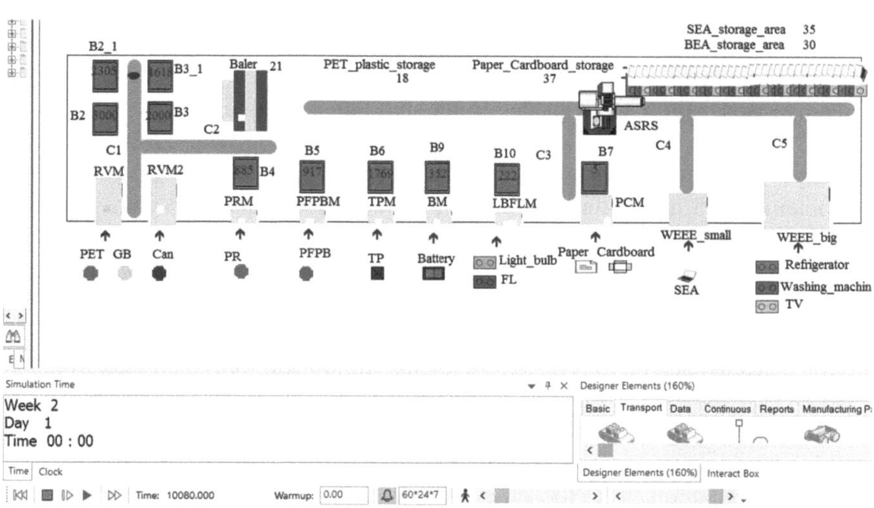

Fig. 5 The optimized system during the simulation in Witness Horizon

approximately 50 kg. will be obtained. The bale is pushed to the C3 conveyor, that will transport it to the to the transfer zone. The ASRS system will take over the bale and will deposit it atPaper_Cardboard_storage. In the new system configuration, the vertical storage of the bales is possible. The design of the Paper and Cardboard module according to the details here presented will be done in the second stage of the project (2017–2018).

In this version of the system, the polystyrene module (PM) was not used.

The battery module (BM) was repositioned, but the components and the flow were not modified.

The Light_bulb and fluorescent lamp module (LBFLM) was repositioned, but the components and the flow were not modified. The storage capacity of the B10 container was increased from 200 to 500 pieces.

The module for collecting electrical equipment and electronic waste was divided in two separate modules WEEE_small and WEEE_big.

The WEEE_small will take over small electronics. After identification and weighing they will be transported to the C4 conveyor to the transfer zone, point from which they will be picked up by the AS/RS system and will be deposited in the SEA_storage_area. Because the storage system was extended vertically, small electronics will be positioned at higher levels because they are lighter.

The WEEE_big module will take over the mobile entities such as refrigerators, washing machines, TV sets and after identification and weighing they will be transported by C5 to the transfer zone, from where they will be picked up by the AS/RS system and deposited in the BEA_storage_area at the lower levels because they are heavier. The designing of the WEEE module according to the here presented details will be done in the second stage of the project (2017–2018).

The results obtained after the simulation of waste flows in the optimized version are the following (see Fig. 5):

- The changes brought to the RVM model dedicated to the collection of PET, aluminium cans and glass bottles lead to a significant increase of the collected quantities. At PET_plastic_storage 18 bales of plastic recipients were stored, in comparison with 7 bales in the preliminary version, as can be seen in Fig. 5. 10,821 plastic recipients were collected, in comparison with 4200 units originally. The B2 container is full (3000 aluminium cans) and the B2_1 container contains 2305 aluminium cans, meaning that 5305 aluminium cans were collected in total, in comparison with 2050 initially. The B3 container is full (2000 glass recipients) and the B3_1 container has 1618 glass recipients, meaning that in total, 3618 glass recipients were collected, in comparison with 1400 units originally.
- The collected quantity for plastic recipients (from PRM) was increased from 350 to 885.
- The collected quantity for plastic film and plastic bags was increased from 3.5 kilos to approximately 9 kilos.
- The collected quantity for different packaging (including Tetra Pack) was increased from 700 to 1769 pieces.

- The collected quantity for batteries was increased from 140 to 352 pieces.
- The collected quantity for light bulbs and fluorescent lamps was increased from 88 to 222 pieces. 134 light bulbs and 88 fluorescent lamps were collected.
- The changes brought to the Paper and cardboard module lead to the increase of the collected quantities from 12 bales to 37 bales. In the Paper_Cardboard_storage 37 bales were deposited, as it can be seen in Fig. 5. Out of these, 24 were cardboard bales and 13 paper bales.
- 35 pieces were stored in the area dedicated to the storage of small electronics (see Fig. 5) as compared with 14 pieces in the preliminary version.
- The report for BEA_storage_area shows that 30 pieces were collected, in comparison with 12 pieces in the preliminary version. In the storage dedicated area 30 mobile entities are positioned, as showed in Fig. 5: 10 blue colour units (symbol used for refrigerators), 10 red coloured units (symbol used for washing machines) and 10 yellow colour units (symbol used for TV sets and monitors).

In the optimized configuration, the system can collect the above mentioned quantities but their collection also depends on external factors: the location of the system, dissemination initiatives of selective waste collection directed at citizens, the public awareness regarding the necessity of selective waste collection etc. The data obtained after the optimization of the system will be used as entry data in the second stage of the project (2017–2018) that will include the design activities for the WEEE module, Paper and cardboard and RVM modules. During this stage (during the design activities) changes can be brought to the system components, and changes can occur regarding the positioning and the composition of some models in comparison with the version shown in the present paper.

6 Conclusion

The presented research is centred on the development of a virtual model for the preliminary architecture of a waste processing system. We used the Witness Horizon software application for modelling the preliminary and the optimized architectures of the system and for making a diagnosis of the performance indicators levels associated with the waste flows for these two architectures. A material flow management algorithm was defined to optimize the system architecture based on the information obtained using this virtual model.

The preliminary architecture is representing the current state of the system as it is now, before optimization. The virtual model of the preliminary system for waste processing was generated, each element was parameterized and the relationships necessary for simulating the waste flow were defined. Then, a simulation was performed on the virtual model of the preliminary architecture to generate various diagnose reports regarding the performance indicators of the system. The waste flow simulation was done in order to make a diagnosis regarding the functioning of the system. Reports were generated and the results obtained for each module were interpreted in order to identify solutions for optimizing the system. The inputs in the system model considered a priority for the optimization were: the RVM module (PET, cans, glass bottles), the Paper and cardboard module and the WEEE module. In order to obtain the optimized version the paper presents the necessary modifications for each module. The virtual model of the optimized architecture obtained after these modifications is also presented. The simulation of the material flow performed on this model confirm the increase productivity and storage capacity for all waste categories.

In our future work, the results generated by the optimization of the system will be used as entry data in the second stage of the project (2017–2018) that will include implementing the new, IoT oriented, WEEE module, Paper and cardboard and RVM modules. During this stage changes should be brought to the system components to convert them into smart components. As a result, changes can also occur regarding the parameters positioning and the composition of some models in comparison with the version shown in the present paper.

Acknowledgements This work was supported by a grant of the Romanian National Authority for Scientific Research and Innovation, CNCS/CCCDI—UEFISCDI, project number PN-III-P2-2.1-BG-2016-0437, within PNCDI III.

References

1. The decision of the Romanian Parliament with reference to the circular economy Package–COM (2015) 614 final, published in the Official Journal of Romania nr. 93/8.02.2016
2. The 2008/98/CE directive of the European Parliament and of the Council from 19 November 2008 regarding waste and the abrogation of certain directives. Available at http://eur-lex.europa.eu/legal-content/EN/TXT/?uri=celex%3A32008L0098
3. The National Strategy for Waste Management—Ministry of Environment and Climatic Change. Available at http://www.mmediu.ro/beta/wp-content/uploads/2013/01/2013-01-11-DGDSP-SNGD.pdf
4. Law nr. 211/2011 regarding the treatment of waste (Art. 17). Available at http://www.mmediu.ro/beta/wp-content/uploads/2012/05/2012-05-17_LEGE_211_2011.pdf
5. Law 249/2015 regarding the management of packaging and waste packaging. Available at http://www.euroavocatura.ro/print2.php?print2=lege&idItem=1204
6. Cotet CE, Popescu D (2014) Material flow management in industrial engineering. In: Khosrow-Pour M (ed) Encyclopedia of information science and technology, category: industrial engineering, 3rd edn, Published in the United States of America by Information Science Reference, pp 3786–3794. ISBN:978-1-4666-5888-2 (hardcover). ISBN:978-1-4666-5889-9 (ebook)
7. Popa CL, Cotet CE, Ionita V, Gavrila St (2015) Modelling processing cell architecture by material flow simulation, Procedia Engineering 100, Published by Elsevier Ltd. pp 334–339 doi:10.1016/j.proeng.2015.01.376
8. Coteț CE, Popa CL, Enciu G, Popescu A, Dobrescu T (2016) Using CAD and flow simulation for educational platform design and optimization, Int J Simul Model 15(1):5–16, ISSN 1726-4529. doi:10.2507/IJSIMM12(2)2.225

9. Ayadi M, Costa Affonso R, Cheutet V, Masmoudi F, Riviere A, Haddar M (2013) Conceptual model for management of digital factory simulation information. Int J Simul Model 2 (12):107–119, ISSN 1726-4529
10. Reverse vending machine from TOMRA. Available at https://www.tomra.com/
11. Ciocoiu CN, Colesca SE, Rudăreanu C, Popescu M-L (2015) Management of waste electrical and electronic equipment in Romania: a mini-review. Waste Manage Res 34(2):96–106, ISSN:0734-242X
12. Pérez-Belis V, Bovea MD, Ibáñez-Forés V (2015) An in-depth literature review of the waste electrical and electronic equipment context. Trends Evol Waste Manage Res 33(1):3–29, Published 18 Nov 2014. ISSN:0734-242X
13. Kumara A, Holuszkoa M, Crocce D, Espinosa R (2017) E-waste: an overview on generation, collection, legislation and recycling practices. Resour Conserv Recycl 122:32–42, ISBN:0921–3449
14. Vimieiro GV, Lange LC (2015) Proposal for an operational performance indicator to evaluate sorting and composting plants for municipal solid waste. Int J Environ Waste Manage 15 (2):114–129. doi:10.1504/IJEWM.2015.068933, ISSN online:1478-9868, ISSN print: 1478-9876

An Advanced CAI Model for Inspection Planning on CMM

Slavenko M. Stojadinovic, Vidosav Majstorovic and Numan M. Durakbasa

Abstract This paper presents an advanced CAI model for inspection planning of prismatic parts on CMM. The model consists from mathematical model, modeling inspection features, sampling strategy, analysis of the probe accessibility, collision-free generation, and optimization of the path using ant's colony and simulation of the path. Needed geometrical information for path planning are taken from IGES and STL file, and input data about tolerance entries from already developed knowledge base. The presented model is an advanced approach oriented towards definition of the intelligent inspection system. Advantage of the model is in reduction the total measurement time by reducing the time needed for the planning of the measurements due to automated generation of certain activities.

Keywords CAI · CMM · Inspection planning · Prismatic parts

1 Introduction

The inspection on CMM is based on a complex software support for different classes of tolerance (location, flatness, perpendicularity, parallelism etc.). Conducting uniform inspection plan on CMM is a special problem, which depends on the metrological complexity of prismatic parts, intuition and skill of the

S.M. Stojadinovic (✉) · V. Majstorovic
Department for Production Engineering, Faculty of Mechanical Engineering, University of Belgrade, Belgrade, Serbia
e-mail: sstojadinovic@mas.bg.ac.rs

V. Majstorovic
e-mail: vmajstorovic@mas.bg.ac.rs

N.M. Durakbasa
Department for Interchangeable Manufacturing and Industrial Metrology, Faculty of Mechanical and Industrial Engineering, Vienna University of Technology, Vienna, Austria
e-mail: numan.durakbasa@tuwien.ac.at

V. Majstorovic and Z. Jakovljevic (eds.), *Proceedings of 5th International Conference on Advanced Manufacturing Engineering and Technologies*, Lecture Notes in Mechanical Engineering, DOI 10.1007/978-3-319-56430-2_5

inspection planner. For this reason, today there is a tendency for lower human participation in certain inspection planning operations which leads to the need to develop new CAI model with elements of artificial intelligence (AI techniques).

In recent decades the CMM inspection planning is developed through the following approaches: (i) manual plan of inspection, (ii) plan generated with CAI software that was held until today, (iii) plan based on expert systems, and (iv) intelligent planning concept, i.e. CAI models with elements of artificial intelligence. The common element of all these approaches is a workpiece, i.e. the touching object used for measurements. In the case of inspection of prismatic parts the objects are basic geometric features (plane, cylinder, cone, etc.). According to the manner of an analysis of workpiece geometry, three approaches could be distinguished: (i) analysis of geometry, (ii) analysis of tolerances, and (iii) combined approach.

On the other hand, inspection process is composed of a few key elements such as path planning, collision avoidance, accessibility analysis and a workpiece setup, as well as configuration of measuring probes. The level of development and implementation of until now developed methods for generating the inspection plan, defines the representation of these elements in inspection plan. A complete system for inspection planning contains all mentioned key elements. Paper [1] presented the elements such as part setup and configuration of the measuring probes. In paper [2–6] the approaches to path planning are given. Research work presented in [7] considers the solutions for collision avoidance, while [8] presents the analysis of the measuring sensor accessibility. The inspection planning may be considered through the division into local and global plan of inspection [9], which clearly group together some elements of the inspection plan.

This paper presents an advanced CAI model of path planning for inspection of prismatic parts on CMM. The model is based on the model presented in [13], while the advances relate to generating and simulation of the optimal probe path based on CAD model of the prismatic part and its tolerances. Experimental example included the measurement path for ten measurement points, but distribution methods allow to enter any number of measurement points, depending on the type and quality of the prescribed tolerances. Based on the defined connection between tolerance and geometry according to [10], this paper presents an advanced CAI path planning model where the output is obtained as point-to-point optimized measurement path for a prismatic part. The path obtained in this form is then simulated to provide a visual check of collisions in the measuring path.

2 CAI Model

The CAI model consists of mathematical model, modeling inspection features, sampling strategy, analysis of the probe accessibility, collision-free generation, and optimization of the path using ant's colony and simulation of the path. This model represents an advanced version of the model given in [13]. Advancement refers to

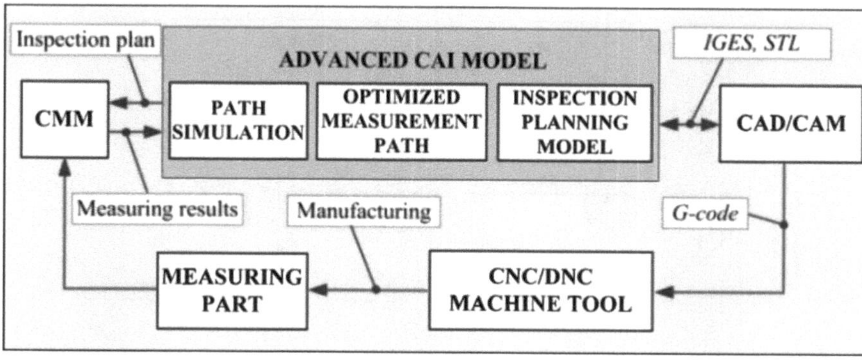

Fig. 1 An advanced CAI model

the developed optimization of the measurement path using ant's colony and simulation of the path. Figure 1 presents the advanced CAI model with its elements and the links between them. The model is suitable for both processes, in-process (manufacturing) measurement and final measurement of the workpieces. In-process measurement can be carried out on the machining centers with the help of measuring heads which are placed as tools. The measuring head has independent communication from the control unit of the machine. In this way, for a given accuracy of the measuring head, it eliminates errors of machine center on the measuring head error. On the other hand final measurement can be carried out by creating a control data list for a concrete CMM.

2.1 Mathematical Model

Figure 2 shows the mathematical model of path planning for inspection of prismatic parts on the CMM. The basic purpose of the model is to define the relationship between the coordinate system of the measuring machine, workpiece and geometrical features of the part.

The model is based on the Eq. (1), which calculates the position vector of measurement point in the coordinate system of the measuring machine as:

$$^{M}r_{P_i} = {}^{M}r_W + {}^{W}r_F + {}^{F}r_{P_i} = {}^{M}r_F + {}^{F}r_{P_i} \tag{1}$$

where $^{M}r_F$ is feature (plane) system's position vector in machine coordinate system, $^{F}r_{P_i}$ is probed point's position vector in feature coordinate system, $^{M}r_W$ is workpiece system's position vector in machine coordinate system, and $^{W}r_F$ is feature system's position vector in workpiece coordinate system.

The matrix T takes into consideration position and orientation of the coordinate systems by [12, 13].

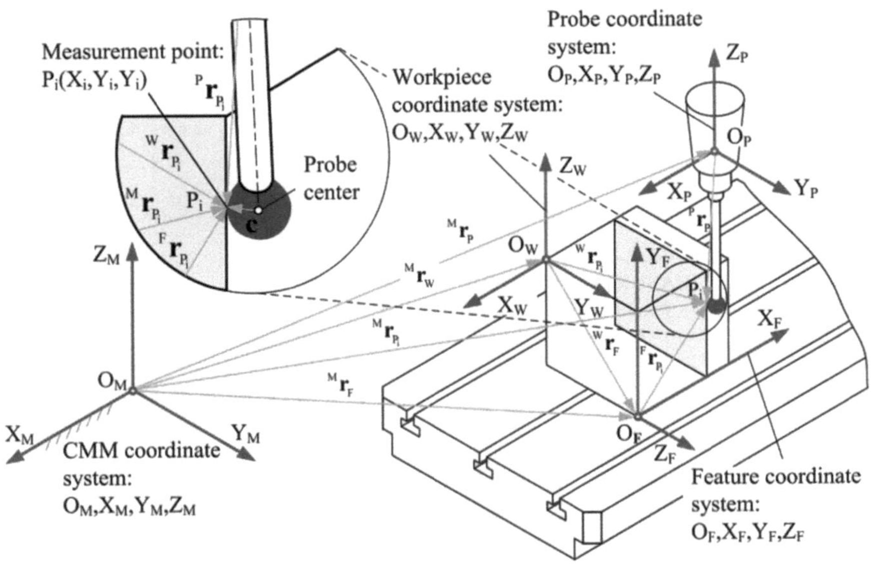

Fig. 2 Mathematical model of the path planning [12, 13]

2.2 *Inspection Features and Parameters*

Inspection feature construction is based on the basic geometric features and their uniquely determined parameters. The geometric features included in this modeling are: (a) point; (b) plane; (c) circle; (d) hemisphere; (e) cylinder; (f) cone; (g) truncated cone; (h) truncated hemisphere. An example of features (hemisphere) with its parameters is presented in Fig. 3.

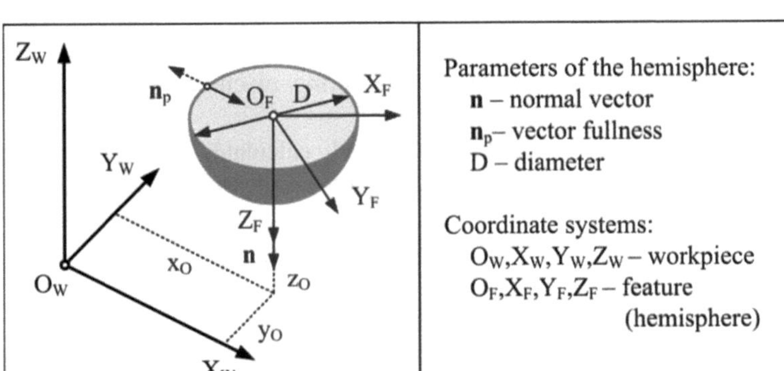

Fig. 3 A hemisphere with its parameters

2.3 Sampling Strategy—Distribution Measurement Points

Sampling strategy is based on *Hemmersly* sequences [11], for the calculation of coordinates of measurement points for two axes of the feature. By modifying the Hemmersly sequences, we define the distribution of measuring points for basic geometric features which are considered in this paper. For these purposes we utilize Cartesian's coordinate system O_F, X_F, Y_F, Z_F and polar cylindrical coordinate system O'_F, X'_F, Y'_F, Z'_F. The coordinates of measuring points in Cartesian's coordinates system are denoted by $P_i(s_i, t_i, w_i)$.

For example, the equations for calculation of measuring point coordinates for hemisphere are:

$$s_i = \sqrt{R^2 - \left(\left(\sum_{j=0}^{k-1}\left(\left[\frac{i}{2^j}\right]\mathrm{Mod2}\right)\cdot 2^{-(j+1)}\right)\cdot R\right)^2}\cdot\cos\left(\frac{i}{N}\cdot 360^\circ\right) \quad (2)$$

$$t_i = \sqrt{R^2 - \left(\left(\sum_{j=0}^{k-1}\left(\left[\frac{i}{2^j}\right]\mathrm{Mod2}\right)\cdot 2^{-(j+1)}\right)\cdot R\right)^2}\cdot\sin\left(\frac{i}{N}\cdot 360^\circ\right) \quad (3)$$

$$w_i = \left(\sum_{j=0}^{k-1}\left(\left[\frac{i}{2^j}\right]\mathrm{Mod2}\right)\cdot 2^{-(j+1)}\right)\cdot R \quad (4)$$

where: $R[\mathrm{mm}]$ is the radius of a hemisphere, N is number of measuring points, $k = \log_2 N$ is constant, Mod2 is mathematical operator whose result is the remainder after dividing the expression by two.

2.4 Collision Free-Generation and Probe Accessibility Analysis

To perform inspection of a primitive, it is necessary to conduct an analysis of the accessibility of the measuring sensor and to determine two new sets of points. Accessibility analysis includes the determination of possible directions of probe approach. Generally, there are five possible directions of probe approach presented in Fig. 4. Based on the necessary approach directions, configuration of the measuring probes will be determined.

The first set of points $P_{i1}(x_{i1}, y_{i1}, z_{i1})$ presents points for the transition from rapid to slow feed of CMM. The probe path between points $P_{i1}(x_{i1}, y_{i1}, z_{i1})$ and $P_i(x_i, y_i, z_i)$ is presented as d_1—a slow feed path, and the probe path between points $P_{i2}(x_{i2}, y_{i2}, z_{i2})$ and $P_{i1}(x_{i1}, y_{i1}, z_{i1})$ is d_2—a rapid feed path. With this approach to

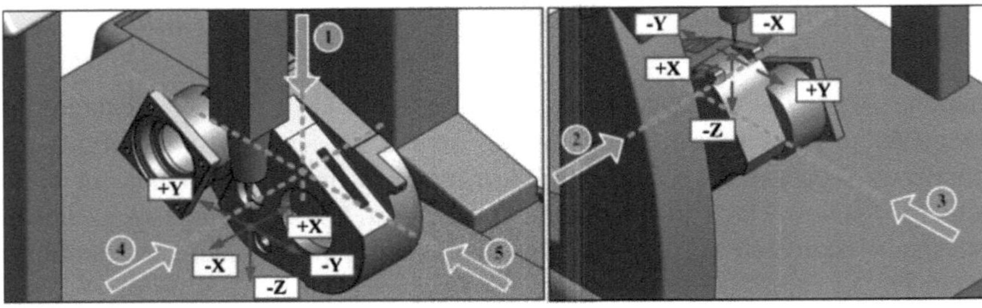

Fig. 4 Five directions of probe approach [12]

defining sets of points and paths, it is possible to avoid collision between the features and probe.

In the inspection of prismatic parts, there are three different cases for the definition of point sets $P_{i2}(x_{i2}, y_{i2}, z_{i2})$ and $P_{i1}(x_{i1}, y_{i1}, z_{i1})$. According to [12, 13], the cases may be classified as follows:

- concave cylindrical surface,
- convex cylindrical surface,
- flat surface.

Defined directions in this section are basic directions of approach. Depending on the number of CMM axes and its path programming, other directions can be derived from the basic directions.

Based on STL model of presenting geometry, the tolerances of workpiece, the coordinates of the last point $P_{(N_{F1})1}$ for inspection of the previous feature and the coordinates of the first point $P_{(N_{F2})1}$ for inspection of the next feature, in [13] the principle of collision avoidance is presented. The principle is iterative and consists of moving line *p* for distance δ until line (segment of the measurement path) becomes collision free.

2.5 Simulation and Optimization of the Measuring Path

As we said the measuring path could be presented as a set of points that a measuring probe passes through during the inspection of a feature. Application of ant's colony optimization in a coordinate metrology is based on the solution of traveling salesman problem, where the set of cities that the salesman should pass through with the shortest possible path corresponds to the set of points of a minimal measuring path length [14]. Precisely, the set of cities corresponds to the set of points, and the salesman corresponds to the measuring probe.

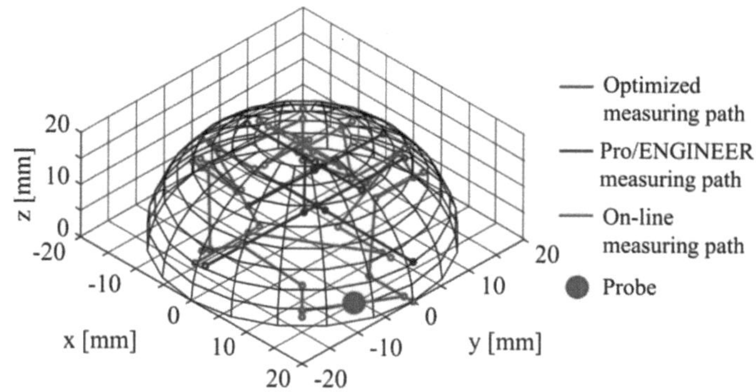

Fig. 5 Simulation three type of measuring path on example hemisphere

The model is based on the following equation for calculation of the measuring probe path during the measurement on N measuring points:

$$D_{tot} = \sum_{i=0}^{N-1} \left(\left| \overrightarrow{P_{i2}P_{i1}} \right| + 2 \cdot \left| \overrightarrow{P_{i1}P_{i}} \right| + \left| \overrightarrow{P_{i1}P_{(i+1)2}} \right| \right) \tag{5}$$

Finding the shortest measuring path is the main criteria for optimization, and it has the influence on the reduction of the total measurement time; it is one of the goals of this research. In application of ant's colony optimization, the ant colony with 1000 ants was used in 100 iterations to obtain the optimal path. The optimized path is shown with red color in Fig. 5.

The simulation provides a visual check two type of the measuring path. First type is path generated in the software Pro/ENGINEER and second on-line programmed CMM's measurement path. The distance ratio for sphere is: $d_{optimized}/d_{Pro/E} = 84.3\%$, $d_{optimized}/d_{on-line} = 62.9\%$. This provides a savings in the path length of at least 15.7% and a maximum of 37.1%.

3 Experiment

The experiment includes manufacturing and measurement of test workpiece produced specifically for this purpose (Fig. 6). The test workpiece consists from all the geometric features covered by the model. The experiment was performed on the measuring machine "ZEISS UMM 500". Manufacturing of the workpiece was performed on 4-axes (X′ Y Z B′) CNC machining center (horizontal milling and drilling machine, type "LOLA HMC 500"), in Laboratory for machine tools of the Faculty of Mechanical Engineering in Belgrade. Surface finish quality is N7 and N8, material is aluminum.

Fig. 6 Manufacturing and inspection of workpiece

The results of the inspection show that all tolerances of the workpiece are within the specified limits. This confirms the efficiency of the proposed advanced CAI model towards intelligent CAI model. The complex geometry of the workpiece by the model changes to the set of points whose sequence defines the initial measuring path of sensors without collision with workpiece.

4 Conclusion

Advanced CAI inspection planning model provides the basis for further development of intelligent planning inspections of prismatic parts on CMM. It is especially reflected in reduce of the preparation measurement time for the process inspection planning based on the automatic generation of measurement protocols and optimizing the probe path for the reduction of measurement time. By reducing the measurement time through these two components total measurement time required for inspection one of the part is reduced. Thus, it is reducing the costs per measurement part.

On the other hand, developed CAI model based on the automatic generation of measurement protocols and path in the point-to-point form leaves the possibility of minimal human participation in the process of inspections planning. Minimal human participation gives the possibility for a minimal error in this part of the inspection process.

The developed CAI model gives answers to industry requirements such as (i) the high geometric variation, (ii) faster product placement on the market based on the reduction of total production time by reducing the time required for inspection planning, (iii) Maintenance a permanent—required level of quality through automation of inspection activities carried out by the inspection planner. The model is suitable for both of the process, in-process (manufacturing) measurement and final measurement of the workpieces.

Limitation of the developed approach is its applicability only for prismatic parts, not for the parts with free-form surfaces for inspection, because the model of inspections planning and optimization is developed only for basic geometric features. For this reason, the subject of future research will be parts with free-form surfaces.

The model is especially suitable for application in case of measuring path planning for geometrically complex prismatic parts with large number of tolerances.

References

1. Hwang YC, Tsai YC, Chang AC (2004) Efficient inspection planning for coordinate measuring machines. Int J Adv Manuf Technol 23:732–742
2. Gu P, Chan K (1996) Generative inspection process and probe path planning for coordinate measuring machines. J Manuf Syst 15(4):240–255
3. Prieto F, Redarce T, Lepage R, Boulanger P (2002) An automated inspection system, Int J Adv Manuf Technol 19:917–925, Springer, London
4. Lin YJ, Murugappan P (1999) A new algorithm for determining a collision-free path for a CMM probe. Int J Mach Tools Manuf 39:1397–1408 (Elsevier Science Ltd.)
5. Stojadinovic S, Majstorovic V, Durakbasa NA (2015) Feature—based path planning for inspection prismatic parts on CMM, XXI IMEKO world congress "Measurement in Research and Industry", In: Proceedings, ISBN 978-80-01-05793-3, pp 1551–1556, Prague, Czech Republic, August 30th–September 4th, 2015, Czech Technical University in Prague, Faculty of Electrical Engineering, Department of Measurement
6. Flack D (2001) CMM measurement strategies, ISSN 1368-6550, National Physical Laboratory Teddington, Middlesex, United Kingdom, TW11 0LW
7. Lin C-Z, Chen C-C (2001) Collision-free path planning for coordinate measurement machine probe. Int J Prod Res 39(9):1969–1992 (Taylor & Francis Ltd.)
8. Lim PC, Menq HC (1994) CMM feature accessibility path generation. Int J Prod Res 32 (3):597–618 (Taylor & Francis Ltd.)
9. Zhao F, Xu X, Xie QS (2009) Computer—Aided inspection planning—The state of the art. Comput Ind 60(7):453–466
10. Majstorovic DV, Stojadinovic MS (2013) Research and development of knowledge base for inspection planning prismatic parts on CMM. In: Cracow-Kielce (ed) 11th international symposium on measurement and quality control, Poland, 11–13 Sept 2013
11. Lee G, Mou J, Shen Y (1997) Sampling strategy design for dimensional measurement of geometric features using coordinate measuring machine, Int J Mach Tools Manuf 37(7):917–934, Great Britain
12. Stojadinovic S (2016) Intelligent concept of inspection planning for prismatic parts on the CMM, Doctoral dissertation (on Serbian language), University of Belgrade, Faculty of mechanical engineering, Belgrade
13. Stojadinovic S, Majstorovic V, Durakbasa N, Sibalija T (2016) Towards an intelligent approach for CMM inspection planning of prismatic parts. Measurement 92:326–339. doi:10.1016/j.measurement.2016.06.037. ISSN:0263-2241
14. Stojadinovic S, Majstorovic V, Durakbasa N, Sibalija T (2016) Ants colony optimization of the measuring path of prismatic parts on a CMM. Metrol Meas Syst 23(1):119–132. doi:10.1515/mms-2016-0011(ISSN0860-8292)

Flexible Control System Used in the Nano-technological Production Flow

Popan Gheorghe and Angelescu Dorin

Abstract The continuous development of new products has increased in pace, with multiple facilities, reduced life cycles of new products and requires automation and robotization of the Nano-technology production processes. However, due to the specificity of the field, the automation of the Nano-technology industry faces a serious challenge. Classic products are made in series and face little change over the years. In the field of Nano-technology, product life cycles were shortened significantly and series production must adapt to the market. The rapid change in production requires a high degree of flexibility. Manufacturing various models and the large number of products requires preparation of the Nano-fabrication to compensate efficiently for fluctuations of the production series. For the automation of Nano-fabrication to be easily and individually adapted to various operations (handling, testing, control, elimination of production flow of parts or defective products) and to be able to react quickly to changes occurring in Nano-technology production processes production must be of modular construction and must provide maximum flexibility in production. The control on the Nano-technology production processes developed by INCDMTM Bucharest offers solutions for the automation of the Nano-technology production processes having both modular special items of flexibilization of the process of handling and interoperable ones, along with the possibility of change facilitated by software that controls the entire process control and selection of parameters admissibility of each item checked. The device is provided with the possibility to ensure laser checking, optical checking via Video Inspection and Atomic Force Microscope. The process control is achieved in the flow, the power is robotic and it has the possibility to check both by sampling and total control.

Keywords Measurement system · Calibrating nano-technology · Nano-metrology · Optical measuring system

P. Gheorghe (✉) · A. Dorin
INCD Mechatronics and Measurement Technique, Bucharest, Romania
e-mail: popangeorge@yahoo.com

A. Dorin
e-mail: dorin.angelescu@incdmtm.ro

V. Majstorovic and Z. Jakovljevic (eds.), *Proceedings of 5th International Conference on Advanced Manufacturing Engineering and Technologies*,
Lecture Notes in Mechanical Engineering, DOI 10.1007/978-3-319-56430-2_6

1 Introduction and Research Problems

The rapid evolution in the field of technologies related to Nano-manufacturing and Nano devices based on electrical, optic, magnetic, mechanic, chemical and biological effects would allow measurements in the length ranges involved. Furthermore, micro sensors, transducers and ultra-accurate machines must be calibrated or verified on production and on reception by the beneficiary because it is through them that the measuring unit is transmitted to users (producers) [7]. Traditional measuring means have proved limited in terms of accuracy because of the physical law [2]. The approach to the field of Nano-technologies, which is worldwide only at the stage of incipient concern, is suggested to be achieved by means of unconventional methods, which mainly allow quantifications in the field of Nano-metrics [3].

Many fields have come to currently employ Nano-technology. The study of biological systems and of certain materials in the Nano-metric field is century-old. The recent opportunity to perform surgery by Nano scale systems has caused a breakthrough in engineering thought and in devising new Nano scale systems [9].

Financial and intellectual investments in semiconductors and optoelectronics have foreshadowed this trend of thought resulting in the progress of Information Technology. The results have brought about a spate of new initiatives and have contributed to defining Nano-technology as confined to structures smaller than 100 nm. Starting from this specification, the semiconductors industry is likely to be espousing the Nano technologies in a few years.

Potential requests for nanotechnology pertain to vast fields, from electronics to optical communication, from new materials to biological systems. By disseminating the results, researchers increase the number of interested scholars, who may be potential large-scale users of Nano-devices, systems and materials.

A few companies have begun assessing such findings, most applications having only reached the research stage and that of laboratory results. The main hindrance standing in the way of using Nano-technologies at a large industrial scale is the lack of an infrastructure.

All production processes, from bolts to aeronautics, display a structure that mandatorily includes a system of control of the characteristics of each landmark, in order to ensure the interchangeability of the landmarks and the functional parameters of the product including them. This is even more important when devising Nano-systems based on optic, electronic, mechanic and biological Nano-devices [5].

In our country and, considerably, abroad, the Nano-systems and the Nano-technologies have reached the research-development laboratory stage. In order to create an efficient production structure, designing them with a warrant within some specific parameters is required. At his dimensional scale tolerances cover a narrow range too. Failing to detect extension beyond the tolerance field may have undesirable effects, ending up in deteriorating the system in which it is used. The production of these systems without rigorous control is unconceivable.

The transition from the laboratory stage to the industrial production stage implies devising a metrological infrastructure which would pursue the reaching of the parameters in the production flow step by step.

2 Experimental Model

According to the paper, the device for calibrating Nano-technology products consists of a rotary feeding system with eight support tables on top of which are placed supports that hold the Nano-devices that needs to be calibrated. Nano-devices that need to be calibrated include electronic Nano-devices, optical Nano-devices, biological Nano-devices, Nano-materials, Nano-sensors, resulting in a phase of Nano-technological process.

For handling without affecting the Nano-device, it is placed in a calibration support for Nano devices. From the handling system, the Nano-device that needs to be calibrated, with the support is transferred to a precision linear displacement system using a manipulator. The precision linear motion system carries the Nano-device to an optoelectronic precision measurement system, a laser measuring system or in front of an AFM calibration system [3]. From this point, the Nano-devices that need to be calibrated are handled by a robot that places it in the special place in the AFM calibration system.

The equipment comprises a rotary feeding system on top of which are placed 8 support tables (Fig. 1). On each support table there is a support is for the Nano-devices that needs to be calibrated.

The equipment (Fig. 1) comprises a rotary supply system endowed with a manipulator, a precision linear mover system, an optical measuring system, a LASER measuring system [1], a measuring system endowed with an atomic force microscope AFM, and a feeding robot for calibration based on the AFM.

The experimental model will include an ultra-accurate, controlled displacement system and will be supplied with optoelectronic, laser and atom level microscope measurement systems displaying the following technical features:

- Accuracy of displacement: 0.2 nm
- Laser measurement resolution: 1 nm
- Optoelectronic measurement resolution: 10 nm

During the calibration process, the device that needs to be calibrated it is placed in order to handle optimally the calibrated Nano-device on the support, which is precisely positioned on the support table, sitting on the plate of the rotary feeding system and on the table of the precision displacement. The transfer of the measurement Nano-device, from the rotating feeding system to a precision moving system is made using a manipulator. The precision displacement measurement

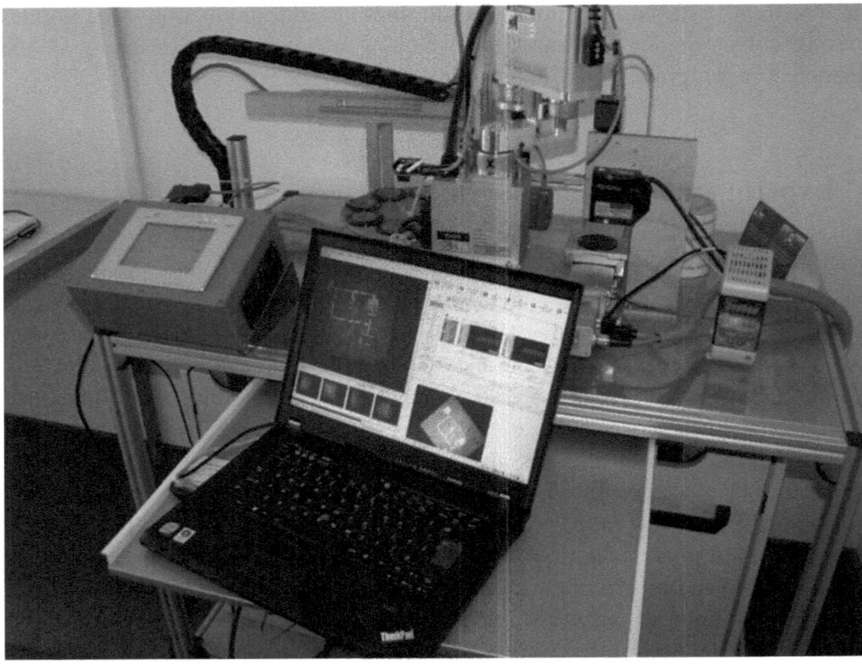

Fig. 1 The experimental model (includes the mechanical, the optical and the optoelectronic sub-ensembles, the optoelectronic measurement sub-ensembles and the algorithms related to (real time) measurement system data acquisition, data processing and measurement protocol presentation)

system takes the Nano-device that needs measuring to the optical laser measuring system or in front of the measuring system endowed with an Atomic Force Microscope (AFM) (Fig. 2). Feeding, for measuring with an AFM is made with a special feeding robot. Depending on the decision of the calibration system (the device is good/ the device is rejected), to and, upon return, the calibrated device is repositioned then in the production flow or it is taken to scrap.

Depending on the decision of the measuring system, the device that needs to be calibrated is either taken to the Nano-technological flow or transferred to scrap.

Depending on the requirements of the Nano-technological process, the equipment has the ability to schedule automatic calibration to one, two or three posts of calibration [11]. For adjustments and experiments, the control software has the option of ordering automatic calibration process or manually controlling a computer system or touch screen.

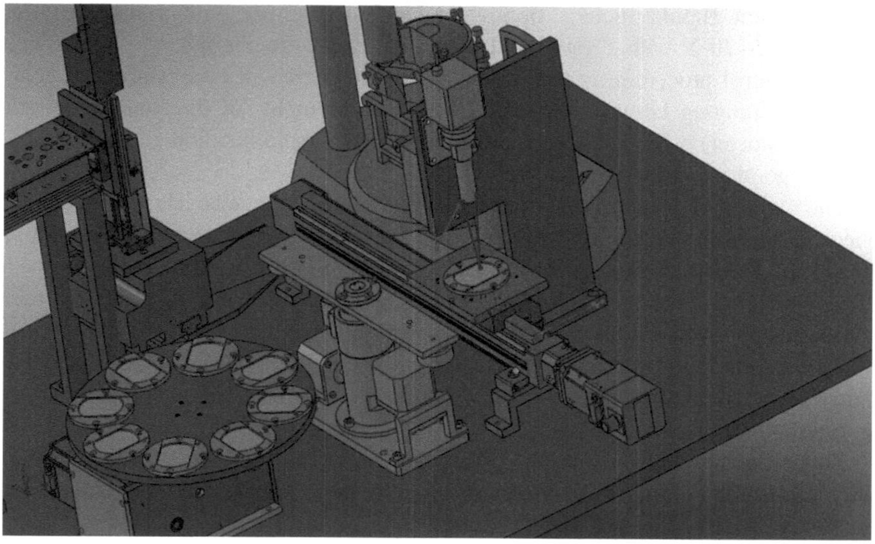

Fig. 2 Optoelectronic control

3 Results and Comments

The equipment developed in research institute INCDMTM, is equipped to perform the flow control by means of three specialized systems:

- Optoelectronic Control (microscope with CCD camera)
- Laser control
- AFM control

In this paper we present only the first specialized control system.

4 Optoelectronic Control (Microscope with CCD Camera)

The purpose of using this method of testing and inspection is to check the conditions of integrity and quality of semiconductor devices (discrete components type SMD), microelectronic circuits, micro machined circuits, printed microcircuits, micro sensors and transducers. This control method can also be used for inspecting defects that may result from handling, assembling or encapsulating devices listed above.

The equipment used in this control must be able to demonstrate the quality conditions of the devices mentioned in accordance with the requirements envisaged in the design.

Equipment should include optics (optical microscope) with a magnification range of 1.5–20 X with a view field accessible and large enough.

The control procedure is the following: the devices will be examined at magnifications ranging between 1.5 and 20 X. Measurements of dimensions (length/ width/ diameter) will be made using the order of magnification of 10–20 X to provide good precision measurements.

The values of dimensions (width/ length routes, contacts, etc.) that can be checked are in the range 10–2000 μm. The optoelectronic measurement resolution is 10 nm. The system must allow linear movement of the sample in two perpendicular directions (corresponding to the test plan) at a distance of at least 5 mm. Measuring resolution must be higher than 100 nm.

Applications that require the optical control are: verification of integrated circuits, verification of printed microcircuits, verification of microsensors based on amorphous magnetic materials.

Following are some examples of application of the optoelectronic control in microelectronic circuits, micromachined circuits, printed microcircuits, and microsensors (Figs. 3, 4, 5, 6, 7, 8, 9 and 10).

Verification of integrated circuits (Fig. 3).

- Verification of routes and establishing dimensional variations from the geometry of the proposed design, comparison with a theoretical form;

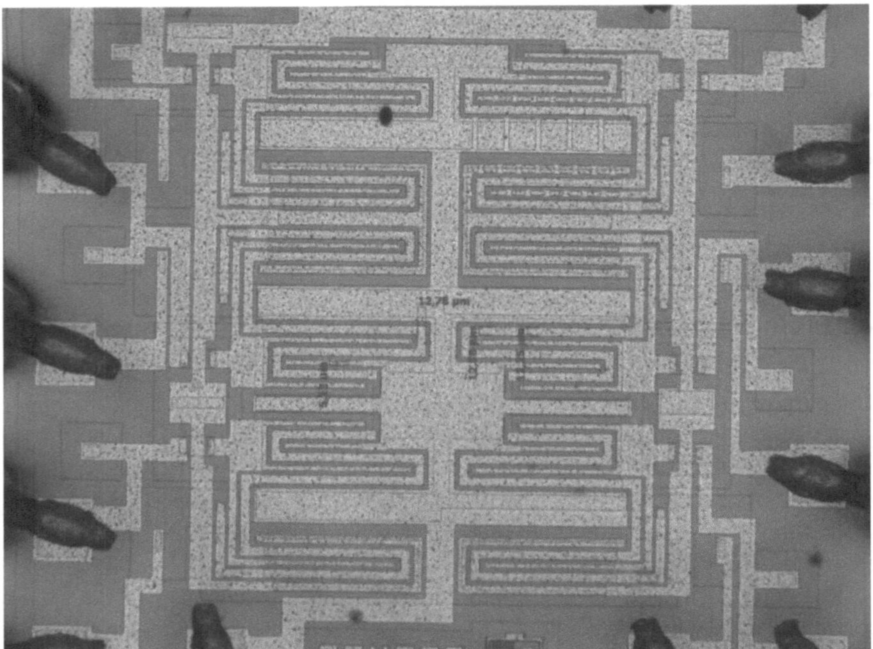

Fig. 3 Integrated circuit

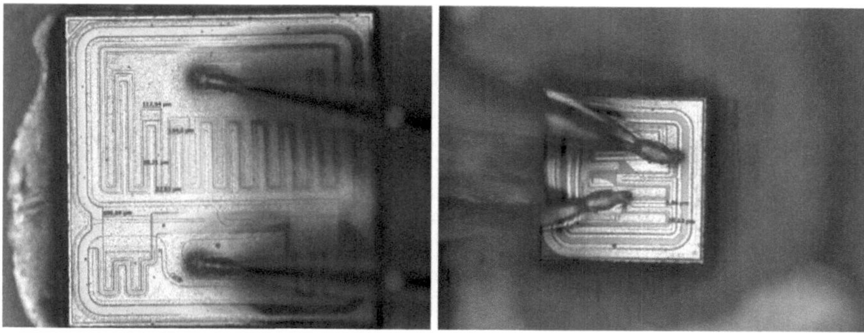

Fig. 4 Micro processed circuits

Fig. 5 Microelectronic circuits (Cable routes, metal plated holes)

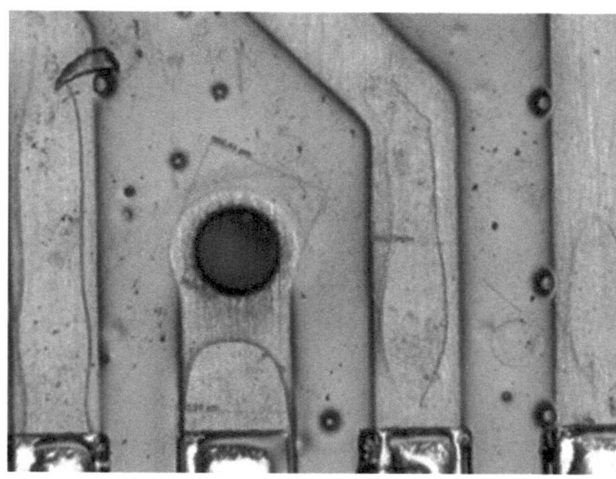

Fig. 6 Micro imprinted circuits

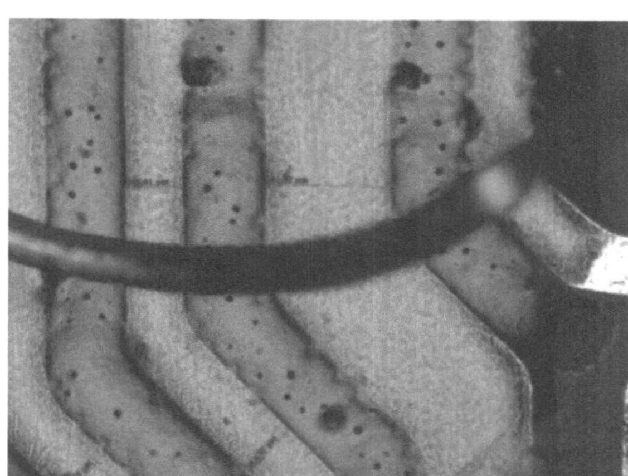

Fig. 7 Microcircuit

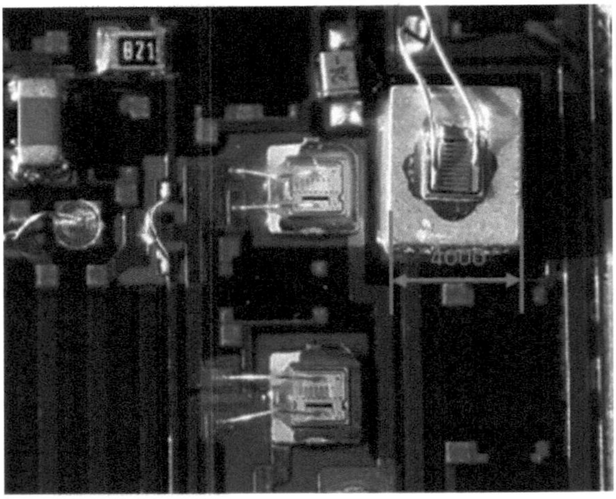

Fig. 8 Microcircuits

Fig. 9 Cable routes, metal plated holes

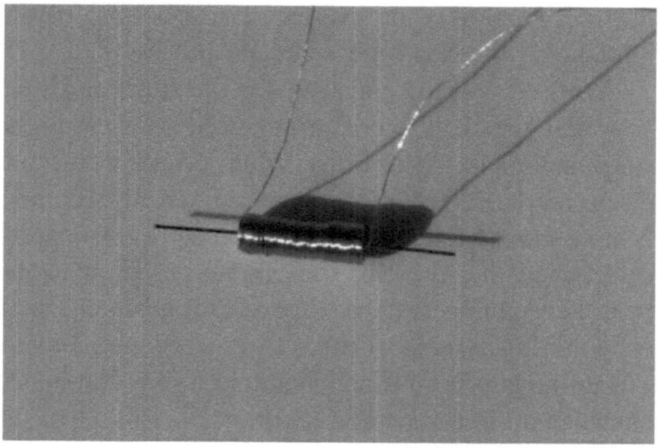

Fig. 10 Inductive sensor

- Verification of junctions and contacts;
- Verification of profiles;

Verification of printed microcircuits (Fig. 6).

- Verification of maximum deviation distance of the routes of the microcircuit;
- Verification of the continuity of routes;
- Verification of the profile of a PCB and deviations from a theoretical shape;
- Verification of positioning (placement) of SMD components on the surface of the printed microcircuits;
- Verification of deviations between the size and position in relation to the size of the printed routes (marks) of the integrated circuits or chips.

Verification of microsensors based on amorphous magnetic materials (Fig. 10).

- Verification of core dimensions (length and diameter);
- Verification of winding dimensions (length and diameter);
- Verification of the distance between sensors placed in a matrix and establishing the distance deviation against the design;
- Verification of the alignment sensors in the sensor matrix;

5 Conclusion and Future Research

In Nano-technology development, the Nano-metrology become an extremely important necessity for correct measurements. It has been shown that the stochastic feature of these structures and their great variety of shapes generate great difficulties in Nano-metrology. In literature ([2–4, 8–10]) it has been discussed in detail about several characteristics such as density, Nano-scale particles, the particle size distribution (PSD), or grain size distribution, particle shape, complex shapes all influencing the Nano-measurement process.

Another issue to be analyzed is the effect of disturbing factors such as noise and vibration which is more pronounced as their higher frequency is closer to the frequency of the measuring system. There were mot highlighted the influences of electromagnetic radiation on measurement accuracy of the system composed of Nano-manipulator/ piezoelectric actuator- sensor for position and movement. It was determined that the electromagnetic radiation in the visible spectrum emanating from the lighting systems (fluorescent tubes and incandescent bulbs) do not influence the functioning of the sensor or the measurement accuracy.

For the transition from laboratory to line Nano-technology production, should be considered the following issues:

- Measurement techniques commonly used for conventional material cannot be applied to Nano-structures.
- Special standards for Nano-structures and Nano-materials must be promoted in order not to induce errors in results evaluation.
- New equipments must be conceived in order to resolve the above mentioned problems.
- Developing such new equipments will allow the production of reproducible Nano-structures creating new study fields regarding Nano-structure proprieties.
- Development of the Nano-metrology, creation of international standards and equipments for calibration of the products and equipments from industrial production, offer the chances of new scientific discoveries regarding new commercialization products.

The experimental model permits optical, laser and AFM microscopic verifications of realized Nano-devices in order to correct possible production errors [6], production calibration and automatic selection of rebuttal in production process.

Future research will aim the development of detailed technologies for various applications in Nano-devices production field that need to be calibrated including electronic Nano-devices, optical Nano-devices, biological Nano-devices, Nano-materials and Nano-sensors.

References

1. Popan Gh, Sorea S, Savu T (2007) Mechatronic system for scavenge the laser ray. Rom rev precis mech Optics Mechatron
2. Chiriac H, Popan G, Ursu D, Gheorghe M (2008) The importance of the production calibration in the Nano domain. Rom rev precis mech optics mechatron Nr:33/2008 - ISSN 1584–5928
3. Chiriac H, Popan G, Ursu D, Gheorghe M Nano-technological measuring system. In: The 2-nd international conference on innovations, recent trend and challenges in mechatronics, High-Tech products development "MECAHITECH' 10"

4. Control procedures in the production flow (2009) International conference MECAHITECH', Bucharest 8–9 October
5. Sorin S, Popan G, Atanasescu A Nanotechnology and solar cells. In: The 2-nd international conference on innovations, recent trend and challenges in mechatronics, High-Tech products development "MECAHITECH' 10"
6. Popan G, Palade DD, Lung I, Tacutu I (2011) Influence of constructive-functional elements of the laser head systems. Accuracy- 6th international working conference—"total quality management. Advanced and intelligent approaches", June 7th–11th, Belgrade, Serbia Rev: 8843 total quality management & business
7. Leão CP, Soares FO, Machado JM, Seabra E, Rodrigues H Design and development of an industrial network laboratory. Int J Emerg Technol Learn 6(2):21–26
8. Barros C, Leão CP, Soares F, Minas G, Machado J (2013) RePhyS: a multidisciplinary experience in remote physiological systems laboratory. Int J Online Eng 9(5):21–24
9. Machado J, Denis B, Lesage J-J (2006) A generic approach to build plant models for des verification purposes. In: Proceedings eighth international workshop on discrete event systems WODES 2006, pp 407–412
10. Silva M, Pereira F, Soares F, Leão CP, Machado J, Carvalho V (2015) An overview of industrial communication networks. Mech Mach Sci 24:933–940
11. Vieira GG, Varela MLR, Putnik GD, Machado JM, Trojanowska J (2016) Integrated platform for real-time control and production and productivity monitoring and analysis. Rom Rev Precis Mech Opt Mechatronics

Hydro-multipoint Forming, a Challenge in Sheet Metal Forming

Viorel Panuoiu and Doina Boazu

Abstract The Hydro-Multipoint Forming is a new technology for sheet metal manufacturing appeared as a response to the demands of the global marketplace toward batch and customized production. The technology is a combination between two metal forming processes multipoint forming and hydroforming. The Hydro-Multipoint Forming reconfigurability is assured mainly by the flexible surface of deformation, generated by the axial moving of a number of pins, from the pins network. Using the hydraulic pressure the necessary force to deform the material along the reconfigurable surface of deformation is obtained. In the paper the elements of the new Hydro-Multipoint Forming technology are presented. A simulation model is developed. The model is created using FEM, and both the interpolator associated with the pressure chamber and the interpolator between the pins and the blank are considered. The process of deformation is analyzed in terms of strains, stresses and prevalent defects, respectively dimpling and wrinkling. Using an experimental tool for Hydro-Multipoint Forming the material deformation is analyzed considering the geometrical variations. Finally, a discussion of the numerical and experimental results is made and limits and possibilities for future process improvements are highlighted.

Keywords Multipoint forming · Reconfigurable technology · Hydroforming · FEM

1 Introduction

The Hydro-Multipoint Forming is a new technology for sheet metal manufacturing appeared as a response to the demands of the global marketplace toward batch and customized production. The technology is a combination between two metal forming processes: multipoint forming and hydroforming.

V. Panuoiu (✉) · D. Boazu
Faculty of Engineering, Dunarea de Jos University of Galati, Galati, Romania
e-mail: viorel.paunoiu@ugal.ro

V. Majstorovic and Z. Jakovljevic (eds.), *Proceedings of 5th International Conference on Advanced Manufacturing Engineering and Technologies*,
Lecture Notes in Mechanical Engineering, DOI 10.1007/978-3-319-56430-2_7

In multipoint forming, both die and punch are two opposite pins networks which materialize at the end of the deformation process the geometry of the part. Because each pin from each network could be vertically moved by numerical control, according to the part profile, the die and the punch have the property of reconfigurability. The starting point of this technology is the concept of reconfigurable discrete dies [1–4], applied in the fields of aerospace and shipbuilding, for obtaining complex and large sized parts by stretch forming. The multipoint forming was studied under different aspects, for avoid dimpling, wrinkling and springback reducing. The springback phenomenon was investigated by Liu et al. [5] which proposed a method for iterative compensation. Paunoiu [6] propose new concepts for springback reduction and compensation. Wang et al. [7] investigated the effect of punch element toward the part quality. Zhang et al. [8, 9], Paunoiu et al. [10], Cai et al. [11] highlighted the influence of discrete pin type toward the springback. The phenomenon of dimpling and the necessity of using an interpolator between the pins and the blank, was investigated by Cai et al. [12], Davoodi et al. [13], Paunoiu et al. [14]. We can conclude that the multipoint forming have some important advantages: the possibility of the geometry control of the formed part, the reduced level of the deformation force, the reconfigurability which lead to the manufacturing of a large variety of parts from the simple ones to complicated 3D sheet metal part.

In hydroforming the pressure assures the force for material deformation. The pressure could be applied directly toward the blank or indirect using a flexible membrane between the fluid and the blank. The advantages of this technology are a better distribution of the thickness, increasing the forming limit and improving the part quality and precision. As a result complex parts are obtained. The process was largely studied, both experimental, Siegert et al. [15], Lang et al. [16], and numerical, Hama et al. [17]. Due to the flexibility of the rubber pad, parts with undercuts and sharp contours can be formed, even with sheets of different initial thicknesses, Del Prete et al. [18], Vollerstsen et al. [19].

According to Paunoiu et al. [20] this new concept of hydro-multipoint forming tool, has the following advantages:

- Reduction of the setup time in half, in comparison to the conventional digitized dies;
- The construction is relatively simple due to the absence of one of the pins network;
- Expansion of parts shapes range due to easy configuration of the multipoint die/punch;
- Data files provided on-line from a simulation program or from a database created on the basis of real or simulated tests of the process, could be used to control the springback phenomenon;
- The dimpling could be avoided because of the presence of the elastic interpolator between the die and the blank;
- The device provides a uniform pressure towards the blank surface due to the presence of the two elastic interpolators leading to the material deformation within large limits.

Together with these advantages, dimpling and wrinkling are major problems in choosing the process parameters.

The present paper introduced the concept of hydro-multipoint forming. Numerical simulations were performed to analyze the effect of interpolator thickness and material toward the dimpling and wrinkling. The influence of blank material characteristics are investigated by simulation to obtain new information about the actual type of deformation process. An experimental setup was built and a series of experiments were provided to validate the hydro-multipoint process.

2 Definition of Hydro-multipoint Forming

Hydro-multipoint forming by combining advantages of the above processes is a potential solution for improving the flexibility of forming technology. The concept simplifies the construction of the multipoint tools with pins by replacing the die/punch with a hydraulic/pneumatic/elastic pressure chamber. The blank sheet is placed between two flexible membranes and is deformed under the action of the hydraulic/pneumatic/elastic pressure applied from the top/bottom of the tool. The part shape is given by the shape of the multipoint, reconfigurable die/punch of the equipment.

There are two research directions for hydro-multipoint forming, depending of the multipoint structure position.

The first direction it is one in which the multipoint configuration is given by the punch. The idea of using an elastic medium in the multipoint forming process appeared for the first time in [21]. Authors proposed a method for hydro-mechanical deep drawing with the punch composed from an array of pins, which are independently shifted in vertical direction, accordingly with the piece profile (Fig. 1). Between the blank and the multipoint punch is interposed an elastic interpolator.

Liu et al. [22], based on hydro-multipoint punch concept, studied the feasibility of the process by investigate the dimpling and parts geometry by experiment and numerical simulation. As novelty they introduced in addition a metallic cover sheet between the elastic interpolator and the multipoint punch. The obtained results reveal the influence of the hydraulic pressure and cover sheet thickness upon the surface quality and part shape error.

The second direction of researches it is the one in which the multipoint configuration is given by the die. An example is presented in (Fig. 2), Paunoiu et al. [23].

The hydro-multipoint forming die concept consists from a superior subassembly provided with an interpolator (2), a lower subassembly also provided with an interpolator (6) and a reconfigurable die (8) consisting of a number of hydraulic cylinders in a network. The blank (11) is deformed accordingly to the geometrical part shape of the die. Thegeometrical shape of the die could be digitally controlled, allowing for the material flow during the deformation process.

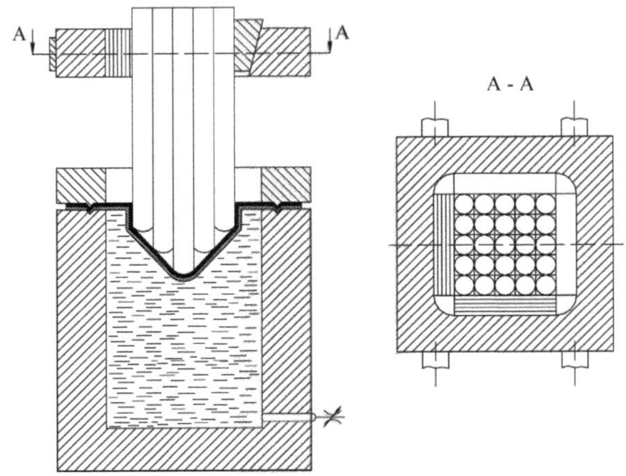

Fig. 1 Hydro-multipoint punch concept [21]

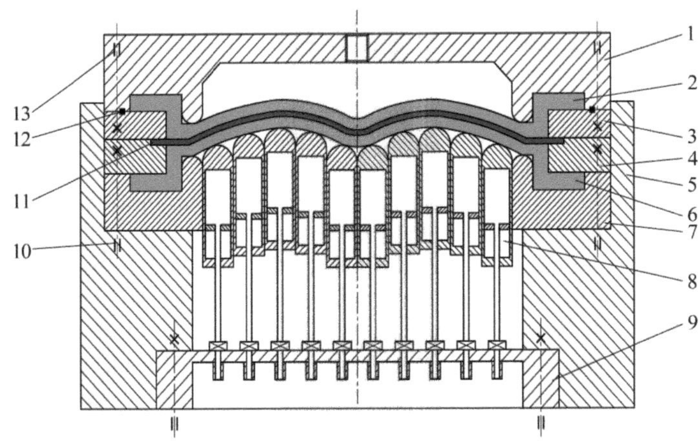

Fig. 2 Hydro-multipoint die concept: *1* superior body; *2*, *6* elastic pad; *3*, *4* fastening plate; *5* inferior body; *7* profiled plate; *8* reconfigurable die; *9* support plate; *10*, *13* screw; *11* blank; *12* sealing ring [23]

Selmi et al. [24] used in their researches a rubber pad for applied the pressure instead of a hydraulic/pneumatic chamber and a metallic sheet instead of the elastomeric interpolator. Their results highlight that the metallic sheet media is more efficient way to eliminate dimpling and edge buckling. Recently, Selmi et al. [25], using a segmented metallic sheet between the pins and the blank developed both a simulation model and conducted a series of experimental investigation. The results validate the efficiency of the new design, in terms of part surface, part shape and forming pressure.

3 FEM Simulations of Hydro-multipoint Forming

The model was developed using the finite element program ANSYS WORKBENCH—Mechanical Explicit Dynamics with AUTODYN as solver. Explicit/Dynamic approach was used due to the high accuracy and the low analysis time. For process simulation the entire geometry of the system was generated in ANSYS WORKBENCH program (Fig. 3). The 253 pins of the die (Fig. 4), the two interpolators (upper rubber and down rubber) and the bank material were generated as solid bodies; all of the 253 steel pins of the die were considered as rigid bodies.

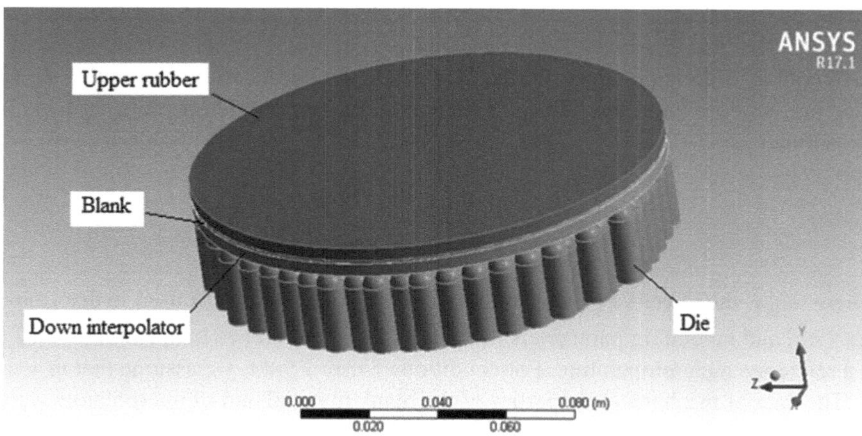

Fig. 3 Hydro-multipoint forming simulation model

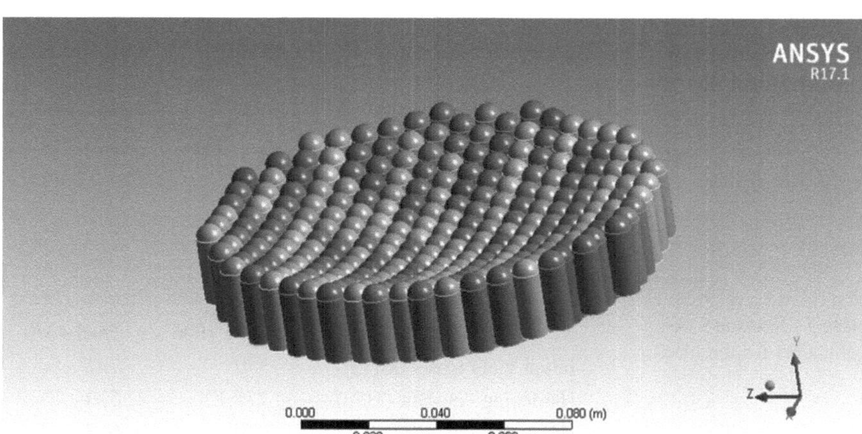

Fig. 4 Hydro-multipoint forming die model

3.1 Materials

The upper interpolator is made from rubber and its thickness was considered of 4 mm. Two materials for the down interpolator were used. One is made from rubber with a thickness of 4 mm and the other one is made from aluminum, 2 mm thick.

The blank is made from mild steel and its thickness was considered of 1 mm.

The blank (density—7850 kg/m^3, Young's modulus—200 GPa, Poisson ratio—0.3) behavior during plastic deformation was modeled using Johnson Cook law given by the formula:

$$\sigma_y\left(\varepsilon_p, \dot{\varepsilon_p}, T\right) = \left[A + B\left(\varepsilon_p\right)^n\right]\left[1 + C\ \ln\left(\dot{\varepsilon_p^*}\right)\right]\left[1 - \left(T^*\right)^m\right] \quad (1)$$

Where ε_p is the equivalent plastic strain, $\dot{\varepsilon}_p$ is the plastic strain-rate and *A, B, C, n, m* are material constants. The normalized strain-rate and temperature in Eq. (1) are defined as:

$$\dot{\varepsilon_p^*} = \frac{\dot{\varepsilon_p}}{\dot{\varepsilon_{p0}}} \text{ and } T^* = \frac{T - T_0}{T_m - T_0}$$

where $\dot{\varepsilon_{p0}}$ is the effective plastic strain-rate of the quasi-static test used to determine the yield and hardening parameters *A,B* and *n*. T_0 is a reference temperature, and T_m is a reference melt temperature. For conditions where $T^* < 1$, we assume that $m = 1$.

The data of mechanical properties of the blank material and of interpolators were obtained from the data program database.

The blank material was considered STEEL 1006, similar with the one used in the experimental investigations, having coefficients of Johnson Cook model presented in the Table 1.

For upper interpolator, during the deformation process the rubber plate will experience large deformation and the stress-strain behavior was modeled by using Mooney–Rivlin law. The 2-parameter Mooney–Rivlin model [27], with the equation for strain energy is:

$$W = C_{10}(\bar{I}_1 - 3) + C_{01}(\bar{I}_2 - 3) + \frac{1}{D_1}(J - 1)^2 \quad (2)$$

Table 1 Johnson-Cook coefficients for the blank

	Steel 1006	Steel 4340
Initial yield stress (MPa)	350	792
Hardening constant (MPa)	275	510
Hardening exponent	0.36	0.26
Strain rate constant	0.022	0.014
Thermal softening exponent	1	1.03

Table 2 Material properties for the rubber 2

Property	Value
Density	1000 kg/m^3
Material constant C_{10}	1.5 * 10^5 Pa
Material constant C_{01}	15,000 Pa
Incompressibility parameter D_1	1.212 * 10^{-9} Pa^{-1}

where C_{10}, C_{01} are material constants characterizing the deviatoric deformation of the material, D_1 is the material incompressibility parameter, J is the determinant of the elastic deformation gradient and $\bar{I}_1$, $\bar{I}_2$ are the first and the second invariant of strain deviatoric tensor. The material constants are given in the Table 2 (we call this material rubber 2). The thickness of the upper interpolator is 4 mm.

For down interpolator were used four types of materials, two rubbers and two aluminum plates (Table 3).

The rubber 1 (Table 3) is a soft rubber and was considered having a linear elastic behavior with properties: density = 100 kg/m^3, Young's modulus = 1 * 10^5 Pa, Poisson ratio = 0.49. The rubber 2 was already defined (see Table 2).

The aluminum material model follows the power flow rule: where $\sigma = K\varepsilon^n$ K = strength coefficient [MPa] and n = strain hardening coefficient.

The material properties used in modeling the two aluminum plates are given in the Table 4.

Table 3 Material variants for the down interpolators

Variant number	Material	Thickness of down interpolator plate (mm)
Case A	Rubber1—soft rubber	4
Case B	Rubber 2—vulcanized rubber	4
Case C	Aluminium 1	2
Case D	Aluminium 2	2

Table 4 Material variants for the aluminum down interpolators

	Aluminum 1 (AL 1100-0)	Aluminum 2 (AL 2024-T4)
Density (kg/m^3)	2707	2770
Young's modulus (GPa)	68.9	71
Poisson ratio	0.33	0.33
Yield strength (MPa)	34.5	276
Strength coefficient (MPa)	180	696
Strain hardening coefficient—n	0.2	0.16

3.2 Numerical Simulation

The model for hydroforming plastic deformation was generated as solid bodies in Ansys Workbench.

For meshing the solid element SOLID164 is used for the three-dimensional modeling of solid structures. The element is defined by eight nodes having the following degrees of freedom at each node: translations, velocities, and accelerations in the nodal x, y, and z directions.

Figure 5 presents the mesh of the system—blank, interpolators and die containing (319,531 nodes and 478,746 solid elements). The element size was considered of 2 mm for the pins of the die and interpolators and 0.5 mm for the blank.

Figure 6 shows the boundary conditions—fixed supports for all 253 pins. A constant pressure of 4 MPa applied from the top of the upper rubber is used.

In numerical simulations contacts interactions were considered with Coulomb friction. The coefficient of friction between the metallic faces was set 0.1, and between the metallic face and rubber surface was set to 0.2.

The contact conditions between surfaces are type Surface_to_Surface. The slave (blank) and master (die) side are defined by segments and as a symmetric contact, both slave and master nodes are checked for penetration.

The pressure applied on the upper interpolator was considered constant of 4 MPa, as we already stated.

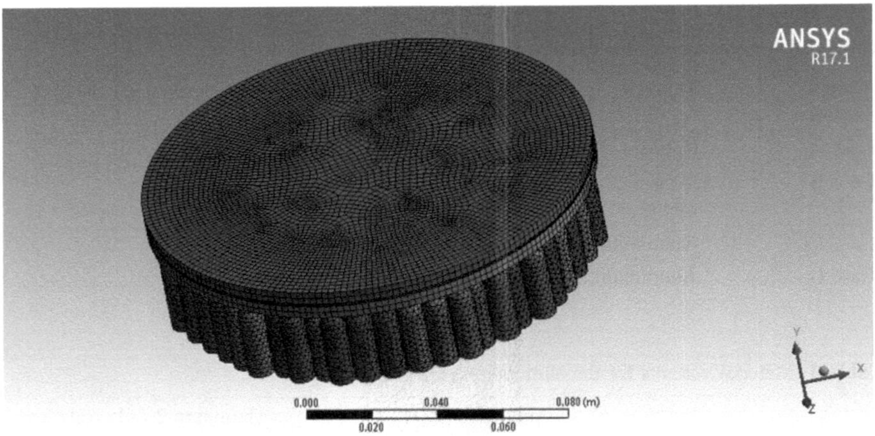

Fig. 5 Hydro-multipoint forming model meshing

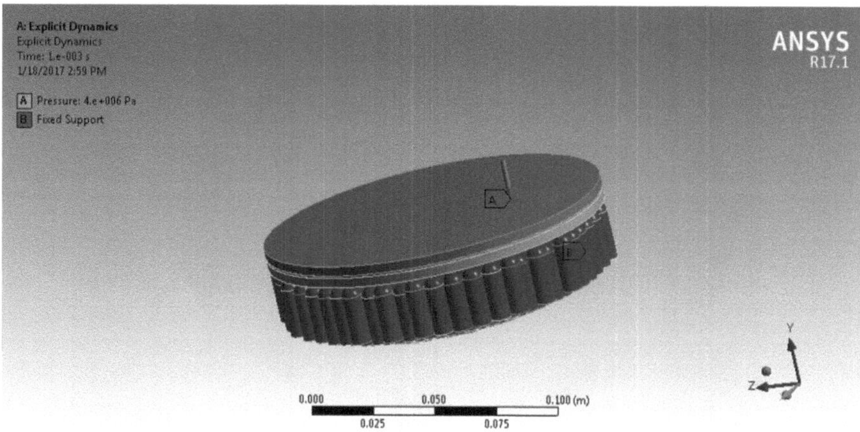

Fig. 6 The boundary condition in simulation

3.3 *Results of Numerical Analysis*

The total deformation of steel plates in cases A, B, C, D (Table 3) are presented in Figs. 7a–d and discussed in Table 5.

The numerical results show the presence both of the two undesirable phenomena: dimpling and wrinkling because of the effect of localized deformation which depends from the contact points positions. If the rubber plate is thinner than dimples could not be eliminated (see case A from Table 5 and Fig. 7a). When the rubber is thicker the accuracy of the specimen is small due to wrinkles (see case B from Table 5 and Fig. 7b). In case of aluminum down interpolator a harder one will decrease the dimples but the wrinkles are more pronounced for the considered parameters (cases C and D from Table 5 and Fig. 7c, d).

An extension of our study shows that the strain rate of the blank material could have an important influence. In this regard we had included in the simulations another steel material, Steel 4340, having a lower strain rate compared to those of the Steel 1006 (Fig. 8). In sheet metal forming, the flow stress of the metal is an important factor for successful operation. The forming pressure and the load are related to the flow stress, and therefore a lower flow stress is desirable in most of the forming operations according to Osakada et al. [26]. The effect of strain rate on the flow stress of metals is not usually so significant at low temperature, but in our study without any restraints of material, a lower strain rate of the blank material becomes important (Table 5).

The maximum equivalent plastic strain (Fig. 9) and maximum equivalent stress von Mises (Fig. 10) can be observed in the case A in the edge region; in cases C and D the maximum equivalent plastic strain and maximum equivalent stress Von Mises appears in the local wrinkling of the steel plate.

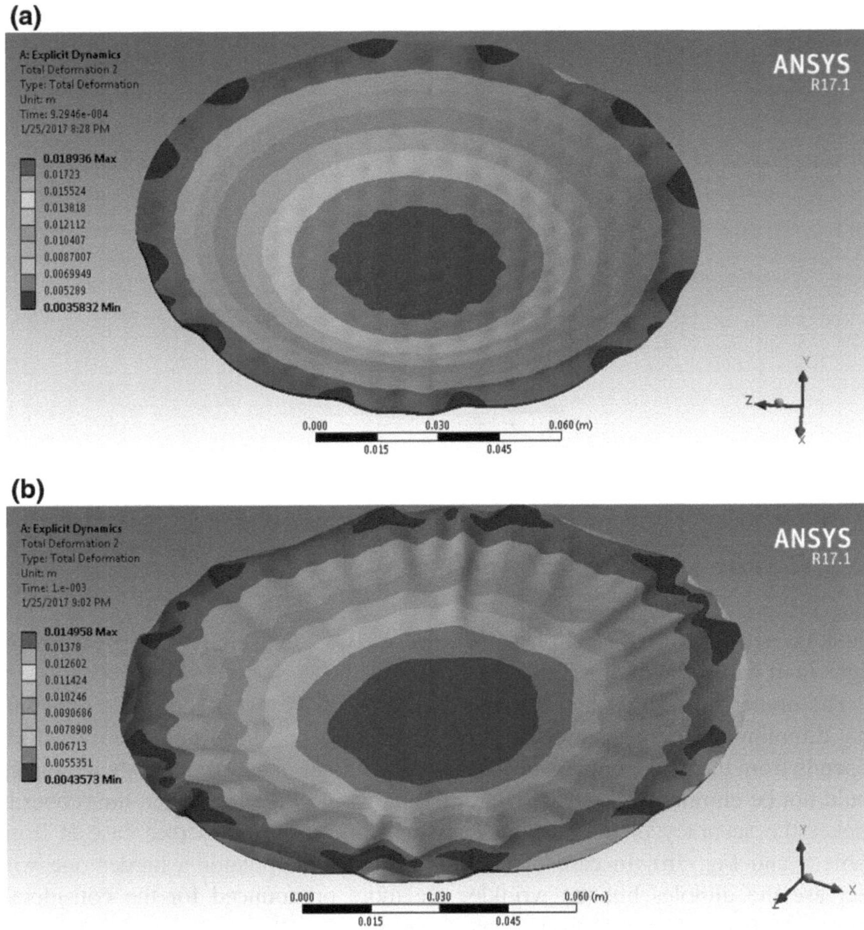

Fig. 7 **a** Total deformation of steel plate, case A. **b** Total deformation of steel plate, case B. **c** Total deformation of steel plate, case C. **d** Total deformation of steel plate, case D

4 Experimental Work

Based on the hydro-multipoint die concept presented above, a hydro-multipoint tool, mechanical actuated was designed. Figure 11a shows the experimental tool. The main subassemblies of the tool are the pressure chamber and the reconfigurable die. The pressure chamber is coupled with a hydraulic system which assures a maximum pressure of 25 MPa. The pressure is transmitted to the blank using a rubber membrane with 4 mm thickness.

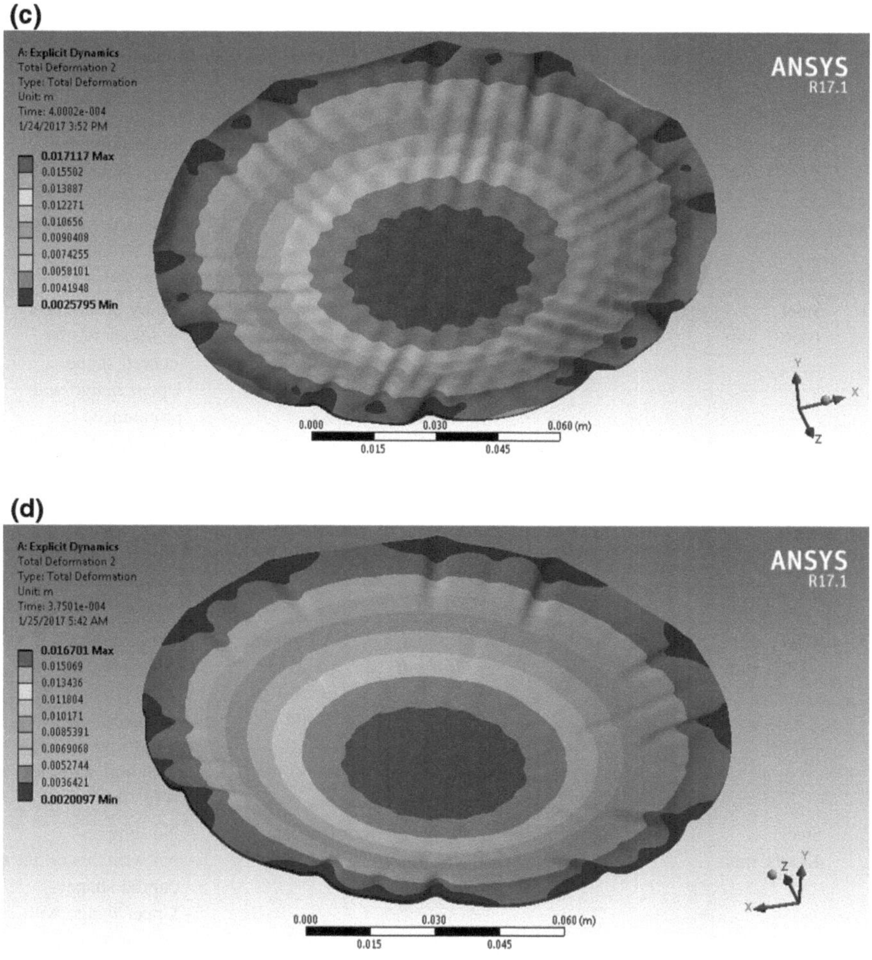

Fig. 7 (continued)

Figure 11b presents the reconfigurable subassembly of the die. The multipoint die was designed with a rows structure inside a metallic circular body; the pins have a diameter of 8 mm each and a hemispherical end. The die is shaped using the piece model. For this, each pin from the network is vertically displaced due to its own threaded area, until it reaches the model surface. The desired workpiece is a spherical sector with a radius of 141.5 mm and a depth of 14 mm.

Circular blanks, having a diameter of 140 mm, made from mild steel sheet, similar with the one used in simulation, were used. The properties of material were: yielding strength—268.8 MPa; ultimate tensile strength—389.6 MPa and hardening exponent—0.21.

Table 5 Discussions about numerical simulations results

Nr. crt.	Blank material thickness	Strain rate of the blank material (Johnson Cook strength)	Upper interpolator/ thickness	Down interpolator/ thickness	Defects
A	Steel 1006/1 mm	0.022	Rubber 2/4 mm	Rubber 2/4 mm	– Deep dimples – No wrinkles on the curved shape – Upper flange with soft wrinkles (Fig. 7a)
B	Steel 1006/1 mm	0.022	Rubber 1/4 mm	Rubber 1/4 mm	– Many pronounced wrinkles on the curved shape - Upper flange with pronounced wrinkles (Fig. 7b)
C	Steel 1006/1 mm	0.022	Rubber 1/4 mm	Aluminum (AL 1100-0)/ 2 mm	– Attenuated dimples – Many wrinkles on the curved shape – Upper flange with pronounced wrinkles (Fig. 7c)
D	Steel 1006/1 mm	0.022	Rubber 1/4 mm	Aluminum (AL 2024-T4)/ 2 mm	– No dimples – Attenuated wrinkles on the curved shape – Short upper flange with soft wrinkles (Fig. 7d)
E	Steel 4340/1 mm	0.014	Rubber 2/4 mm	Rubber 1/4 mm	– No dimples – No wrinkles on the curved shape – Upper flange with very soft wrinkles (Fig. 8)

First, a number of experiments were done without any interpolator between the blank and pins. The deformation was made at a pressure of 4 MPa. Figure 12a pre-sents the deformed part. Both the dimpling and wrinkles appear. The part surface is affected by dimples produced due to localized deformation of pins ends profiles. The edge region is highly affected by wrinkles because the outer contour of the pins network is not uniform in this case. The blank squeezed into the space of pins on the contour. The image is similar with the one obtained in simulation (see case A), with rubber interpolators.

The obtained experimental part diameter is not uniform and varies between 134.7 and 136.7 mm, measured along the middle of the part, and the experimental height is 13.5 mm, measured in the part center.

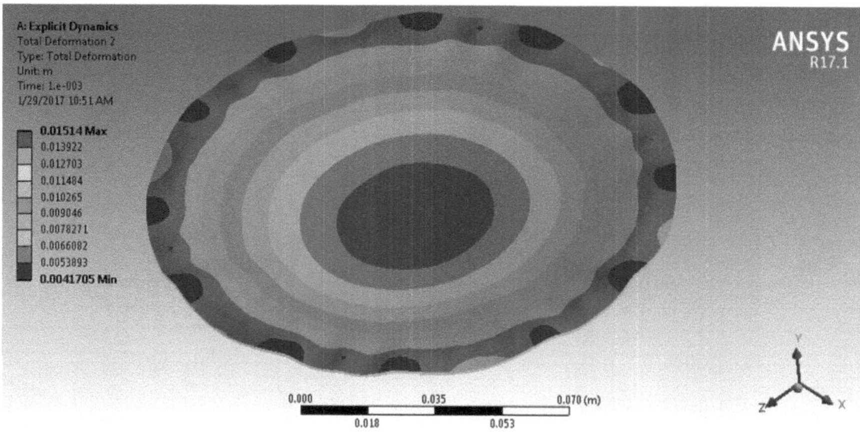

Fig. 8 Total deformation of steel plate, steel 4340

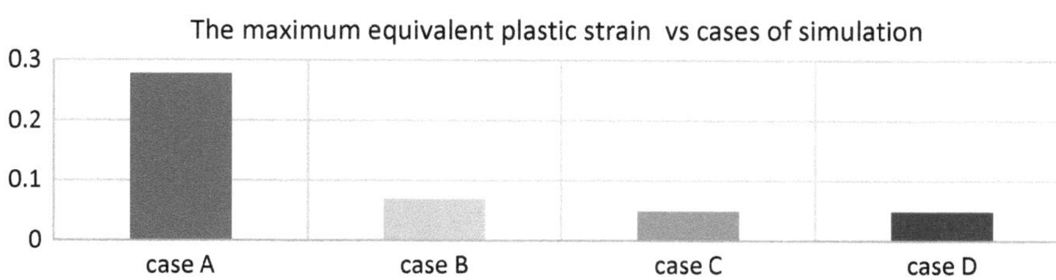

Fig. 9 The maximum equivalent plastic strain for the four study cases

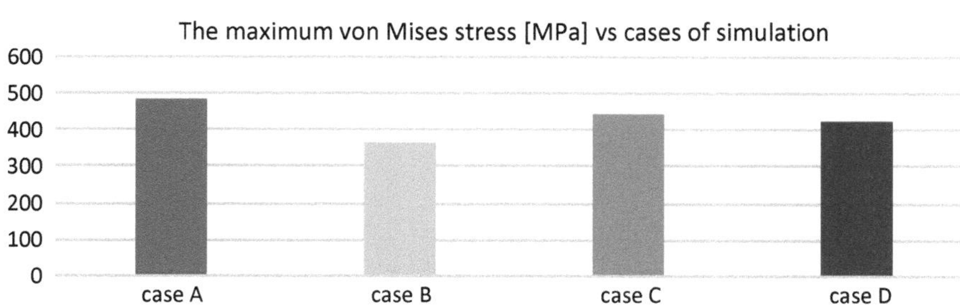

Fig. 10 The maximum von Mises stress in (MPa) versus study cases

Next, a number of experiments were done using a metallic interpolator between the blank and pins, at the same pressure of 4 MPa. 2 mm thick Al sheet was used as down interpolator. Figure 12b presents the deformed part. The inside part surface is of good quality even some imperfections in die set-up and pins manufacturing exists. The edge region is also affected by the non-uniformity of the pins network, but the size of wrinkles is smaller. We have the same observations about dimensions as above.

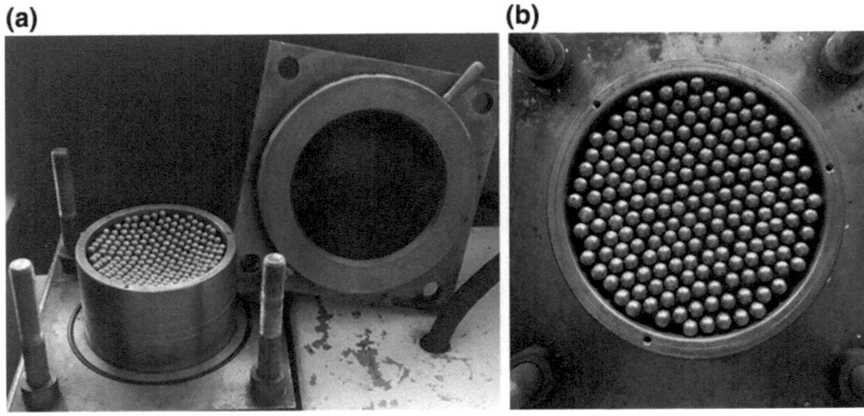

Fig. 11 **a** General image of the hydro-multipoint forming tool. **b** Die subassembly of the hydro-multipoint forming tool

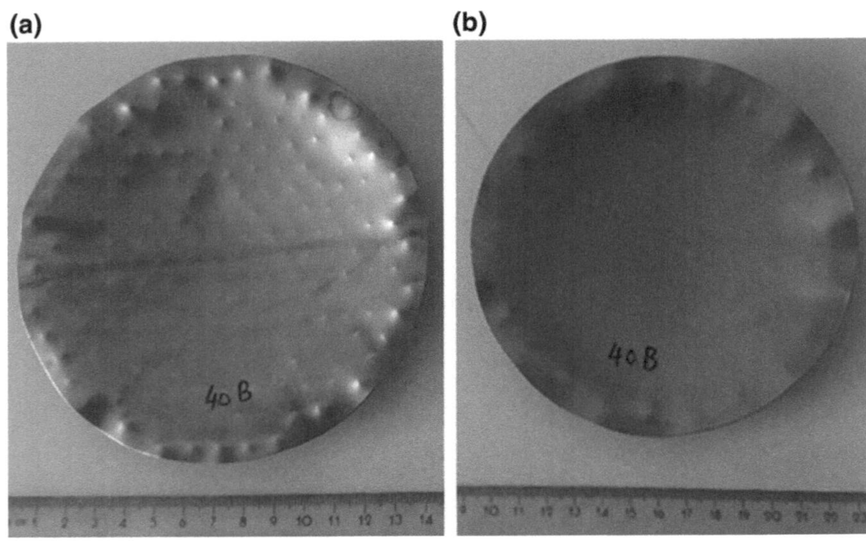

Fig. 12 **a** Hydro-multipoint part deformed without interpolator. **b** Hydro-multipoint forming part deformed with interpolator

5 Conclusions

The hydro-multipoint forming tool presented in this work could be used for manufacturing complex sheet metal parts. The main advantages of this type of equipment are the reduction of the setup process time at a half in comparison with the conventional digitized dies and the combining of hydroforming and multipoint forming advantages. Based on this concept a series of numerical simulations and experimental investigations were performed in terms of avoidance of dimpling and wrinkling.

- The dimples appear in the case A when a soft rubber for down interpolator are used and is attenuated in the case C when a metallic interpolator is used. In the case A the hydraulic pressure causes dimpling and squeezes the down interpolator into the gap of pins. The deformed shape in the case A has no wrinkled region.
- The wrinkling of the deformed shape appears in the cases B, C and attenuated in the case D. In case B the wrinkled regions cause geometrical error that cannot be compensated. These wrinkled regions appear also in the cases C and attenuated in case D. A start point in the future investigation of hydro-multipoint forming process can be the case D (down interpolator of aluminum AL2024-T4) with the main purpose to remove the wrinkled region on the flange.
- The increase of the thickness of the down interpolator made of soft rubber could have a good influence on the quality of the pieces (in the future work this remark will be investigated) and a starting point is the simulation case B in the Table 5.
- It is interesting to note that using a harder material for blank in simulations lead to obtaining quality parts, without dimples and wrinkles, the simulation case D in the Table 5.
- A binder will be needed for dimpling and wrinkling phenomena avoidance because the pressure at the outer contour is not uniform due to irregular pins network in this region.
- A new structure of the pins network inside the die body will be the next step in the improving of the design construction of the tool for hydro-multipoint forming.
- Because of the complexity of the deformation every imperfection in manufacturing and die set-up should be avoided.

References

1. Papazian J (2002) Tools of change. Mech Eng 31–40
2. Hardt DE, Norfleet NA, Valentin VM, Parris A (2001) In-process control of strain in a stretch forming process. Trans ASME 123
3. Valjavec M, Hardt DE (1999) Closed-loop shape control of the stretch forming process over a reconfigurable tool: precision airframe skin fabrication. In: Proceedings of the ASME symposium on advances in metal forming, Nashville, November 1999
4. Walczyk DF, Hardt DE (1998) Design and analysis of reconfigurable discrete dies for sheet metal forming. J Manuf Syst 17(6):436–454
5. Liu W, Yang Y-Y, Li M-Z (2010) Numerical simulation of multipoint stretch forming and controlling on accuracy of formed workpiece. Int J Adv Manuf Technol 50:61–66. doi:10.1007/s00170-009-2501-1

6. Paunoiu V (2014) New approaches for springback-based offline dimensional control in sheet metal forming. Indian J Eng Mater Sci 303–310. ISSN: 0971-4588
7. Wang S, Cai Z, Li M, Numerical investigation of the influence of punch element in multi-point stretch forming process. Int J Adv Manuf Technol 49:475–483. doi:10.1007/s00170–009–2420-1
8. Zhang Q, Dean TA, Wang ZR (2006) Numerical simulation of deformation in multi-point sandwich forming. Int J Mach Tools Manuf 46:699–707
9. Zhang Q, Wang ZR, Deana TA (2007) Multi-point sandwich forming of a spherical sector with tool-shape compensation. J Mater Process Technol 194:74–80
10. Paunoiu V, Cekan P, Gavan E, Nicoara D (2008) Numerical simulations in reconfigurable multipoint forming. Int J Mater Form Suppl 1:181–184
11. Cai ZY, Li MZ (2006) Digitized die forming system for sheet metal and springback minimizing technique. Int J Adv Manuf Technol 28:1089–1096
12. Cai ZY, Wang SH, Li MZ (2008) Numerical investigation of multipoint forming process for sheet metal: wrinkling, dimpling and springback. Int J Adv Manuf Technol 37(9–10):927–936
13. Davoodi B, Zareh-Desari B (2014) Assessment of forming parameters influencing spring-back in multipoint forming process: a comprehensive experimental and numerical study. Mater Des 59:103–114
14. Paunoiu V, Cekan P, Banu M, Epureanu A, Nicoara D (2008) Simulation of the combined reconfigurable multipoint forming and rubber forming. In: Special edition metal forming conference 2008, Steel Research International, vol 1 no 79. ISSN 1611-3683
15. Siegert K, Haussermann M, Losch B, Rieger R (2000) Recent developments in hydroforming technology. J Mater Process Technol 98:251–258
16. Lang LH, Wang ZR, Yuan SJ, Zhang SH, Danckert J, Nielsen KB (2004) Hydroforming highlights: sheet hydroforming and tube hydroforming. J Mater Process Technol 151:165–177
17. Hama T, Hatakeyama T, Asakawa M, Amino H, Makinouchi A, Fujimoto H, Takuda H Finite-element simulation of the elliptical cup deep drawing process by sheet hydroforming. J Finite Elem Anal Des 43(3):234–246
18. Del Prete A, Anglani A, Primo T, Spagnolo A (2008) Computer aided simulation as valid tool for sheet hydroforming process development. Int J Mater Form 1:317–322
19. Vollertsen F, Breede R, Lange K (1999) Method for deep drawing with multiple elastomer membranes. CIRP Ann Manufact Technol 48(1):221–226 (1999)
20. Paunoiu V, Baroiu N, Maier C, Epureanu A, Marinescu V (2011) Reconfigurable deep drawing equipment, Patent number RO 128719-A2
21. Finckenstein EV, Kleiner M (1991) Flexible numerically & controlled tool system for hydro-mechanical deep drawing. Ann CIRP 40:311–314
22. Liu W, Chen YZ, Xu YC, Yuan SJ, Evaluation on dimpling and geometrical profile of curved surface shell by hydroforming with reconfigurable multipoint tool. Int J Adv Manuf Technol doi:10.1007/s00170-015-8264-y
23. Paunoiu V, Teodor V, Baroiu N (2015) The Hydro-Multipoint forming process of complex sheet metal parts. J Mach Eng 15(3):16–116, ISSN 1895-7595
24. Selmi N, BelHadjSalah H (2013) Finite element and experimental investigation of the multipoint flexible hydroforming. Key EngMater 554–557:1290–1297. doi:10.4028/www.scientific.net/KEM.554-557.1290
25. Selmi N, BelHadjSala H Ability of the flexible hydroforming using segmented tool. Int J Adv Manuf Technol. doi:10.1007/s00170-016-9160-9
26. Osakada K (1997) Effects of strain rate and temperature in forming processes of metals. J Phys IV Colloq 07(C3), C3-XXXVII–C3-XLIV
27. Ansys/Autodyn version 11.0, User documentation (2011)

Study on Hard Turning Process Versus Grinding in Manufacturing Some Bearing Inner Rings

Mitica Afteni, Ion Terecoasa, Cezarina Afteni and Viorel Paunoiu

Abstract Advanced manufacturing technologies are based on-time-delivery principles. The customer requirements regarding quality, cost and delivery terms become more and more important, both for supplier and sub-supplier. Taking into account these considerations, all companies started the implementation of some new strategies and technologies based on continuous improvement principle. One of these is hard turning, as a cost efficient alternative to grinding. The technology is analyzed in the case of a bearing component, as a new process step versus the existing grinding process. Experimental tests were performed in order to describe the advantages of this technology. Cubic Boron Nitride grades (CBN) are used for cutting. The main influence of this technology in process flow is analyzed from different approaches. Experimental results were analyzed in terms of the ring surface quality, roughness and profile, and also on process time and production cost.

Keywords Quality · Hard turning · Grinding · Bearings components

1 Introduction

The customer requirements regarding quality, cost and delivery terms became more and more important, both for supplier and sub-supplier. Taking into account these all companies started to define and implement new strategies based on continuous improvement principles.

The development of the manufacturing technologies was observed in the last period. This trend affects all industrial areas.

M. Afteni (✉) · I. Terecoasa
Rulmenti SA, Barlad, Romania
e-mail: afteni.mitica@urb.ro

C. Afteni · V. Paunoiu (✉)
Faculty of Engineering, Dunarea de Jos University of Galati, Galați, Romania
e-mail: viorel.paunoiu@ugal.ro

V. Majstorovic and Z. Jakovljevic (eds.), *Proceedings of 5th International Conference on Advanced Manufacturing Engineering and Technologies*, Lecture Notes in Mechanical Engineering, DOI 10.1007/978-3-319-56430-2_8

The investment plans are based on new technologies or up-grating the existing ones. Advanced manufacturing technologies [1] are based on-time-delivery principles. On time delivery is a measure of process and supply chain efficiency which measures the amount of finish goods or services delivered to customers on time and in full [2]. Hard turning is a technique that can be used to substitute grinding in the finishing operations for hardened steel (HRC 45 and above) [3].

One of the advantages of hard turning is that the quality of surface finish produced is equivalent to the one produced in grinding [4].

New approaches were observed in manufacturing processes, different studies were performed in order to analyze the results in terms of surface quality, process parameters as cutting forces [5], cutting speed, depth of cut, feed rate, material removal rate [6, 7] and also on the effect of this parameter on the surface integrity [8, 9]. Good surface quality of roughness about 0.42 μm obtained during hard turning operation was reported by [10], surface profile, namely: form, waviness, roughness and microroughness were studied in [11].

A study on the hard turning process versus grinding process results was developed by [12] using hard turning (HT) conditions as follows: cutting speed of 150 m/min, feed rate of 0.1 mm/rev, depth of cut of 0.15 mm.

Dry hard turning produced initial surface profiles with regular tool nose traces and surface roughness with the Ra = 0.5 mm. As a result, the replacing of ground surfaces by hard turned surfaces based on 2 Droughness data can be highly risk yin case of the bearing steel processing material.

Surface roughness and dimensional accuracy play very important role while performing machining operation [13]. Surface roughness is mainly a result of process parameters such as tool geometry and cutting conditions [14]. Surface finish is very important feature of any machining process and the main requirement of many manufacturing, automotive and aerospace applications [4]. The functional behavior of machined parts is decisively influenced by the fine finishing process which represents the last step in the process chain and can as well be undertaken by cutting as grinding [15].

In case of a grinding process rough and finish operations are performed using two set-up. In case of a hard turning process rough and finish operations can be performed with one clamping using a CNC lathe. The cutting speed varies from 100 to 200 m/min in case of a hard turning operation performed for a material with 60HRC with CBN inserts [16].

In [17, 18] was reported that the surface roughness is significantly influenced by feed and cutting speed; depth of cut has very less effect during hard turning of AISI 4340 alloy steel. Most optimal hard turning parameters were obtained from Taguchi analysis using signal to noise ratio as cutting speed (180 m/min), feed (0.125 mm/rev), and depth of cut (0.8 mm).

Experimental tests were performed in [19] with the following cutting regime cutting speed (180,130,160 m/min), feed (0.05, 0.15, 0.25 mm/rev), and depth of cut (0.1, 0.35, 0.6 mm). Different studies were developed for modeling methods of cutting processes [20, 21]. Was concluded that cutting speed is the parameter with insignificant influence on surface roughness after the application of the Taguchi method for optimizing cutting parameters in hard turning operations of AISI 4142 steel. The materials such as tool steels, die steels, bearing steel [22], alloys steels, case-hardened steels, white cast irons, and alloy cast irons are widely machining in hard machining operations [18].

In this study the main purpose was to analyze the experimental data results of the cutting regime for hard turning, in terms of surface quality, accuracy and cutting time in order to be able to replace the grinding semi-finished and finished steps in case of an inner ring surface. The inner ring raceway was chosen to perform the experimental tests.

2 Problem Definition

In the last period were developed different technologies for manufacturing of different components especially for automotive industry. One component used usually in a car is a bearing. Bearings are components manufactured from different metallic or non-metallic materials.

A standard bearing is composed by two rings named outer and inner ring, a number of rollers depending on the application conditions, one or two brass, plastic or metal sheet cages and rivets.

Some experimental tests were performed in case of a special radial roller bearing, Fig. 1. This bearing is a special design used in wind energy applications. This bearing has one inner ring (1), 14 rollers (2) and a plastic cage (3).

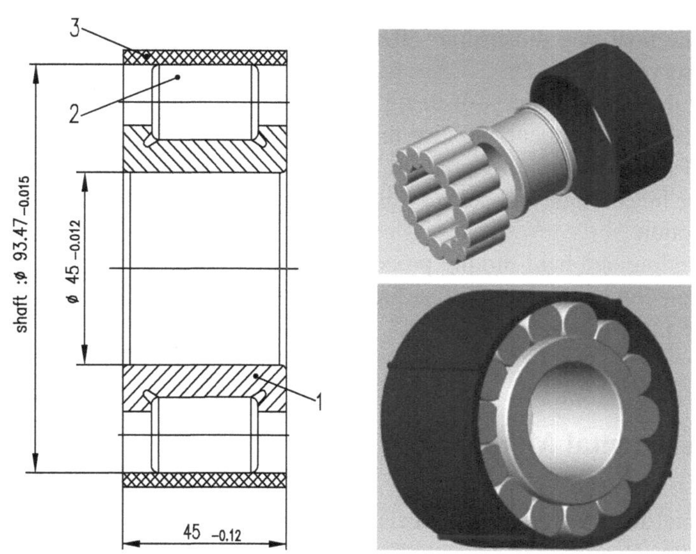

(a) Bearing design-2D view (b) Bearing design-3D view

Fig. 1 Bearing components drawing

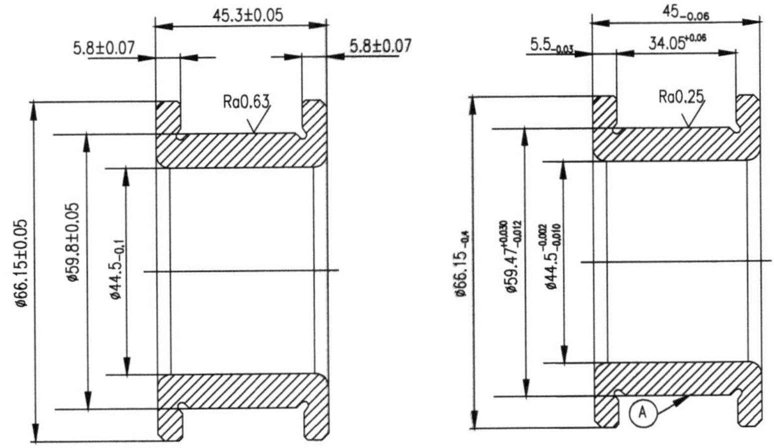

(a) Turned sample (b) Grinded sample

Fig. 2 Inner ring shape and dimensions

Different mechanically processes were homologated in order to obtain the final shape of rings, cages and rollers. For the inner ring the diameter of the race way is an important data in bearing construction, surface A in Fig. 2b. Two manufacturing processes could be applied to obtain the accuracy and the surface quality of the raceway.

One is grinding and the other one is hard turning. Schematically dimensions of the rings before grinding operations, soft turned and heat treated, can be observed in Fig. 2a. The final ring dimensions after grinding or hard turning of the inner ring can be observed in Fig. 2b.

One of the major impacts of using hard turning is that this process has small impact on the environment and also the consumption of the energy is less than in the case of a normal grinding process. From this point of view an important effect also will be in lead time and delivery time improvement with impact on the internal costs reduction.

The grinding and hard turning process were performed in order to compare the two cutting solutions in terms of surface quality, accuracy and cutting time in the case of the inner ring raceway.

3 Experimental Methodology

Experimental tests were carried out on a turning machine, type GLS 2000, using CBN inserts. Nose radius of cutting insert was 0,4 mm. Grinding process was developed on a normal grinding machine type, SAW 4.

In order to compare the results of the hard turning process implementation with grinding process results, some rings from the same production lot were manufactured. Cutting regime for both processes is described as: cutting speed (vc), feed rate (f), depth of cut (ap). Considering different values for the above mentioned cutting process parameters was analyzed the effect in terms of quality of the surface, rough, profile and roundness. On the other side using hard turning process the cutting time was analyzed. The experimental tests were carried out in case of a raceway diameter, outer surface; the length of cut was 20 mm. This surface has an important role in the bearing function because the contact area between ring and rollers is assured during the function. A continuous oil film must exist on the raceway surface.

The value of the roughness for this surface must be under 0.25 μm after super-finish operation as indicated in Fig. 2b.

The surface roughness measured in the paper is the arithmetic mean deviation of surface roughness of profile Ra.

3.1 Process Flow Description

The rings for experimental tests were prepared according to the flow charts depicted in Fig. 3.

Grinding process is developed in different phases such as rough, semi-finish, finish and super-finish.

Considering that one hard turning phase can replace grinding semi-finish and finish phases some improvements could be obtained in terms of machining efficiency, high equipment share, and the cost decrease due to the time saving which include electronic cost, auxiliary materials cost, manpower cost.

The allowance after quenching is 0.3 mm for both types of process, hard turning and grinding rings were prepared and checked before experimental tests. All surfaces were grinded as follow: finish grinding faces, inner and outer diameter rough grinding.

In order to be able to process and measure the results of grinding versus hard turning rings were measured before using a control device as depicted in Fig. 4.

3.2 Materials

Raw material tested was 100CrMnSi6-4. This steel is used in bearing production for rings and rollers. In this study the material was approved for series production and the chemical composition is shown in Table 1.

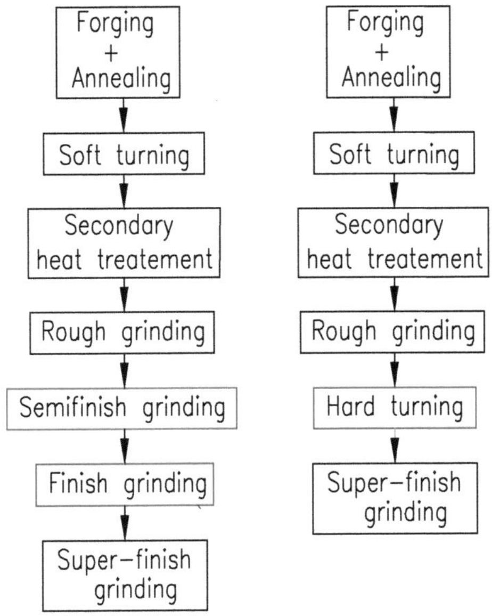

Fig. 3 Technological chains: Grinding process and hard turning process

Table 1 Chemical composition of 100CrMnSi6-4

%C	%Si	%Cr	%Mn	%Ni	%Cu	%Mo
0.98	0.64	1.42	1.13	0.15	0.1	0.07

3.3 *Testing Procedure*

Grinding operations were performed using a grinding wheel. The dressing of the wheel was performed with a dressing diamond. The characteristics of the Bakelite bounded grinding wheel used to process this kind of rings are: white fused aluminum oxide with fine grits size, soft hardness and vitrified bond. The symbol of the wheel is 66A80K4V1E.

Coolant liquid is used for both types of process. RELUBRO TRM is a semi-synthetic cutting fluid. This liquid is suitable for cutting, milling, general machining and grinding. Grinding process cutting regime can be observed in Table 2.

Table 2 Grinding cutting regime

Process	Cutting regime			
Grinding	vc, (m/s)	f, (mm/rev)	ap, (mm)	Speed, (rev/min)
	31	0.08	0.03–0.05	1058

Fig. 4 Control device

All cutting parameters were controlled via CNC part program. All data were processed and the average values were considered for analysis.

Carbide insert tip used was VBGW 16040T00815ME, Fig. 5. CBN dimensions and angles are described in Table 4.

Before grinding all rings are supposed to quenching and tempering heat treatments, temperature 845 °C, and time 60 min. The hardness is 60HRC.

The experimental findings were compared to the requirement of the surface finish on the basis of the design drawing of the ring and also with the grinding results.

Actual values showed in Table 3 used in current production lead to a cutting time about 0.68 min/ring in case of hard turning and in case of grinding process with the regime showed in Table 2 the average time is 0.6 min/piece/process step.

Fig. 5 Experimental device

Table 3 Hard turning actual cutting regime

Process	Cutting regime			
Hard turning	vc, (m/s)	f, (mm/rev)	ap, (mm)	Speed, (rev/min)
	130	0.07	0.05	500–1000

Table 4 Insert dimensions

d, (mm)	l, (mm)	s, (mm)	r, (mm)	d1, (mm)
9.52	16.6	4.78	0.4	4.4

Based on this observation was defined an experimental plan and implemented through 75 test specimens, in order to study the cutting time improvement and also parameters related to quality of surface.

Table 5 depict the experimental plan and the value of the cutting parameters.

The tests were carried out into 3 groups (25 samples/group). Each group was machined using the following procedure:

Table 5 Experimental plan

Process: hard turning number of group/modified parameter	No. of measured samples/regime	Cutting regime			
		vc, (m/s)	f, (mm/rev)	ap, (mm)	Speed, (rev/min)
Group 1 (25 samples) Cutting speed (v)	5	100	0.07	0.05	693
	5	130	0.07	0.05	693
	5	140	0.07	0.05	693
	5	150	0.07	0.05	693
	5	160	0.07	0.05	693
Group 2 (25 samples) Feed rate(s)	5	130	0.08	0.05	693
	5	130	0.09	0.05	693
	5	130	0.1	0.05	693
	5	130	0.15	0.05	693
	5	130	0.2	0.05	693
Group 3 (25 samples) Depth of cut (ap)	5	130	0.07	0.075	693
	5	130	0.07	0.1	693
	5	130	0.07	0.125	693
	5	130	0.07	0.15	693
	5	130	0.07	0.2	693

Group 1: 25 samples were machined with five different cutting speed, the feed rate (0.07 mm/rev) and depth of cut (0.05 mm) were mentioned constant.

Group 2: 25 samples were machined with five different feed rates, the cutting speed (130 m/min) and depth of cut (0.05 mm) were mentioned constant.

Group 3: 25 samples were machined with five different depths of cut, the cutting speed (130 m/min) and feed rate (0.07 mm/rev) were mentioned constant.

4 Results and Discussions

4.1 Hard Turning Process

4.1.1 Cutting Speed Influence on Process Time

All data were analysed based on the principle of cutting time reduction. Tests were performed with different cutting speed. The actual cutting speed used to process the bearing steel by hard turning is 130 (m/min). Five different cutting regimes were defined till the cutting speed was 160 (m/min). The feed rate and depth of cut were maintained in these cases at constant values.

The minimum value of the cutting time was measured for a value at cutting speed of 160 m/min (Fig. 6).

An increase with about 30% of cutting speed lead to a decrease of cutting time with about 35%.

4.1.2 Feed Rate Influence on Process Time

In this case the tests were developed for constant value of cutting speed and depth of cut. The feed rate was increased till 0.2 mm/rev (Fig. 7).

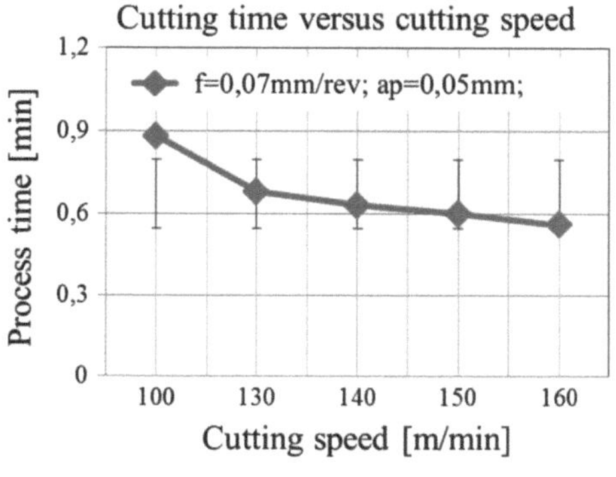

Fig. 6 Influence of the cutting speed on the process time

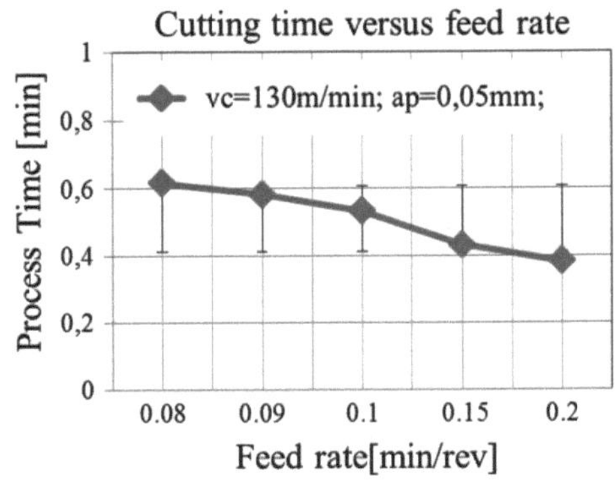

Fig. 7 Influence of the feed rate on the cutting time

Analysis of results lead to the conclusion that the minimum cutting time was obtained for a feed rate value about 0.2 mm/rev.

4.1.3 Depth of Cut Influence on Process Time

In the last tests the depth of cut value was increased from 0.05 till 0.2 (Fig. 8). In this case the results shown that the value of the cutting depth in case of bearing steel has insignificant influence on process time.

The scope of these experiments was to define the optimum working regime theoretically taking into account that the cutting time depend on the cutting regime.

4.1.4 Roughness and Surface Profile

The measurements of surface accuracy, profile, roughness and roundness for each cutting regime were performed on Talysurf and Talyrond (Figs. 9, 10, 11, 12, 13, 14, and 15).

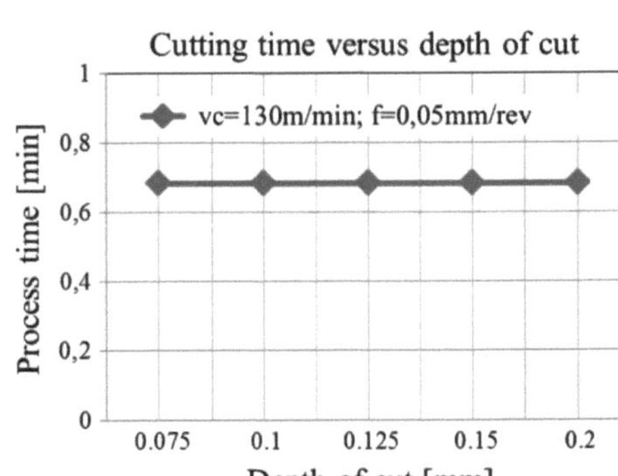

Fig. 8 Influence of the depth of cut on the process time

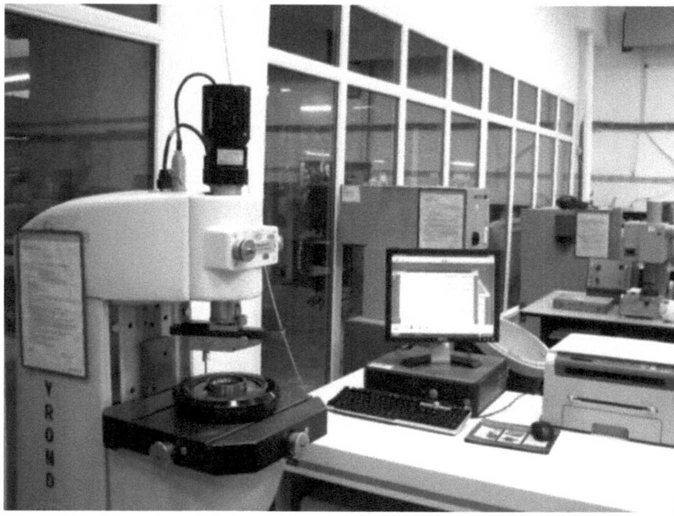

Fig. 9 Surface quality control equipment

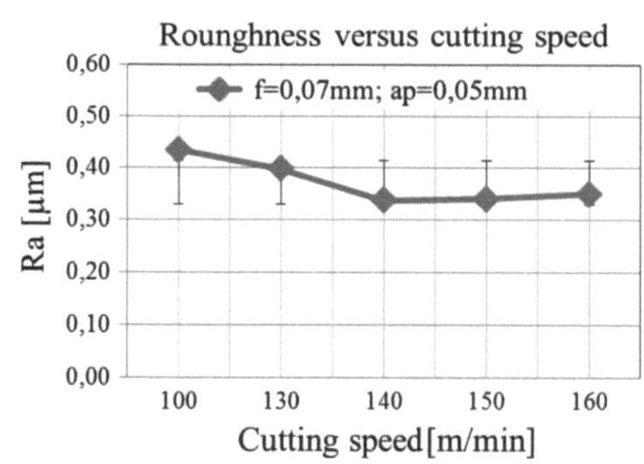

Fig. 10 Influence of cutting speed on roughness

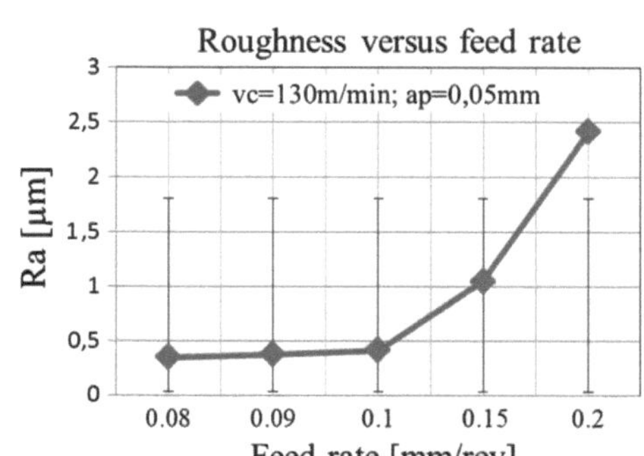

Fig. 11 Influence of feed rate on roughness

Fig. 12 Influence of depth of cut on roughness

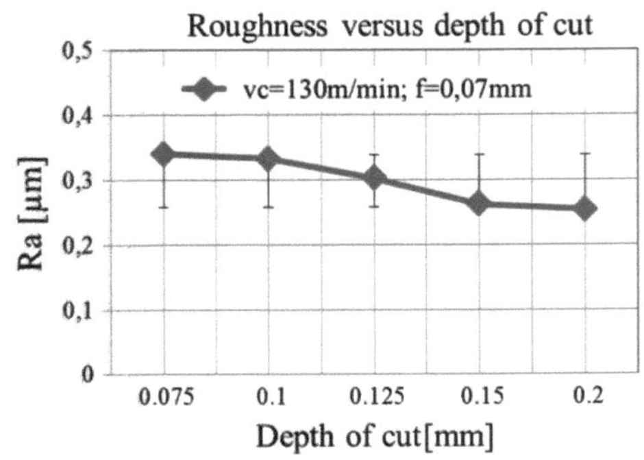

Fig. 13 Influence of cutting speed on surface profile

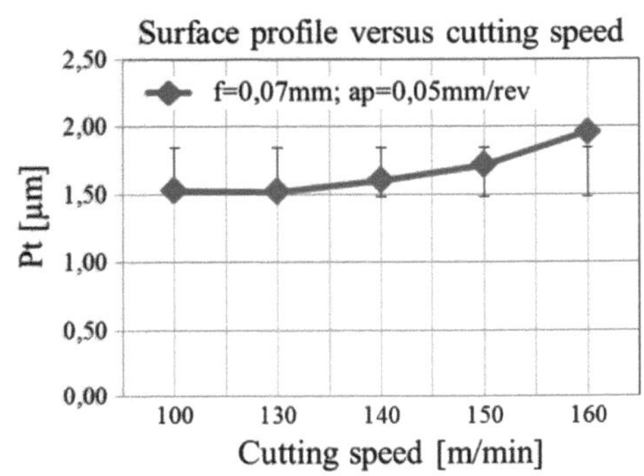

Fig. 14 Influence of feed rate on surface profile

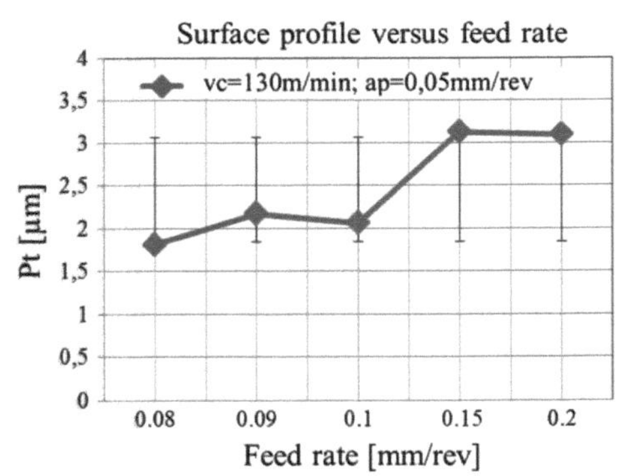

4.2 *Grinding Process*

4.2.1 Roughness and Surface Profile

Results of the experimental tests in case of grinding process can be observed in Table 6.

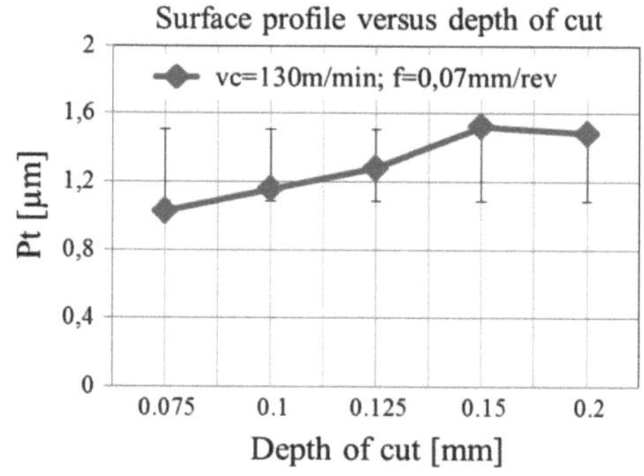

Fig. 15 Influence of depth of cut on surface profile

Table 6 Grinding results

Grinding	Ra, (µm)	Pt, (µm)	Tb, (min)
	0.34	2.12	0.6

Analysis of grinding results shows that the quality of surface after finish grinding process is satisfactory.

4.2.2 Roundness

The most important effect of the roundness is that the working condition and also the loading condition of the bearing will be affected if this parameter is out of specification. In Figs. 16 and 17 are depicted the results regarding the out-of-roundness for both processes grinding and hard turning.

An important effect of the out-of-roundness is that the fit and the clearance between the components of the bearing are affected. Analyzing the valley and the peaks of the measurement results could be concluded that in case of a bearing processed with the regime recommended in Table 7 the hard turning process can be used with good results in place of grinding.

In case of hard turning results showed in Fig. 17 a more constant shape of roundness peaks and valley can be observed, this means that a constant oil film rest between rollers and raceway during the function of the bearing.

All experimental data were analyzed considering that the identified values of the cutting regime parameters can be applied in case of series production. A summary with intervals of measured values versus the cutting regime parameters used to obtain the minimum values of roughness, roundness, surface profile and cutting time is depicted in Table 7.

Taking into account that different factors have influence on the process results, experimental data analysis show that the minimum values for process parameters can be obtained using cutting regimes characterized as described in Table 7.

Fig. 16 Grinding process

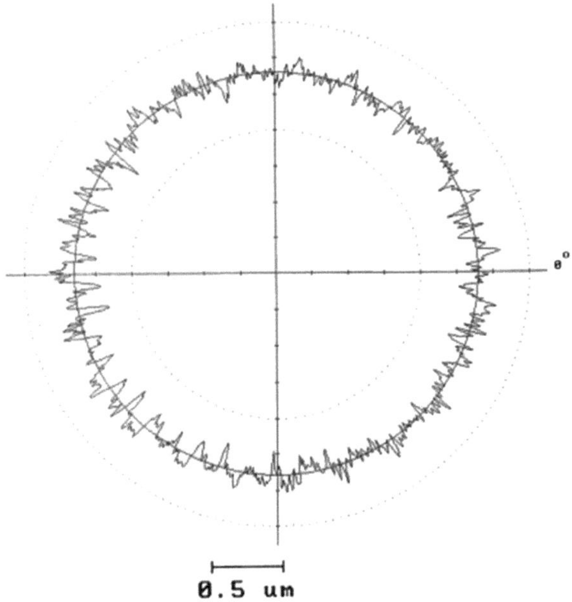

Fig. 17 Hard turning process

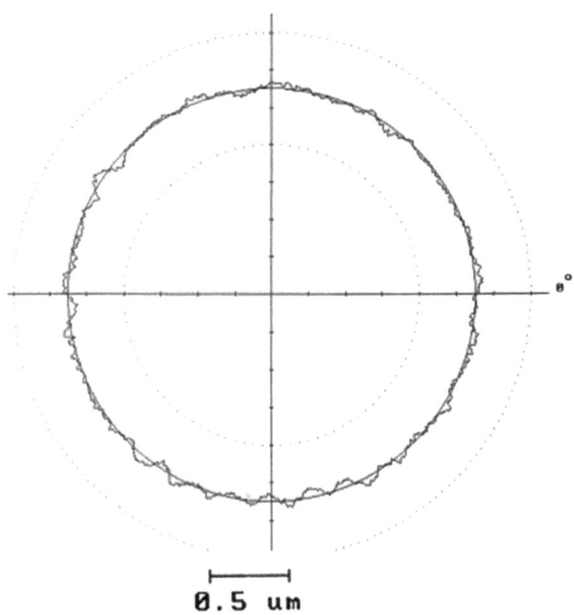

Table 7 Summary of experimental results in case of hard turning process

Experimental results		Recomemded cutting regime for minimum of measured results		
Parameter	Interval of measured values	vc (m/min)	f (mm/rev)	ap (mm)
Roughness	(0.25–2.42) (μm)	130	0.07	0.2
Surface profile	(0.74–3.09) (μm)	130	0.07	0.075
Roundness	(0.25–0.9) (μm)	150	0.07	0.05
Cutting time	(0.383–683) (min)	130	0.2	0.05

5 Conclusions

An experimental study has been conducted in case of bearing steel manufacturing process versus with the purpose to be able to replace some grinding process phases with hard turning process.

The study included experimental analysis of cutting parameters as cutting speed, depth of cut and feed rate, with a total of 75 test samples in case of hard turning process. All experimental tests were performed in industrial environment.

After the experimental tests results analysis it was concluded:

1. The cutting time was improved with about 35% and a feed rate increase with 35%, in case of hard turning process, caracterised by a cutting speed increased with 30% compared with the existing process regime.
2. In case of an increase with 35% of the depth of cut was observed no significant influence on the cutting time.
3. Surface quality regarding roughness show that the hard turning process lead to a smooth surface.
4. Analysis of profiles surfaces lead to the conclussion that the profile after hard turning process is almost equal with the one obtained after grinding process. Hard turning processes could replace grinding process steps, semifinish and finish in case of this kind of bearing for mass production, only if the last process step is super finishing.
5. An important advantage of the hard turning process is that the preponderence of the set-up time in total production lead time is smaller than in case of a grinding process, due to the fact that two set-up operation of grinding machines are replaced with one in case of hard turning process, this advantage has influence on the internal supply chain on one side and on customer supply chain on the other hand.
6. The improvement of cutting time considering the influence of the hard turning operation is around 35%.

Future extensions of this work may explore avenues for further improvement in surface quality and productivity via implementation on mass production of the resulted cutting regimes, as well as modeling using a multi-scale numerical analysis and validation of the proposed cutting regimes considering different production lot size.

Acknowledgements This work was supported by Rulmenti SA Barlad.

References

1. Dangayach GS, Pathak SC, Sharma AD (2006) Advanced manufacturing technology: a way of improving technological competitiveness. Int J Global Business Competitiveness 2(1):1–8
2. Urs RR, Mahesh BP, Sandesh S (2014) On-Time Delivery Improvement Using Lean Concepts—A Case Study of Norglide Bearings. Int J Innovative Res Sci Eng Technol 3 (6):13349–13354
3. Seveen K, Muniswaran K (2007) Experimental study on hard turning of hardened tool steel with coated carbide cutting tools. Masters thesis, Universiti Teknologi Malaysia, Faculty of Mechanical Engineering
4. Gosiger MC (2012) Fundamentals of hard turning—an depth look at the process, (http://cdn2.hubspot.net/hub/139128/file-17761415-pdf/docs/gos_wp_hardturning_f.pdf)
5. Piska M, Forejt M (2006) Theory of machining, forming and cutting tools. UST FSI VUT, Brno, Czech Republic. ISBN 80-214-2374-9
6. Tóth T, Kundrák J, Gyáni K (2004) The removal rate as a parameter of qualification for hard turning and grinding. Tools Methods Competitive Eng 1(2):629–639
7. Kundrak J, Karpuschewski B, Gyani K, Bana V (2008) Accuracy of Hard Turning. J Mater Process Technol 202:328–338
8. Abrao AM, Aspinwal DK (1996) The surface integrity of turned and ground hardened bearing steel. Wear 196:279–284
9. Choi Y (2009) A comprehensive study of residual stress distribution induced by hard machining versus grinding. Tribological Lett 36:277–284
10. Das DK, Sahoo AK, Das R, Routara BC (2014) Investigations on hard turning using coated carbide insert: Grey based Taguchi and regression methodology. Procedia Mater Sci 6:1351–1358
11. Zawada-Tomkiewicz A (2011) Analysis of surface roughness parameters achieved by hard turning with the use of PCBN tools. Est J Eng 17(1):88–99
12. Grzesik W, Żak K, Kiszka P (2014) Comparison of surface textures generated in hard turning and grinding operation. Procedia CIRP 13:84–89
13. Waydande P, Ambhore N, Chinchanikar S (2016) A review on tool wear monitoring system. J Mech Eng Autom 6(5A):49–53
14. Singh D, Rao PV (2008) Improvement in surface quality with solid lubrication in hard turning. In: Proceeding of the world congress on engineering, vol 3, London
15. Tönshoff HK, Arendt C, Amor RB (2000) Cutting of hardened steel. CIRP Annals-Manufact Technol 49(2):547–566
16. Davim J (2008) Machining, fundamentals and recent advances. Br Libr Cataloguing. doi:10.1007/978-1-84800-213-5
17. Basil MP, Raju T, Biju B (2014) Optimization of cutting parameters in hard turning of AISI 4340 steel. Int J Innovative Res Adv Eng (IJIRAE) 1(8): 93–98. ISSN: 2349-2163,
18. Kaçal A, Yıldırım F (2013) High speed hard turning of AISI S1 (60WCrV8) cold work tool steel. Acta Polytechnica Hungarica 10(8):169–186
19. Puh F, Šegota T, Jurković Z (2012) Optimization of hard turning process parameters with pcbn tool based on the taguchi method. Technical Gazette 19(2):415–419

20. Umer U, Butt SI, Askari SJ, Danish SN, Xie L (2008) Comparative analyses for different modeling methods in high speed turning operations for hardened steel. J Mech Eng 54 (12):850–854
21. Davim JP, Maranhão C, Faria P, Abrão A, Rubio JC, Silva LR (2009) Precision radial turning of AISI D2 steel. Int J Adv Manuf Technol 42(9):842–849
22. Bhadeshia HKDH (2012) Steels for bearings. Prog Mater Sci 57:268–435
23. Grzesik W, Zak K (2013) Comparison of surface textures produced by finish cutting, abrasive and burnishing operations in terms of their functional properties. J Machine Eng 13(2):46–58
24. Sahin Y, Motorcu AR (2008) Surface roughness model in machining hardened steel with cubic boron nitride cutting tool. Int J Refractory Metals Hard Mater 26:84–90
25. Derakhshan ED, Akbari AA (2009) Experimental investigation on the effect of workpiece hardness and cutting speed on surface roughness in hard turning with CBN tools. In: Proceeding of the world congress on engineering, vol 2, London
26. Sahin Y (2009) Comparison of tool life between Ceramic and Cubic Boron Nitride (CBN) cutting tool when, machining hardened steels. J Mater Process Technol 209(7):3478–3489
27. D'Addona DM, Raykar SJ (2016) Analysis of surface roughness in hard turning using wiper insert geometry. Procedia CIRP 41:841–846

On the Advanced Milling Technology of Thin-Walled Parts for Aerospace Industry

Miroslav Piska and Petra Ohnistova

Abstract This paper is focused on advanced analysis of a progressive milling technology of high strength alumina alloy 7475-T7351 that is used in modern aerospace industry today. The analyses comprises a study of the material, its mechanical properties, use of alternative monolithic-carbide tool designs, micro-geometries of the milling cutters, coatings and a study of cutting parameters on dynamics of cutting, its productivity, quality of the surfaces and other specific cutting phenomena. This work includes an experimental verification of the proposed technology using 5-axis CNC machining centre, evaluation of 2D/3D surface structures using optical high-resolution 3D surface device. Dynamometer Kistler 9575B/SW DynoWare were used for measuring of instantaneous force loading in long time series. The results are relevant for milling of thin-walled parts and the up and down milling strategies without any other surface treatment of the parts. Some fatigue studies are in progress.

Keywords Tool geometry · Force loading · Surface topography

1 Introduction

Although the current trend in aerospace industry is to produce thin-walled structural parts made from composite materials some applications still require conventional machining operations [1]. As described in the paper of Starke [2] the most frequently machined materials in aerospace industry are high strength alumina alloys which are supplied in either cast or wrought conditions. Application of these alloys further differs in the type of application: wrought alloys are widely used for so-called *primary structures* (a failure of such components is directly affecting operability of the aircraft, e.g. wing structures or the fuselage) and cast or wrought

M. Piska (✉) · P. Ohnistova
Institute of Manufacturing Engineering, BUT FME Brno,
Technicka 2, 616 69, Brno, Brno-Královo Pole, Czech Republic
e-mail: piska@fme.vubtr.cz

V. Majstorovic and Z. Jakovljevic (eds.), *Proceedings of 5th International Conference on Advanced Manufacturing Engineering and Technologies*,
Lecture Notes in Mechanical Engineering, DOI 10.1007/978-3-319-56430-2_9

aluminum alloys are used for the *secondary structures*, that do not immediately affect the safety of the aircraft upon their premature failure.

2 State of the Art

Components used in the aircraft industry should be therefore differentiated not only according to the manufacturing process and used material but also according to the workpiece geometry and needed mechanical or physical properties. In order to respect weight limitations of an aircraft, every component must be fully optimized in terms of its weight and material excess. For this reason, many of the aerospace components are having thin-walled structures [3]. These structures vary significantly according to the operational use: complex thin-walled structures such as turbine blades used in propulsion systems or simplified thin-walled components such as directional flap situated on the wing, but probably the largest use of the thin-walled structures are structural and wing components of the aircrafts such as ribs and spars. These structures have to be lightened and they therefore consist of deep pockets which are separated by thin-walls in the highest removal rates.

Milling of thin-walled components involves several different issues due to deflections of the walls caused by unilateral cyclic loading of the workpiece [4] which results in geometrical errors and a deteriorated surface structure. Another issue occurring during machining is self-excited oscillation of the tool and also mechanical vibrations of the machined part [5]. The literature provides a wide range of analytical, numerical and experimental methods how to deal with the every issue mentioned above [6]. However not many methodologies are dealing with the influence of the shoulder milling process and its cutting conditions on the surface topography and fatigue performance of the thin-walled components.

Resistance to fatigue is one of the main requirements for primary and secondary aircraft structures and it is already well established by many studies, However, if any loading mode is applied, the micro-crack initiations and their propagation leading to the fracture usually begin on the real material surfaces. The reason for this phenomenon is understandable because many loadings such as torque, tensile or bending induce the maximum stress concentration located on the surface of the exposed workpiece. It is therefore logical that the slightest nonconformity of the surface can lead to the crack initiation on the surface [7]. So to examine the effect of the shoulder milling as finishing operation affecting the surface quality of the thin-walled parts (usually one of the most loaded areas of the structures) makes sense.

However, the problem of the influence of the cutting conditions for shoulder milling strategy and then on the surface integrity and related fatigue performance of thin-walled components is not very available in the literature. This paper closely examine the influence of machining parameters of shoulder milling finishing operation and the generated force loading, surface topography and especially the Abbott-Firestone curves of the parts made of the high strength aluminum alloy

7475-T7351. These results will be correlated with the fatigue performance studies consequently.

2.1 Study of the Material 7475-T7351

The material used in this study is graded as 7475-T7351 and is the one of the most frequently used aluminum alloys in the aerospace industry today [8]. Aluminum alloy 7475, known also as the alloy AlZn5.5MgCu, belongs to the series 7xxx where the main alloying element to aluminum is zinc and copper—Table 1. The heat treatment includes heating to the solid solution of selected phases, cooling, stress-relieving by a controlled stretching (permanent set from 0.5 to 3% for a sheet) and then artificially aging for achievement of a higher strength, a better corrosion resistance and a better resistance to the stress corrosion cracking [9, 10].

The aluminum alloy 7475 is an alloy with a so-called controlled toughness, developed by the producer ALCOA, providing a combination of high strength, good fracture toughness and high resistance to the fatigue crack propagation [11]—Table 2. The alloy 7475 is a refinement of the alloy 7075 and its fracture toughness for plates are almost 40% greater than for the previous version 7075 at the same temperature. The prevalence in some properties is a result of the reduction of the Fe, Si and Mg contents and is as influenced by thermo-mechanical and heat treatments procedures also [8]. The material 7475 is recommended when a high fracture toughness of a part (typically the aircraft wings or wing spars) are considered. The application of T7351 heat treatment provides also optimal resistance to stress corrosion cracking.

Table 1 Chemical composition of 7475-T7351 [8]

Element	Weight content (%)
Si	0.10 max.
Fe	0.12 max.
Cu	1.20–1.90
Mn	0.06 max.
Mg	1.90–2.60
Cr	0.18–0.25
Zn	5.20–6.20%
Ti	0.06 max.
Others, each	0.05 max.
Al	Balance

Table 2 Mechanical properties of 7475-T7351 [8]

Thickness of the sheet (mm)	25–38	50–63	75–89
Tensile strength (MPa)	490	476	448
Yield strength (MPa)	414	393	365
Elongation (%)	9	8	8

2.2 Theory of the Force Loading at Shoulder Milling

In aircraft industry are typical demands on parts with high strength to weight ratios that lead to rampant use of thin-walled aluminum or titanium parts that should be machined with multi-axes CNC centers. However, cutting forces generated when machining cause deformations due to low rigidity of the components what affects their accuracy and quality [5]. These variable cutting forces affect the workpiece and accurate position of the tool. Static loadings cause the geometrical inconformity related to the deflection of a wall or of a tool. Dynamic loadings result in wall or tool vibrations, deflection of the spindle or deflection either or both of them. The surface topography and the fatigue performance of the component may be affected so [4].

Determination of the force loading during machining operation can be predicted theoretically and verified experimentally.

Theoretical calculations are derived from the specification of the chip cross section—corresponding to the analysis of a helical conoid and specific cutting force [12]. Cross section area of the chip can be derived for each cutting flute (tooth) and its geometry. It is possible for each milling operation to calculate how the cutting edge passes through the material and the chip cross section varies at each time moment. The loading of the milling tooth in cutting can be defined as (1).

$$F_c = \int_{\phi_1}^{\phi_2} \mathrm{d}F_c = c_o. \int_{\phi_1}^{\phi_2} \sin^{1-mc} \phi \, . \mathrm{d}\phi \tag{1}$$

where the constant $\underline{c_0}$ represents a constant regarding the geometrical and material parameters, $\underline{\phi}$ the angle of rotation and $\underline{mc}$ the coefficient expressing the effect of variable chip thickness on specific cutting force. This equation leads to the integration using hypergeometric function. However, for engineering applications a numerical integration can be used so Eq. (1) can be extended then by the number of teeth in the cut $\underline{z}$ (depending on the cutting conditions and tool geometry) to relation (2).

$$F_c = \sum_{i=1}^{z_z} A_{Di} \, . kc_i. \tag{2}$$

Workpiece loading and workpiece deflection depend not only on this cutting force F_c but also on the force perpendicular to the cutting force acting in the radial direction of the cutting tool F_{cN}. Its prediction is rather difficult because of its high influence of the tool macro-geometry and micro-geometry, tribology of the interface tool-chip-workpiece and tool wear. The magnitudes of F_{cN} and passive force F_{cN} can vary with the tool wear significantly.

So far, there is a few different variations how to derive forces acting in the technologically important directions (e.g. following the direction of cutting speed). For example Tlusty [13] proposes relations for the cutting force $\underline{F_c}$, feed force $\underline{F_f}$ and passive force $\underline{F_p}$ in the form (3).

$$
\begin{aligned}
F_C &= C_{Fc} \cdot a_P^{x_{Fc}} \cdot f^{y_{Fc}} \\
F_f &= C_{Ff} \cdot a_P^{x_{Ff}} \cdot f^{y_{Ff}} \\
F_p &= C_{Fp} \cdot a_{p1}^{x_{FP}} \cdot f^{y_{FP}} \\
F &= \sqrt{F_C^2 + F_f^2 + F_p^2}
\end{aligned}
\tag{3}
$$

The resultant force should be equal to the resultant force from the measured force components in Cartesian and technological system, respecting the cutting force $\underline{Fc}$, normal force $\underline{Fcn}$ an axial passive force $\underline{Fp}$ (4).

$$F = F_M = \sqrt{F_x^2 + F_y^2 + F_z^2} = \sqrt{F_c^2 + F_{cN}^2 + F_p^2}, \tag{4}$$

or for one cutting edge engaged in cutting in the form (5).

$$F_1 = F_{M1} = \sqrt{F_{x1}^2 + F_{y1}^2 + F_{z1}^2} = \sqrt{F_{c1}^2 + F_{cN1}^2 + F_p^2}. \tag{5}$$

However, these forces are time-dependent on the machining time so exact experimental measurement of the cutting forces in engineering application is inevitable. For the testing and evaluation of the force loading the time positioning is used in order to detect the maximal force loading corresponding to the position of the tool at the maximal cutting cross-section (including maximal chip thickness). Maximal cutting cross-section (maximal chip thickness while milling) differs with every type of the milling strategy. Conventional shoulder milling is basically divided into the up and down milling a chip thickness is significantly varying along the cutting trajectory of the tool—Fig. 1.

2.3 Surface Integrity of Machined Surface

Machined surface can be defined as a border between a machined component and an environment commonly. Machining process influences the state of the surface therefore the term surface integrity was introduced in the late 70s. Surface integrity conceals several aspects: surface topography and surface microstructure, mechanical properties and residual stresses [14, 15], but fatigue properties should be included also.

The surface texture, also denoted as surface topography or surface finish, is the characteristic of the surface defined by several different parameters with the basic

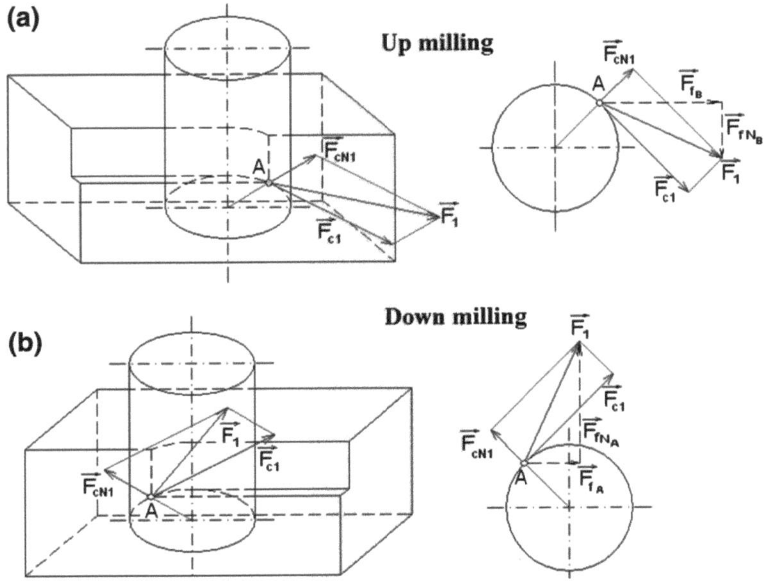

Fig. 1 Force loading for up milling (**a**) and down milling (**b**) force distribution [13]

differencing into the 2D profile parameters and 3D surface parameters [16]. More precisely the surface texture is the three-dimensional set of repetitive and random deviations from the micro-geometry of the machined surface [17].

According to the European Standard EN ISO 4287 [18], all parameters have to be defined in the same coordinate system. Most commonly used coordinate system (also integrated into the International Organization for Standardization) is the Cartesian orthogonal coordinate system. Axis X coincides with direction of measurement which is parallel to the mean line of the profile, axis Y also belongs to the measured profile and axis Z is perpendicular to the surface and oriented away from the measured surface.

A suitable direction of the reference measurement plane is the one in which the values of maximum height parameters are evident. In the other words, the roughness of evaluated surface is determined by its profile in a section perpendicular to the machining direction. In the case of shoulder milling, the appropriate direction for surface evaluation is identical with the direction of the feed. For the purpose of this work, surface topography after shoulder milling operation was measured in two directions: (1) in the direction of the feed according to the standards EN ISO 4287 and (2) in the direction of the tool axis.

The profile parameters are evaluated from the part of a profile called sampling length (l). They correspond to the distribution of profile heights z(x) along the measuring direction x. The surface roughness can be described according to the ISO 4287 with the basic parameters including the 3 sets of 13 parameters: R-parameters derived from the profile roughness and P-parameters and W-parameters derived respectively from the surface basic profile and the profile waviness. For the purpose of this work only some R and W parameters were analyzed and Firestone-Abbot curve describing the material ratios was defined for every machining strategy and corresponding cutting conditions.

2.4 *Effect of Machining on the Surface Topography and Fatigue*

Several phenomena can be observed during machining operation and plastic deformations. In simple terms, the surface layer is subjected to the elastic and plastic deformation and is influenced by heating produced during machining which results in structural changes in the material, micro-cracks, strain hardening, imposed residual stresses and especially in surface irregularities constituting the surface roughness. These effects can highly affect mechanical properties of the workpiece such as fatigue strength. The selection of the appropriate cutting condition is essential for the desired mechanical properties of the workpiece because the improper cutting condition develop the substandard surface texture which, together with cyclic loading, leads to the fatigue failure of the material [19].

In various early publications and fatigue models the average surface roughness parameter Ra is used to evaluate the influence of the surface quality on the fatigue life. But relatively large variance in fatigue results is observed for the same Ra values. This raises the question of whether this parameter is suitable for evaluation of the influence of the surface topography on fatigue or not. Therefore in many recent studies it has been recognized that the measure of maximum surface irregularities may be better indicator of the expected fatigue life [20].

According to Siebel and Gaier [21] the maximum depth of surface deviation after machining is the most significant key factor influencing the fatigue properties of the material. In their work they compared fatigue strength with the maximum depth of a surface groove Rt in steels and non-ferrous metals (Al–Cu and Al–Mg alloys). Output from their work was following: above critical grove depth (R0), the reduction in fatigue endurance limit is proportional to log Rt and R0 values for aluminum alloys was 4–6 μm. It has been lately assumed that not only the maximum profile height Rt but also the shape of the groove can be critical factor influencing the fatigue life because they can be source of the local stress concentrations which can initiate cracking of the material [20].

An example from engineering perspective can be taken from the Renault automotive manufacturer using the maximum profile irregularities (locally marked as Rx) as the primary aspect for the measure of the fatigue performance. Further, the secondary aspects are mean depth and mean spacing (marked as R and AR) [20].

According to Gómez [22] who studied the influence of cutting parameters of the turning process and the fatigue performance of the aluminum alloy A92024-T351, the most affecting cutting parameters of the surface topography (especially Ra) is the feed. His work confirms an increase in Ra with increase of the feed, on the other hand the influence of the cutting speed on the surface quality was remarkable only for higher feeds where the higher cutting speed showed higher Ra values.

Ojolo [19] presented a study of the machining parameters influence on the fatigue life of end milled aluminum alloy 2024. Their work presented a decrease of the surface roughness with an increase of the cutting speed, which may be a result of the thermal softening effect. The increase in the feed rate significantly influences and increases the surface roughness. Ojolo [19] supposes that at higher feed rates the teeth of a cutter doesn't perform perfect swiping on the entire surface of the machined zone to make a smooth surface. It was as well observed that an increase in rake angle of the cutter leads to a better surface quality.

3 Experimental

A number of cutting experiments was carried out on the 5-axis milling center MCV 1210 controlled with Sinumerik 840D. All experiments were performed using three different SECO tools with different micro-geometries and coatings. Cutting experiments were performed for different cutting conditions: variable cutting speed and variable feed. Force loading testing was performed using stationary KISTLER 9575B/SW DynoWare and machined surface was analyzed using high resolution optical device ALICONA-IF G4. Aluminum alloy 7475-T7351 was selected as a workpiece material because of its real use in aerospace industry today especially for structural parts which are subjected to the cyclic loading during flight. The experimental workflow for an execution of the experimental procedure can be seen in Fig. 2.

3.1 Tool Micro-geometry

The following SECO end-milling tools recommended for a thin-wall machining were used: $\phi 16 \times 55 \times 115$ JS513160D3C.0Z3-NXT (SMG N11, (Ti,Al)N coating), $\phi 16 \times 55 \times 115$ JS514160D3C.0Z4-NXT (SMG N11, (Ti,Al)N based coating), and $\phi 6 \times 32 \times 125$ JS453160E3R100.0Z3-HEMI (SMG N11, with the PVD HEMI polished coating). Tool micro-geometries were analyzed using particular software subprogram *Alicona Edge Master* compatible with optical microscope ALICONA-IF G4, Fig. 3. The principle of this measurement was based on the gradual positioning of the reference plane perpendicularly to the cutting edge of each tooth. Measured values of intersection of the reference plane with the cutting edge were statistically processed to obtain final result of the mean radius of mean cutting edge and the helix inclination angle.

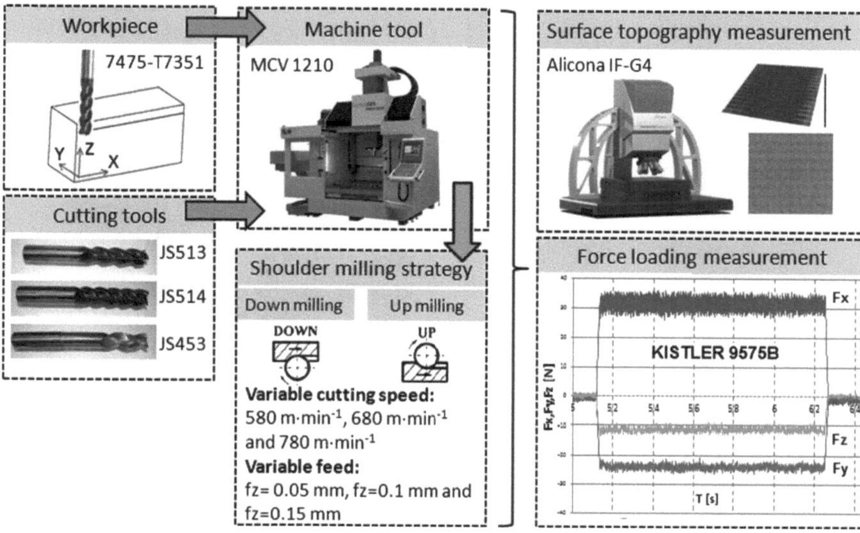

Fig. 2 Experimental workflow

3.2 *Force Generation When Shoulder Milling*

The dimensional accuracy of a thin-walled structure depends on the machining operation resp. the generated cutting forces. For a need of a very real data, the experimental approach for determination of cutting forces was selected instead of the FEM simulations or predictions.

The force loading during machining was measured between tool and a solid workpiece with the stationary KISTLER 9575B/SW dynamometer. Machining and force loading testing was performed using the 5-axis CNC center MCV 1210, Tajmac-ZPS, a.s., Zlin, Czech Republic.

The Kistler dynamometer was fixed with screws on the machine tool table in the direction corresponding to the coordinate system of the machine. The precise vise was mounted to the dynamometer in order to have the workpiece placed in the middle of the dynamometer and to avoid measurement errors.

Force loading was measured at different cutting conditions (variable feed and cutting speed) in order to examine the influence of the conditions on the direction and value of the cutting forces for both up and down milling. Values measured with the KISTLER dynamometer were processed and evaluated with Dynoware software —Fig. 4.

Force loading in the technologically important directions ($\underline{F_c}$ and $\underline{F_{cN}}$) followed the expected force distribution, presented in Fig. 1. The mean values of the maximal instantaneous force acting in $\underline{X}$, $\underline{Y}$ and $\underline{Z}$ direction was used for the determining the measured resultant force $\underline{F_{IM}}$.

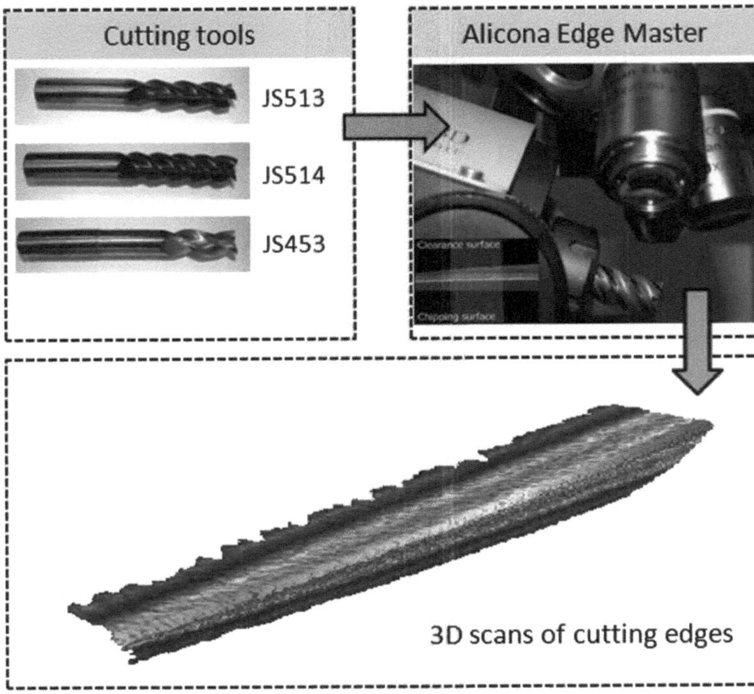

Fig. 3 A workflow of a tool micro-geometry measurement (by means of Alicona Edge Master)

3.3 Measurement of the Surface Topography When Shoulder Milling

Based on the previous research of the influence of the cutting conditions on the surface integrity and the fatigue performance of the components several different selections had been made. First, as a reference roughness parameters specifying the surface topography the Ra, Rt and Rv were chosen to highlight, because of their wide use in industry.

Consequently, the waviness parameters as Wa, Wq, Wt and Wz, and Wt, were chosen as the reference parameters for further investigations.

The Abbott-Firestone curve that can be found from a profile trace by drawing lines parallel to the datum and measuring the fraction of the line which lies within the profile was defined for every machined surface. Parameters defining the Abbott-Firestone curve are: Rk describing core roughness depth (height of the core material), parameter Rpk characterizing mean depth of the peaks above the core material, parameter Rvk describing the mean depth of the valleys below the core material, parameter Rmr1 (or Mr1) characterizing the fraction of the surface which consists of peaks above the core material and Rmr2 (or Mr2) charactering the fraction of the surface which will carry the load [26].

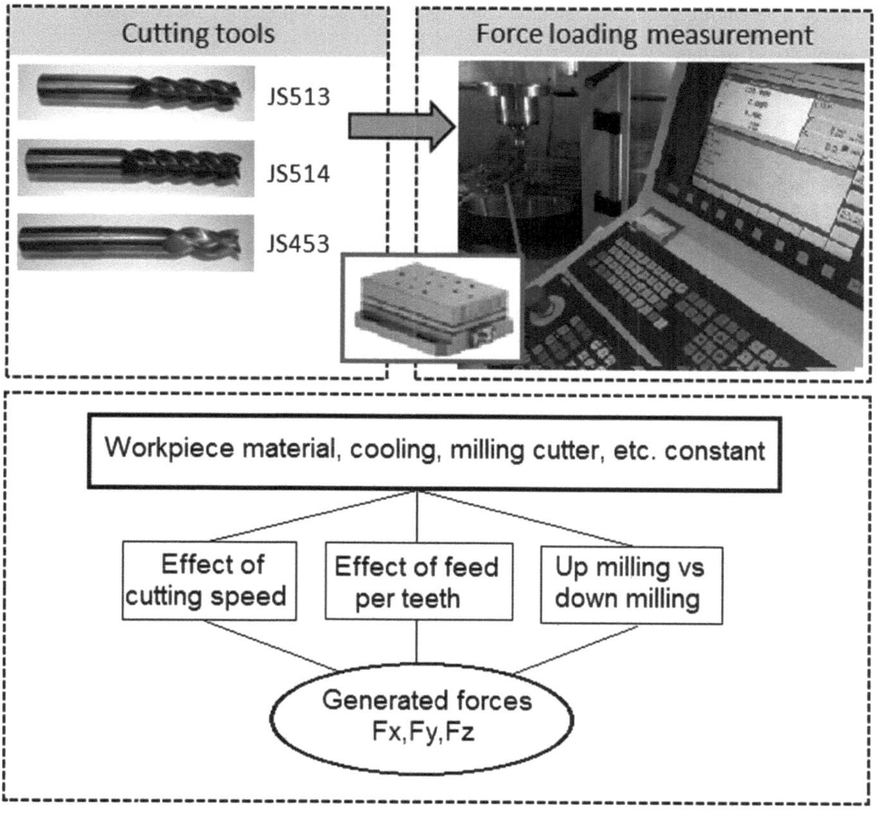

Fig. 4 Configuration of the force measurement with Kistler dynamometer

Literature review dealing with different machining processes and their influence on the surface quality and fatigue life regards turning [22, 28], ball-end milling [29] and end-milling [19] that showed that the most frequently analyzed cutting parameters are feed, cutting speed and the rake angle of a cutting edge. In this study the effect of feed and cutting speed on the surface topography were chosen as reference variable parameters for three standard SECO tools—Fig. 3. The cutting conditions with respect to the SECO tool recommendation are summarized in the Table 3.

Complex measurement of the surface topography was performed on the rigid sample using ALICONA-IF G4. The surface of the dimensions 5×5 mm^2 was measured containing about 30 millions of data characterizing the surface roughness and surface waviness of every specimen (Fig. 5).

Surface topography was measured in two directions X and Y corresponding to the coordinate system of the measurement device ALICONA- IF G4, see Fig. 6. Axis X corresponded to the direction of the tool axis, direction Y is perpendicular to

Table 3 The cutting conditions shoulder milling and surface topography measurement

Cutting speed (m·min^{-1})		
580	680	780
Feed per tooth (mm)		
0.05 mm	0.10 mm	0.15 mm
Axial and radial depth of cut (mm)		
a_p = 25		a_e = 0.2

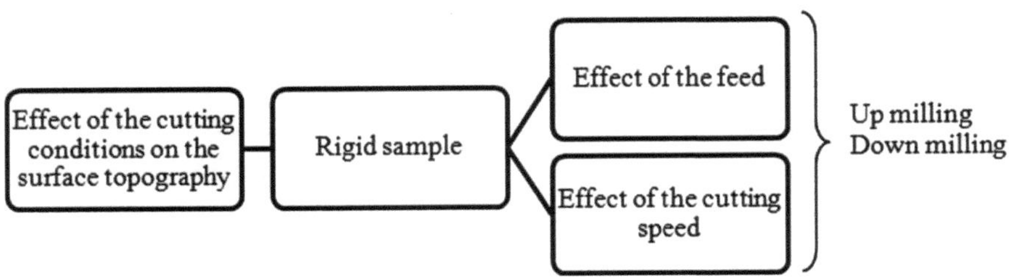

Fig. 5 Methodology of surface topography measurement

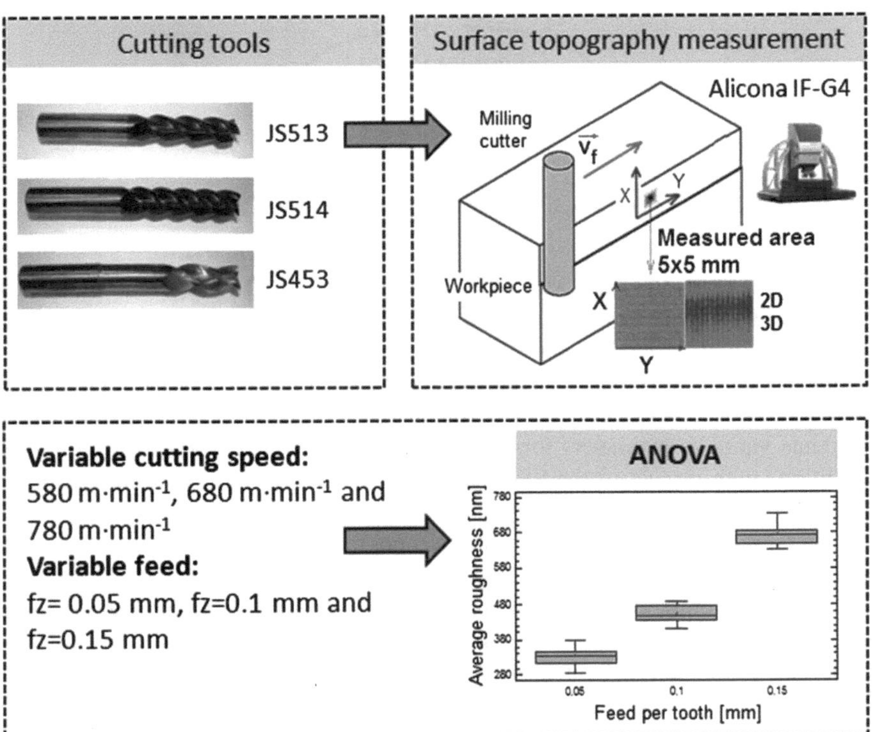

Fig. 6 Surface topography measurement, experimental set-up

the previous and corresponds to the direction of the feed. The surface topography was investigated for the three SECO tools.

4 Results and Discussion

4.1 Tool Analyses

Analysis of micro-geometry of each tool was performed in order to focus on the potential factors affecting the surface topography. As a main monitored parameter the mean radius of cutting edge, r_n, effective length, le, and inclination of helix angle, λs, were measured for every flute (tooth) of the milling tools.

Measured values are summarized in the Table 4. Tools JS453 and JS514 show similar radii of cutting edge (16.14 and 17.83 μm) however for JS453 (Fig. 7) the cutting edge was the lowest. JS513 has the cutting edge rounded to the dimension of radii more than 26 μm with no other modification of the cutting edge. Measurement of the tool geometry showed slight differences in the measured tools geometry for each tooth. For the tools JS5453 and JS514 the difference between mean radii of the cutting edge did not exceed 1.5 μm, the standard deviation of 0.13 μm and maximal difference between minimal and maximal value is 0.24 μm specifically for tool JS453 were found. The results for the tool JS513 were slightly bigger in standard deviation 0.61 μm and maximal difference of 1.24 μm. The maximal difference for tool JS513 was almost 5 μm (standard deviation 2.44 μm and maximal difference of 4.67 μm) which might be caused with the grinding. The results could not be compared with the designed tool geometry proposed by SECO, because these values undergo non-disclosure agreements.

4.2 Cutting Conditions and Forces

Force loading in technologically important directions (F_c, F_{cN} and F_{1M}) was determined for cutting tools JS514, JS453 and JS513. Mean values of the maximal measured instantaneous force loadings in X and Y direction served for determining the resultant force F_{1M} based on which the force loadings F_c and F_{cN} were analyzed —Fig. 8. The determined values of F_c, F_{cN} and F_{1M} served also for a comparison and the investigation of the influence the feed per tooth and cutting speed on the force loading of the cutter or workpiece.

Table 4 The micro-geometry of SECO tools

Tool designation	d (mm)	le (mm)	λs (°)	Tooth nr.	Mean radius of the edge r_n (μm)	Statistical data (μm)	
JS453	16	35	35°	1	16.10	Average	16.14
				2	16.28	Standard deviation	0.13
				3	16.04	Range (max-min)	0.24
JS514	16	55	46°	1	17.92	Average	17.83
				2	18.66	Standard deviation	0.61
				3	17.32	Range (max-min)	1.34
				4	17.41		
JS513	16	50	65°	1	23.39	Average	26.13
				2	26.95	Standard deviation	2.44
				3	28.50	Range (max-min)	4.67

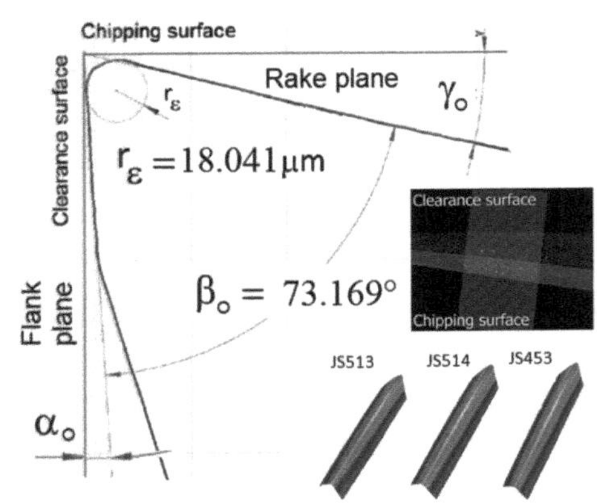

Fig. 7 Geometrical parameters of the SECO JS453 tool Alicona—Edge Master

4.2.1 Cutting Speeds and Forces

Analysis of the influence of a cutting speed on the force loading is summarized in the Fig. 9, including the corresponding statistical regression curves.

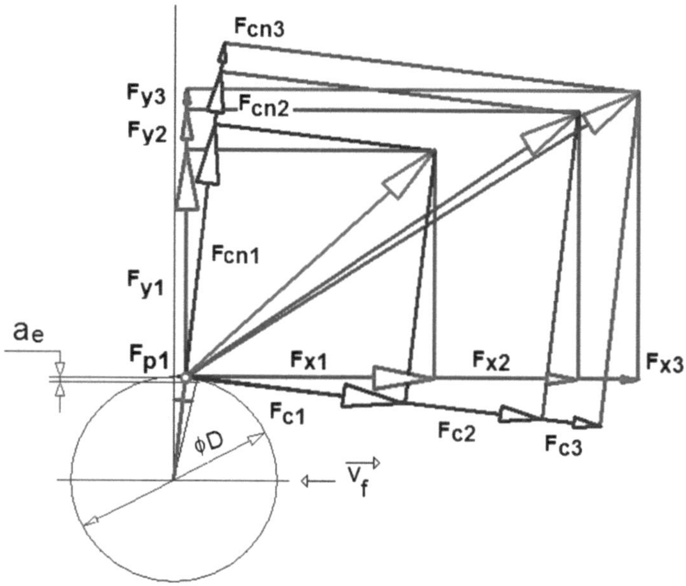

Fig. 8 A change of force loading (F_c, F_{cN}, F_{1M}) with increasing feed per tooth (*down* milling, the tool JS513)

Feeds and Force

Analysis of the influence of a feed on the force loading is summarized in the Fig. 10.

Evaluation of the effect of the feed and the cutting speed on the force loading of the workpiece confirmed that:

- the higher feed per tooth was used (with other cutting conditions kept constant), the higher cutting force and higher ratio of cutting force to perpendicular cutting force were measured (three times increase of the feed lead to 60% increase of the force loading);
- the increasing cutting speed did not resulted significantly (probability 95%) in the increase of cutting forces;
- the down milling technology made higher pressure loadings of the tools onto machined surfaces.

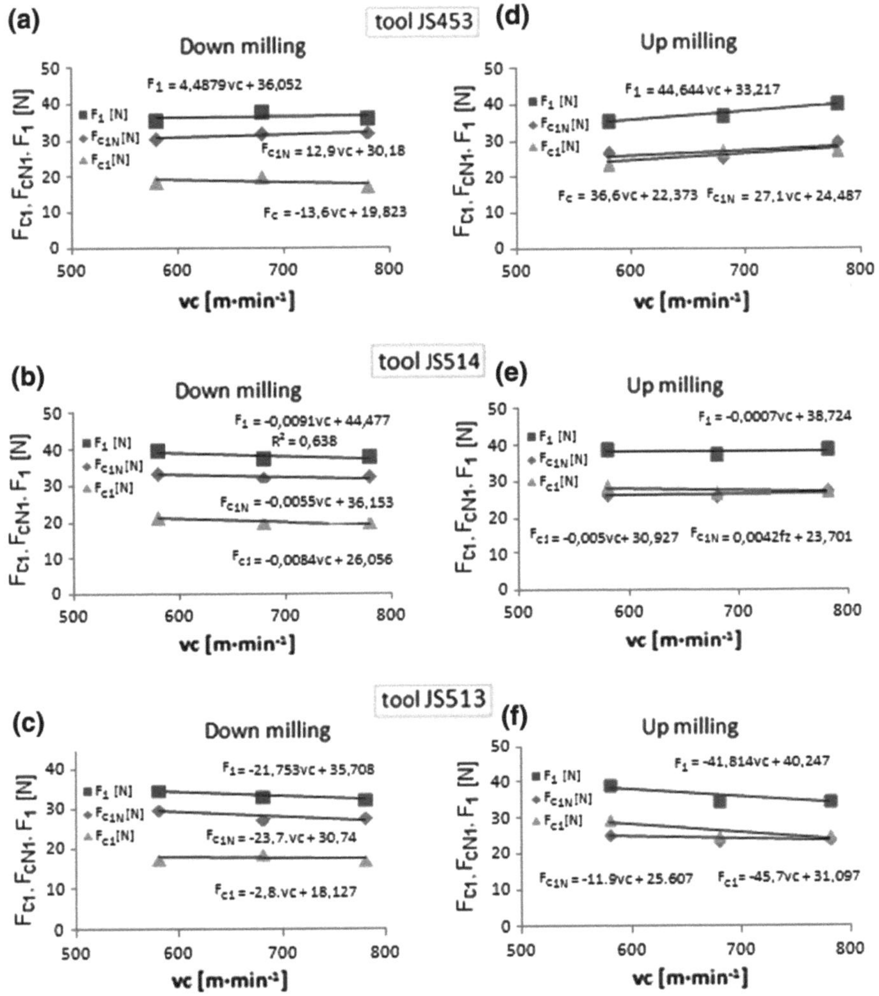

Fig. 9 Force loading (F_c, F_{cN}, F_{1M}) as a function of increasing cutting speed

4.3 Cutting Conditions and Surface Topography

Measured roughness (Ra, Rt and Rv) and waviness (Wt) values were processed using statistical tools ANOVA in order to determine the influence of cutting conditions on the surface topography (see Figs. 11, 12).

Procedure of statistical analysis of measured values consisted of a one-way analysis of variance for R and W surface parameters, for up and down milling.

Measured values were analyzed first using F-test for the significant difference between the mean values of roughness parameters from the levels of the cutting speed at the 95% confidence level for both types of the milling operation and both

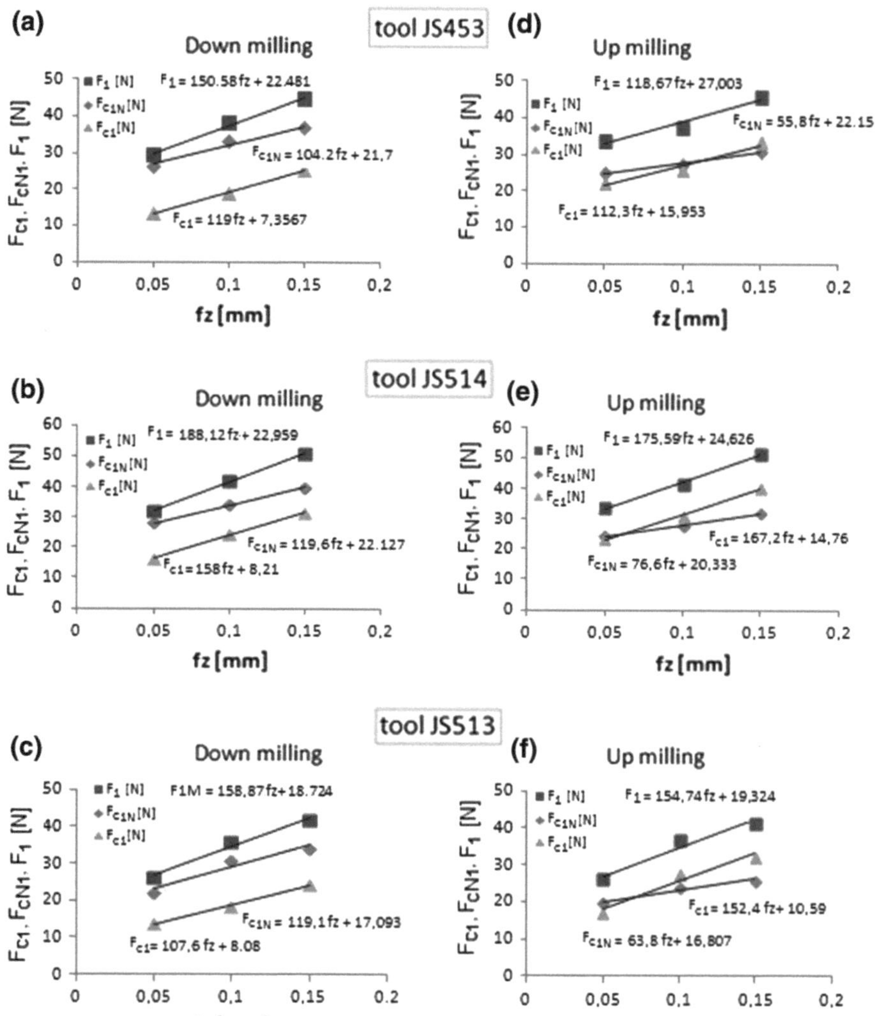

Fig. 10 Force loadings as functions of increasing feeds

directions of the measurement (X direction parallel to the tool axis, Y parallel direction of the feed).

Multiple range test was used for multiple comparisons to determine the significance of differences between each pair of the cutting speeds. Finally, Leven's test for equality of variances was used for testing the null hypothesis (the standard deviations of roughness values within each of the three levels of cutting speed or feed are the same).

For the milling tool JS513, the impacts on the surface topography can be summarized as following:

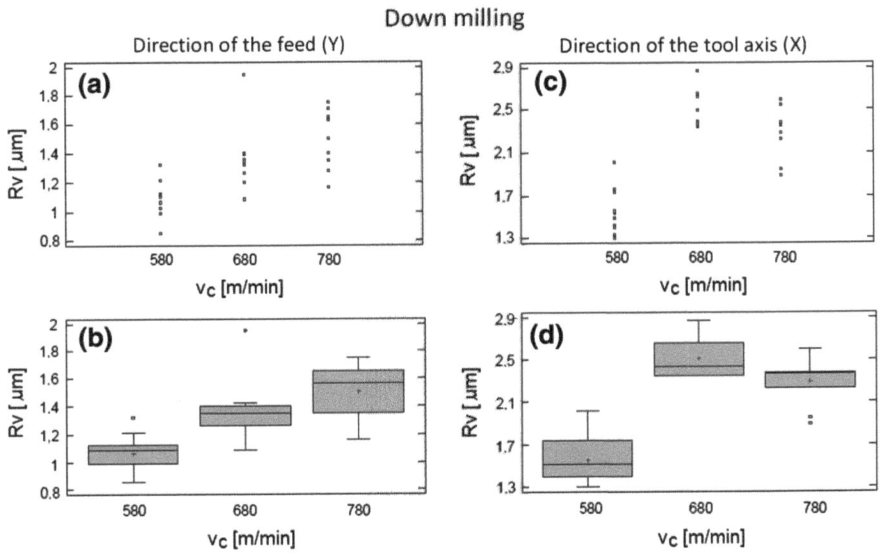

Fig. 11 An effect of cutting speed on maximum height of the profile (*down* milling)

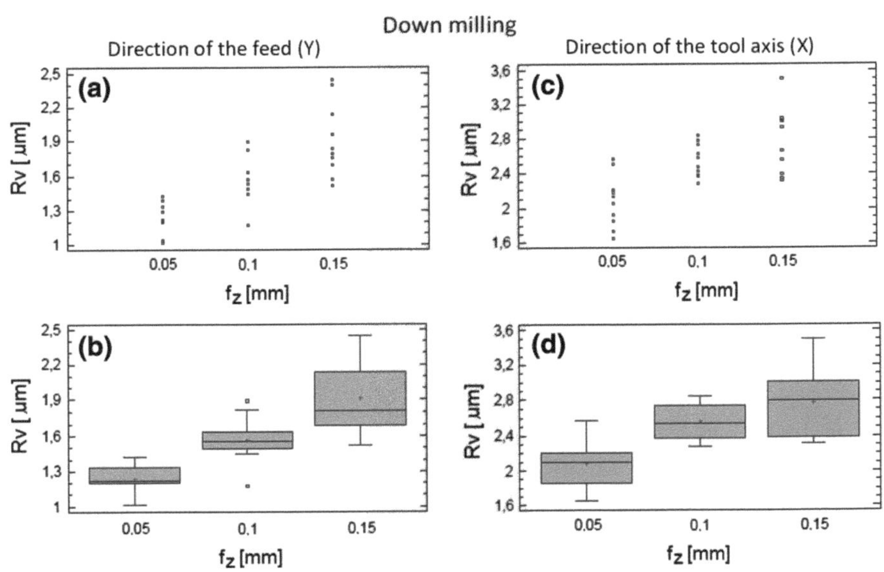

Fig. 12 Study on effect of feed on maximum height of the profile (*up* milling)

- an increase in feed per tooth leads to a deterioration of the surface topography in both directions of the measurement (F-test showed a significant difference of the mean values for all roughness parameters),

- the increase of the feeds by 300% resulted in 30% increase of Ra magnitude and 33% increase of Rv in the direction of the tool axis X,
- the increase of the feeds by 300% was followed with approximately 160% increase of Ra and 50% increase of Rv in the direction of the feed Y,
- a surface quality is inferior in the direction of the tool axis (X) for lower values of the feed per tooth (fz = 0.05 mm) but a steeper rise in the roughness values was observed for the direction of the feed Y,
- the regular repetition of the roughness profile shape was observed only in Y direction where the repetition corresponds to the feed movement of the tool,
- a measured roughness parameters after up and down milling do not exhibit any predominance of the strategy in terms of a surface topography,
- the analysis of residual stresses imposed by the machining has to be performed consequently.

Surface topography for the tool JS513

Furthermore, the results for the effect of the cutting speed (vc = 580 m/min, vc = 680 m/min and vc = 780 m/min) on the surface roughness can be summarized as:

- cutting speed has only the modest effect on the surface topography in X and Y directions (F-test did not show a significant difference of the mean values of every roughness parameter),
- the measured roughness parameters after up and down milling do not exhibit any predominance of one strategy in terms of a surface topography,
- the analysis of residual stresses imposed by the machining has to be performed also.

Finally, regarding the tool and the same cutting conditions:

- a waviness increases with an increase in feed per tooth (F-test showed a significant difference of the mean values of every waviness parameter),
- a profile of the waviness has a regular shape in the direction of the tool axis X with the repetition for each value of the feed per tooth.

Surface topography for the tools JS453 and JS514

For the tool JS453, the effect of the cutting conditions on the surface topography was found as following:

- in a direction of the tool axis X—the highest roughness for up milling were observed for the mean value of the feed per tooth (fz = 0.1 mm), for down milling strategy these values were the lowest,
- in a direction Y parallel to the feed speed—the roughness values increase with increase of the feed per tooth (for both strategies),
- cutting speed has almost no effect on the surface quality in X and Y direction,
- a profile of the waviness has a regular shape in the direction of the tool axis X with the repeatability for each feed per tooth.

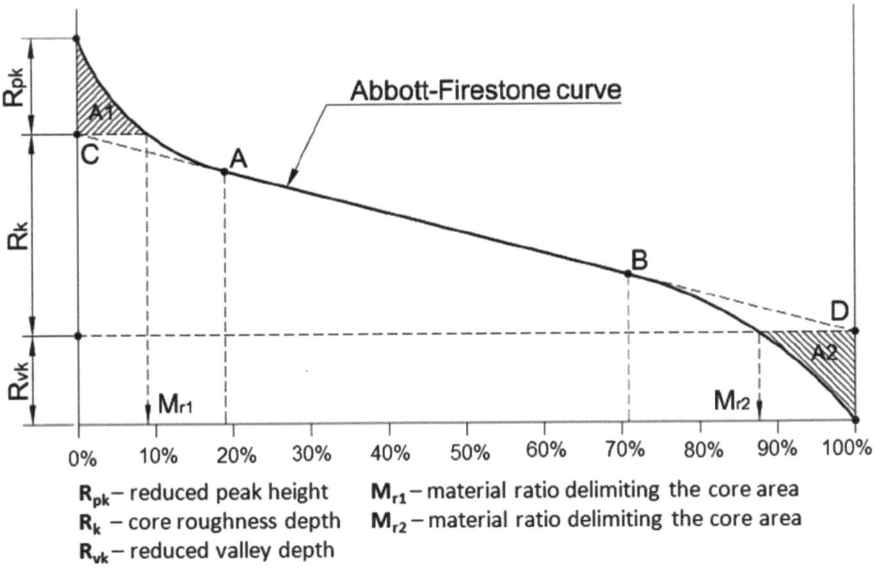

Fig. 13 The Abbott-Firestone curve [27]

For the tool, the effect of the cutting conditions on the surface topography was following:

- in a direction Y of the feed: the roughness values increase with increase in the feed per tooth for both strategies,
- cutting speed has no effect on the surface quality in X and Y direction,
- a profile of the waviness has a regular shape in the direction of the tool axis X with the repetition for every value of the feed per tooth.

4.3.1 Surface Bearing Ratio, the Abbott-Firestone Curve

Many engineering applications use the 2D profile parameters as Ra, Rt, Rv according to standard ISO EN 4287. However, increasing quality and functional requirements lead to implementation and use of other more advanced specific parameters, describing the surface characteristics and affecting their functionality.

The basic roughness and waviness parameters describe the distribution of profile heights z(x) along the measuring direction X [23], so they do not provide any information about the shape, slopes or sizes of irregularities of the surface. However, booming use of optical measuring devices enables to measure whole complex surface parameters in order to obtain proper qualitative criteria of the machined surface [24, 25]. Surface can be characterized by the Abbott-Firestone curve (Figs. 13 and 14) specifying the ratio air to material. Abbott-Firestone curve expresses percentage of the material of the profile elements at a defined height level relative to the evaluation length. Otherwise said the analysis of the material distribution allows understanding the air to material ratio at any level, starting at the highest peak as a function of level.

The surface criteria in the Abbott-Firestone curve can be described by several different parameters as Rk describing core roughness depth (height of the core material). This parameter is very important because it is giving information about volume of the material which is possible to wear during operational use. Parameter Rpk is characterizing mean depth of the peaks above the core material, more specifically Rpk express volume of the material which will be removed during initial wearing process. Lower value of Rpk is preferable because of the higher dimensional accuracy during operational use. Parameter Rvk is describing the mean depth of the valleys below the core material, otherwise said Rvk is giving information about ability of the machined surface to retain liquids. Rmr1 (or Mr1) is characterizing the fraction of the surface which consists of peaks above the core material and Rmr2 (or Mr2) is characterizing the fraction of the surface which will carry the load [26] (Fig. 14).

4.3.2 Feed and the Surface Bearing Ratio

An increase of the feed during shoulder milling process leads to the modification of the shape of the Abbott-Firestone curve and its parameters (Figs. 15, 16, and 17). Rk and Rpk parameters are increasing in both direction of the measurement however the parameter Rvk is remain the same or slightly decreasing. Base on these facts several different conclusions can be derived.

- use of the higher feeds leads to a more pronounced wear of the workpiece in the operation,
- use of the higher feeds do not affect (or decrease) ability of the surface to retain liquids,
- no statistically significant difference (reliability of 95%) between down milling and up milling strategies was found.

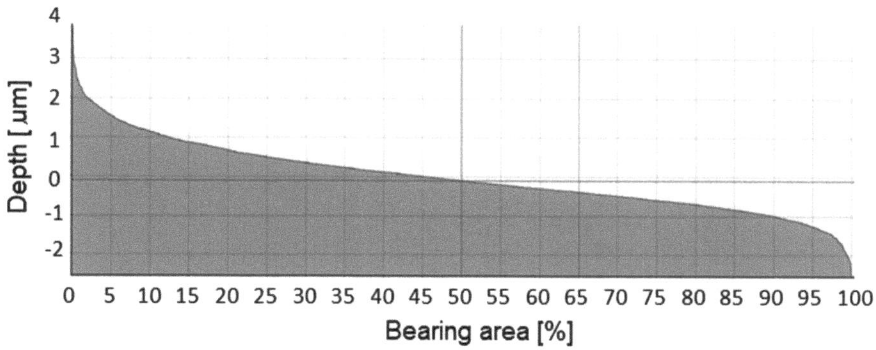

Fig. 14 An example of the Abbott-Firestone curve for milled material 7475-T7351

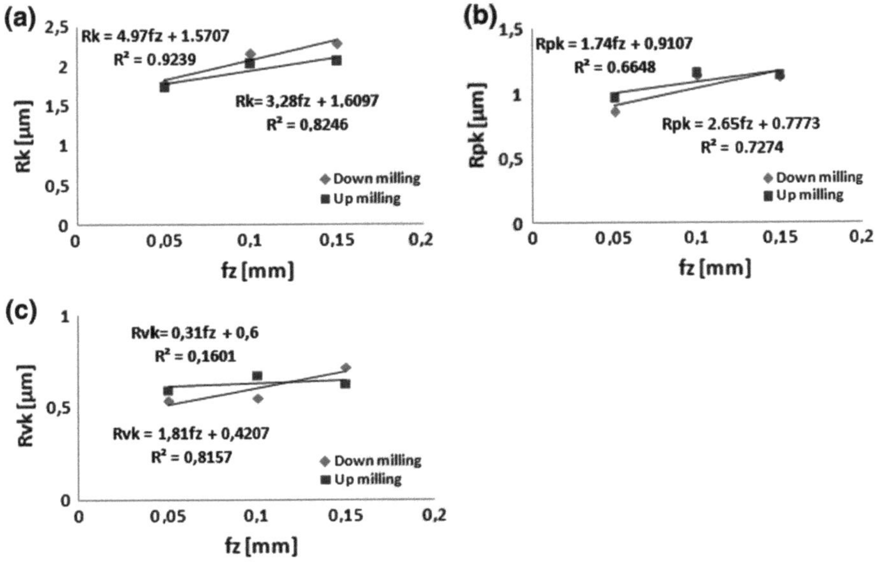

Fig. 15 The effect of feed on the Abbott-Firestone parameters (*down* milling, up milling; measurement in the X—direction)

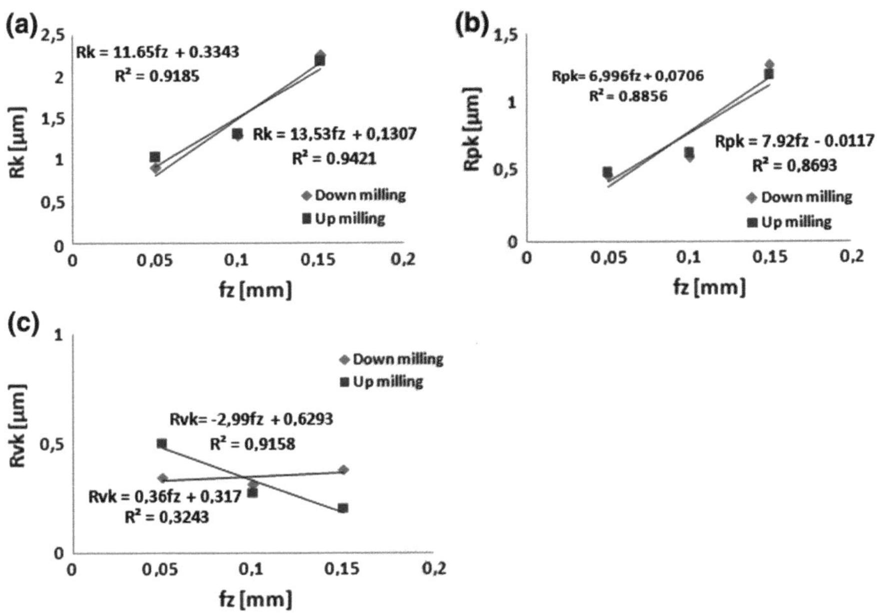

Fig. 16 The effect of feed on Abbott-Firestone parameters (*down* milling, up milling; measurement in the Y—direction)

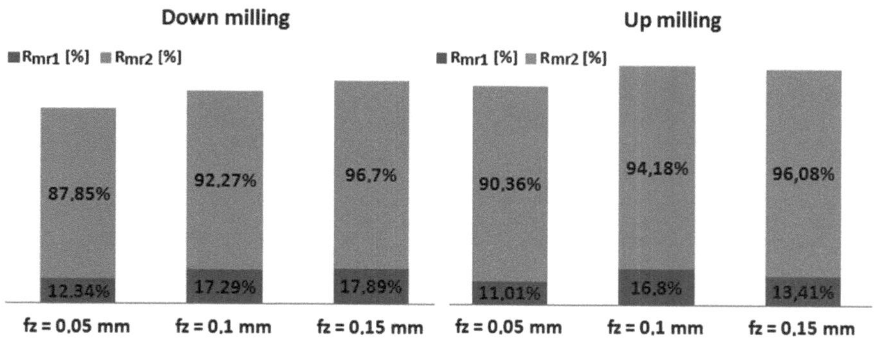

Fig. 17 Material parameters Rmrl and Rmr2, influence of the feed, measurement in the Y-direction

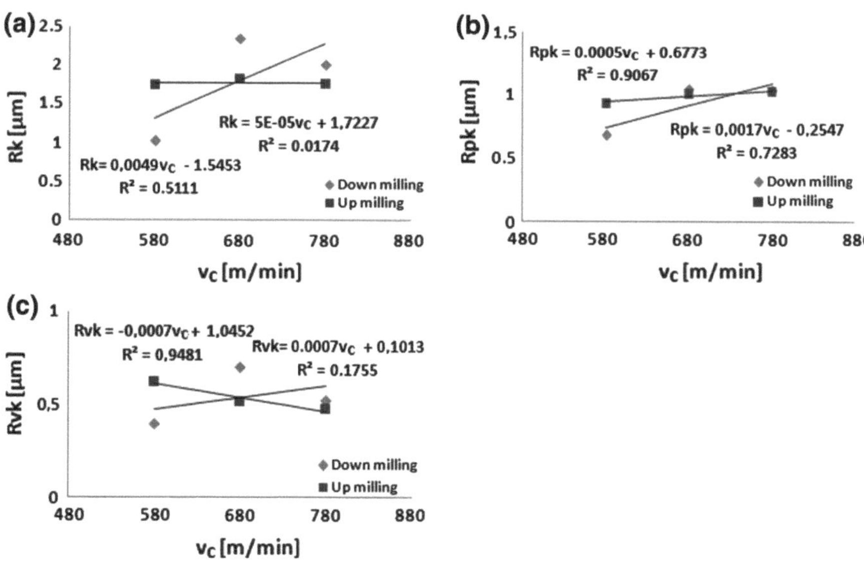

Fig. 18 The effect of cutting speed on Abbott-Firestone parameters (*down* milling and up milling; measurement in the X—direction)

4.3.3 Cutting Speed and the Surface Bearing Ratio

An increase of the cutting speed during shoulder milling process leads to the modification of the shape of the Abbott-Firestone curve and its parameters only for down milling strategy. Parameters and shape of the curve remains the same for up milling strategy (Figs. 18, 19, and 20). The highest values of the Rk and Rpk

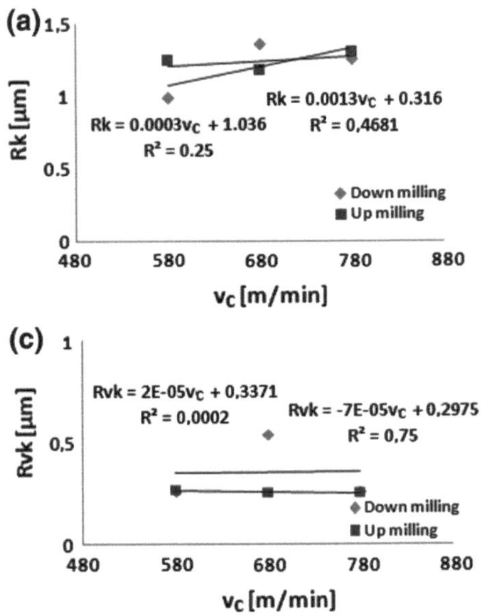

Fig. 19 The effect of cutting speed on the Abbott-Firestone parameters (*down* milling and up milling; measurement in the Y-direction)

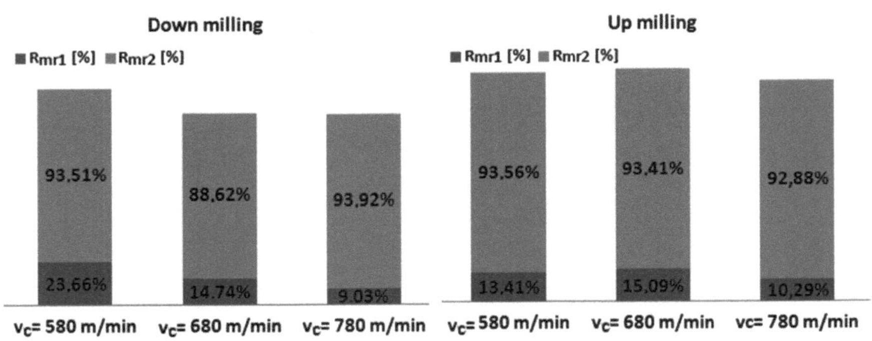

Fig. 20 The effect of cutting speed on the Abbott-Firestone parameters Rmr1 and Rmr2, measurement in the Y-direction

parameters were observed for mid value of the cutting speed (cutting speed recommended by the tool manufacturer). Use of the mid value of the cutting speed according to the manufacturer recommendation leads to a more pronounced wear of the workpiece in the operation and increased ability of the machined surface to retain liquids.

5 Conclusions

The main goal of the work was to analyze progressive shoulder milling technology of thin-walled components for aerospace industry and determine influence of its cutting parameters on the surface topography and force loading of the workpiece. Based on the cutting conditions, force loading, tool micro-geometry, surface topography and Abbott-Firestone curve analysis of the shoulder milling of aluminum alloy 7475-T7351 the following conclusions can be made:

- the increase of the feed leads to the higher force loading of the thin-walled components (the higher cutting force and higher ratio of cutting force to perpendicular cutting force were measured),
- the effect of increasing cutting speed on cutting forces in the tested interval of cutting speeds wasn't statistically significant with reliability of 95%,
- the down milling technology made higher pressure loadings of the tools and machined surfaces (resultant force of the force loading is oriented more perpendicularly to a machined surface),
- the lowest force loading was observed of the tool with the highest helix angle $\underline{\lambda s}$ (tool JS513),
- an increase in feed per tooth leads to an increase of the roughness parameters in the direction of the feed movement and in the direction of the tool axis (F-test showed a significant difference of the mean values of Abbott-Firestone roughness parameter),
- the effect of the cutting speed on the surface topography wasn't statistically significant on the tested interval of cutting speeds with reliability of 95% (F-test did not show a significant difference of the mean values of every roughness parameter),
- the measured roughness parameters after up and down milling do not exhibit any predominance of one strategy in terms of a surface topography on the tested interval however the analysis of residual stresses imposed by the machining has to be performed in order to fully determine its influence on the surface integrity,
- use of the higher feeds leads to higher Rpk parameter values so more pronounced wear of the workpiece in the operation can be expected while using higher feeds,
- use of the higher feeds do not affect (or decrease) the value of the Rvk parameter, so ability of the surface to retain liquids (and the risk of the higher stress corrosion cracking) won't be affected while using higher feeds,
- there is no significant difference with reliability of 95% between down milling and up milling strategies in terms of ability to resistant to the wear in operational use or retain liquids.

The current research is devoted to the effect of physical, technological and surface parameters on fatigue with statistical data assessment, including the study of material plasticity and matrix-inclusions interactions.

Acknowledgements This research work was supported by the Brno University of Technology, Faculty of Mechanical Engineering, Specific research 2016, with the grant "Research of modern production technologies for specific applications", FSI-S-16-3717.

References

1. Raymer D (2012) Aircraft design: a conceptual approach (Aiaa education series), 5th edn. American Institute of Aeronautics and Astronautics, Reston
2. Starke EA, Staley JT (1996) Application of modern aluminum alloys to aircraft. Prog Aerospace Sci 32(2–3): 131–172. ISSN 03760421
3. Shanmugam NE, Liew RJY, Thevendran AV (1998) Thin-walled structures: research and development. In: Second international conference on thin-walled structures. Elsevier, New York
4. Ratchev S, Liu S Becker AA (2005) Error compensation strategy in milling flexible thin-wall parts. J Mater Process Technol 162–163: 673–681. ISSN 09240136
5. Aijun T, Zhanqiang L (2008) Deformations of thin-walled plate due to static end milling force. J Mater Process Technol 206(1–3): 345–351. ISSN 09240136
6. Ratchev S, Liu S, Huang W, Becker AA (2004) Milling error prediction and compensation in machining of low-rigidity parts. Int J Mach Tools and Manuf 44(15): 1629–1641. ISSN 08906955
7. Shahzad M (2011) Majid. Influence de la rugosité et des traitements d'anodisation sur la tenue en fatigue des alliages d'aluminium aéronautiques 2214 et 7050. TOULOUSE, Thesis. L'UNIVERSITÉ DE TOULOUSE. Supervisor M. Farhad Rézaï-Aria
8. Alloy 7475 Plate and Sheet. Iowa: ALCOA. (vid. 2016-08-20). Available from: https://www.arconic.com/mill_products/catalog/pdf/alloy7475techplatesheet.pdf
9. Alloy designations. Alumeco (online). Odense, 2016. (vid. 2017-02-09). Available from: http://www.alumeco.com/Knowledge-and-Technique/Aluminium-data/Temper-descriptions.aspx
10. Al-Rubaie K, Barroso EKL, Godefroid LB (2006) Fatigue crack growth analysis of pre-strained 7475–T7351 aluminum alloy. Int J Fatigue. 28: 934–94
11. Static and dynamic fracture properties for aluminum 7475 T7351: Final report. University of Dayton. Research Institute. Ohio, 1975. (vid. 2017-02-09). Available from: http://www.dtic.mil/dtic/tr/fulltext/u2/a014353.pdf
12. Forejt, M, Píška M (2006) Teorie obrábění, tváření a nástroje (Theory of metal cutting and forming). Brno. Akademické nakladatelství CERM s.r.o.
13. Tlusty J (1999) Manufacturing Process and Equipment. 1st ed. Prentice Hall, p 928. ISBN 10-0201498650
14. Grzesik W (2011) Wit. Podstawy skrawania materiałów konstrukcyjnych. Wydawnictwa Naukowo Techniczne. Opole
15. Davim JP (2010) Surface integrity in machining. Springer, London
16. Degarmo, PE, Black JT, Kohser RA 2003 Materials and processes in manufacturing. 9th edn. update ed. Wiley. Hoboken. (vid. 2017-02-09). Available from: http://dcetind.weebly.com/uploads/9/1/6/3/9163431/solutions_manual_-_materials___processing_in_manufacturing__demargo_.pdf
17. Jiang XJ, Whitehouse DJ (2012) Technological shifts in surface metrology. CIRP Annals—Manuf Technol 61(2): 815-836. ISSN 00078506
18. EN ISO 4287 (1997) Surface texture: profile method—terms, definitions and surface texture parameters
19. Ojolo SJ (2014) Machining variables influence on the fatigue life of end-milled aluminium alloy. Int J Mater Sci Appl 3(6): 391–398. ISSN 2327-2635

20. Novovic D, Dewes RC, Aspinwall DK, Voice W, Bowen AP (2004) The effect of machined topography and integrity on fatigue life. Int J Mach Tools Manuf 44(2–3): 125–134. ISSN 08906955
21. Siebel E, Gaier AM (1957) Influence of surface roughness on the fatigue strength of steels and non-ferrous alloys. Metal Fatigue: Eff Small Defects Nonmetallic Inclusions Eng Digest 18 (3): 109–112
22. Gómez A, Sanz A, Marcos M (2012) An analysis of the influence of cutting parameters on the turning process on the fatigue life of aluminum alloy UNS A92024-T351. Adv Mater Res 498:19–24
23. EN ISO 4287 (1997) Surface texture: profile method—terms, definitions and surface texture parameters
24. Abbott EJ, Firestone FA (1933) Specifying surface quality. Mech Eng 55:569–572
25. Rîpă M, Tomescu L, Hapenciuc M (2003) Tribological characterisation of surface topography using Abbott-Firestone curve. University "Dunărea de Jos" of Galati, România. ISSN ISSN 1221-4590
26. Taylor H (2002) Form Talysurf Intra: operator's handbook
27. Zhang T, Bartolo MP, Vasco J, Silva B, Galo C (2006) Laser micromachining for mould manufacturing: I the Influence of operating parameters. Assembly Auto 26(3): 227–234. ISSN 0144-5154
28. Sasahara H (2005) The effect on fatigue life of residual stress and surface hardness resulting from different cutting conditions of 0.45% C steel. Int J Mach Tools Manuf 45(2): 131–136. ISSN 08906955
29. Souto-Lebel A, Guillemot N, Lartigue C, Billardon AR (2011) Characterization and influence of defect size distribution induced by ball-end finishing milling on fatigue life. Procedia Eng 19: 343–348. ISSN 18777058

Reducing of Scuffing Phenomenon at HCR Spur Gearing

Milan Rackov, Maja Čavić, Marko Penčić, Ivan Knežević, Miroslav Vereš and Milan Tica

Abstract The article deals with possibility of increasing the resistance of HCR spur gearing from a scuffing point of view. Scuffing is the process that occurs when the surfaces of two contacting bodies are joined by localized welding and then pulled apart. A material transfer occurs between the two contacting surfaces due to high metal-to-metal contact and hence produces a weld. Since there are great pressures between teeth flanks and the load is higher, the scuffing is the most important damage of teeth flanks of HCR involute gears. The scuffing traces appear in the form of streaks or scratches with rough bottoms and sides, often appearing as bands of variable depth width oriented in the direction of the height of the tooth, and affect isolated zones or their whole width. In the case of warm scuffing; the combination of high pressure exists between teeth surfaces, high sliding speeds, and excessive contact temperature, resulting from pressure and sliding speed values, which cause oil film rupture between the teeth flanks. HCR profiles are more complicated than standard involute profiles, they have greater predisposition for occurring interference, pointed tip thickness, but also undercut of teeth during the

M. Rackov (✉) · M. Čavić · M. Penčić · I. Knežević
Faculty of Technical Sciences, University of Novi Sad, Novi Sad, Serbia
e-mail: racmil@uns.ac.rs

M. Čavić
e-mail: scomaja@uns.ac.rs

M. Penčić
e-mail: mpencic@uns.ac.rs

I. Knežević
e-mail: ivanknezevic@uns.ac.rs

M. Vereš
Faculty of Mechanical Engineering, Slovak University of Technology in Bratislava, Bratislava, Slovakia
e-mail: miroslav.veres@stuba.sk

M. Tica
Faculty of Mechanical Engineering, University of Banja Luka, Banja Luka, Bosnia and Herzegovina
e-mail: milan.tica@unibl.rs

V. Majstorovic and Z. Jakovljevic (eds.), *Proceedings of 5th International Conference on Advanced Manufacturing Engineering and Technologies*, Lecture Notes in Mechanical Engineering, DOI 10.1007/978-3-319-56430-2_10

production (primary production interference). Due to increased addendum height, there is larger possibility of occurring some interference or pointed tooth tip. Therefore it should prevent these errors and check if all equation and constraints are satisfied. This paper describes finding optimal solutions for geometry of the tooth curve profile. It will be defined certain values addendum heights for meshing wheel according to criteria of specific slips and corrected head shape of the teeth of both wheels. In the same time, this optimization is joined with assessment and theoretical analysis of the impact of the HCR tooth profile resistance to scuffing on the basis of integral temperature criterion according to Winter-Michaelis criterion. A significant benefit in a theoretical area is generalization of the integral temperature criterion for involute HCR gearing.

Keywords HCR gearing · Geometry · Scuffing · Optimization

1 Introduction

High contact ratio (HCR) gear pair is a contact between gears with at least two pairs of teeth in contact. High contact ratio is obtained with increased addendum height, so larger than in standard gearing. Proposed geometry of HCR gearings is much complicated due to the fact there is larger possibility of occurring meshing and during the production interference, much larger than interference happening in standard involute profiles. Also here there is a higher risk of too small thickness of a tooth tip and significantly less favorable values of specific slips into the flanks [1].

It is well known that increasing the average number of teeth in contact leads to excluding or reduction of the vibration amplitude. First, it was established experimentally that dynamic loads decrease with increasing contact ratio in spur gearing [2]. Moreover, in order to get a further reduction of the vibration, HCR gear profiles can be optimized. Sato et al. [3] found that HCR gears are less sensitive with respect to manufacturing errors. In particular, such kind of gears allows larger tolerance in the tip relief length. Moreover, they found that, in the absence of pressure angle error, the best contact ratio should be about 2; otherwise, it is better to have a contact ratio about 1.7 or higher than 2.3. Kahraman and Blankenship [4] published an experimental work on HCR gear vibration; they found that the best behavior is obtained with an integer contact ratio, even though other specific non integer (rational) contact ratios can minimize the amplitude of some specific harmonics of the static transmission error. It is important to note that in Ref. [4] HCR gears were obtained by modifying the outside diameter; the other macro-geometric parameters, e.g. the number of teeth, were left unchanged.

Contact ratio is increased by increasing tooth height. Dynamic loads and noise are reduced by using high contact ratio gears. According to results of different measurements of gear pair, reduction of noise proved to be the best using HCR gearing with the value of contact ratio $\varepsilon_\alpha = 2$. Decrease in noise is caused by $\varepsilon_\alpha = 2$ because there are always two pairs of teeth in contact, which means when one pair of teeth go out from the contact, another pair of teeth is coming in contact and applied force is considerably smaller since it is divided on two pairs of teeth. Therefore, gearing in automotive industry should be done with $\varepsilon_\alpha = 2$ in order to reduce the noise and dynamic forces [5].

Scientific and technological progress in engineering leads to the increasingly better use of resources, reducing energy intensity of production and increasing the reliability and efficiency during operation. This trend is also reflected in the field of the development of gears, which in recent years have increased the transmitted power, while in the same time reduced the volume and weight of the gear.

Increasing the transmitted power of gear relative to the unit of volume is, however, associated with the increase of thermal load transfer. Apart from usual kinds of teeth damages (fracture in the heel and formation of pitting), the increased heat load has discovered another kind of damage, the so-called scuffing.

Problems with teeth scuffing mainly occur in such transfers, which cannot be used for lubrication oils with EP additives (EP—extreme pressure). Naturally, scuffing may arise in other cases, even assuming that the gear lubrication oils have additives. In such cases, scuffing occurs in overloading transfers that operate at the higher peripheral speed of transfers with poorly chosen tooth geometry or that influence an excessive increase in oil temperature operation (e.g. high ambient temperature, etc.).

In any case, progressive scuffing means serious damage to the tooth profile shape, which subsequently means removing the teeth from the service. Increased noise transmission over the standard border may refer in some cases to the reason for exclusion from the transfer operation [6, 7].

The problem related to the involute gear with the extended duration of the contact period, known in the literature as HCR (High Contact Ratio), is discussed in this paper. Due to the larger path of action and the higher peripheral speed, the scuffing can occur more frequently in HCR gear teeth. The objective of this paper is to analyze and discuss the possibilities to reduce the occurring of scuffing as a teeth damage on external HCR gears.

2 HCR Gearing and Possibilities of Optimization of Geometrical Parameters

The main indicator of HCR gearing, which differs from the commonly used standard involute profiles, is higher contact ratio, at least two pair of teeth in contact.

According to the results of different measurements of gear pair, the reduction of noise proved to be the best using HCR gearing with the value of the contact ratio $\varepsilon_\alpha = 2$. Decrease in noise is caused by $\varepsilon_\alpha = 2$ because there are always two pairs of teeth in contact, meaning that, when one pair of teeth leaves the contact, another pair of teeth is coming in contact; hence, the applied force is considerably smaller since it is divided between two pairs of teeth.

Contact ratio depends on the length of line of action and tooth pitch $\varepsilon_\alpha = f\,(g_\alpha, p_{bt})$. Tooth pitch on the base circle of LCR gearing is equal to the base pitch on HCR gearing, and it is considered as constant. This means that, achieving the greatest value of the contact ratio ε_α has to be obtained by the greatest possible increase of the length of line of action g_α [8]. The length of line of action g_α is calculated in following equation:

$$g_\alpha = \sqrt{r_{a1}^2 - r_{b1}^2} + \sqrt{r_{a2}^2 - r_{b2}^2} - a_w \sin \alpha_{wt} \tag{1}$$

Consequently, according Ref. [8] the contact ratio is the objective function of both addendum heights and the addendum modification factor of pinion $\varepsilon_\alpha = f(h_{a1}^*, h_{a2}^*, x_1) = \max$, i.e. optimization parameters h_{a1}^*, h_{a2}^*; x_1 makes a non-linear optimization of triple constraint, with limitation requirements defined for [8, 9]:

- removal of the meshing interference,
- minimum arc thickness of the tooth tip $s_{a1,2}$,
- distribution x_c to x_1, x_2 has to be performed through balancing specific slips, strength, or a particular condition, respectively compromising their combinations [8, 9].

2.1 *Interference During the Production*

This interference occurs in the production process of gear forming when the tooth of the rack tool is in collision with a produced transition curve of the gear wheel, resulting in a so-called undercut tooth (Fig. 1).

This phenomenon largely depends on the method of manufacturing process. Unfavorable conditions arising with manufacturing by tool rack, so if it is not known in advance the means of production, always should check the production interference of gearing for production by tool rack. After satisfying conditions not to occur the interference during the production, following relations are obtained [8, 9]:

$$h_{a1}^* \leq \frac{r_{b1} \cdot \sin^2 \alpha_t}{m_n \cdot \cos \alpha_t} + x_1 \tag{2}$$

$$h_{a2}^* \leq \frac{r_{b2} \cdot \sin^2 \alpha_t}{m_n \cdot \cos \alpha_t} + x_2 \tag{3}$$

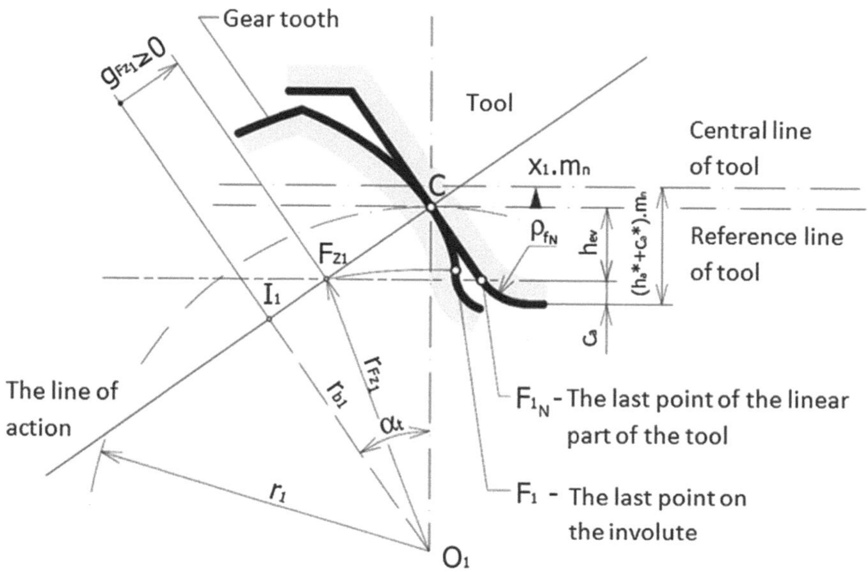

Fig. 1 Interference during the production with tooth rack tool

2.2 *Interference During the Production*

Meshing interference is referred in the case of a collision between curves of teeth profiles as interference between these curves. It means that the meshing interference may occur as a collision of head of gear and the transition curve of pinion (Fig. 2) and/or head of pinion and the transition curve of gear wheel [8, 9].

Addendum heights h_{a1}^* and h_{a2}^* can be expressed:

$$h_{a1}^* \leq \frac{1}{m_n}\left[a_w \frac{\sin \alpha_{wt}}{\sin \alpha_{at1}} - r_{b2}\frac{\tan \alpha_t}{\sin \alpha_{at1}} + \frac{m_n}{\sin \alpha_t \cdot \sin \alpha_{at1}}\left(h_{a2}^* - x_2\right) - r_1\right] - x_1 \quad (4)$$

$$h_{a2}^* \leq \frac{1}{m_n}\left[a_w \frac{\sin \alpha_{wt}}{\sin \alpha_{at2}} - r_{b1}\frac{\tan \alpha_t}{\sin \alpha_{at2}} + \frac{m_n}{\sin \alpha_t \cdot \sin \alpha_{at2}}\left(h_{a1}^* - x_1\right) - r_2\right] - x_2 \quad (5)$$

2.3 *Minimum Thickness of the Tooth Head Circle*

Changing addendum heights will certainly influence the total thickness of the tooth on tip circle. Greater tooth height, as well as a positive correction factor, may affect the thickness of the tooth on tip circle under the permissible value (Fig. 3).

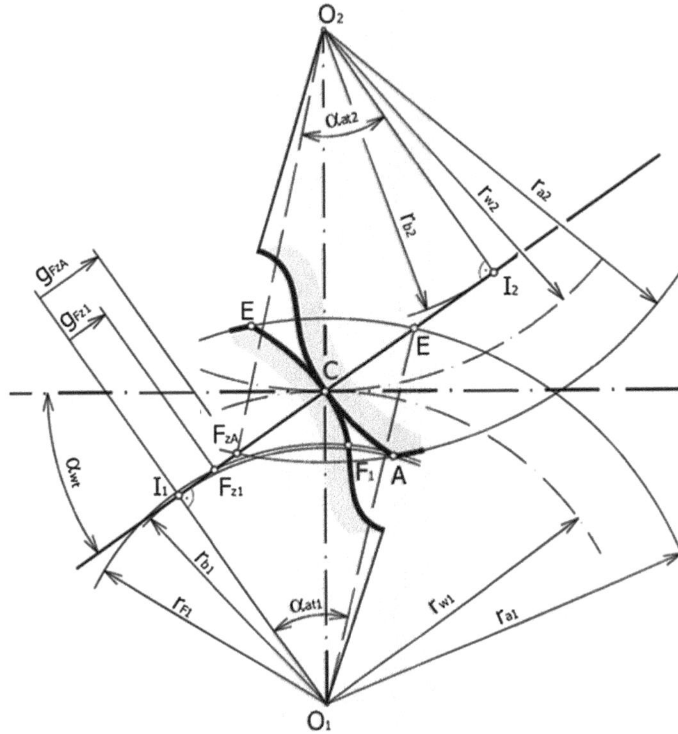

Fig. 2 Meshing interference between head of gear wheel and transition curve of pinion

Fig. 3 Determination of the tooth thickness on the tip circle da

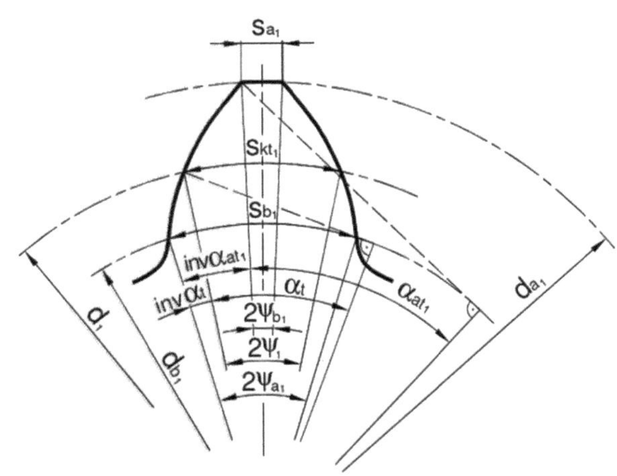

After these conditions [8, 9], addendum heights h^*_{a1} and h^*_{a2} can be expressed:

$$h^*_{a1} \leq \frac{0.2}{\frac{p_t + 4x_1\, m_n\, \tan\alpha_t}{d_1} + 2\cdot(\mathrm{inv}\alpha_t - \mathrm{inv}\alpha_{at1})} - \frac{d_1}{2\,m_n} - x_1 \tag{6}$$

$$h^*_{a2} \leq \frac{0.2}{\frac{p_t + 4x_2\, m_n\, \tan\alpha_t}{d_2} + 2\cdot(\mathrm{inv}\alpha_t - \mathrm{inv}\alpha_{at2})} - \frac{d_2}{2\,m_n} - x_2 \tag{7}$$

2.4 *Optimization of HCR Geometry*

After different obtained conditions, the goal is to achieve that contact ratio has value two ($\varepsilon_\alpha = 2$), so that always two pairs of gears are engaged. With this value of contact ratio it is expected to reduce the vibration and gear noise. In order to achieve high contact ratio, addendum height is made higher in order to obtain larger line of action. The equation of contact ratio represents goal function with the aim it should have the value of two:

$$\varepsilon_\alpha = \frac{z_1}{2\pi}\left[\tan\alpha_{at1} - \tan\alpha_{wt} - i\cdot\tan\alpha_{at2} + i\cdot\tan\alpha_{wt}\right] = 2 \tag{8}$$

3 Scuffing as the Most Important Damage of HCR Tooth Flanks

Scuffing is not a fatigue phenomenon and it may occur instantaneously. Based on the severity of the damage, scuffing can be categorized as mild, moderate, or severe (Fig. 4). Scuffing is classified as mild if it occurs only on small areas of the teeth and is confined to the peaks of the surface asperities. It is generally nonprogressive. Moderate scuffing occurs in patches that cover significant portions of the teeth. If the operating conditions do not change, moderate scuffing may be progressive. Severe scuffing occurs on significant portions of the gear tooth (for example, the entire addendum, the entire dedendum, or both). In some cases the surface material may be plastically deformed and displaced over the tip of the tooth or into the root of the tooth. Unless corrective measures are taken, severe scuffing is usually progressive [6, 7].

Scuffing is severe adhesion that causes transfer of metal from one tooth surface to another due to welding and tearing. The damage typically occurs in the addendum, dedendum, or both, away from the operating pitchline, in narrow or broad bands that are oriented in the direction of sliding. Scuffing may occur in localized patches if it is due to load concentrations. The scuffed area appears to have a rough or matte texture.

(a) (b)

Fig. 4 Light scuffing (**a**), severe scuffing (**b**) [7]

Scuffing is the process that occurs when the surfaces of two contacting bodies are joined by localized welding and then pulled apart. A material transfer occurs between the two contacting surfaces due to high metal-to-metal contact and hence produces a weld. The high metal-to-metal contact is the result of a local failure of the gear lubricant which has been caused by frictional heating due to high sliding speed and high surface pressure. This type of scuffing is called warm scuffing.

The scuffing traces appear in the form of streaks or scratches with rough bottoms and sides, often emerging as bands of variable depth and width oriented in the direction of the height of the tooth, affecting isolated zones or their whole width. The scuffing traces are generally more clearly marked at the tooth tip and root of the teeth in the high sliding zones.

In the case of warm scuffing, it is in fact the combined action of high pressure between surfaces, high sliding speeds, and excessive contact temperature resulting from pressure and sliding speed values, which causes the rupture of oil film between the teeth flanks. During the start-up or running-in of certain gears, some local scuffing of lesser importance, which is characterized by shallow traces and very fine roughness, may appear in certain points of the teeth in the zone where the contact pressure is maximum. In general, after a certain time of operation at reduced load, these localized traces of scuffing diminish by wear. Once this happens, the gear may operate under its nominal load. In this case, a slight increase in the lubricant viscosity will allow for better safety in service. On ground gears, localized scuffing can be observed at the tip and root of the teeth as the result of insufficient tip relief or too great a deviation in the profile. Identical phenomena can also appear near tooth ends due to insufficient longitudinal correction or too great helix deviation.

Scuffing damage can be prevented through the design changes or through the operational/break-in changes. Design-related changes include optimizing the gear geometry/accuracy and the use of nitrided steel. Operational and break-in changes include the use of high viscosity lubricants with anti-scuff additives, reduced lubricant temperatures, and the running-in of new gearboxes at reduced loads.

4 Calculating the Tooth Scuffing Resistance According to the Criterion of Integral Temperature

Temperature criterion is one of the hypotheses about developing the scuffing phenomena. It is based on the assumption that there is a limit temperature of the surface of the teeth, at which the scuffing occurs. These criteria can be divided into criteria based on local instantaneous surface temperature and the criteria based on the mean temperature of the surface of the teeth.

Theoretically, the best criterion among these criteria is the sophisticated Blok criteria (1937). Blok was one of the earliest researchers to study scuffing and to propose a hypothesis to explain the experimental observations. In his study, he suggested that scuffing would only take place if a critical temperature is reached at the sliding interface [10].

For practical calculation of safety against scuffing of gearing, more favorable temperature criteria are the criteria based on the mean surface temperature. Winter-Michaelis criterion [11] of integral temperature (median average contact temperature) is among the most modern criteria for assessing resistance to teeth scuffing. It is based on defining the contact surface temperature of teeth in relation to the Blok theory. The calculation consists of identifying the instantaneous temperature along the mating line under the Blok and adding tooth bulk to a constant temperature. Comparative average temperature of the surface is then determined as the quotient of the integral thus calculated during the unsteady temperature along the mating line and a length image. The calculation of safety against scuffing the teeth is a relatively simple and provides good agreement with measured or detected data in practice, both for pure mineral, but also for doped and synthetic oils. For these reasons, integral temperature criterion appears to be the best so far published comprehensive method for calculating gearing resistance to scuffing. Therefore, in addition to Blok criterion, it became a part of the standard DIN 3990 [10].

4.1 General Principles of the Integral Temperature Criterion in HCR Involute Gearing

Integral temperature criterion is based on the middle temperatures of surface calculated on the basis of Blok theory according to the Eq. (10). Comparative medium temperature of surfaces ϑ_i (the integral temperature) consists of two temperatures: the medium value of local instantaneous flash temperature along the action line ϑ_{Blm} and tooth bulk temperature ϑ_O [10].

Therefore, integral temperature is calculated from the following relationship:

$$\vartheta_i = \vartheta_0 + \mathbf{B}\vartheta_{Blm} \tag{9}$$

where B is a weighting factor which takes into consideration qualitatively different temperatures influencing the actual tooth bulk temperature ϑ_O and temperature ϑ_{Blm}, which is defined only as a comparative temperature, and it does not reflect the actual size of the temperature in the contact points.

4.2 Calculating Medium Temperature of the Tooth Surface ϑBlm for HCR Involute Gearing

The size of the local instantaneous flash temperature ϑ_{BlE} is explained and determined in [9]. To compare the resistance of particular types of gearing against the scuffing, it is necessary to identify local flash temperatures ϑ_{BlE} in the reference point. Reference point is defined by the integral temperature for HCR involute gearing as the endpoint of contact line on the head of pinion, i.e. the point E on the contact line. The temperature at the reference point E can be expressed:

$$\vartheta_{BlE} = 0.62\,\mu_m \left(\frac{F_{nE}}{b}\right)^{0.75} \left(\frac{E_r}{\rho_r}\right)_E^{0.25} \frac{\left|v_{\rho 1E} - v_{\rho 2E}\right|}{\sqrt{B_{M1}\, v_{\rho 1E}} + \sqrt{B_{M2}\, v_{\rho 2E}}} \tag{10}$$

With referring to [9] medium temperature value can be obtained as

$$\vartheta_{Blm} = \mu_m\, w^{0.75}\, v_0^{0.5}\, a^{-0.25}\, X_M\, X_G\, X_\varepsilon \tag{11}$$

where: ϑ_{BlE}—medium temperature value;
μ_m—mean coefficient of friction;
w—normal unit load ($w = F_n/b$);
v_o—peripheral speed.

The values E_r and B_M depend only on the material of meshing gear wheels. Then, the coefficient of material X_M can be defined by the expression [13]:

$$X_M = \frac{E_r^{0.25}}{B_M^{0.5}} \tag{12}$$

where, for standard steel it follows [13]:

$$\sqrt{B_M} = \sqrt{\lambda\rho c} = \sqrt{50 \cdot 7800 \cdot 10^{-9} \cdot 482 \cdot 10^3} = 13.7\,\mathrm{Nmm^{-1}s^{-0.5}K^{-1}}$$

$E_r = 2.26 \cdot 10^{11}$ $\mathrm{Nm^{-2}}$, $\lambda_{1,2}$—coefficients of thermal conductivity of wheel materials, ρ—specific densities of gear material.

Then, the coefficient of material is calculated as: X_M = 50 K·N$^{-0.75}$s$^{0.5}$m$^{-0.5}$ mm.

X_G is a dimensionless coefficient of gear geometry, which can be modified in the following form [9, 13]:

$$X_G = 1.24 \sqrt[4]{\frac{\text{tg}\alpha_{wt}}{\cos^2 \alpha_{wt}}} \cdot \frac{\sqrt{u+1}}{2} \cdot \frac{\sqrt{\rho_{1E}} - \sqrt{\rho_{2E}}}{\sqrt[4]{\rho_{1E}\rho_{2E}}} \quad (13)$$

Factor of load distribution is obtained [9] in the following form:

$$X_\varepsilon = \frac{1}{2\ \varepsilon_1 \varepsilon_\alpha} \left[-0.543\varepsilon_1\varepsilon_2 + 0.204\varepsilon_2^2 + 0.123\varepsilon_1^2 + 0.759\varepsilon_2 + 1.001\varepsilon_1 - 0.539\right] \quad (14)$$

If the medium value of the coefficient of friction μ_m along the contact line is considered as constant, then the general influence on the medium temperature value ϑ_{Blm} will be expressed with the product of the coefficients X_G and X_ε. Therefore, the occurrence of scuffing depends on the factor of load distribution (X_ε) and the factor of gear geometry (X_G).

The product of coefficients $X_G \cdot X_\varepsilon$ is crucial for the assessment of impact on the size of the temperature ϑ_{Blm} and thus on the resistance to scuffing of HCR involute gearing. It is clear that there is strong dependence of the temperature ϑ_{Blm} on the shape of the tooth [12]. So, it is very important what profile is used for teeth meshing (involute, cycloidal, Novikov, etc.) since it determines the temperature in contact and the occurrence of scuffing.

5 Joined Optimization of Geometrical and Scuffing Parameters of HCR Gearing

The goal of the optimization is to obtain HCR involute gearing with the contact ratio factor $\varepsilon_\alpha = 2$; yet, at the same time, the occurrence of scuffing must be avoided.

Describing geometrical and manufacturing constraints, there are several constraints that should be satisfied (interference during the production, meshing interference, minimum thickness of the tooth head circle, slide-conditions in the HCR involute gearing). According to [8, 9], several constraints are provided and they come from the limitation condition, Eqs. (2–7), but there is an equation of contact ratio which represents the goal function with the aim that it should have the value of two, Eq. (8).

Beside geometric condition, there is another requirement to avoid the occurrence of scuffing and consequently reduce the rapid damage of tooth flanks. Using the same variable parameters h_{a1}^*, h_{a2}^* and x_1, flash temperature at the reference point E (Eq. 10) is considered as the highest temperature at the mating line. Since scuffing occurrence depends on flash temperature, the value ϑ_{BIE} should be an additional constraint that has to be minimized. Since the occurrence of scuffing depends on the factor of load distribution (X_ε) and the factor of gear geometry (X_G), the product of the coefficients $X_G \cdot X_\varepsilon$, is crucial for the assessment of impact on the flash temperature and thus on the resistance to scuffing of HCR involute gearing. According to the Eqs. (10), (11) and (13):

$$\vartheta_{BIE} = \mu_m w^{0.75} v_0^{0.5} a^{-0.25} X_M X_{GE} X_{\varepsilon E} = f(X_{\varepsilon E} \cdot X_{GE}) \\ \Rightarrow f\left(\frac{\left|\sqrt{v_{\rho 1E}} - \sqrt{v_{\rho 2E}}\right|}{\left(\frac{1}{\rho_{rE}}\right)^{0.25}} \cdot \frac{\left|\sqrt{\rho_{1E}} - \sqrt{\rho_{2E}}\right|}{(\rho_{1E}\rho_{2E})^{0.25}}\right) \Rightarrow \min \tag{15}$$

and for both gears made of standard steel (i.e. $B_{M1} = B_{M2} = B_M$), the equation for ϑ_{BIE} depends on variable parameters only in tangential velocities ($v_{\rho 1E}$ and $v_{\rho 2E}$), reducing the radius of curvature at the reference point E (ρ_{rE}) and the radii of curvature of profiles in the point E (ρ_{1E}, ρ_{2E}). Therefore, additional tooth parameters should be calculated:

- tangential velocity at the reference point E:

$$v_{\rho 1E} = \rho_{1E}\,\omega_1, \quad v_{\rho 2E} = \rho_{2E}\,\omega_2, \quad u = \frac{\omega_1}{\omega_2} \tag{16}$$

- radius of curvature of profiles in the reference point E:

$$\rho_{1E} = r_{b1} \cdot \tan\alpha_{at1}; \quad \rho_{2E} = a_w \cdot \sin\alpha_{wt} - r_{b1} \cdot \tan\alpha_{at1} \tag{17}$$

- reduced radius of curvature at the reference point E:

$$\frac{1}{\rho_{rE}} = \frac{1}{\rho_{1E}} + \frac{1}{\rho_{2E}} \tag{18}$$

However, in order to demonstrate this hypothesis about obtaining minimal flash temperature at the reference point E, the conditions for achieving maximal temperature will be found:

$$\theta_{BIE} = \mu_m w^{0.75} v_0^{0.5} a^{-0.25} X_M X_{GE} X_{\varepsilon E} = f(X_{\varepsilon E} \cdot X_{GE}) \\ \Rightarrow f\left(\frac{\left|\sqrt{v_{\rho 1E}} - \sqrt{v_{\rho 2E}}\right|}{\left(\frac{1}{\rho_{rE}}\right)^{0.25}} \cdot \frac{\left|\sqrt{\rho_{1E}} - \sqrt{\rho_{2E}}\right|}{(\rho_{1E}\rho_{2E})^{0.25}}\right) \Rightarrow \max \tag{19}$$

The factor of load distribution (X_ε) and the factor of gear geometry (X_G) depend on the variables $v_{\rho 1E}$, $v_{\rho 2E}$, ρ_{1E} and ρ_{2E} which are also the functions of variable parameters h^*_{a1}, h^*_{a2} and x_1, for the same centre distance a_w. Therefore, it is possible to simultaneously satisfy both these functions: the objective function which must be equal two and additional constraints according to scuffing. That is the reason for the strong dependence of the temperature ϑ_{Blm} on the shape of the tooth [12].

Obtaining the parameters for HCR involute gearing for maximal temperature in the reference point E, the condition for rapid scuffing occurrence on teeth flanks is achieved. Experimentally, it is proved on a test rig according to FZG methodology [9]. Searching numerical optimization results of involute HCR gearing is given in Ref. [9]. There is no just one solution of this task; there is infinite number of optimization results. All other results obtained by this numerical optimization give the same result: contact ratio equals two, Eq. (8) and maximum condition, Eq. (19). Therefore, all these optimization parameters are located in 3D field defined by parameters h^*_{a1}, h^*_{a2} and x_1. Choosing any combination inside the field of obtained parameters gives conditions that satisfies the aim function and gives maximum condition [9].

Applying any of these parameters obtained by numerical optimization to the real pinion and gear wheel, it will cause condition for rapid damaging HCR teeth flank due to scuffing. Scuffing occurrence can be visible by naked eye, but limiting condition when scuffing occurs can be recognized by several ways. One of the ways of scuffing occurrence is gravimetric criterion, when the pinion and gear wheels are measured after certain period of operating. Depending of their mass and volume, when they lost particular mass quantity, they can be considered as damaged and not more for use. Also, there is another method for recognizing scuffing and it is criterion of maximal damage depth due to scuffing. This method is also doing after certain periods of operating, when the depth due to scuffing is controlled and when it amounts 7 μm or more, the gears are damaged due to scuffing [13]. Obtained information are noticed there and given as information about occurring scuffing phenomenon. Using the same HCR gears ($z_1 = 21$, $z_2 = 51$, $m_n = 4$ mm, $a_w = 144$ mm) and transmission oil PP90, scuffing damage occurred for the torque $M_{k1} = 256$ Nm and rotation speed 760 min^{-1}, temperature of oil in that moment was 70 °C [8, 9].

Of course, if this scuffing phenomenon occurs after short period of time, it means that using opposite constraint, i.e. that Eq. (15) tends to be minimal, longer operating of gears is expecting without scuffing occurrence. In that way scuffing damaging of gear teeth can be avoided. In this way, the hypothesis is proved about obtaining minimal flash temperature and thus obtaining parameters for minimal ϑ_{BlE} and $\varepsilon_\alpha = 2$ as the solution to this problem.

6 Conclusions

High contact ratio (HCR) gear pair is a contact between gears with at least two pairs of teeth in contact. High contact ratio is obtained with increased addendum height, so larger than in LCR gearing. Proposed geometry of HCR gearings is much complicated due to the fact there is larger possibility of occurring meshing and during the production interference, much larger than interference happening in standard LCR profiles. Also, there is a higher risk of too small thickness of a tooth tip and significantly less favorable values of specific slips into the flanks. Contact ratio is increased by increasing tooth height. Dynamic loads and noise are reduced by using high contact ratio gears. According to the results of different measurements of gear pair, the reduction of noise proved to be the best using HCR gearing with the value of contact ratio $\varepsilon_\alpha = 2$. Decrease in noise is caused by $\varepsilon_\alpha = 2$ because there are always two pairs of teeth in contact, meaning that, when one pair of teeth leaves the contact, another pair of teeth arrives in the contact, while the applied force is considerably smaller since it is divided between two pairs of teeth. Therefore, gearing in automotive industry should be performed with $\varepsilon_\alpha = 2$ in order to reduce the noise and dynamic forces.

Due to the increased addendum height, there is a larger possibility of some interference or pointed tooth tip to occur. Consequently, these errors should be prevented by verifying whether all equations and constraints are satisfied. Conditions for teeth on pinion and gear wheel are related to the following: conditions for non-occurrence of interference during the production, conditions for non-occurrence of meshing interference, conditions for minimal thickness of the tooth head circle of both gears, and additional condition related to the balance of specific sliding at the beginning and the end of meshing in order to reduce losses during meshing [8, 9]. Also, there is a possibility of scuffing occurrence due to higher tangential velocities and high pressure between teeth. Using the criterion of integral temperature, in order to avoid scuffing occurrence the relation must be minimized.

The main contribution of this paper is a generalization of the integral temperature criterion for involute HCR gearing. It has been shown that relations for integral temperature criterion for involute HCR gearing need to have the calculation of the factor of load distribution for the case $\varepsilon_\alpha \geq 2$. In the continuation of this researching, the criterion for integral temperature of involute HCR gearing can be optimized and the minimal flash temperature can be obtained. For that case, the factor of load distribution (X_ε) and the factor of gear geometry (X_G) were derived for the case of involute HCR gearing. Summary results of the article show that tooth shape significantly affects the resistance of the teeth to the scuffing. It can be expressed by analytical relations derived from the integral temperature criterion and confirmed by scuffing tests. Based on the derived relationships it can be derived for the special occasion when $\varepsilon_\alpha = 2$ to optimize HCR gearing which is used to increase their resistance to scuffing.

References

1. Kuzmanović S, Vereš M, Rackov M (2010) Product Design as the Key Factor for Development in Mechanical Engineering, In: Proceedings of international conference mechanical engineering in XXI Century. Niš, Serbia, pp 113–116
2. Kasuba R (1981) Dynamic loads in normal and high contact ratio spur gearing. In: International symposium on gearing and power transmissions. Tokyo, Japan, pp 49–55
3. Sato T, Umezawa K, Ishikawa J (1983) Effect of contact ratio and profile correction of spur gears on the rotational vibrations. Bull JSME 26(221):2010–2016
4. Kahraman A, Blankenship GW (1999) Effect of Involute Contact Ratio on Spur Gear Dynamics. ASME J Mech Des 121:112–118
5. Vereš M, Kuzmanović S, Rackov M (2012) Experimental research of HCR gearing from pitting damage point of view, In: Proceedings of 7th international symposium about mechanical and industrial engineering—KOD. Balatonfüred, Hungary, pp 317–320
6. Michaelis K Gear failures—scuffing (course at the university of Ljubljana), Forschungsstelle für Zahnräder und Getriebebau Gear Research Centre, FZG TU München
7. Gears—Wears and Damage to Gear Teeth—Terminology (1995) International organization for standardization, ISO 10825
8. Rackov M, Vereš M, Kanovi Ž, Kuzmanović S (2013) HCR gearing and optimization of its geometry. Adv Mater Res 633:117–132
9. Rackov M (2014) Optimization of HCR gearing geometry from scuffing point of view, PhD thesis (in English), slovak university of technology in Bratislava, faculty of mechanical engineering
10. Blok H (1937) Theoretical study of the temperature rise at surfaces of actual contact under oiliness conditions. Inst Mech Eng General Discuss Lubr 2:222–235
11. Winter H, Michaelis K, Oct/Nov (1984) Scoring load capacity of gears lubricated with EP-Oils, technical university of Munich, gear technology
12. Tragfähigkeitsberechnung von Stirnrädern—Berechnung des Freßtragfähigkeit, DIN 3990–4
13. Vereš M (1987) Odolnost ozubenia voči zadieraniu z hladiska jeho tvaru. Kandidatska dizertačna praca, Slovenska vysoka škola technicka v Bratislava, Strojnicka fakulta, Bratislava

Parametric Design of an Electric Driven Reduction Gear for Anchor and Mooring Winches

Florin Susac, Nicuşor Baroiu, Bogdan Liviu Marcu, Daniel Ganea and Virgil Teodor

Abstract The paper presents an Inventor-based technique for designing an electric driven reduction gear for anchor and mooring winches. The technique consists in a parametric design so that the final product to be automatically model in correlation with input parameters. For that, simple logical rules are used in order to define and use the parameters and properties of the designed model through *iLogic* programming language which is included in *Autodesk Inventor* package. A *iLogic* rule can be considered as a short *Visual Basic* program created and used to automatically monitor and control, as *Inventor* objects, the model parameters values depending on input data. The aim of this is to extend and consolidate the capacity and automation of designing tasks. The *iLogic* rules may be incorporated directly in *Inventor* files or may be externally saved in order to be accessed by any files of the designing software packages.

Keywords Parametric design · ILogic · Visual basic · Reduction gear · Anchor and mooring winch

1 Introduction

The reduction gears with toothed wheels are mechanisms structured as independent assemblies and having a constant ratio gear, which are running in closed and sealed cases. They are used for speed reduction and, at the same time, for amplifying the transmitted torque [1, 2]. Sometimes, reduction gears are used for reversing the

F. Susac · N. Baroiu (✉) · B.L. Marcu · D. Ganea · V. Teodor
Department of Manufacturing Engineering, "Dunarea de Jos" University of Galati, Galati, Romania
e-mail: Nicusor.Baroiu@ugal.ro

F. Susac
e-mail: florin.susac@ugal.ro

V. Teodor
e-mail: virgil.teodor@ugal.ro

V. Majstorovic and Z. Jakovljevic (eds.), *Proceedings of 5th International Conference on Advanced Manufacturing Engineering and Technologies*, Lecture Notes in Mechanical Engineering, DOI 10.1007/978-3-319-56430-2_11

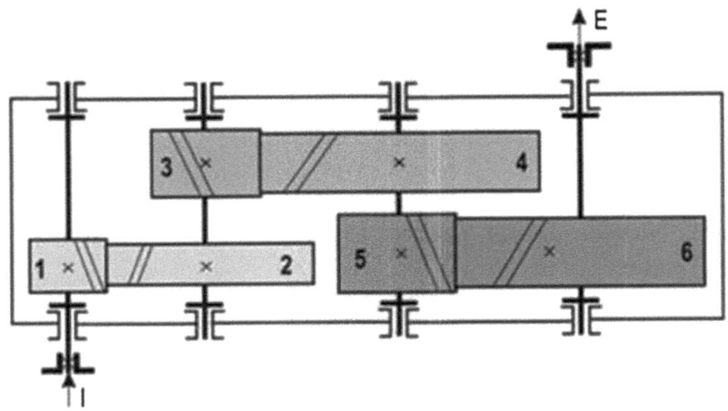

Fig. 1 Reduction gear with cylindrical toothed wheels [5]

direction of motion. Rarely, speed amplifiers which are structured as reduction gears are used in order to increase the output speed [3, 4]. The main characteristics of a reduction gear are: reduction gear type, electric power (in kW), engine shaft speed and ratio gear.

The reduction gears have mainly cylindrical, conical and worm gears mounted in series or series-parallel, which represents the reduction gear steps, Figs. 1, 2 and 3 [5].

In order to set-up a faster and more efficient design of technical products, *Autodesk Inventor* has the possibility of creating and managing „intelligent" parts and assemblies (*iParts, iAssemblies*) and allows to use and create libraries of standard parts through integrated concept called Content Center [6, 7]. At the same time, *Autodesk Inventor* develops efficient designing environment and advanced programming features as *iLogic* [8].

iLogic uses text-type parameters, as true/false (*True/False*) and multi-values type list (*Multi-value list*), which allow to modify the standard numerical parameters belonging to Inventor models.

Fig. 2 Reduction gear with conical toothed wheels [5]

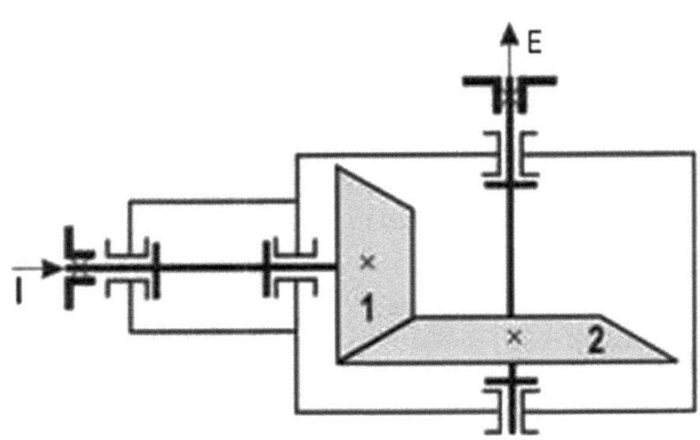

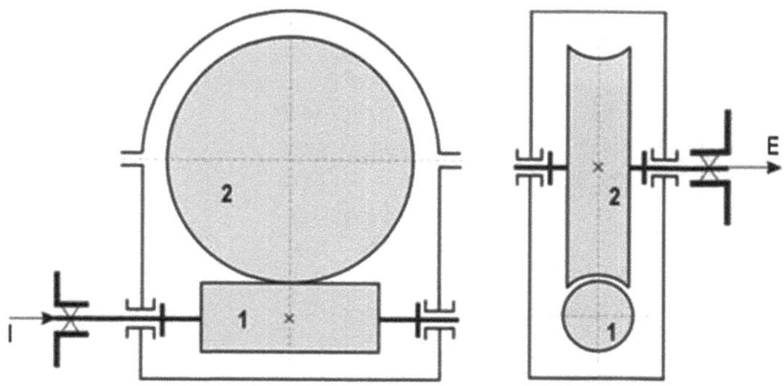

Fig. 3 Reduction gear with worm toothed wheels [5]

2 Graphical Modelling of Reduction Gear Elements

When modelling the reduction gear of an anchor winch, pinion shaft and toothed wheel are the starting point. These are multiplied depending on the number of reduction gear steps, which in this case is three, Fig. 4. The parameters of both the pinion and toothed wheel are defined according to *iLogic* rules.

The pinion is created using a *iFeature*, Fig. 5, which contains the specific parameters of gear teeth: number of teeth, number of teeth of mating gear, hand of teeth, tip diameter of gear wheel, root circle diameter of gear wheel, pitch circle diameter, reference center distance, helix angle, wheel width.

After defining the specific parameters of teeth gear, the ends of pinion shaft are modeled using a sketch and function *Revolve*. The list of parameters belonging to any modeled component can be accessed through option *Manage →fxParameters*, Fig. 6.

Fig. 4 The schema of a reduction gear of an anchor winch

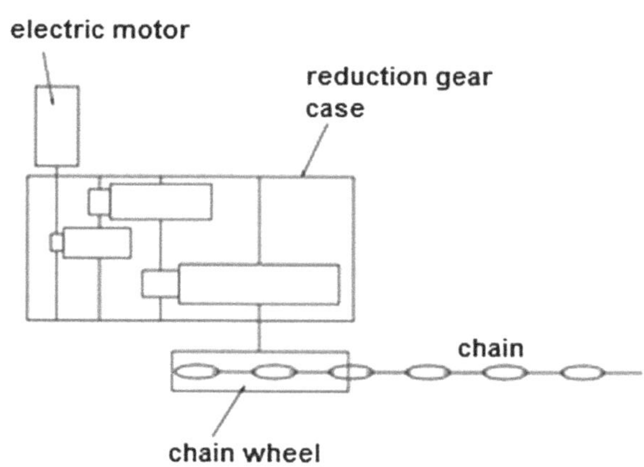

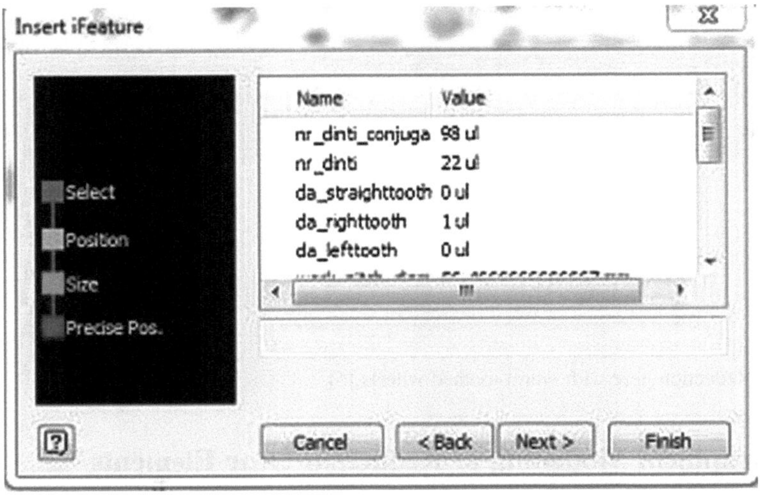

Fig. 5 *iFeature*–specific parameters of gear teeth

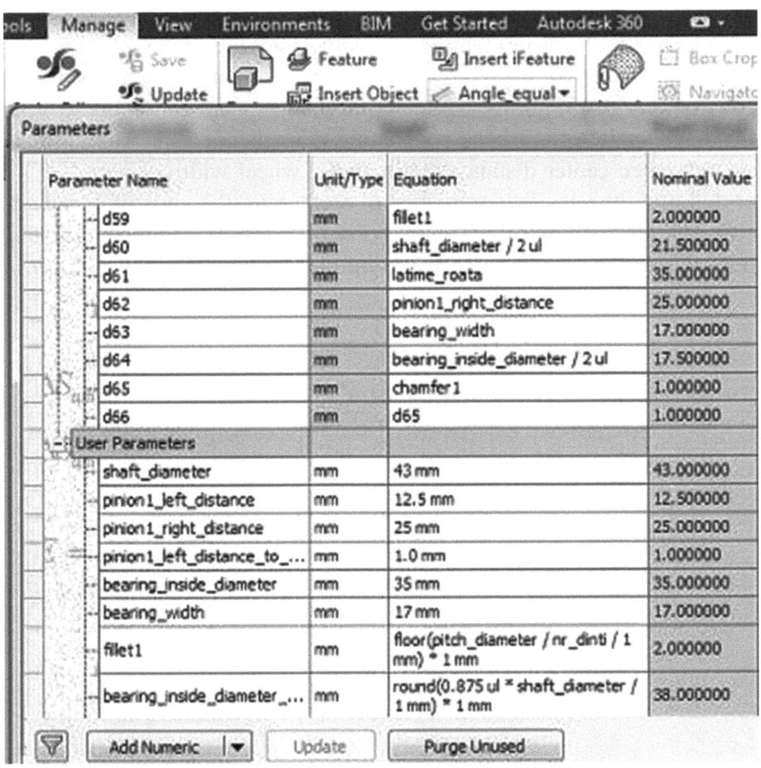

Fig. 6 The parameters list of *Pinion 1* component

These parameters can be manually created using option *Add numeric.*

At the beginning, the parameters will receive approximate numerical values, the exactly values used for modelling the pinion shaft being later calculated by *iLogic* rules.

It is important to mention that the dimensions in the sketch are not absolute values, parametric values are given to them, which are defined by generic names (for example, instead of a number corresponding to the value of the area where the bearing is mounted, the *bearing-width* parameter is used).

Considering the issues previously described, the 3D model of the pinion shaft belonging to first step of the reduction gear, Fig. 7, was generated.

For modelling the toothed wheel, the same procedure is followed. There are two differences when modelling toothed wheel and pinion shaft. These refers to modifying the cylinder shape for wheeling the teeth and machining of equidistant holes for reducing the toothed wheel weight, Fig. 8.

This component has its own defined list of parameters, too. After generating *Pinion1.ipt*and*Wheel1.ipt* components, these will be multiplied in order to be used for components belonging to step 2 and step 3 of the reduction gear.

For *Wheel 3*, due to the fact that in the most cases it has large dimensions, in order to reduce the manufacturing costs, a structure consisting in a welded hub and disc and an assembly gear rim-hoop are used. In order to generate this toothed the following components were created in *Parts* folder: hub, disc, hoop, which will form the assembly *Wheel 3.iam.*

Figure 9 presents both the assembly components and *iLogic* list of rules. Depending on overall dimensions of the toothed wheel, some constructive options may be used: with one or two discs, the width of the disc, the number of the holes in the disc, the holes diameter may vary [1].

3 Algorithm for Reduction Gear Calculation

In order to carry out a correct design of the reduction gear components, rigorous calculations are necessary so that the resulted components should with stand the tension gear but, at the same time, it is essential to avoid over dimensioning of components. The calculations are made according to *iLogic* rules, grouped in windows (*Forms*), which contain many working areas: parameters, rules and

Fig. 7 3D model of pinion shaft of reduction first step

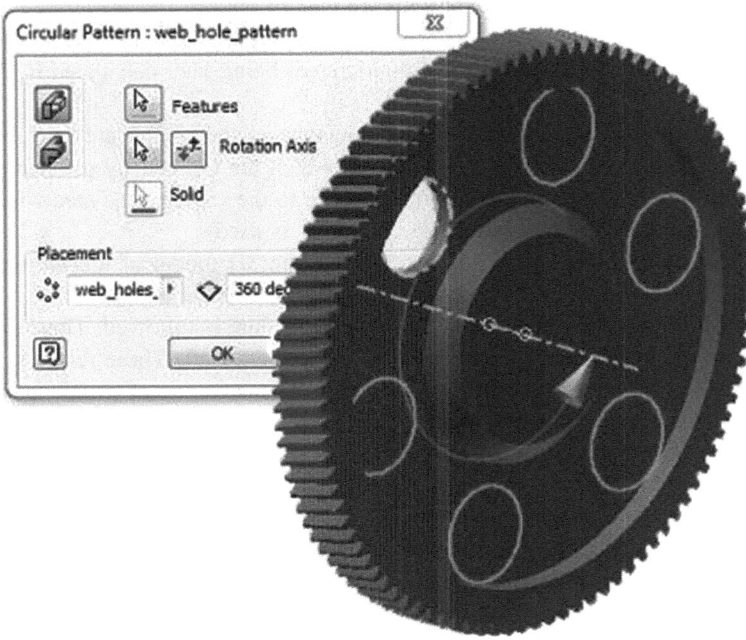

Fig. 8 The modelling of reduction gear basic toothed wheel (Wheel 1)

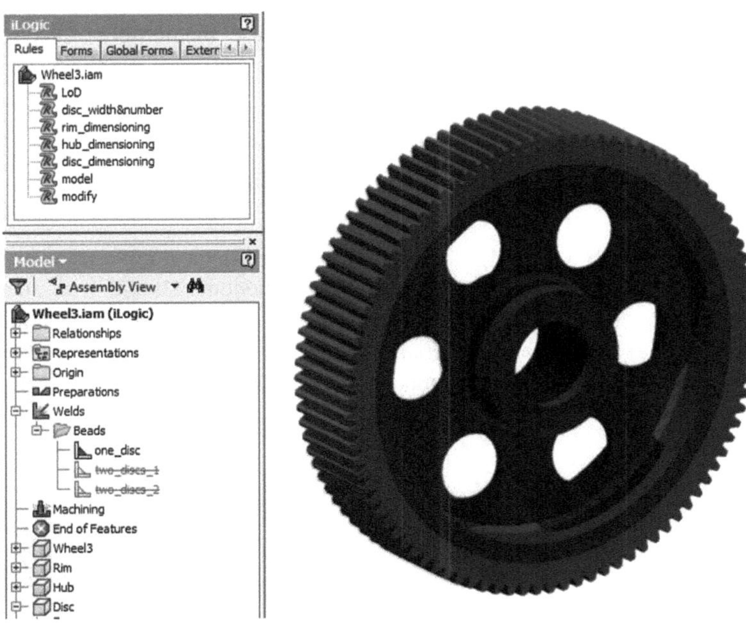

Fig. 9 Assembly *Wheel 3*

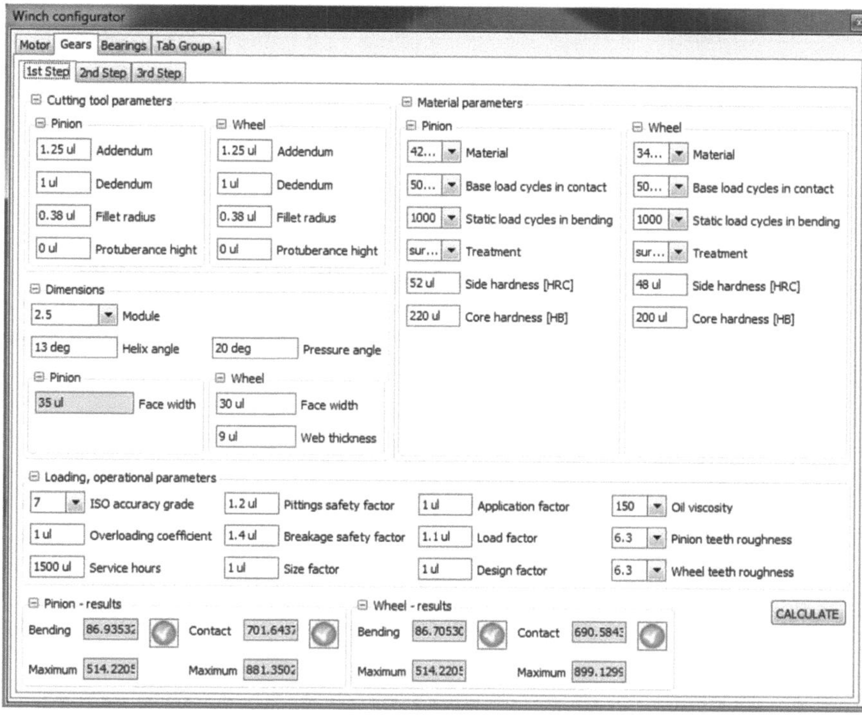

Fig. 10 The window *Winch configurator* in section *Gears*

properties, working tools (*Tab Group*)—*Motor, Gears, Bearings, Tab Group 1* etc., area for positioning or for properties.

As an example, in section *Motor* many groups of data were created, Fig. 10:

(1) *Input data*—input data delivered by clients, as nominal pulling force, pulling speed of the chain, chain (chain link diameter);
(2) *DMT data*—input data which are not delivered by clients, but delivered by the producer, as chain wheel socket etc.;
(3) *Chosen motor*—necessary calculated power of electric motor and fields for filling the electric motor data from producer catalogue which are necessary for running calculations;
(4) *Gear ratio*—a group of data divided in three areas, one area for each step of the reduction gear, for the number of teeth belonging to gear toothed wheels;
(5) *Output data*—area for comparing and verifying the data delivered by client and calculated data;
(6) The listing area for actual gear ratio and *Calculate* command, which runs *Motor* rule.

3.1 Choosing the Electric Motor

The necessary power of the electric motor is calculated considering the output shaft power and the reduction gear efficiency:

$$P_{nec} = \frac{P_{arb}}{\eta} = \frac{F_n \cdot v}{\eta} \tag{1}$$

Therefore, an electric motor with a power of 3 kW and a speed of 1405 rpm is chosen [9].

The code sequence of rule *Motor* which calculates the necessary power of the electric motor is as follows:

```
''NECESSARY POWER FOR MOTOR
Parameter("necessary_power")=Parameter("nominal_force") * Parame-
ter("chain_speed") /(60000*0.97^Parameter("number_of_steps"))
If Parameter("motor_power")<Parameter("necessary_power") Then
MessageBox.Show("Motor power must be greater then the calculated
necessary power!" &vbCrLf&"Setting motor power to
neccesary power!", "Error")
Parameter("motor_power")=Parameter("necessary_power")
End If
```

3.2 Choosing the Gear Ratio

The minimum and maximum gear ratio of the reduction gear are calculated with following relations:

$$i_{\min} = \frac{M_{AP}}{M_{ME} \cdot \eta}; \quad i_{\max} = \frac{n_{ME}}{n_{AP}} \tag{2}$$

where:

- M_{AP}, n_{AP} are the torque, respectively the speed of the main shaft;
- M_{ME}, n_{ME}- the torque, respectively the speed of the electric motor shaft.

The code sequence of the rule *Motor* is as follows:

```
''NECESSARY GEAR RATIO
Parameter("i_min")=(Parameter("nominal_force")*a/2000) /
(30000*Parameter("motor_power")/(PI*Parameter("motor_rotational_spe
ed"))*0.97^Parameter("number_of_steps"))
Parameter("i_max")=Parameter("motor_rotational_speed") /
(1000*Parameter("chain_speed")/(PI*a))
Parameter("i_mediu")=(Parameter("i_min")+Parameter("i_max"))/2
```

3.3 Choosing the Teeth Number of Toothed Wheels

For a good functioning of the gear, it is recommended that the teeth number of the pinions to be between 17 and 23 and gear ratios to be a number with as many decimals as possible. These will lead to a uniform wear of all teeth of the toothed wheel.

The teeth number of the toothed wheels will be chosen so that the resulted gear ratios to have a value as close as possible to the previously calculated gear ratios. The three gear ratios are chosen using equations:

$$i_2 = \sqrt[3]{i_{med}}; \quad i_1 = 1,05 \cdot \sqrt{\frac{i_{med}}{i_2}}; \quad i_3 = \frac{i_{med}}{i_1 \cdot i_2}. \tag{3}$$

To automate this stage a suggestion algorithm was created. This algorithm will automatically suggest the teeth number of each toothed wheel (rule *suggest_-teeth_number*). After running this rule, the following are obtain:

$$Z_{11} = 22;\ Z_{12} = 98,\ with\ i_1 = 4.455;$$
$$Z_{21} = 21;\ Z_{22} = 89,\ with\ i_2 = 4.238;$$
$$Z_{31} = 23;\ Z_{32} = 93,\ with\ i_3 = 4.043.$$

Finally, $i = i_1 \cdot i_2 \cdot i_3$.

3.4 The Toothed Wheel Calculation of Resistance

For resistance calculation of the gear, the following verifications are mandatory: the contact checking of tooth flanks and bending checking of the tooth base. The necessary equations for these verifications are available in rules *step1_calcul*, *step2_calcul*, respectively *step3_calcul*.

The fatigue contact stress is calculated using [10]:

$$\sigma_H = Z_B \cdot Z_H \cdot Z_E \cdot Z_\varepsilon \cdot Z_\beta \cdot \sqrt{\frac{F_t \cdot K_A \cdot K_g \cdot K_{V\beta} \cdot K_{H\beta} \cdot K_{H\alpha}}{d_1 \cdot b_w} \cdot \frac{i_1 + 1}{i_1}} \quad (4)$$

where:

Z_B is the contact factor for a single pear of teeth and is given by:

$$Z_B = M_{1(2)} - \varepsilon_\beta \cdot \left(M_{1(2)} - 1\right) \quad (5)$$

The code sequence of rule *step1_calcul* for calculating Z_B (pin1_261) and factors M_1 (pin1_262) and M_2 (wh1_262) is as follows:

```
Dim pin1_262 as Double=Tan(pin1_86*PI/180) / Sqrt(
(Sqrt(((pin1_101/pin1_94)^2) - 1) - 2*PI/pin1_66) *
(Sqrt(((wh1_101/wh1_94)^2) - 1) - (pin1_125-1)* 2*PI/wh1_67 ) )
Dim wh1_266 as Double=Tan(pin1_86*PI/180) / Sqrt(
(Sqrt(((wh1_101/wh1_94)^2) - 1) - 2*PI/wh1_67) *
(Sqrt(((pin1_101/pin1_94)^2) - 1) - (pin1_125-1)* 2*PI/pin1_66 ) )
Dim pin1_261 as Double=0
Dim wh1_265 as Double=0
If pin1_126=0 Then
pin1_261=Max(pin1_262,1)
wh1_265=Max(wh1_266,1)
Else If pin1_126>=1 Then
pin1_261=1
wh1_265=1
Else
pin1_261=pin1_262-pin1_126*(pin1_262-1)
wh1_265=wh1_266-pin1_126*(wh1_266-1)
```

M_1 is a factor; ε_β- axial covering rate and is given by relation:

$$\varepsilon_\beta = \frac{b_\omega \cdot \sin\beta}{\pi \cdot m} \quad (6)$$

where:

b_ω is toothed width; β- helix angle; m–normal modulus.

$$M_1 = \frac{\tan \alpha_{wt}}{\sqrt{\left[\sqrt{\left(\frac{d_{a1}}{d_{b1}}\right)^2 - 1} - \frac{2\pi}{Z_1}\right] \cdot \left[\sqrt{\left(\frac{d_{a2}}{d_{b2}}\right)^2 - 1} - (\varepsilon_\alpha - 1) \cdot \frac{2\pi}{Z_1}\right]}} \quad (7)$$

α_{wt}—transverse pressure angle at pitch cylinder; $d_{a1\,(2)}$—tip diameter of the gear wheel; $d_{b1\,(2)}$—base circle diameter; ε_α—transverse contact ratio.

Z_H—***contact zone factor*** [10]:

$$Z_H = \sqrt{\frac{2\cos\ \beta_b}{(\cos\ \alpha_t)^2 \cdot\ \tan\ \alpha_{wt}}} \tag{8}$$

β_b- base helix angle; α_t- transverse pressure angle at the pitch.
The code sequence of rule *step1_calcul* for calculating Z_H (pin1_267) is:

```
Dim pin1_267 as Double=Sqrt(2*Cos(pin1_88*PI/180) /
    ((Cos(pin1_82*PI/180)^2)*Tan(pin1_86*PI/180)) )
```

Z_E—***elasticity factor of the material*** [10]:

$$Z_H = \sqrt{\frac{1}{\pi \cdot \left(\frac{1-\nu_1^2}{E_1} + \frac{1-\nu_2^2}{E_2}\right)}} \tag{9}$$

$\nu_{1\ (2)}$—Poisson ratio; $E_{1\ (2)}$—Young modulus.
The code sequence of rule *step1_calcul* for calculating Z_E (pin1_268) is:

```
Dim pin1_268 as Double=Sqrt( 1/ (PI*((1-(pin1_152^2))/pin1_151 + (1-
(wh1_152^2))/wh1_151 )) )
```

Z_ε—***contact ratio factor*** [10]:

$$Z_\varepsilon = \sqrt{\frac{(4-\varepsilon_a)\cdot(1-\varepsilon_\beta)}{3} + \frac{\varepsilon_\beta}{\varepsilon_a}} \tag{10}$$

The code sequence of rule *step1_calcul* for calculating Z_ε (pin1_270) is:

```
If pin1_126 <1 Then
pin1_270=((4-pin1_125)*(1-pin1_126)/3+pin1_126/pin1_125)^0.5
Else pin1_270=(1/pin1_125)^0.5
End If
```

Z_β—***helix angle factor:***

$$Z_\varepsilon = \sqrt{\cos\ \beta} \tag{11}$$

F_t—***transverse force:***

$$Z_\varepsilon = \sqrt{\cos\ \beta} \tag{12}$$

M_t—torsional moment:

$$M_t = \frac{30000}{\pi} \cdot \frac{P_{nec}}{n} \tag{13}$$

The code sequence of rule *step1_calcul* for calculating F_t (pin1_144) and M_{t1} (pin1_142) and M_{t2} (wh1_142) is:

```
Dim pin1_142 as Double=30000/PI*Parameter("motor_power") /
Parameter("motor_rotational_speed")
Dim wh1_142 as Double=30000/PI*Parameter("motor_power") /
(Parameter("motor_rotational_speed") /pin1_68)
Dim pin1_144 as Double=pin1_142 * 2000 / pin1_92
SharedVariable("step1_transverse_force")=pin1_144
```

K_A ***application factor***–adjusts the nominal load in order to compensate the external applied loads; $K_A = 1$ (according to ISO 6336-6-2004)

K_g ***load distribution factor***–$K_g = 1.1$ (according to ISO 6336-6-2004)

$K_{V\beta}$ ***dynamic factor***

$$K_{V\beta} = N \cdot K + 1 \tag{14}$$

where:

N is the reference factor of speed.

There are four functioning domains depending on the reference factor of the speed: under critical (N < 0.85), critical, of resonance (0.85 $\leq$ N $\leq$ 1.15), intermediate (1.15 < N < 1.5), over critical (N $\geq$ 1.5), according to ISO 6336-1-2006.

$$N = \frac{n_1}{n_{cr1}} = \frac{n_1}{\frac{30 \cdot 10^3}{\pi \cdot Z_1} \cdot \sqrt{\frac{c_\gamma}{m_{red}}}} \tag{15}$$

where:

n_1 is pinion shaft speed; n_{cr1}—resonance speed of the pinion; c_γ—average rigidity of simultaneously gearing teeth.

The code sequence of rule *step1_calcul* for calculating factor N (pin1_187) is:

```
Dim wh1_186 as Double=30*(10^3)*((pin1_164/pin1_183)^0.5) /
(PI*pin1_66)
Dim pin1_186 as Double=wh1_186
Dim pin1_187 as Double=Parameter("motor_rotational_speed")/pin1_186
```

$K_{H\beta}$—***face load factor*** (***contact stress***) is defined as the intensity of the maximum load comparing with average load [11]:

$$K_{H\beta} = \sqrt{\frac{2 \cdot F_{\beta y} \cdot c_{\gamma\beta}}{\frac{F_{tm}}{b_w}}} \tag{16}$$

$F_{\beta y}$—actual longitudinal tooth flank deviation after grinding, without corrections; $c_{\gamma\beta}$—gear rigidity; F_{tm}—mean transverse tangential load at the reference circle; b_w-working face width.

$K_{H\alpha}$—***transverse load factor*** (***contact stress***) is the effect at no uniform load distribution between more simultaneously pears of teeth in contact [11].

$$K_{H\alpha} = \frac{\varepsilon_\gamma}{\varepsilon_\alpha \cdot Z_\varepsilon^2} \tag{17}$$

Z_ε—contact ratio factor.

The code sequence of rule *step1_calcul* for calculating Z_ε (pin1_270) is:

```
If pin1_126 <1 Then
pin1_270=((4-pin1_125)*(1-pin1_126)/3+pin1_126/pin1_125)^0.5
Else pin1_270=(1/pin1_125)^0.5
End If
```

The fatigue contact stress for first gear step can be determined if all factors are already calculated. Therefore, the following results are obtained:

$$\sigma_{H\,pinion} = 701.644\ [\text{MPa}] \text{ and } \sigma_{H\,wheel} = 690.584\ [\text{MPa}]$$

The previously calculated values must be lower than the maximum admissible fatigue contact stress [10]:

$$\sigma_{H\,adm} = \frac{\sigma_{H\,\lim} \cdot Z_N \cdot Z_L \cdot Z_V \cdot Z_R \cdot Z_W \cdot Z_X}{S_H} \tag{18}$$

Z_N—***life factor of contact stress*** (is calculated according to ISO 6336-2-1996), depending on N_{HE}—number of running cycles, N_{Hst}—static number of load cycles in contact, N_{HB}—base number of load cycles in contactq_H—Wohler curve exponent for contact:

$$Z_N = \left(\frac{N_{HB}}{N_{HE}}\right)^{\frac{1}{q_H}} \tag{19}$$

Z_L—***lubrication factor.*** The lubrication factor for mineral oils can be determined as a function of nominal viscosity at 40 °C and the value $\sigma_{H\ \lim}$ of the softest material for the two toothed wheels of the gear and can be calculated according to ISO 6336-2-1996:

$$Z_L = C_{ZL} + \frac{4(1 - C_{ZL})}{\left(1,2 + \frac{134}{v_{40}}\right)^2} \tag{20}$$

v_{40}—oil viscosity at 40 °C.

The code sequence of rule *step1_calcul* for calculating C_{ZL} (pin1_278) is:

```
Dim pin1_278 as Double=0
If Min(Parameter("pinion1_material_contact_endurance"),
Parameter("wheel1_material_contact_endurance")) <850 Then
pin1_278=0.83
Else If min(Parameter("pinion1_material_contact_endurance"),
Parameter("wheel1_material_contact_endurance")) >1200 Then
pin1_278=0.91
Else
pin1_278= 0.83+0.08*(Min(Parameter("pinion1_material_
contact_endurance") ,Parameter("wheel1_material_contact_
endurance")) - 850)/350
End If
```

Z_V—***peripheral speed factor*** [10]:

$$Z_L = C_{ZV} + \frac{2(1 - C_{ZV})}{\sqrt{0,8 + \frac{32}{v}}} \tag{21}$$

$C_{ZV} = C_{ZL} + 0,02$; v is peripheral speed of pitch circle diameter.

Z_R—***roughness factor*** (is calculated according to ISO 6336-2-1996):

$$Z_R = \left(\frac{3}{R_{z10}}\right)^{C_{ZR}} \tag{22}$$

$$R_{z10} = R_z \cdot \sqrt[3]{\frac{10}{\rho_{red}}} \tag{23}$$

R_z is teeth roughness; ρ_{red}—relative radius of curvature.

The code sequence of rule *step1_calcul* for calculating Z_R (pin1_283) is:

```
Dim pin1_283 as Double=( 3/
(0.5*(Parameter("pinion1_teeth_roughness")+
Parameter("wheel1_teeth_roughness"))* ((10/pin1_287)^(1/3)) ))^pin1_291
```

Z_W—***hardness ratio factor.*** This factor considers the increasing of surface hardness due to gearing of a steel toothed wheel with a pinion which has an increased hardness and a lower roughness. Z_W is determined according to ISO 6336-2-1996 standard. Therefore, $Z_W = 1$.

The code sequence of rule *step1_calcul* for calculating Z_W (pin1_294) is:

```
Dim pin1_294 as Double = 0
If Min(Parameter("pinion1_material_contact_endurance"), Parame-
ter("wheel1_material_contact_endurance")) >130 And
Min(Parameter("pinion1_material_contact_endurance"), Parame-
ter("wheel1_material_contact_endurance"))<470 Then
pin1_294 = 1.2-(Min(Parameter("pinion1_material_contact_endurance"),
Parameter("wheel1_material_contact_endurance"))-130)/1700
Else If Min(Parameter("pinion1_material_contact_endurance"), Parame-
ter("wheel1_material_contact_endurance"))<=130 Then
pin1_294=1.2
Else
pin1_294=1
End If
```

Z_X—***size factor.*** The size factor is defined according to ISO 6336-2-1996, considering statistics which indicates that the stress level when fatigue occurs is decreasing with the increasing of the components dimensions. Usually,$Z_X = 1.S_H = 1.2$—safety factor.

From calculation, the following results are obtained:

$$\sigma_{H\ adm\ pinion} = 881.351\ [\mathrm{MPa}] \text{ and } \sigma_{H\ adm\ wheel} = 899.129\ [\mathrm{MPa}]$$

It is obviously seen that both for pinion and toothed wheel, the contact resistance condition is $\sigma_H \leq \sigma_{H\ adm}$, Fig. 11.

Based on the same calculation reasons, the bending checking is carried out both for pinion shaft and toothed wheel, Fig. 12.

$$\sigma_{F\ pinion} = 86.935\ [\mathrm{MPa}] \text{ and } \sigma_{H\ wheel} = 86.705\ [\mathrm{MPa}]$$

$$\sigma_{F\ adm\ pinion} = 514.221\ [\mathrm{MPa}] \text{ and } \sigma_{H\ adm\ wheel} = 514.221\ [\mathrm{MPa}]$$

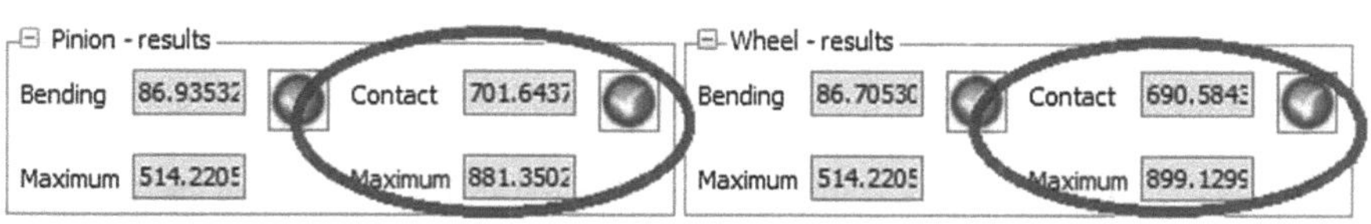

Fig. 11 Contact stress at pinion and toothed wheel

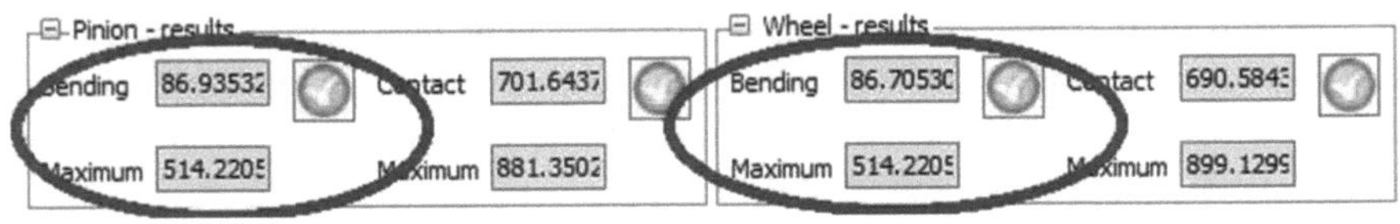

Fig. 12 Bending stress at pinion and toothed wheel

4 Methodology for Parameters Configuration

The parametric design in *Autodesk Inventor* is based on rules application over some components which are already 3D modelled or over some components which will be extracted from Inventor predefined components data base.

In this manner, a centralized file is created (*Content Centre*), which contains a package of rules having well defined goals, some of them being activated by other rules or runs within other rules. At the same time, there are rules which are set to automatically run when one of the parameters controlled by them is changing its value. Table 1 and Fig. 13 shows some examples of rules and their specific goals.

When opening file*assembley1.iam*the *winch configurator* window will be shown, see Fig. 10, due to *start_form* rule, which is chosen to run when opening the file with *Event Trigers* tool from *iLogic* framework. This window is structured in four sections: *Motor, Gears, Bearings and Tab Group 1,* Fig. 14. Some fields allow

Table 1 Examples of rules from centralized file

LoD	Sets the level of predefined iLogic detail when opening fileassembley1.iam (centralized file)
Delete_parameters	Automatically erases all parameters of the file
ISO	Sets the tolerances values for a specific precision level according to ISO-1328-1:1995, ISO-1328-1:1996, ISO-1328-1:1997 [12]
Add_parameters	Automatically adds all parameters later used by the centralized file
Motor	Calculates the electric motor power in accordance to input data
Sugest_teeth_number	Recommends the optimum number of teeth so that the total gear ratio should be as close as possible to calculated average gear ratio
Step1_calcul, step2_calcul, step3_calcul	Calculates the stresses in the gear and calculates the contact and bending tensions in the gear teeth
Constrains	Applies the position constraints of the pinion and toothed wheel
Shaft_dimensioning	Calculates the necessary parameters for dimensioning the shafts
Add_bearings_code	Chooses a proper pair of bearings for each shaft
Bearings_dynamic_load_calcul	Calculates the dynamic forces which stresses the bearings

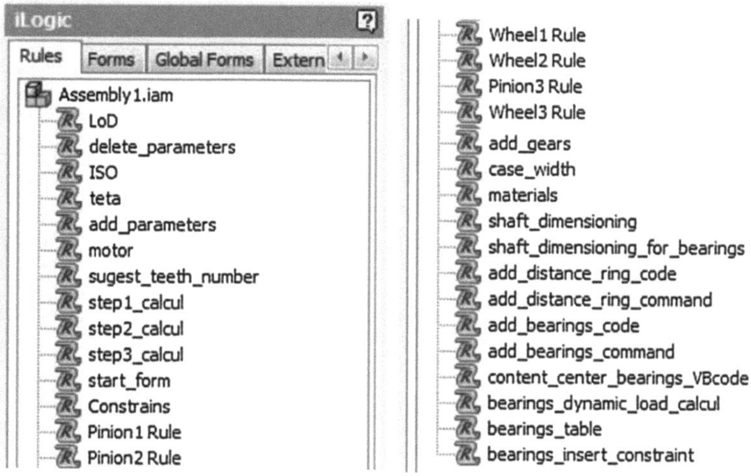

Fig. 13 List of rules in the centralized file (*assembley1.iam*)

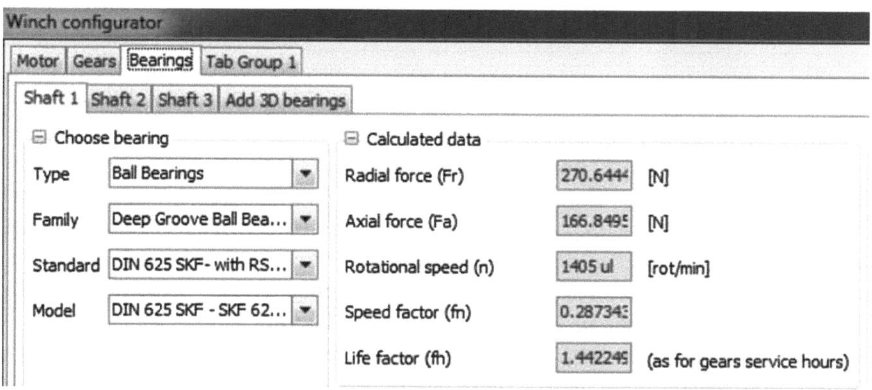

Fig. 14 Winch configurator window in section bearings

to fill in or modify data and others allow only to visualize data which were previously calculated and do not allow any modifications of this data in order to prevent overwriting or errors which may propagate in future calculations.

After selecting the data in *Motor* section, the necessary elements for designing the toothed wheels and calculating the contact and bending stresses are filled in section *Gears*. This section has one sub-section (*sub-tab*) for each reduction gear step: *1st Step, 2nd Step and 3rd Step*. Similar to *Motor* section, many groups of data were created in section *Gears*. These groups of data contains the parameters of the cutting tool both for pinion shaft and toothed wheel, material characteristics which counts for resistance calculation, pinion and toothed wheel dimensional parameters such as normal modulus, helix angle $\beta°$, pressure angle $\alpha°$, teeth and disc width,

Fig. 15 3D view of the designed 3-steps reduction

safety coefficients, precision level, total running time of the reduction gear, oil viscosity, teeth surface roughness etc.

Similar, the pairs of bearing are selected in section *Bearings,* Fig. 14. Considering that the bearings type are of a large variety, it was chosen a more accurate designing when introducing them in the assembly of the *Content Centre* data base. Thus, in order to diminish the designing time, this stage was automated by creating a rule in *Visual Basic* language, *content_center_bearings_VBcode*, which extracts from *Content Center* the necessary bearing.

After finalizing the resistance calculations for toothed wheels, the functions for running the rules *add_gears, case_width, Constrains, shaft_dimensioning, add_distance _ring_command* rules are called from section *Tab Group 1.* The pinion shafts, distance rings and bearings are shown in Fig. 15.

5 Conclusions

The use of powerful graphical modelling tools, such as *Autodesk Inventor*, correlated with programming languages as *iLogic*, included in Autodesk Inventor package, consolidates the designing capacities. At the same time, they allow the automation of the designing tasks, giving the possibility of integrating components, assemblies and subassemblies in very well structured projects, by managing very easy and accurate the tool. This is possible simultaneously with carrying-out all interdependencies between components, simulating the movements and interactions and carrying-out the resistance calculations.

This paper presents a technique for using *Autodesk Inventor* graphical package, sustained by *iLogic* rules in order to design a reduction gear for an anchor and mooring winch. The designing was completely parametric so that the final product can be automatically modelled in strong relation to the input data.

This may be extended for automatically calculating other reduction gears through monitoring and controlling the *Inventor* objectives, such as parameters, characteristics and components, by modifying the input data. At the same time, the parametric design can be used for any other engineering application.

Acknowledgements This work was supported by a grant of the Romanian National Authority for Scientific Research and Innovation, CNCS—UEFISCDI, project number PN-II-RU-TE-2014-4-0031.

References

1. Rădulescu (1986) Gh., Îndrumar de proiectareînconstrucția de mașini, Ed. Tehnică, București
2. Matejic M, Blagojevic M, Djordevic Z, Marjanovic N, Petrovic N (2013) Comparative analysis of different type reducers for winch drum driving unit. In: Proceedings of 7th international quality conference, pp 255–262
3. Kedrowski D, Slimak S (1993) Wobbling gear drivetrain for cordless screwdriver, Honors Theses,. http://scholarworks.wmich.edu/honors_theses/index.23.html#year_1993
4. González-Palacios MA, Angeles J (1999) The design of a novel mechanical transmission for speed reduction. J Mech Des 121(4):538–543
5. http://www.scritub.com/tehnica-mecanica/Reductore-cu-roti-dintate2015211321.php
6. Stăncescu C (2010) Modelareaparametricășiadaptivă cu Inventor—Vol I și II, Editura Fast, ISBN 978-973-86798-4-9, București, 2009
7. http://usa.autodesk.com/autodesk-inventor/
8. http://adndevblog.typepad.com/files/cp2544_ilogic_inventor_api.pdf
9. https://www.rotor.nl/Home.aspx
10. ISO 6336-2-1996—Calculation of load capacity of spur and helical gears. Calculation of surface durability (pitting)
11. ISO 6336-1-2006—Calculation of load capacity of spur and helical gears. Basic principles, introduction and general influence factors
12. ISO 1328-1-1997—Definitions and allowable values of deviations relevant to corresponding flanks of gear teeth

Experimental Determination of Brittle Fracturing Appearance During Static Indentation of Materials Based on Stone

Miloš Pjević, Ljubodrag Tanović and Filip Vučetić

Abstract During micro cutting process, the presence of both ductile and brittle mode is observed. The appearance of one or the other, or the value of critical penetration depth, when transition is happening, depends on the material itself, cutting condition, tool geometry and its motion as well. Widely used method for determining critical penetration depth is static indentation with diamond indenter. This paper presents experimental determination of critical penetration depth for material based on stone. Static indentation is used with Vickers indenter. It is shown that, with increase of indentation load, first phase lacks the appearance of micro cracks. After exceeding the threshold force (penetration depth), micro cracks appear, which is then followed by severe destruction of the material. Due to its material properties, penetration depth is not constant. It varies from grain to grain.

Keywords Micro cutting · Indentation · Brittle materials · Ductile mode

1 Introduction

Advancements in technology are facing the production industry with constant increase of variety of materials used for manufacturing. Technological development is directed towards miniaturization of products, with saving materials and improving energy efficiency as base motives.

M. Pjević (✉) · L. Tanović
Department of Production Engineering, Faculty of Mechanical Engineering, University of Belgrade, Belgrade, Serbia
e-mail: mpjevic@mas.bg.ac.rs

L. Tanović
e-mail: ltanovic@mas.bg.ac.rs

F. Vučetić
Innovation Centre of the Faculty of Mechanical Engineering, University of Belgrade, Belgrade, Serbia
e-mail: fvucetic@mas.bg.ac.rs

V. Majstorovic and Z. Jakovljevic (eds.), *Proceedings of 5th International Conference on Advanced Manufacturing Engineering and Technologies*, Lecture Notes in Mechanical Engineering, DOI 10.1007/978-3-319-56430-2_12

Micro-cutting has taken a significant and stable place among widely spread technologies nowadays. Micro machining of brittle materials is present in optical, electrical and many other industries. However, although brittle materials have extraordinary mechanical properties, their machinability can be quite an issue. Understanding of phenomena occurring during the process of micro machining is essential for further application of this technology.

Two different regimes are present during the micro-cutting of brittle materials, i.e. ductile regime and brittle regime [1–5]. Proper functionality of micro parts is often conditioned by a fine machined surface without any cracks present. For instance, camera lens with cracks in its surface will have image sharpness problems. Having this in mind, it is very important to determine the boundary between these two regimes.

Idea of this paper is to present the experimental determination of the parameters at which the transition from ductile to brittle regime of stone-based materials occurs. For this purpose the method of static indentation is used.

2 Static Indentation

Static and dynamic indentation presents one of the methods used for explaining the phenomena that occur during micro-cutting [6, 7]. Literature overview [8] shows that various indenters are used for this purpose.

Scheme of indentation process is shown on Fig. 1. During indentation, in the case when normal load (F_n) or depth of the indent (a) is lower than its critical value (a_c) for tested material, process will be followed by elastic and plastic deformation of material (Fig. 2a).

With the increase of F_n, over the critical value, uncontrolled growth of radial and median cracks inside material occur (Fig. 2b). These cracks will not directly lead to the detachment of material pieces, but the fracture strength in zones with cracks will be significantly decreased. New fracture strength can be calculated as:

$$\sigma_f = \frac{1}{\gamma(\xi)^{1/3}(\cos\psi)^{2/9}}\left(\frac{H}{E}\right)^{1/6}\left(\frac{K_c^4}{F_n}\right)^{1/3} \tag{1}$$

Fig. 1 Indentation process

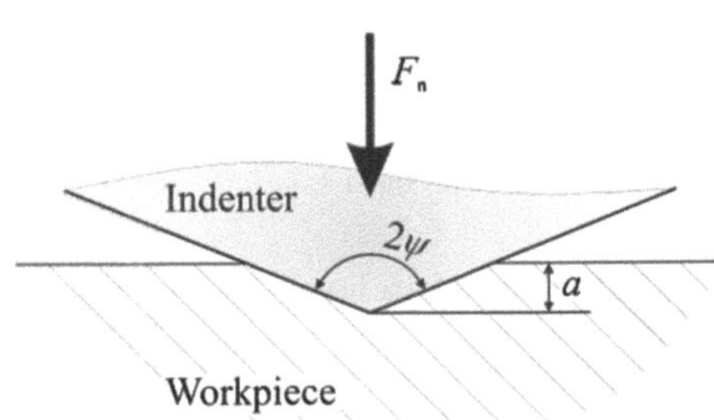

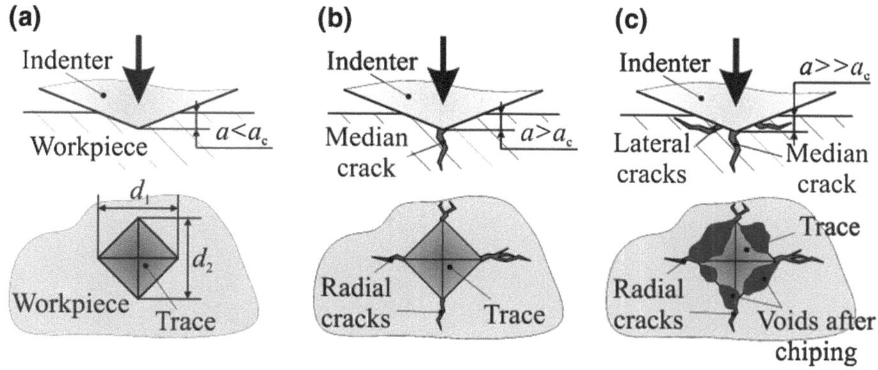

Fig. 2 Indentation phases

where H represents hardness, E elasticity modulus, Kc fracture toughness, F_n indentation normal load, γ a constant which depends on crack geometry and orientation, 2ψ the included angle between opposite indenter edges, and ξ a constant.

Further increase of normal load introduces the lateral cracks into the material with tendency of further growth during unloading due to residual stresses (Fig. 2c). Their appearance could be used to describe the removal of the material that is proportional to their size.

3 Experimental Setup

Experiments are conducted on Micro Vickers Hardness Tester TIME-TH710, shown in Fig. 3. The plan of experiment was based on possibility to vary normal load (F_n = 0.098÷9.807 N) and indentation dwell time (T = 0÷15 s). Tests were conducted for three materials: marble “Plavi tok” and two granites, “Bukovik” and “Jošanica”.

Multiple measurements were done for every F_n–T pair in order to achieve a higher confidence level in test results. Also, measurements were conducted on all of the present phases (different grains) in each material with the aim of obtaining a proper overview of possible/emerging phenomena. In addition to calculating the hardness value, dimensions of indent diagonals (d_1 and d_2), along with the angle between opposite indenter edges (2ψ = 136°), can be used to determine the depth of indentation:

$$a = \frac{d_1 + d_2}{4} \operatorname{ctg} \psi \tag{2}$$

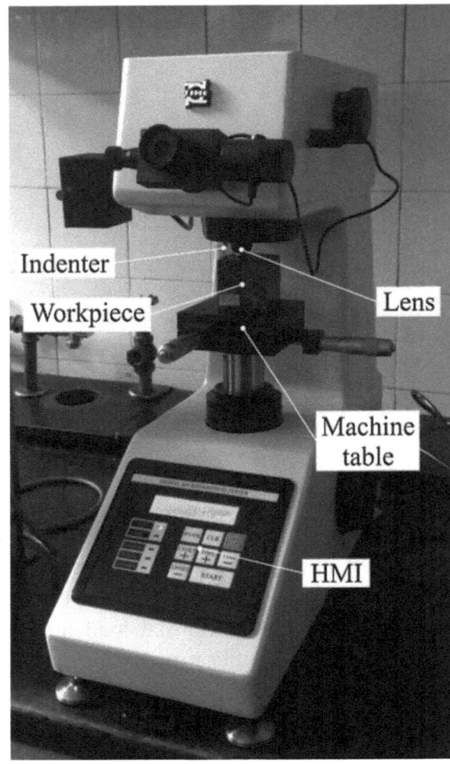

Fig. 3 Experimental setup

4 Results and Discussion

Some results of experiments of all three materials (marble "Plavi tok" and two granites, "Bukovik" and "Jošanica") are given in Table 1.

Indentation of granite "Bukovik" is shown in Fig. 4. Large difference between measured hardness of "Bukovik" (Table 1) can be attributed to the high level of its heterogeneity. Variation of material hardness leads to variation of penetration depth. Indent depths are in the range from 2 to 29 μm, depending on the normal load intensity and material's hardness. Phase 3 (Fig. 2c) can be observed in Fig. 4a–c, whilst the indentation trace in Fig. 4d represents the clear case of plastic deformation. Differences between indent depths from Fig. 4b, d, on one side, and from Fig. 4c, on the other, with respect to differences in normal load intensity, can be also explained by heterogeneity of the material (indents were made on different grains).

Static indentation has shown that granite "Bukovik" can be machined in ductile regime if the depth of cutting does not exceed 2 μm. Since the test apparatus does not have the option to vary the normal load between 0.49 and 0.98 N, this critical depth of 2 μm can be considered to be a conservative value and that the exact value may be somewhat higher.

Table 1 Experimental conditions and results

	Fig. no.	F [N]	T [s]	Hardness [HV]	Diagonals	a [μm]	
					d_1 [μm]	d_2 μm]	
Bukovik	4a	9.807	0	HV1 = 89.9	144.5	142.75	29.0
	4b	4.904	0	HV0.5 = 919.8	31.13	32.38	6.4
	4c	0.98	0	HV0.1 = 102.7	43.82	41.19	8.6
	4d	0.49	10	HV0.05 = 927.7	10.38	9.69	2.0
Jošanica	5a	9.807	10	HV1 = 695.8	54.44	48.82	10.4
	5b	1.96	0	HV0.2 = 1041.0	19.25	18.57	3.8
	5c	1.96	0	HV0.2 = 811.8	21.94	20.82	4.3
	5d	0.98	0	HV0.1 = 897.4	15.07	13.69	2.9
Plavi tok	6a	9.807	0	HV1 = 107.1	129.75	133.50	26.6
	6b	0.98	0	HV0.01 = 122.4	30.69	29.00	6.0
	6c	0.245	0	HV0.01 = 127.5	19.38	19.50	3.9
	6d	0.098	15	HV0.025 = 122.8	12.44	11.69	2.4

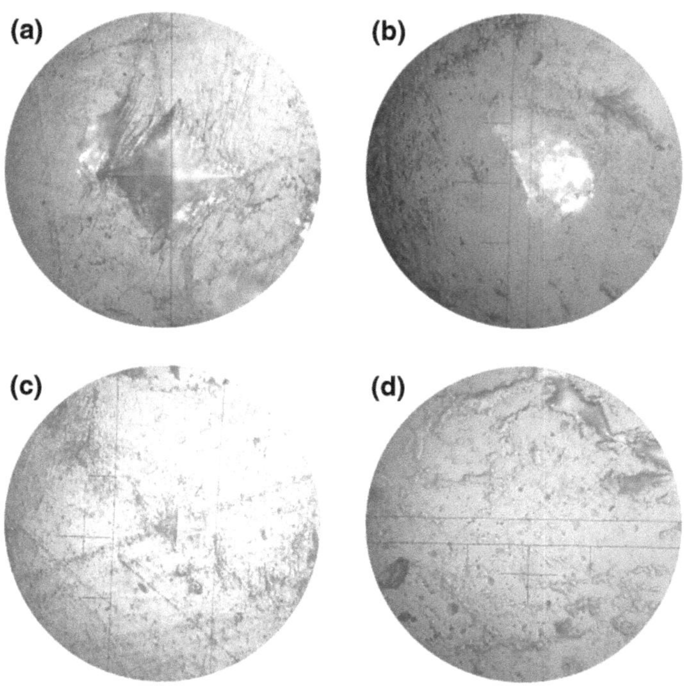

Fig. 4 Static indentation on Granite "Bukovik" with Vickers indenter

Granite "Jošanica" has higher mean value of hardness compared to "Bukovik", and its variation of hardness is much lower.

The depth of indents is in the range from 2.9 to 10.4 μm (Fig. 5). In case when normal load intensity was 9.807 N (Fig. 5a), phase 3 (Fig. 2c) was present. Decreasing the indentation load to 1.96 N leads to transition from phase 3 to phase 2 (Fig. 2b, c). There were no cracks in the material when the normal load of 0.98 N was applied (Fig. 5d).

From around 3.8 to 4.3 μm of indent depth, only radial/median cracks were present in the material, but, due to the lateral motion of the tool during the micro-cutting, lateral cracks are to be expected. In order to obtain micro machining of "Jošanica" in the ductile regime, cutting depth should be lower than 2.9 μm.

Marble "Plavi tok" has a different structure than previously mentioned materials. Besides structure, its mechanical properties also differ significantly. The lowest deviation of measured hardness is present in this material. It can be seen from Fig. 6a that all possible cracks are present (3rd phase in Fig. 2c). Intensity of normal load has maximum of 9.807 N, which corresponds to the indent depth of 26.6 μm. Lowering load intensity gives the indent depth of 6 μm (Fig. 6b) with radial/median cracks present, but in this case the lateral cracks are in their initial stage.

With indent depth of around 4 μm lateral cracks are no longer present (Fig. 6c). Further lowering of the indent depth to the 2.6 μm, makes the full transition to the ductile regime, without the presence of cracks in the material (Fig. 6d).

Figure 7 gives a comparison of the expected machining regimes for all of the tested materials with respect to the cutting depth.

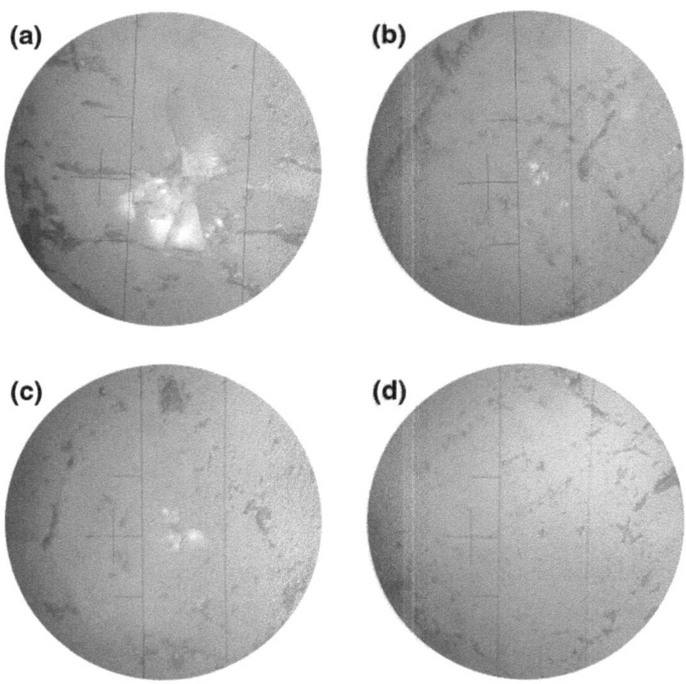

Fig. 5 Static indentation on Granite "Jošanica" with Vickers indenter

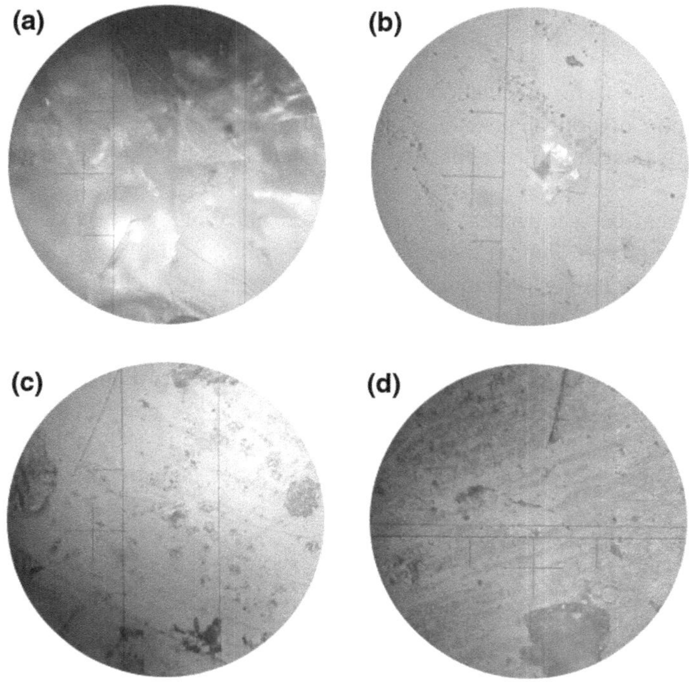

Fig. 6 Static indentation on Marble "Plavi Tok" with Vickers indenter

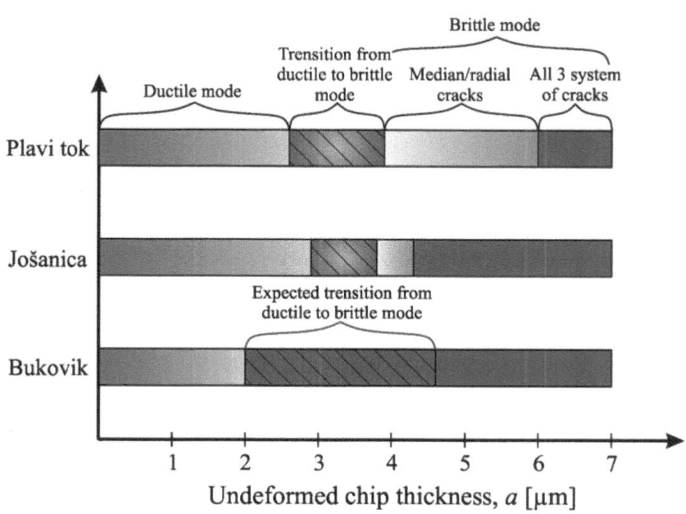

Fig. 7 Cutting regimes in the function of cutting depth for granite "Bukovik" and "Jošanica" and marble "Plavi tok"

5 Conclusion

Once again, the static indentation proved itself to be a useful and fast method for determination of critical penetration depth for micro cutting of brittle materials. Main disadvantage of used apparatus is its inability of finer normal load intensity adjustment.

Experiment has shown that all of the tested materials can be machined in the ductile regime, even though they are brittle by nature.

The best machining potential in ductile regime, if we include transition phase between ductile and brittle regime, had the marble "Plavi tok".

Granites "Jošanica" and "Bukovik" exhibit similar behavior, but this statement should be taken with caution, primarily due to the lack of hard phases in the surface of "Bukovik". Values of critical penetration depth can't be compared precisely, but it's expected that both granites exhibit similar behavior. This should be confirmed with future tests on a static indenter that has better adjustment of normal load intensity.

References

1. Nakasuji T et al (1990) Diamond turning of brittle materials for optical components. CIRP Ann Manuf Technol 39(1):89–92
2. Fang FZ, Liu XD, Lee LC (2003) Micro-machining of optical glasses—a review of diamond-cutting glasses. Sadhana 28(5):945–955
3. Liu K, Li XP, Liang SY (2007) The mechanism of ductile chip formation in cutting of brittle materials. Int J Adv Manuf Technol 33(9-10):875–884
4. Zhou M et al (2002) Brittle–ductile transition in the diamond cutting of glasses with the aid of ultrasonic vibration. J Mater Process Technol 121(2):243–251
5. Yan J et al (2001) On the ductile machining of silicon for micro electro-mechanical systems (MEMS), opto-electronic and optical applications. Mater Sci Eng: A 297(1):230–234
6. Anton RJ, Subhash G (2000) Dynamic vickers indentation of brittle materials. Wear 239 (1):27–35
7. Ghosh D, Subhash G, Sudarshan TS, Radhakrishnan R, Gao XL (2007) Dynamic indentation response of fine grained boron carbide. J Am Ceram Soc 90(6):1850–1857
8. Malkin S, Hwang TW (1996) Grinding mechanisms for ceramics. CIRP Ann Manuf Technol 45(2):569–580

Dynamic Optimization of the Cam-Lever Mechanism for Thermoforming Machine Tool Driving

Maja Čavić, Marko Penčić, Milan Rackov, Ivan Knežević and M. Zlokolica

Abstract Considering that the most important use of thermoforming is in the production of plastic packaging for the food and pharmaceutical industry, it is essential that formed products remain sterile through the entire thermoforming process. Most machines used for thermoforming have a tool holder with one degree of freedom—DOF, which allows only vertical motion of the tool. After the thermoforming process, the formed products are ejected from the tool with compressed air, which may cause contamination and/or deformation of the products. We propose a working mechanism for driving the tool, which, compared to conventional machines, guarantees both a shorter working cycle and sterility of the formed products during the entire process. Products are punched out after forming and accepted and transported with an adequate mechanism to the manipulation module, where they are sorted and packed. This paper presents a dynamic optimization of the thermoforming machine working mechanism with 2 DOFs which consists of two cam-lever mechanisms that enable translation, rotation and complex motion of the tool. Based on the set of technical requirements, kinematic synthesis of the cam-lever mechanism is performed. SVAJ diagrams for the cams and the dimensions of the lever mechanism links are defined. Based on the kinetostatic analysis, a dynamic model of the cam-lever mechanism is formed and the driving torque for both lifting and rotation of the tool is determined. The optimization problem is formed and the objective function is defined as the minimization of the required

M. Čavić · M. Penčić (✉) · M. Rackov · I. Knežević · M. Zlokolica
Faculty of Technical Sciences, University of Novi Sad, Novi Sad, Serbia
e-mail: mpencic@uns.ac.rs

M. Čavić
e-mail: scomaja@uns.ac.rs

M. Rackov
e-mail: racmil@uns.ac.rs

I. Knežević
e-mail: ivanknezevic@uns.ac.rs

M. Zlokolica
e-mail: mzlokolica@uns.ac.rs

V. Majstorovic and Z. Jakovljevic (eds.), *Proceedings of 5th International Conference on Advanced Manufacturing Engineering and Technologies*, Lecture Notes in Mechanical Engineering, DOI 10.1007/978-3-319-56430-2_13

driving torque. Based on the set constrains, a dynamic optimization is performed using the method of genetic algorithm. By comparing the results before and after optimization, it is concluded that the driving torque is lower by 50.3%.

Keywords Thermoforming machine · Cam-lever mechanism · Dynamic optimization · Genetic algorithm

1 Introduction

Thermoforming is one of the oldest and most common methods for processing plastic materials [1]. It is an industrial process in which thermoplastic sheets or foils are shaped into a new form by using heat and pressure or vacuum [2]. Thermoplastic materials are easily formed by heating, so thermoforming is very widely used [3]. The most important use is in the food and pharmaceutical industry for packaging of food and medicine—blister packs, food containers [4], in electrical industry for enclosures, anti-static trays [5], in medical industry for radiotherapy and pressure masks [6], prosthetic parts [7], thermoplastic aligners [8], in automotive industry for wheel covers, door interiors [9], in aircraft industry for interior trim panels, cowlings [10], in buildings for door panels, bathtubs [11], in furniture industry for cutlery trays, kitchen panels [12], in nautical for boat hulls, dashboards [13], in graphic design and arts for coloring 3D printed surfaces [14, 15] etc.

This paper shows the dynamic optimization of the thermoforming machine working mechanism, with the aim of reducing the driving torque.

2 Thermoforming Process

Thermoforming consists of three basic operations: foil manipulation, product forming and product manipulation [16]. Each operation is performed on a separate module of the thermoforming machine. The product forming is carried out within the working module with a special tool that consists of two parts—Fig. 1. The upper part of the tool is stationary, while the lower part moves in accordance with the work process. When the movable part of the tool is in the lowest position, foil is retracted into the tool. Lower part of the tool is lifted and used to press the foil against the upper—stationary part of the tool. This is followed by a standstill period during which the product forming is carried out. Afterwards, the products are separated from the foil with a tool that performs an operation of separation by punching. The foil leftovers are removed from the machine while the products remain in the tool. At the end of the process, the lower part of the tool moves downwards and the products are taken out from it with the appropriate mechanism and transported to the module for product manipulation, where they are sorted and packed.

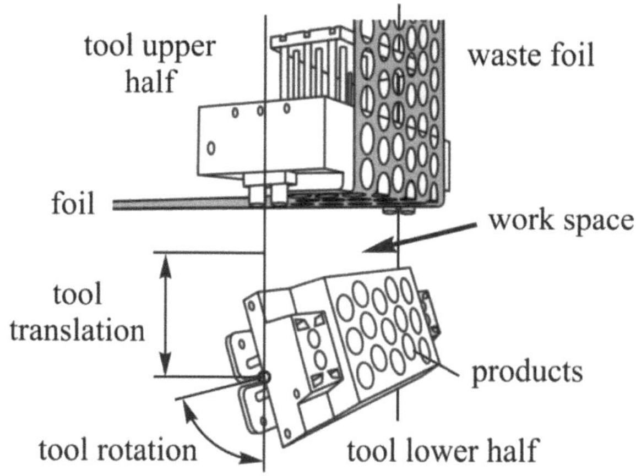

Fig. 1 Thermoforming tool

Conventional thermoforming machines [17] have a tool holder with 1 DOF which allows only vertical motion of the tool. Formed products are ejected from the tool with compressed air—blown out, which may cause contamination of the products. Therefore, a mechanism for product manipulation is necessary. Such a mechanism should enable the removal of formed products from the lower part of the tool through the working area—the space between the upper and the lower part of the tool [18], as well as the transport of products to the manipulation module, while taking care to keep the products sterile throughout the entire thermoforming process.

The space within the working area is relatively small, so access to the products is limited. This problem is solved by an additional degree of freedom, by rotating the tool [19, 20]. When the movable part of the tool is in the lowest position, it rotates by a certain angle—usually 80°, and then stands still. During this standstill, products are removed from the tool by an appropriate mechanism and transported further to the manipulation module.

Figure 2 shows a working cycle of the tool of a thermoforming machine. At the beginning of the cycle, the tool rotates by a certain angle (1). Then it performs a combination of rotation and translation until it reaches a vertical position (2). From this point, the tool continues to perform only translation (3), until it reaches a certain point and remains still (4). The products are formed during this interval. After the thermoforming, the tool performs a short translational motion upwards in order to separate the products from the foil by punching (5). The tool then translates downwards (5), and then starts to rotate as well (6). In the lower end position, translational motion stops and the rotation continues (7) until the tool reaches the position for product removal. Then a standstill period follows (8), during which the products are removed from the tool and transported to the manipulation module, where they are then sorted and packed.

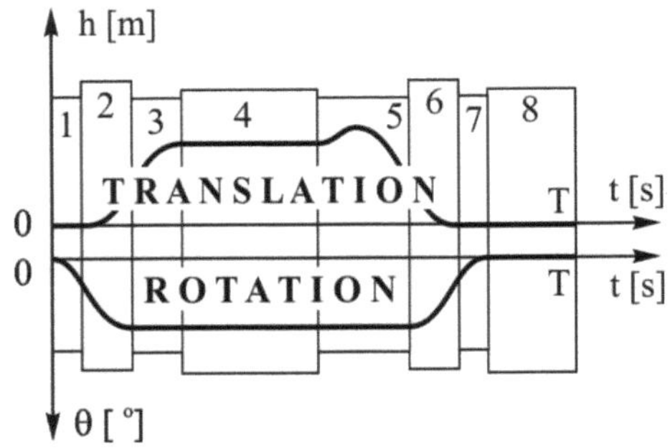

Fig. 2 Working cycle of the tool: *1*—rotation of the tool, *2*—complex motion of the tool, *3*—translation of the tool, *4*—tool standstill and thermoforming process, *5*—translation of the tool and cutting process, *6*—complex motion of the tool, *7*—rotation of the tool and *8*—tool standstill and extraction of the products

3 Working Mechanism

The motion of the tool is carried out by the working mechanism which consists of two cam-lever mechanisms—Fig. 3. One is for the vertical motion of the tool holder and the second is to change the angular position of the tool. The working mechanism has 2 DOFs and allows translation, rotation and complex motion of the tool. Different ways for driving the tool and controlling the cycle are analyzed and based on that, the mechanism structure for the vertical motion of the tool is adopted [21]. Synthesis of the lifting mechanism [22] and the mechanism for the tool rotation was performed [23, 24].

The kinematic scheme of the thermoforming machine working mechanism is shown in Fig. 4. The cam mechanisms are used to control the working cycle, i.e. to transform the continuous rotation of the actuator into the appropriate, exactly defined, periodic rotary—oscillatory unequal motions or periodic standstill of input links of the lever mechanism 3_R and 3_L. The kinematic parameters of links 3_R and 3_L are φ_{3R}, ω_{3R}, ε_{3R}, φ_{3L}, ω_{3L}, ε_{3R}. The SVAJ diagrams for the cam mechanisms are defined—for the rotation and lifting of the tool, which represent the input

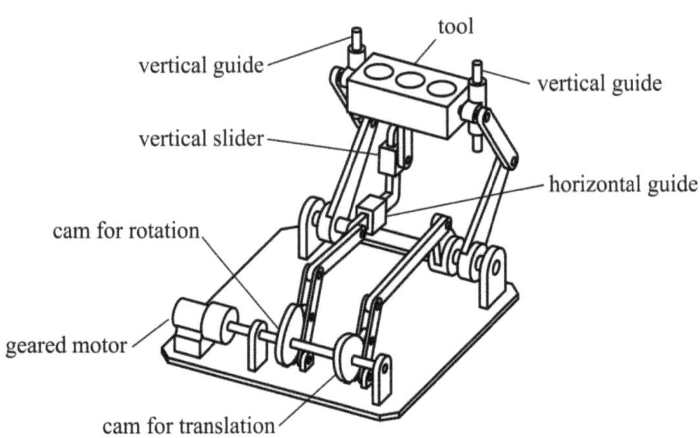

Fig. 3 Thermoforming machine working mechanism

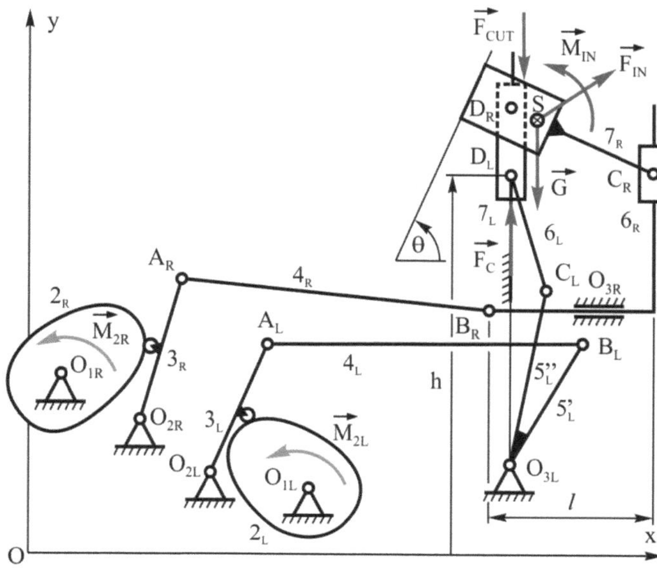

Fig. 4 Kinematic scheme of working mechanism

parameters for the lever mechanisms. The cams are set on the same shaft for motion synchronization—one actuator drives both mechanisms. Therefore, the input kinematic parameters of the cam mechanisms are equal, so $\varphi_2 = \varphi_{2R} = \varphi_{2L}$—angle of rotation, $\omega_2 = \omega_{2R} = \omega_{2L}$—angular speed and $\varepsilon_2 = \varepsilon_{R2} = \varepsilon_{L2}$—angular velocity.

The lever mechanism for the lifting/lowering transforms the rotary oscillatory motion which the input link 3_L receives from link 2_L of the cam mechanism into the output rectilinear motion of link 7_L—the tool holder. The output parameter of the lever mechanism is the vertical position of the tool holder—the parameter *h*. The lever mechanism for lifting/lowering consists of a four-bar linkage—links 3_L, 4_L and $5_L'$ and a slider-crank mechanism—links $5_L''$, 6_L and 7_L. The four-bar linkage distances the drive from the working parts of the machine and allows for better power and motion transmission, while the slider-crank mechanism transforms the rotary motion of link 5_L into the rectilinear motion of the tool 7_L.

The lever mechanism for the rotation of the tool consist of the eccentric slider-crank mechanism—links 3_R, 4_R and 5_R, and an additional mechanism—links 6_R and 7_R, where link 7_R represents the tool. The eccentric slider-crank mechanism transforms the input rotary oscillatory motion of link 3_R into the translational rectilinear motion of link 5_R. Input link 3_R receives the motion from output link 2_R of the cam mechanism. One of the requirements is that the mechanism for rotation enables only rotational motion of the tool—the tool is rotated only when the mechanism for rotation is active and that the lifting mechanism enables only translational motion of the tool—the tool is lifted/lowered only when the mechanism for lifting is active. This requirement was realized by using an additional mechanism—links 6_R and 7_R. Link 5_R has the possibility of horizontal translational motion, while link 6_R can slide vertically on link 5_R. If link 5_R is stationary, then the rotating mechanism is passive—link 6_R does not move horizontally relative to link 7_L and therefore only the vertical motion of links 6_R, 7_R and 7_L is possible—the tool moves vertically. If the mechanism for rotation is active, then link 5_R moves horizontally and therefore so does link 6_R, which results in the rotation of link 7_R—the tool rotates. The rotation of the tool is not dependent on the vertical motion of the tool holder, meaning the tool can rotate even when the tool holder is motionless. When the lifting mechanism is passive, there is no vertical motion and therefore only the rotation of the tool is realized. If links 5_R and 7_L are moving, then both mechanisms are active and the tool performs complex motion. The output parameter of the rotating mechanism of the tool is angle θ.

4 Dynamic Optimization

Based on the analysis of thermoforming machine solutions [19, 20], the following technical parameters are adopted: productivity of 30 cycles per minute, vertical stroke of the tool $h_L = 150 \div 170$ mm, punching stroke $h_C = 1.5 \div 3.5$ mm, rotation angle of the tool $\theta_R = 80°$, maximum rotation angle of the output links of the cam mechanisms $\varphi_{3Lstroke}$ and $\varphi_{3Rstroke}$ is 45°, maximum punching force $F_{CUT} = 200$ kN, mass of the movable tool part m = 550 kg and dimensions of the space in the vertical plane 1 × 1.6 m.

Based on the set requirements the kinematic analysis is performed, the working cycles of tool motion are defined—Fig. 5 as well as SVAJ diagrams for the cam

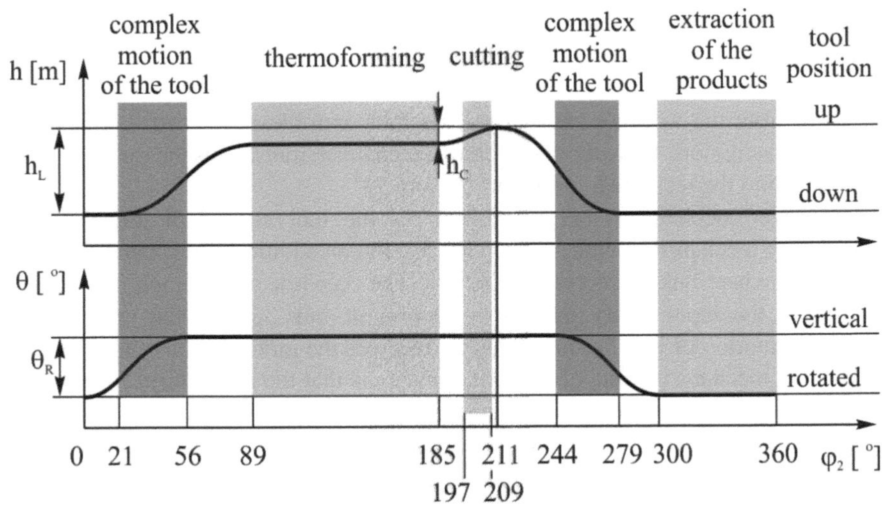

Fig. 5 Working cycle of the tool

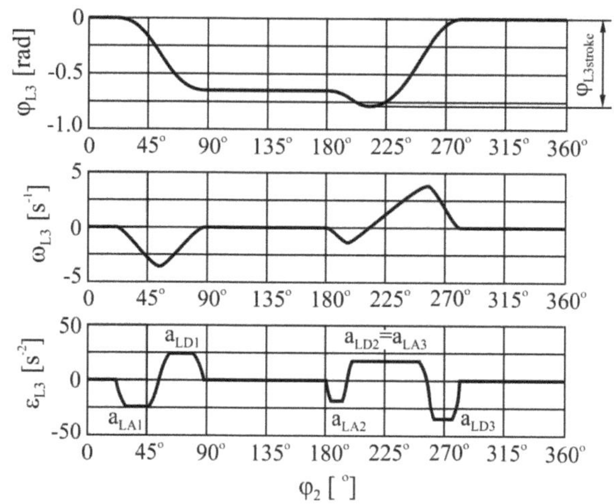

Fig. 6 SVAJ diagrams for tool lifting

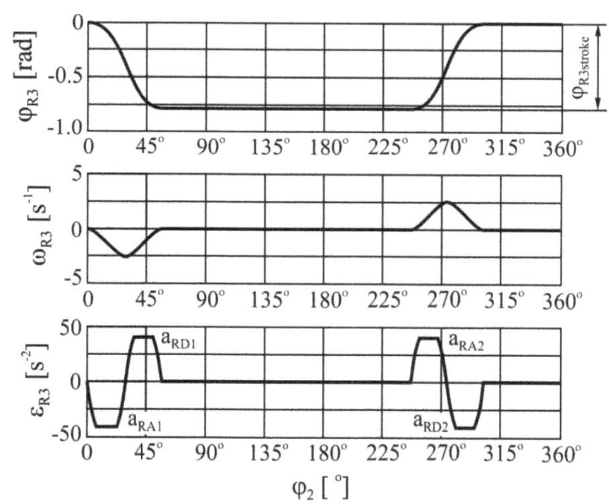

Fig. 7 SVAJ diagrams for tool rotation

mechanisms—Figs. 6 and 7. A modified trapezoidal law is used because it provides a minimum value of the maximum acceleration during the working cycle. Skew factors for lifting p_{L1} and p_{L2}, and rotation p_{R1} and p_{R2} all equal 1, and follower strokes for lifting $\varphi_{3Lstroke}$ and rotating $\varphi_{3Rstroke}$ both equal 45°. These factors significantly influence the change in the intensity of acceleration a_{LA1}, a_{LD1}, a_{LA2}, a_{LD2}, a_{LA3}, a_{LD3}, a_{RA1}, a_{RD1}, a_{RA2}, a_{RD2}, which is shown in [25, 26].

Based on [22–24] the preliminary design parameters of the mechanism for lifting and rotation of the tool are determined—Tables 1 and 2.

The total stroke of the tool is 163.64 mm, punching stroke is 1.56 mm and the total rotation angle of the lower part of the tool is 80°. The procedure by which the driving torque M_M and the reactions in the joints are defined, is presented in [27]. The diagram of the driving torque for one cycle is shown in Fig. 8.

Table 1 Values of design parameters for non-optimized mechanism

$O_{2L}A_L$ [mm]	A_LB_L [mm]	$O_{3L}B_L$ [mm]	$O_{3L}C_L$ [mm]	C_LD_L [mm]	α [°]	x_{O3L} [mm]	y_{O3L} [mm]	$\varphi_{2Lstart}$ [°]
212	874	235	350	350	5	1000	−100	90

Table 2 Values of design parameters for non-optimized mechanism

$O_{2R}A_R$ [mm]	A_RB_R [mm]	C_RD_R [mm]	y_{O3R} [mm]	l [mm]	$\varphi_{2Rstart}$ [°]
310	656	185	216	298	102.5

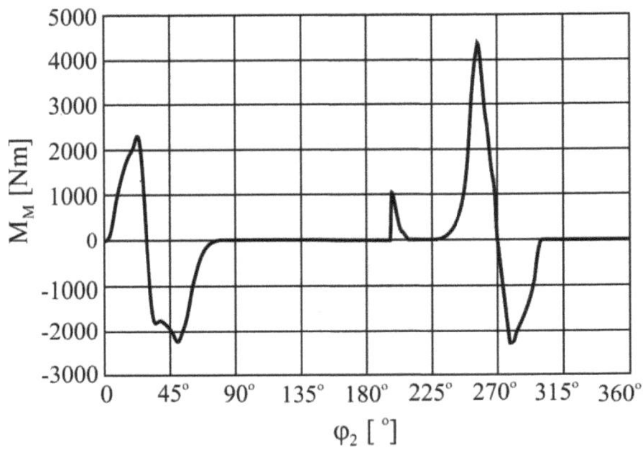

Fig. 8 Driving torque for non-optimized mechanism

The main parameter which determines the dynamic characteristics of the lifting mechanism is the total vertical force acting on the tool. It consists of punching force F_{CUT} that acts on the tool during punching, inertial load of the tool F_{IN}, weight of the tool G and force from the cylinder F_C that supports the slider with the tool and acts all the time. The punching force acts downwards and is 200 kN—technical requirement. Although punching lasts from 185° to 211° of the cycle, technological tests show that material resistance during punching does not exist at the beginning of the punching. That is why the punching force is taken to act from 196° to 211° of the cycle. The inertial load of the tool depends on the mass of the movable part of the tool which is 550 kg—technical requirement, as well as on the acceleration of the tool. The cylinder that supports the slider with the tool is adopted in order to eliminate the influence of the tool's mass and to reduce the middle value of total vertical force acting on the tool—the force generated by the cylinder is 7500 N.

The main load of the rotating mechanism is the inertial torque M_{IN} which occurs as a result of tool rotation and depends on the angular acceleration and inertia of the tool, which equals 17 kgm^2. The tool's center of gravity is located on the axis of tool rotation—points S and D_R are overlapping.

The driving unit consists of a motor and gear reducer—geared motor, by which power is transferred to the cams of both mechanisms. Therefore, when choosing the motor power and the nominal torque of the gear reducer, one must take into consideration the combination of torques for lifting and rotation. It should be noted that positive torque indicates that power is taken from the motor, respectively to accelerate mass or to overcome the load. Conversely, a negative torque indicates that the motor performs mass deceleration. During positive torque, three maximums occur: 2238 Nm at 22°, 1170 Nm at 198° and 4340 Nm at 256° of the cycle. During negative torque, two maximums occur: 1817 Nm at 35° and 2275 Nm at 279° of the cycle. Generally observed, the maximum torque is 4340 Nm which is the basic parameter for the selection of the drive unit.

4.1 Optimization Problem

The optimization problem is the minimization of the objective function—*MIN f(x),* $x \in D$ for the set constraints, where $x = (x_1, x_2, ..., x_m)$—vector of variables, $D = \{x \in R^n | \; g(x) \leq 0 \wedge h(x) = 0\}$—a set of solutions that fulfils the defined constraints and $g(x) \leq 0$ and $h(x) = 0$—vectors of constraints.

The optimization of a complex cam-lever mechanism is rarely found in literature. In the papers that do exist, there are two approaches to this problem. In the first one, only the dimensions of the lever mechanism are optimized for the prescribed parameters and motion. In the second [28–30], the synthesis of the cam mechanism is performed according to the given motion of the working link of the mechanism. The second approach could lead to a non-standard profile of the cam which can cause problems during manufacturing and during the exploitation of the cam mechanism.

In this paper, optimization of the complete mechanism is performed—the standard cam mechanism with recommended law of motion is adopted, while the parameters of the law of motion—skew factor p and follower stroke $\varphi_{3stroke}$, are subjected to the optimization along with the dimensions of the mechanism.

The first group of optimization variables are the parameters of the lever mechanism, the lengths of the levers for lifting/lowering $O_{2L}A_L$, A_LB_L, $O_{3L}B_L$, $O_{3L}C_L$, C_LD_L, the angle α between the levers $O_{3L}B_L$ and $O_{3L}C_L$, support O_{3L} position, lengths of the levers for rotation $O_{2R}A_R$, A_RB_R, C_RD_R, vertical position of the support O_{3R}, the horizontal dimension l of the link 5_R, the initial values of the angle position of links 3_L and 3_R. The second group of variables are the parameters of the cam mechanism, and those are skew factors p_{L1} and p_{L2} for lifting, p_{R1} and p_{R2} for rotating, and strokes $\varphi_{3Lstroke}$ for lifting and $\varphi_{3Rstroke}$ for rotating. Since the output strokes $\varphi_{3stroke}$ are standardized by the manufacturer at 30°, 45° or 60°, in this procedure they are not optimized, but are adopted: $\varphi_{3Lstroke}$ and $\varphi_{3Rstroke}$ will have the same angle of 45°. The lower and upper bounds of the design parameters are presented in Tables 3 and 4.

The aim of the optimization is the minimization of the driving torque. Considering that the driving torque is variable in time, its maximum value during the cycle will be minimized. The objective function is defined as:

Table 3 Lower and upper bounds of design parameters

	$O_{2L}A_L$ [mm]	A_LB_L [mm]	$O_{3L}B_L$ [mm]	$O_{3L}C_L$ [mm]	C_LD_L [mm]	α [°]	x_{O3L} [mm]	y_{O3L} [mm]	$\varphi_{2Lstart}$ [°]
Lower	100	700	100	250	250	0	900	−100	70
Upper	300	900	300	450	450	5	1100	100	110

Table 4 Lower and upper bounds of design parameters

	$O_{2R}A_R$ [mm]	A_RB_R [mm]	C_RD_R [mm]	y_{O3R} [mm]	l [mm]	$\varphi_{2Rstart}$ [°]	p_{L1}	p_{L2}	p_{R1}	p_{R2}
Lower	200	500	100	100	200	100	0.7	0.7	0.7	0.7
Upper	400	700	300	300	300	140	1	1	1	1

$$f = abs(M_{M\max}) \tag{1}$$

In accordance with the lower and upper bounds for the variables, constrains are defined that ensure the possibility of assembly and the functionality and efficiency of the mechanism. The possibility of assembly is a constraint that makes sure the mechanism can be assembled in all possible positions:

$$abs\left(\frac{\overline{A_LC_L}^2 + \overline{A_LB_L}^2 - \overline{B_LC_L}^2}{2\overline{A_LC_L}\,\overline{A_LB_L}}\right) \leq 1 \tag{2}$$

where:

$$\overline{D_LE_L}^2 \geq (x_{CL} - x_{DL})^2 \tag{3}$$

$$\overline{A_RB_R}^2 \geq (y_{O3R} - y_{AR})^2 \tag{4}$$

$$\overline{D_RE_R}^2 \geq (x_{CR} - x_{ER})^2 \tag{5}$$

Functional constrains enable fulfillment of the defined technical requirements. Vertical stroke of the tool h_L, punching stroke h_C and rotation angle of the tool θ_R:

$$0.17 > h_L > 0.15 \text{ m} \tag{6}$$

$$0.003 > h_C > 0.0015 \text{ m} \tag{7}$$

$$\theta_R = 80° \tag{8}$$

Mechanism efficiency is provided through the values of the pressure angle. The lifting mechanism consists of a four-bar linkage and the slider-crank mechanism. The most critical position—the period of punching, when the largest vertical force is acting upon the mechanism, links 6_L and $5_L''$ of the slider-crank mechanism must be collinear with the direction of the acting load and therefore:

$$\varphi_{6L\,for\,\varphi_2=211^\circ} = 90^\circ \tag{9}$$

$$\varphi''_{5L\,for\,\varphi_2=211^\circ} = 90^\circ \tag{10}$$

During the period of punching, the angle between links 4_L and $5_L'$ of the four-bar linkage should be as close to an angle of 90° as possible, and therefore:

$$\varphi'_{5L\,for\,\varphi_2=211^\circ} - \varphi_{4\,for\,\varphi_2=211^\circ} \leq 90^\circ \tag{11}$$

During the cycle, the pressure angles must have values that are smaller than those prescribed for the lever mechanisms. The position angles of links 6_L, $5_L'$ and $5_L''$ of the lever mechanism must fulfill the following conditions:

$$abs(\varphi_{6L} - 90^\circ) < 45^\circ \tag{12}$$

$$abs(\varphi''_{5L} - 90^\circ) < 45^\circ \tag{13}$$

$$\varphi'_{5L} - \varphi_4 \geq 45^\circ \tag{14}$$

The mechanism for rotation consists of a slider-crank mechanism and an additional mechanism. For efficient functioning of the slider-crank mechanism during the cycle, the pressure angle must have a smaller value than the prescribed value for the lever mechanisms:

$$abs(\varphi_{4R}) < 45^\circ \tag{15}$$

4.2 Results

Optimized parameters of the cam-lever mechanisms of the thermoforming machine working mechanism, are shown in Tables 5 and 6.

Table 5 Values of design parameters for optimized mechanism

$O_{2L}A_L$ [mm]	A_LB_L [mm]	$O_{3L}B_L$ [mm]	$O_{3L}C_L$ [mm]	C_LD_L [mm]	α [°]	x_{O3L} [mm]	y_{O3L} [mm]	$\varphi_{2Lstart}$ [°]
221	839.5	237.5	298	298.5	0.885	917	93	110

Table 6 Values of design parameters for optimized mechanism

$O_{2R}A_R$ [mm]	A_RB_R [mm]	C_RD_R [mm]	y_{O3R} [mm]	1 [mm]	$\varphi_{2Rstart}$ [°]	p_{L1}	p_{L2}	p_{R1}	p_{R2}
200	682.5	112.5	299.5	212.5	101.5	1	0.7	0.88	0.88

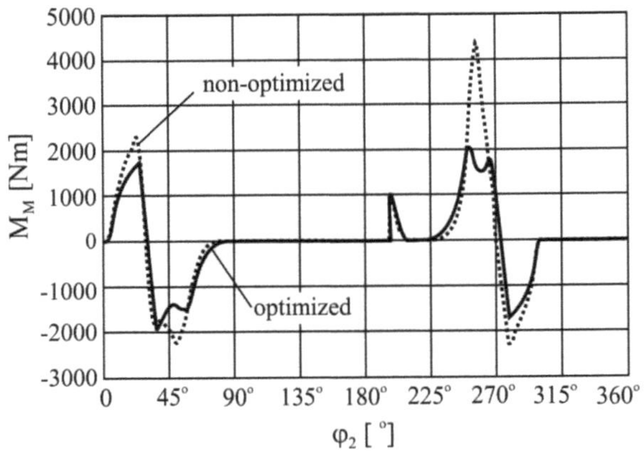

Fig. 9 Driving torque for optimized and non-optimized mechanism

The total stroke of the tool is 154.7 mm, punching stroke is 3.3 mm, and the total rotation angle of the tool is 80°. The diagram of driving torque M_M for lifting and rotation of the tool is shown in Fig. 9.

During positive torque, three maximums occur: 1795 Nm at 24°, 1064 Nm at 198° and 2155 Nm at 250° of the cycle. During negative torque, two maximums occur: 1970 Nm at 36° and 1789 Nm at 278° of the cycle. Generally observed, the maximum torque is 2155 Nm which is the basic parameter for the selection of the drive unit.

5 Conclusions

This paper presents a dynamic optimization of the thermoforming machine working mechanism with 2 DOFs which consists of two cam-lever mechanism that enable translation, rotation and a combination of these two motions—complex motion of the tool. Based on the analysis of thermoforming machines designs, the following technical parameters are adopted—productivity, tool stroke, punching stroke, the angle of rotation of the tool, punching force and mass of the tool. Based on the set requirements, kinematic synthesis of the cam mechanism is performed. SVAJ diagrams for cams are defined—for lifting and rotation of the tool. The type of motion law for the follower is defined—modified trapezoidal law. Within the kinematic synthesis the preliminary design parameters of the lever mechanism are determined. By applying the kinetostatic approach, the dynamic model of the cam-lever mechanism is formed and the driving torque for lifting and rotation of the tool is determined. The optimization problem is formed with the objective function defined as the minimization of the driving torque. The upper and lower bounds for optimization variables—parameters of the cam and lever mechanisms, are prescribed. Based on the set constrains—assembly, functionality and efficiency of the mechanism, dynamic optimization was performed using the genetic algorithm method. By comparing the results before and after optimization, it is concluded that the driving torque is 50.3% less. At the end, it should be noted that the analysis of the kinematic behavior of an optimized mechanism was also performed, but was omitted from the results because it overcomes the scope of this paper. Character changes of velocity and acceleration of the links of the lever mechanism are similar to the input links (3_L and 3_R), while the intensity is different due to the motion range of the particular link. By comparing the obtained values, it is concluded that the velocities and accelerations are within the boundaries recommended for industrial plants of this type.

References

1. Lynch JR (1960) Thermoforming—its processes and applications. Defense Technical Information Center
2. Crawford RJ (1998) Plastics engineering. Butterworth-Heinemann
3. Klein P (2009) Fundamentals of plastics thermoforming: synthesis lectures on materials engineering. Morgan and Claypool Publishers
4. Rosato D, Rosato D, Rosato M (2004) Plastic product material and process selection handbook. Elsevier
5. Herman FM (2014) Encyclopedia of polymer science and technology. Wiley
6. Modjarrad K, Ebnesajjad S (eds) (2013) Handbook of polymer applications in medicine and medical devices. Elsevier
7. Chimento J, Highsmith J, Crane N (2011) 3D printed tooling for thermoforming of medical devices. Rapid Prototyping J 17(5):387–392
8. Sander C, Wiethoff HC, Sander FM (2009) Orthodontic treatment with thermoplastic aligners. Informationen aus Orthodontie & Kieferorthopädie 41(3):183–188
9. Gruenwald G (1998) Thermoforming: a plastics processing guide. CRC Press, Taylor & Francis Group
10. Ashter SA (2014) Thermoforming of single and multilayer laminates: plastic films technologies, testing and applications. William Andrew, Waltham
11. Throne JL (2008) Understanding thermoforming. Carl Hanser Verlag
12. Lokensgard E (2008) Industrial plastics: theory and applications. Cengage Learning
13. Florian J (1996) Practical thermoforming: principles and applications. CRC Press
14. Zhang Y, Tong Y, Zhou K (2016) Coloring 3D printed surfaces by thermoforming. IEEE Trans Vis Comput Graph. doi:10.1109/TVCG.2016.2598570
15. Schüller C, Panozzo D, Grundhöfer A, Zimmer H, Sorkine E, Sorkine-Hornung, O (2016) Computational thermoforming. ACM Trans Graphics 35(4), article 43
16. Engelmann S (2012) Advanced thermoforming: methods, machines and materials, applications and automation. Wiley, New York
17. Machine TQC-650D (HONETOP Machinery). http://www.honetop.com/products/plastic-cup-thermoforming-machine-tqc-650d-ID135.html. Accessed Jan 2007
18. Packaging Thermoforming Solutions (AMUT). http://www.amutgroup.com/downloads/pages/BROCHURENEWCOMIitalow.pdf. Accessed Jan 2017
19. Automatic Pressure Forming Machines (ILLIG Maschinenbau). http://www.illig.de/uploads/tx_illig/RDM_50K54K70K75K_5-2010_EN_web_04.pdf. Accessed Jan 2017
20. GABLER Thermoform. http://gabler-luebeck.de/en/thermoform/loesungen. Accessed Jan 2017

21. Kostić M, Čavić M, Zlokolica M, Veselinović Č (2006) About driving-transmission systems in thermoforming machines. In: Proceedings of the international conference on power transmissions (BAPT), Novi Sad, pp 509–514
22. Kostić M, Čavić M, Zlokolica M, Veselinović Č (2007) Characteristics of mechanism synthesis procedures in design. In: Proceedings of the PSU-UNS international conference on engineering and technology (ICET), Hat Yai
23. Kostić M, Zlokolica M, Čavić M (2005) Certain application of two-D.O.F. lever mechanism. In: Proceedings of the international research/expert conference on trends in the development of machinery and associated technology (TMT), Antalya, pp. 965–968
24. Kostić M, Čavić M, Zlokolica M, Kovačević D, Matušić S (2013) Optimization of two degrees of freedom complex mechanism. In: Proceedings of the PSU-UNS international conference on engineering and technology (ICET), Novi Sad, pp 1–4 (T.10-1.6)
25. Norton LR (2002) Cam design and manufacturing handbook. Industrial Press
26. Chen YF (1982) Mechanics and design of cam mechanisms. Pergamon Press
27. Čavić M, Penčić M, Rackov M, Knežević I, Zlokolica M (2016) Dynamic analysis of the thermoforming machine working mechanism. In: Proceedings of the international scientific conference on mechanical engineering technologies and applications (COMETa), Jahorina, pp 177–184
28. Belforte G, Colombo F, Raparelli T, Maffiodo D (2013) Study of the press forming mechanism of a thermoforming machine. Proc Inst Mech Eng Part C: Journal of Mechanical Engineering Science 228(10):1715–1723
29. Mundo D, Danieli GA, Yan HS (2006) Kinematic optimization of mechanical presses by optimal synthesis of cam-integrated linkages. Trans Can Soc Mech Eng 30(4):519–532
30. Mundo D, Liu JY, Yan HS (2006) Optimal synthesis of cam-linkage mechanisms for precise path generation. J Mech Des 128(6):1253–1260

Cyber-Physical Manufacturing Systems (CPMS)

Zivana Jakovljevic, Vidosav Majstorovic, Slavenko Stojadinovic, Srdjan Zivkovic, Nemanja Gligorijevic and Miroslav Pajic

Abstract Increased product variety that market needs impose to manufacturers, requires high level adaptability of manufacturing systems that can be achieved through introduction of reconfigurable manufacturing systems composed of interoperable devices with ever-changing architecture. Control and management of such a complex system of systems requires fast and reliable real-time virtualization of real world applications as well as real-time feedback from virtual (cyber) model to the real world. The border line between real-world manufacturing system and its cyber representation is characterized by extremely high information permeability thus composing these two systems into a unique system—Cyber-Physical Manufacturing System (CPMS). Recent advances in the fields of Cyber-Physical Systems (CPS) and Internet of Things (IoT) enable creation of CPMS. In this paper we provide an overview of the research works that are currently conducted in the field of CPMS, and we outline the interconnection between CPMS and Industry 4.0. The motivation of this overview is the identification of the R&D activities that are necessary for industry-wide application of CPMS.

Z. Jakovljevic (✉) · V. Majstorovic · S. Stojadinovic · N. Gligorijevic
University of Belgrade, Faculty of Mechanical Engineering, Kraljice Marije 16,
11000 Belgrade, Serbia
e-mail: zjakovljevic@mas.bg.ac.rs

V. Majstorovic
e-mail: vidosav.majstorovic@sbb.rs

S. Stojadinovic
e-mail: sstojadinovic@mas.bg.ac.rs

N. Gligorijevic
e-mail: nemanja.gligorijevich@gmail.com

S. Zivkovic
Military Technical Institute, Coordinate Metrology Lab, Belgrade, Serbia
e-mail: srdjan.vti@gmail.com

M. Pajic
Duke University, Department of Electrical and Computer Engineering,
100 Science Dr Hudson Hall Room 206, Durham, NC 27708, USA
e-mail: mp275@duke.edu

V. Majstorovic and Z. Jakovljevic (eds.), *Proceedings of 5th International Conference on Advanced Manufacturing Engineering and Technologies*,
Lecture Notes in Mechanical Engineering, DOI 10.1007/978-3-319-56430-2_14

Keywords Cyber-physical manufacturing systems · Industry 4.0 · Internet of things and services

1 Introduction

The developments in the field of Information and Communications Technologies (ICT) for industrial practice leveraged the research and development activities in Cyber-Physical Systems (CPS) applications [1, 2]. Using CPS, real and virtual worlds are integrated in internet environment in industry thus generating a foundation for Cyber-Physical Manufacturing (CPM) [3, 4]. Further developments in ICT on the one hand, and manufacturing sciences and technologies on the other, bring about 4th industrial revolution, known as Industry 4.0.

Cloud computing and Internet of Things (IoT) represent the basic hardware-software platform for development and implementation of CPM [5]. There are a number of partial IoT solutions in the fields of services [6] that can be utilized for improvement of some elements and functions within CPM.

IoT has changed the world, and therefore the application of ICT in manufacturing through research, development and application of CPM [7]. Multilevel data transfer and process control, starting from single entities up to the complex systems in a single chain, make IoT the key element for the success of CPM application. On the other hand, these models utilize or generate massive data thus making cloud computing indispensable tool for CPM support.

IoT for CPM can be defined as a future in which the machines work 24/7, and humans and systems within CPM are connected by Internet through which they control services and processes related to manufacturing. Today, digital factory and digital manufacturing are the reality, and immediate future is the smart factory-of-things. On the road towards CPM, the starting point is: "the integration between IoT and PLM platforms using semantic web technologies and Open Services for Lifecycle Collaboration (OSLC) standard on tool interoperability" [3].

Service-oriented manufacturing (SOM) represents a new worldwide manufacturing paradigm, and CPM represents a solution ideally suited for further development of SOM concept [8]. Within CPM, one of the key issues represents the data quality: their source, generation, transfer and utilization. For solving these problems, formal semantics of workflow nets (WF-nets) based on construct process-oriented ontology is utilized along with the processes optimization.

Within CPM, it is extremely important to create consistent data exchange, mostly using open and global information network with two-way data transfer [9]. Knowledge bases represent a special research area within CPM.

In this paper we present an overview of the current research in the field of Cyber-Physical Manufacturing Systems (CPMS) as well as its interconnection with the Industry 4.0. Since CPMS represent one of the most important enabling factors for implementation of Industry 4.0, we dedicate one section to Industry 4.0 model from the perspective of CPMS. Finally, in conclusion we summarize the findings of the review and provide some remarks regarding the future research in the field of CPMS.

2 The Development of Cyber-Physical Manufacturing Concept

Extensive research and development activities in embedded systems over the last decades resulted in generation of smart sensors and actuators with integrated computation and communication capabilities. These devices represent CPS—systems that integrate physical processes and computing elements through their real-time interaction [10, 11]. Smart sensors and actuators are nowadays widely implemented in manufacturing facilities. Nevertheless, the notion of CPS is wider than smart actuators and sensors—CPS can be regarded as systems of systems [12], and as such they can be implemented at all levels of manufacturing process control (manufacturing resources, their elements, manufacturing cells, manufacturing lines, up to the whole enterprise and network of enterprises). However, the potentials that CPS devices offer through interconnection of cyber and physical world are yet to be appropriately exploited in manufacturing.

CPMS represent the highest level of CPS application in manufacturing [2]. Within CPMS real manufacturing system and its cyber (virtual) representation are fused into one CPS through the extensive exchange of data (Fig. 1). Implementation of CPMS requires the design of reliable networks of smart resources with common data semantics. New methods for big data analysis and

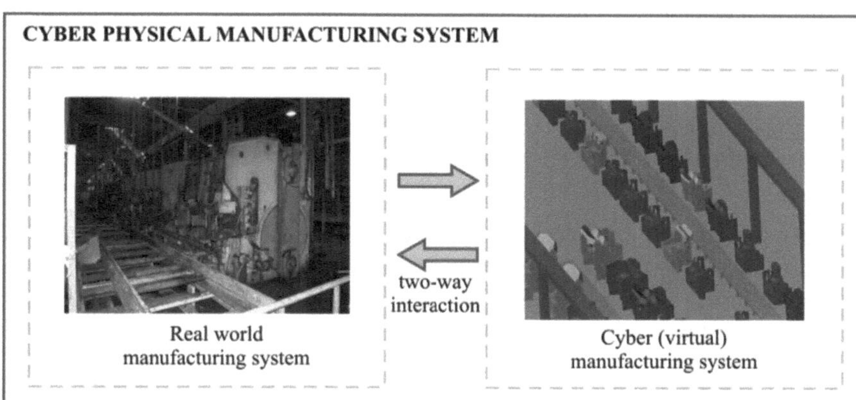

Fig. 1 Cyber physical manufacturing system

Table 1 Examples of application of smart sensors and actuators in manufacturing

Manufacturing resource	Sensor/actuators
Machine tools, robots	Motors (with integrated encoders and drivers)
	HMI
	Vision systems
	Limit switches
	Encoders
	Spindles
	Grippers
	Various sensors for process monitoring, etc.
Pneumatic systems (e.g. manipulators)	Cylinders (with integrated control valves)
	Limit switches and other sensors
Transport systems	Motors
	Vision sensors
	Limit switches, etc.
Inventory	RFID readers
	Limit switches, etc.
Measurement equipment	Measuring heads
	Laser sensors
	Structured light systems, etc.

information representation and retrieval are necessary. The issues of data protection and security are also very important.

As a basis for CPMS integration, the smart sensors are utilized to monitor the physical environments, while the smart actuators change the physical parameters of the system [13]. These devices are connected into the control network that realizes the objective control function. Table 1 provides an overview of sensors and actuators that can be employed for the realization of control networks in manufacturing applications.

3 Towards Real CPMS Application

To reach the goal of real-world industry-wide implementation, Cyber Physical Manufacturing Systems have been in the focus of a number of research projects worldwide. In this section, we provide basic information about selected research projects financed by European Commission (EC) under Horizon 2020 and FP7, as well as the projects recently financed by National Science Foundation (NSF) in USA.

Through the topic related to digital automation within Horizon 2020 Factories of the Future (FoF) call [14] EC has recognized the need for introduction of cyber physical systems in manufacturing. Most of the EC financed Horizon 2020 and FP7 projects in the field of CPMS are submitted under the FoF calls, but there are also a number of projects within ICT (Information and Communication Technologies), SMEINST (Horizon 2020 dedicated SME instrument), and TWINN (Twinning) calls. Some of the projects related to CPMS are [15]:

- MAYA, Multi-disciplinArY integrated simulAtion and forecasting tools, empowered by digital continuity and continuous real-world synchronization, towards reduced time to production and optimization (FoF): research on synchronization of digital and real world factory and development of simulation methodologies for optimization of production activities during factory lifecycle;
- sCorPiuS, European Roadmap for Cyber-Physical Systems in Manufacturing (FoF): the objective of this project was to provide support for research and innovation activities in the field of CPS application in manufacturing;
- GO0D MAN, aGent Oriented Zero Defect Multi-stage mANufacturing (FoF): this project investigates the architecture of agent based distributed system to support Zero Defect Manufacturing;
- EXCELL, Actions for Excellence in Smart Cyber-Physical Systems applications through exploitation of Big Data in the context of Production Control and Logistics (TWINN): the objective of this project is the strengthening of scientific and technology capacities of participating academic institutions in the field of Big Data analysis for manufacturing;
- SKILLPRO, Skill-based Propagation of 'Plug and Produce'—Devices in Reconfigurable Production Systems by AML (FoF): The research in the field of generation of interfaces for seamless integration of heterogeneous devices enabling 'plug and produce' capabilities within reconfigurable manufacturing systems;
- PERFoRM, Production harmonizEd Reconfiguration of Flexible Robots and Machinery (FoF): The objective of this research is generation of solid manufacturing middleware for easy reconfiguration of manufacturing resources and facilitation of 'plug and produce' paradigm.
- DISRUPT, Decentralized architectures for optimized operations via virtualized processes and manufacturing ecosystem collaboration (FoF): Architecture for decentralized control of manufacturing facilities through generation of cyber representation of manufacturing resources and decision support system;
- IOTI4.0, Integral Open Technology for Industry 4.0 (SMEINST): Creation of open source programmable logic controllers based on Arduino that will enable enhanced connectivity of industrial devices.

Similarly, in the USA, NSF has emphasized the need for the research in CPMS through introduction of the program within NSF Division of Civil, Mechanical and Manufacturing Innovation that refers to Cyber manufacturing systems [16]. In addition, there is a program within Directorate for Computer & Information Science & Engineering that refers to CPS. Within these programs, a number of projects referring to cyber physical manufacturing were awarded [16] as Synergy or Frontiers grants, EAGER (EArly-concept Grant for Exploratory Research) or Grant Opportunities for Academic Liaison with Industry (GOALI) grants. Some of them are [16]:

- An Integrated Simulation and Process Control Platform for Distributed Manufacturing Process Chains (Synergy grant): the research on the manufacturing platform that will integrate physical process and numerical simulation via a distributed network.
- Software Defined Control for Smart Manufacturing Systems (Frontiers grant): the objective of this project is the generation of methodology for detection of anomalies within manufacturing systems by comparing the behavior of the system obtained from sensors and behavior of the system modeled in virtual environment.
- Architecture and Protocols for Scalable Cyber-Physical Manufacturing Systems (EAGER): research on service oriented architecture, protocols that enable plug-n-play of manufacturing machines, real-time Ethernet protocols for scalable cyber physical manufacturing systems, and methods for virtualization of manufacturing machines.
- Robust Quality Control Tools for Cyber-Physical Manufacturing Systems: Assessing and Eliminating Cyber-Attack Vulnerabilities (GOALI): research on data security issues in quality control within cyber physical manufacturing systems.
- Real-Time Scheduling in Networked 3D Printers (EAGER): the objective of this project is the development of cloud based routing, batching, and scheduling algorithms for decentralized 3D printing.
- Abstractions and Architectures for Open Composable Services (EAGER): research on open communication and interoperability protocols for exchange of data between software tools and manufacturing applications.
- Modular System Design for CyberManufacturing of Customized Apparel (EAGER): research on end-to-end software systems architecture for personalized manufacturing of individualized clothing.

The fact that the most of grants are EAGER points out that CPMS are far away from mature technology and that there is a lot of research work to be done in this field.

In addition to Horizon 2020 and NSF going on projects, there is a number of reported research works in the field of CPMS. Some of them are summarized in the sequel.

In CPMS, the functionality of the system as a whole is obtained through communication and interoperability of sensors and actuators. The complexity of CPMS integration through interconnection of low level devices requires a systematic approach to system integration. In [17] the authors proposed 5-level architecture that represents the guidelines for CPMS developing and deploying.

The architecture is known as 5C architecture (Fig. 2) and it consist of the following [17] levels:

- **Connection**—acquiring accurate and reliable data from machines and their components;
- **Conversion**—extraction of information from acquired data;

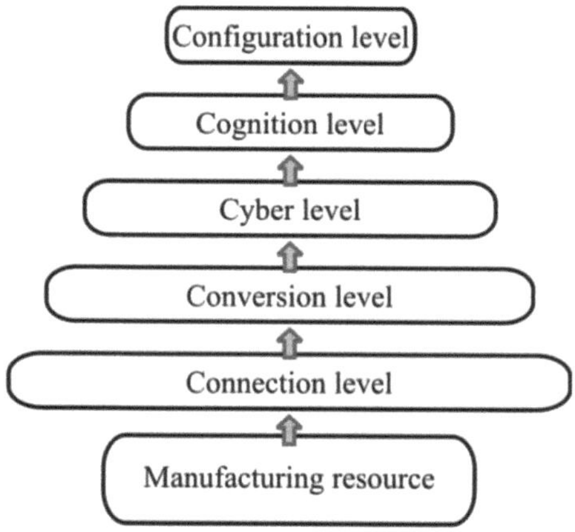

Fig. 2 5C CPS architecture—adapted from [17]

- **Cyber**—carries out the changes in cyber (digital) representation of the machine/component according to the information from previous step and compares new with previously obtained data;
- **Cognition**—generation of knowledge from obtained information;
- **Configuration**—feedback from cyber to physical space.

5C architecture is composed in such a way that sensory information generates digital (cyber) twins of machines' components that further enable creation of machines' digital twins through information aggregation. Finally, at the production system level, information about machines can be utilized for self-configurability and self-maintainability of the factory.

The implementation of CPMS requires easy reconfiguration of manufacturing resources and profuse communication between components. Consequently, one of the most important issues for the implementation of CPMS represents smart infrastructure that will enable configuration and connection of CPS devices to the power and communication networks with little or no human assistance [18]. For these purposes a multi-vendor project SmartFactory^{KL} was launched with the goal to define the standards and concepts for interoperability of devices manufactured by different companies [19]. SmartFactory^{KL} represents modular and scalable production line that is characterized by plug and produce capabilities. The high level modularity is enabled by generation of electrical, mechanical and communication interfaces for multi-vendor production resources. Special infrastructure boxes are designed for power supply and connection to communication network.

In our recent research at Faculty of Mechanical Engineering in Belgrade [20], we have developed a Cyber-Physical Manufacturing Metrology Model for integration of Coordinate Measuring Machines in CPMS. The goal of this model is intelligent planning of parts inspection on CMM that exploits the benefits of cyber representations of part and CMM. The model integrates the following modules: (1) module for automatic recognition of geometrical features from part's CAD model, (2) intelligent inspection process planning module for extraction of

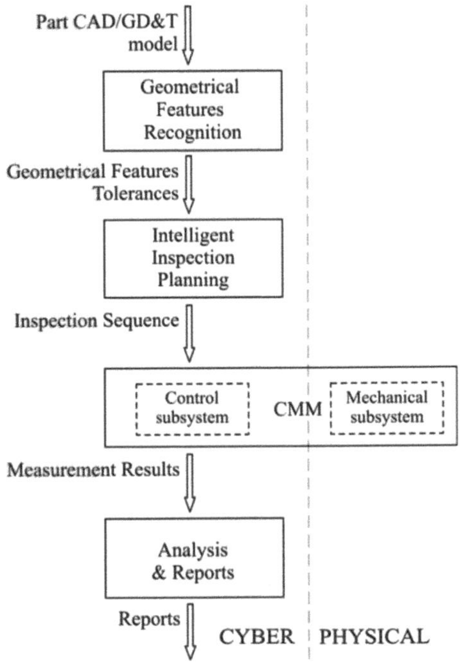

Fig. 3 Cyber-physical manufacturing metrology model

metrological features and generation of inspection sequence, (3) CMM that carries out the inspection, and (4) module for analysis of results and creation of reports (Fig. 3).

A prerequisite for the development and implementation of CPMS is the development of the Industry 4.0 concept. According to [19] in Industry 4.0, all manufacturing resources and products represent CPS that autonomously exchange information, trigger actions based on events and control each other independently. Therefore, the application of CPMS goes hand in hand with the implementation of Industry 4.0 concept.

4 Industry 4.0 Model

The developments in the fields of CPS and internet of things and services will bring about radical changes in the way the goods are manufactured. Namely, the implementation of CPS in manufacturing opens up new possibilities for introduction of highly flexible production resources through generation of reconfigurable manufacturing systems (RMS). RMS, along with the application of internet of things and services for product variety management, are foreseen as the enabling technology for the shift of manufacturing companies to the new manufacturing paradigm—mass customization and personalization [21].

Table 2 Industrial revolutions

Industrial revolution	Enabling technology	Characteristic equipment	Market needs
Industry 1.0	Water and steam power	Machine tools	Customized products
	Mechanization	Steam engine	
Industry 2.0	Electricity	Transfer lines	Low cost products
	Division of labor		
	Interchangeable parts		
Industry 3.0	Microcomputers	PLC—based control	Products variety
		CNC machine tools	
		Robots	
	Information technologies		
Industry 4.0	Cyber physical systems	Reconfigurable manufacturing systems	Low cost customized/individualized products
	Internet of things and services		

The significance of the effects that these changes will have to the industry is comparable with the effects that introduction of steam engine, electricity, division of labor, microcomputers and information technologies had in the past [22–24]—Table 2. They will bring about new industrial revolution—Industry 4.0.

To timely meet Industry 4.0 challenges, the German professional associations BITKOM, VDMA and ZVEI established The Industrie 4.0 Platform that provided the recommendations for implementing the strategic initiative INDUSTRIE 4.0 in Germany [24]. In these recommendations, the following cross-cutting issues for Industry 4.0 are identified [24, 25]:

- Vertical integration of resources within the shop-floor/factory,
- Digital end to end engineering across the whole value chain and product life cycle,
- Horizontal integration through creation of value adding networks of enterprises.

Computational and communication capabilities of CPS resources will bring about the radical changes in the manufacturing process control systems. Current vertical integration through IEC 62264 [26] automation pyramid that can be illustrated through the following hierarchical line of centralized control resources: (1) sensors and actuators, (2) Programmable Logic Controller, (3) Supervisory Control and Data Acquisition, (4) Manufacturing Execution System, (5) Enterprise Resource Planning will give the way to truly distributed control systems (Fig. 4). The functional hierarchy of control tasks will remain in the sense of IEC 62264, but instead of individual control resources in charge of performing certain high level functions, these functions will be distributed to a number of low level resources, each representing a CPS. Functions will be performed through communication and interoperation of different autonomous/intelligent resources that can be smart

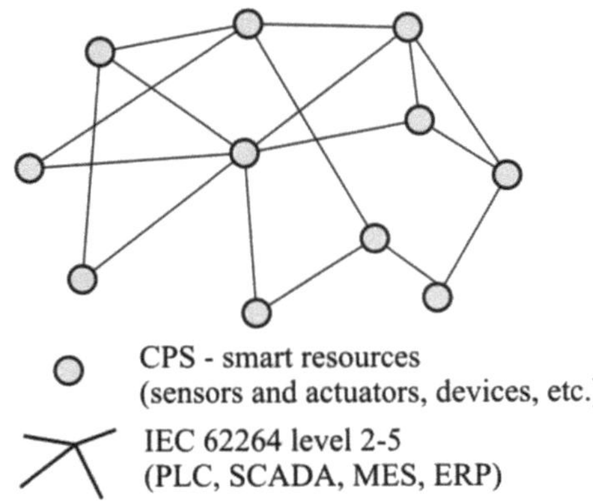

Fig. 4 Distributed control systems

sensors and actuators, machine components or the whole machines and production lines. To enable easier integration into the smart factory, certain functions can be encapsulated into different functional modules [27].

The distribution of control tasks enables high level modularity which is necessary for achieving the reconfigurability of manufacturing systems. Nevertheless, fast and reliable reconfiguration requires that the resources are characterized by high level autonomy, self-configurability and that they are self diagnosable [28].

Introduction of scalable RMS within Industry 4.0 will have significant influence not only on meeting the customers' needs, but also on the resources consumption. The configuration of manufacturing system when needed and as needed [29] aims to increase the resources productivity along with the decrease of energy consumption and invested human labor.

Digital end to end engineering implies [25]:

1. the digitalization of all processes during manufacturing value adding chain from product design, production planning, production to sales, and
2. the life cycle management of products and manufacturing resources.

Connection of manufacturing processes with digital world (manufacturing processes digitalization) requires real time acquisition of all relevant data from manufacturing facilities. On the other hand, decomposition of manufacturing resources into smaller functional units [30] (CPS) up to the level of sensors and actuators paves the way for this kind of data acquisition. Namely, small functional CPS units are capable of receiving and transferring relevant data to the virtual world thus enabling the digital representation with high level of detail. This gives the opportunity for generation of digital twin of the real world manufacturing facility [31] and implementation of the highest level of CPS in manufacturing—Cyber Physical Manufacturing System.

In this context, besides manufacturing resources, the products represent important elements. Tracking not only product types, but also product instances during their life cycle within manufacturing facility, and beyond, is an important issue within Industry 4.0. Industry 4.0 product is a smart product that is uniquely identifiable and can be always traced [24]. The unique product ID can be obtained using passive RFID [32] or it can be included in higher level Product Embedded Information Devices (PEID) that have computation and communication capabilities [33]. Thus, depending on the level of embedded intelligence, smart products can represent autonomous agents capable to provide information about their history, target states and alternative paths to reach these states [24, 33]. Smart products and internet of things make it possible to manufacturing companies to stay in contact with their products through the whole product lifecycle [34]. Obtained information regarding product instance performances can be utilized for its maintenance or for the improvements in product type design. Thus, Industry 4.0 Product Lifecycle Management system will be closed loop system that closes information loop between all phases in product lifecycle: Beginning-of-Life, Middle-of-Life and End-of-Life.

Obviously, one of the main pillars of Industry 4.0 with respect to vertical integration and digital end to end engineering is communication network that will enable intensive data exchange. Starting prerequisite is availability of reliable (stable), scalable and affordable communication infrastructure. Furthermore, real-time exchange of data between CPS is crucial for distributed control systems as well as for real-time generation of manufacturing facility digital twin. Thus, one of the most important issues in Industry 4.0 is communication latency.

To meet the requirements of Industry 4.0 control system, it is necessary to develop fast communication characterized by sensor to actuator latency of 1 ms (Fig. 5). The network that can enable such communication is known as tactile internet. The term tactile internet is chosen since end to end speed of 1 ms corresponds to the reaction of human to tactile sense, which has the shortest reaction time of all human senses [35]. To enable this kind of latency, tactile internet will require new system architecture, protocols, as well as authentication that is embedded in the physical transmission [35]. For wireless communication, significant innovations will be necessary in signal modulation with the objective to enable fast access network. In addition, a special attention will be needed in order to cope with possible interference, since it is expected that a large number of devices

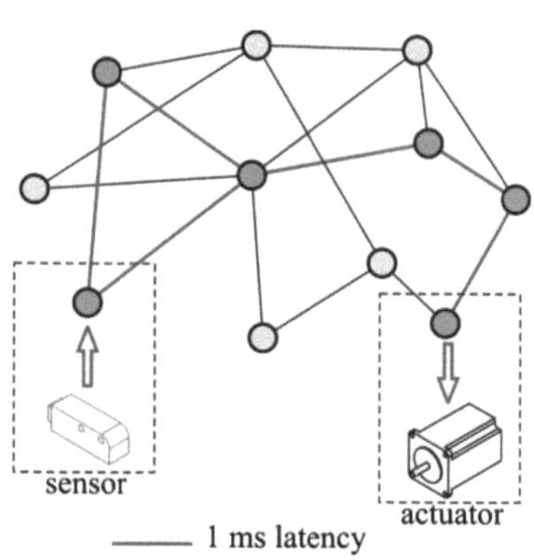

Fig. 5 Tactile internet latency

(sensors, actuators, products and other manufacturing resources such as robots or machine tools) will ask the simultaneous access to the network.

According to the Forbes Magazine [36] the number of IoT devices will reach up to 30.7 billion in 2020 and 75.4 billion in 2025, along with the potential of up to $11.1 trillion annual turnover by 2025. Internet protocol version 6 (IPv6) is suitable for addressing such a large number of devices. Namely, its 128 bit address space enables addressing 2^{128}, or ca. 3.4×10^{38} devices, and it is suitable for generation of scalable networks that are necessary in Industry 4.0 manufacturing facility [37].

During generation of manufacturing system digital twin, manufacturing resources leave digital shadow that contains big data [38] regarding the system performances. These data carry profuse information that can be effectively utilized for product redesign, production planning, management decisions and similar purposes. To convert big data in useful information, there is a need to develop efficient, intelligent techniques for data mining and information representation. These techniques are based on implementation of machine learning and pattern recognition. It is expected that cloud computing will have a significant role in storing, handling and analysis of big data obtained from manufacturing processes [39].

Widespread exchange of confidential information (data, know how, etc.) between resources usually using wireless communication brings about additional issues regarding unauthorized access to various data and its potential misuse. The security and data protection become one of the hottest research and development topics within Industry 4.0 [24].

In addition to technical changes, Industry 4.0 will bring about significant changes in the society. The changes at workplace are inevitable. Industry 4.0 will go a step forward with respect to Industry 3.0 in removing workers from routine, monotonous, repetitive work tasks. Industry 4.0 employees will be involved in monitoring and control of intelligent equipment, or will work in collaboration with it. Their main work tasks will be in the field of knowledge based decision making. To answer these needs the interdisciplinary (manufacturing engineering, automation engineering, IT) training along with deep transferable skills especially in communication will be necessary. The employees will have to attend lifelong learning and companies will have to enable continuous professional development and training on the job for their workforce. The changes in the way things are manufactured will raise the issue of work-life balance; they will lead to changes in education system and will highly influence the society as a whole.

Considering all needs for introduction of industry 4.0, VDI/VDE Society Measurement and Automatic Control (GMA) provided Reference Architecture Model Industrie 4.0 (RAMI4.0) in its Status Report [25]. The objective of RAMI4.0 is to integrate all aspects necessary for three types of integration within Industry 4.0 (vertical, end-to-end and horizontal) into one reference model. RAMI4.0 is represented through three axes (Fig. 6). First axis considers IT representation of Industry 4.0 component (CPS with communication capability) and contains six layers: Asset, Integration, Communication, Information, Functional and Business layer. Second axis takes into account the automation hierarchy of IEC 62264. To address all aspects of Industry 4.0 including end-to-end digital integration of products and

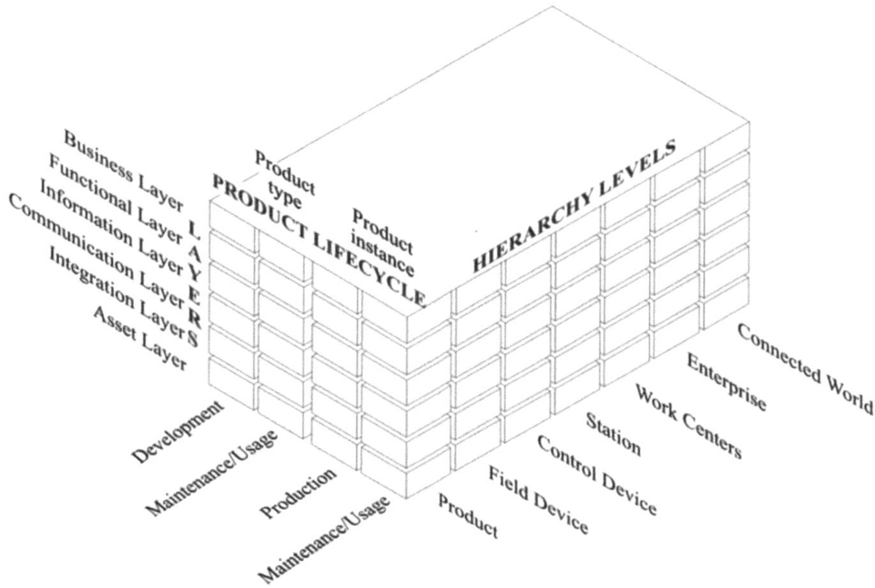

Fig. 6 RAMI—adapted from [25]

horizontal integration of enterprises, RAMI4.0 adds Product and Connected world to IEC 62264 hierarchy thus defining a total of 7 Hierarchy Levels: Product, Field Device, Control Device, Station, Work Centers, Enterprise, and Connected World. Furthermore, for modeling the Life Cycle and Value Stream, RAMI4.0 Introduces the Development and Maintenance/Usage of product type, and Production and Maintenance/Usage of product instance.

5 Conclusion

CPMS are based on implementation of CPS and IoT in manufacturing, and they represent an enabling means for introduction of mass customization and personalization manufacturing paradigm. However, a lot of research and development work has to be done on the road to the real-world industrial application of CPMS. The research fields can be summarized as follows. New methods for distribution of control tasks between smart resources along with the methods for integration of event-based distributed control systems are essential. The common semantics and data models for different devices have to be achieved through the development of interfaces, i.e., middleware and data protocols that will enable plug and produce capabilities of reconfigurable manufacturing systems. Furthermore, fast and reliable methods for manufacturing systems configuration according to the current needs are necessary. Real-time data acquisition and exchange has to be enabled by introduction of fast and reliable communication networks with tactile internet capabilities. In addition, there is a need for intelligent methods for big data analysis and information retrieval and representation. PLM systems have to be improved in order to exploit the benefits of smart products. In addition, intensive communication between smart resources, products and information systems raises the issue of data security and privacy.

Currently, a lot of research efforts are invested in creation of framework and strengthening of R&D institutions' capacities for introduction of CPMS, such as H2020 projects sCorPiuS and EXCELL and recommendations from [17, 24, 25]. The necessity for generation of distributed systems is recognized, and a number of projects (H2020 projects GO0D MAN, DISRUPT and IOTI4.0, as well as NSF EAGER and Synergy grants) and research works such as [28] are conducted in this field. Research on interfaces that will enable interoperability of multi-vendor devices (H2020 projects SKILLPRO and PERFoRM, NSF EAGER grants, research from [19]) and automatic information retrieval and representation (H2020 project MAYA, NSF Frontiers grant, research from [11, 20]) is in progress. In addition, in the field of data security, there is a NSF GOALI grant.

Indisputably, CPMS are enabling the implementation of Industry 4.0 concept. Although the research in the field is evident, a lot of additional R&D efforts are necessary to get to the industry-wide application of this concept.

Acknowledgements This work is a part of research which is supported by the Ministry of Science and Technological Development of Serbia, with Grant Numbers TR35007, TR35020 and TR35022.

References

1. Monostori L et al (2016) Cyber-physical systems in manufacturing. CIRP Ann Manuf Technol 65:621–641. doi:10.1016/j.cirp.2016.06.005
2. Monostori L (2014) Cyber-physical production systems: roots, expectations and R&D challenges. Procedia CIRP 17:9–13. doi:10.1016/j.procir.2014.03.115
3. Shariatzadeha N, Lundholma T, Lindberga L, Sivarda G (2016) Integration of digital factory with smart factory based on Internet of Things. Procedia CIRP 50:512–517. doi:10.1016/j.procir.2016.05.050
4. Lee J, Bagheri B, Jin C (2016) Introduction to Cyber Manufacturing. Manuf Lett 8:11–15. doi:10.1016/j.mfglet.2016.05.002
5. Botta A et al (2016) Integration of cloud computing and internet of Things: a survey. Future Gener Comput Syst 56:684–700. doi:10.1016/j.future.2015.09.021
6. Mineraud J, Mazhelis O, Su X, Tarkoma S (2016) A gap analysis of Internet-of-Things platforms. Comput Commun 89–90:5–16. doi:10.1016/j.comcom.2016.03.015
7. Thramboulidis K, Christoulakis F (2016) UML4IoT—a UML-based approach to exploit IoT in cyber-physical manufacturing systems. Comput Ind 82:259–272. doi:10.1016/j.compind.05.010
8. Song Z, Sun Y, Wan J, Liang P (2016) Data quality management for service-oriented manufacturing cyber-physical systems. Comput Electr Eng 48:1–11. doi:10.1016/j.compeleceng.2016.08.010

9. Barthelmeya A, Störklea D, Kuhlenköttera B, Deusea J (2014) Cyber physical systems for life cycle continuous technical documentation of manufacturing facilities. Procedia CIRP 17:207–211. doi:10.1016/j.procir.2014.01.050
10. Lee EA, Seshia SA (2015) Introduction to Embedded Systems—A Cyber-Physical Systems Approach. Second Edition. LeeSeshia.org
11. Jakovljevic Z, Markovic V, Puzovic R, Majstorovic V (2016) Recognition of one class of quadrics from 3D point clouds. Procedia CIRP 57:292–297. doi:10.1016/j.procir.2016.11.051
12. Rajkumar R, Lee I, Sha L, Stankovic J (2010) Cyber-physical systems: the next computing revolution. Proceedings of 47th ACM/IEEE design automation conference. Anaheim, CA, USA. 13–18 June 2010, pp 731–736
13. Hu F, Lu Y, Vasilakos A, Hao Q, Ma R, Patil Y, Zhang T, Lu J, Li X, Xiong N (2016) Robust cyber–physical systems: concept, models, and implementation. Future Gener Comput Syst 56:449–475. doi:10.1016/j.future.2015.06.006
14. Horizon 2020 Work Programme 2016–2017:17. Cross—cutting activities (Focus Areas) http://ec.europa.eu/research/participants/data/ref/h2020/wp/2016_2017/main/h2020-wp1617-focus_en.pdf. Accessed 27.02.2017
15. http://cordis.europa.eu/projects/home_en.html. Accessed 27.02.2017
16. www.nsf.gov. Accessed 27.02.2017
17. Lee J, Bagheri B, Kao H-A (2015) A cyber-physical systems architecture for industry 4.0-based manufacturing systems. Manuf Lett 3:18–23. doi:10.1016/j.mfglet.2014.12.001
18. Lapp Group AG (2014) Die Zukunftsfabrik. Kabelwelt: Industrie 4.0. Revolution in der Fabrikhalle 2:6–10
19. Weyer S, Schmitt M, Ohmer M, Gorecky D (2015) Towards industry 4.0-standardization as the crucial challenge for highly modular, multi-vendor production systems. IFAC-PapersOnLine 48(3):579–584. doi:10.1016/j.ifacol.2015.06.143
20. Majstorovic V, Stojadinovic S, Durakbasa N (2016) Path planning for inspection prismatic parts on CMM as a part of cyber-physical manufacturing metrology model. J Proc Manuf Syst 11(1):3–8
21. ElMaraghy H, Schuh G, Elmaraghy W, Piller F, Schönsleben P, Tseng M, Bernard A (2013) Product variety management. CIRP Ann Manuf Technol 62(2):629–652
22. Jovane F, Koren Y, Boër CR (2003) Present and future of flexible automation: towards new paradigms. CIRP Ann Manuf Technol 52(2):543–560
23. Pilipovic M, Jakovljevic Z (2017) Automatizacija proizvodnje (Manufacturing automation). University of Belgrade, Faculty of Mechanical Engineering, Belgrade
24. Kagermann H, Wahlster W, Helbig J (2013) Recommendations for implementing the strategic initiative industrie 4.0. acatech—National Academy of Science and Engineering. http://www.acatech.de/fileadmin/user_upload/Baumstruktur_nach_Website/Acatech/root/de/Material_fuer_Sonderseiten/Industrie_4.0/Final_report__Industrie_4.0_accessible.pdf. Accessed 14 Feb 2017
25. VDI/VDE-Gesellschaft Mess- und Automatisierungstechnik (2015) Reference architecture model Industrie 4.0 (RAMI4.0). http://www.zvei.org/Downloads/Automation/5305%20Publikation%20GMA%20Status%20Report%20ZVEI%20Reference%20Architecture%20Model.pdf. Accessed 14 Feb 2017
26. International Organization for Standardization (2013) IEC 62264–1:2013 enterprise-control system integration—Part 1: models and terminology
27. Vyatkin V (2012) IEC 61499 Function blocks for embedded and distributed control systems design. ISA. ISBN: 978-1-9360007-93-6
28. Lesi V, Jakovljevic Z, Pajic M (2016) Towards Plug-n-Play numerical control for reconfigurable manufacturing systems. In: IEEE international conference on emerging technologies and factory automation, ETFA, Nov 2016, Art. no. 7733524
29. Koren Y, Shpitalni M (2010) Design of reconfigurable manufacturing systems. J Manuf Syst 29(4):130–141
30. Holtewert P, Wutzke R, Seidelmann J, Bauernhansl T (2013) Virtual fort knox federative, secure and cloud-based platform for manufacturing. Procedia CIRP 7:527–532

31. Schroeder GN, Steinmetz C, Pereira CE, Espindola DB (2016) Digital twin data modeling with automationML and a communication methodology for data exchange. IFAC-PapersOnLine 49(30):12–17
32. IEEE Internet Initiative (2015) Towards a definition of the Internet of Things (IoT), Revision-1, on-line: http://iot.ieee.org/images/files/pdf/IEEE_IoT_Towards_Definition_Internet_of_Things_Revision1_27MAY15.pdf. Accessed 27.02.2017
33. Kiritsis D (2011) Closed-loop PLM for intelligent products in the era of the internet of things. J Comput Aided Des 43(5):479–501. doi:10.1016/j.cad.2010.03.002
34. Stark J (2015) Product lifecycle management (Volume 2): the devil is in the details (decision engineering), 3rd edn. Springer, ISBN-13: 9783319244341
35. The Tactile Internet. ITU-T Technology watch report. https://www.itu.int/dms_pub/itu-t/oth/23/01/T23010000230001PDFE.pdf. Accessed 03.03.2017
36. Louis Columbus (2016) Roundup of internet of things forecasts and market estimates.https://www.forbes.com/sites/louiscolumbus/2016/11/27/roundup-of-internet-of-things-forecasts-and-market-estimates-2016/#135ae5af292d. Accessed 27.02.2017
37. Teare D, Paquet C (2004) CCNP Self-Study: advanced IP addressing. Cisco Press. http://www.ciscopress.com/articles/printerfriendly/174107. Accessed 03.03.2017
38. Colangelo E, Bauernhansl T (2016) Usage of analytical services in industry today and tomorrow. Procedia CIRP 57:276–280
39. Gilchrist A (2016) INDUSTRY 4.0: The Industrial Internet of Things. APRESS(eBook). doi:10.1007/978-1-4842-2047-4

Graphical Method in CATIA for Side Mill Tool Profiling Using the Generating Relative Trajectories

Virgil Teodor, Florin Susac, Nicusor Baroiu, Viorel Păunoiu and Nicolae Oancea

Abstract A new graphical solution is proposed, for profiling the side mill which generates a cylindrical helical surface with constant pitch. The method uses a specific complementary theorem: the generating trajectories theorem. This is a new express form of the theorems of surface generation. A graphical solution is presented. This solution is developed in the CATIA design environment. Examples of this new method application are presented, for profiling side mill tools which generate cylindrical helical surfaces with constant pitch used in industry.

Keywords Slide mill profiling · CATIA · Generating relative trajectories · Helical surface

1 Introduction

The issue of side mill tools' profiling, tools bounded by revolution peripheral primary surfaces, for generation of cylindrical helical surfaces with constant pitch (e.g: helical flutes of the cutting tools; gear worms; driven screws or worms for pumps) has analytical solutions, [1], based on Olivier theorems, Gohman theorem, Nikolaev condition, as so as based on complementary theorem: the method of "minimum distance" or the method of "substitutive circles" [2].

V. Teodor · F. Susac (✉) · N. Baroiu · V. Păunoiu · N. Oancea
Department of Manufacturing Engineering, University of Galati,
Domneasca Street 111, 800201 Galati, Romania
e-mail: florin.susac@ugal.ro

V. Teodor
e-mail: virgil.teodor@ugal.ro

N. Baroiu
e-mail: Nicusor.Baroiu@ugal.ro

V. Păunoiu
e-mail: viorel.paunoiu@ugal.ro

V. Majstorovic and Z. Jakovljevic (eds.), *Proceedings of 5th International Conference on Advanced Manufacturing Engineering and Technologies*,
Lecture Notes in Mechanical Engineering, DOI 10.1007/978-3-319-56430-2_15

Also, concerns for graphical solution of this problem exist: the grapho–analytical method, V.M. Vorobiev, for profiling side mill which generate the flute of the helical drill. This approach proposes a graphical representation of the in–plane sections curves of the drill's flute and, from here, the graphical representation of the future side mill.

The development of the graphical design environment allows elaborating some graphical methods, as the method of the solid model, Baicu [3], which creates a solid model of the worm to be generated. Subsequent, this model is revolved around the axis of the future side mill. In this way, the determining of the solid model of tool is permitted and, from here, the axial section of the tool, as base for profiling the secondary order tool, which will generate the primary peripheral surface of the side mill.

Similarly solutions are proposed by Veliko and Nankov [4], regarding the profiling of tools for generation of helical involute tooth.

Also, the authors use the solid modelling for determining the shape of actually surface generated with a side mill of which model is determined [4, 5].

Berbinschi [6], using the capabilities of the CATIA software, developed an original methodology for modelling the side mill for generating a cylindrical helical surface with constant pitch—the normal projection of the side mill tool's axis method, basic, the Nikolaev theorem. This method is very precise, intuitive and easy to apply.

The method allow determining the characteristic curve—the contact curve between the revolution surface and the helical surface to be generated—avoiding the uncertainties regarding the existence of the singular points onto composed profiles, as so as, the "shadow" zones in the relative position of side mill's axis regarding the surface to be generated.

Also, the 3D modelling was used by Li et al. [7] for modelling the thread generation using the side mill tools. The contact with blank is modelled and the generating problems are analyzed for various positions of the tool regarding the blank's axis.

In this paper, is proposed a new methodology, developed in the CATIA design environment, for profiling side mill which generate a cylindrical helical surface with constant pitch. The method is developed based on the complementary theorem of the "relative generating trajectories" method [8], as new approach for the enwrapping surfaces' study.

The method is a graphical one and avoids the complicated equations which can be difficult to handle in case of analytical profiling for tools which generate certain piece's profile.

The 3D tool's models can be used for machining the tools using CNC machine–tools.

2 The Relative Generating Trajectories Method

The contact between the cylindrical helical surface with constant pitch and the revolution surface is analyzed regarding the primary peripheral surface of the side mill (revolution surface) in planes perpendicular to the axis of the side mill, see Fig. 1.

They are defined the reference systems:

XYZ is the reference system joined with the Σ surface, with *Z* axis overlapped to the axis $\vec{V}$ of the helical surface;

$X_1Y_1Z_1$—reference system joined with the axis of the future side mill, the Z_1 axis, overlapped with the axis $\vec{A}$ of the side mill tool's primary peripheral surface;

a—distance, measured on the common normal between the axes *Z* and Z_1 (constructive value).

It is accepted that the Σ surface, in its own reference system, *XYZ*, is described by parametrical equations, in form:

$$\Sigma \quad \left| \begin{array}{l} X = X(u,v); \\ Y = Y(u,v); \\ Z = Z(u,v), \end{array} \right. \tag{1}$$

with *u* and *v* independent variable parameters.

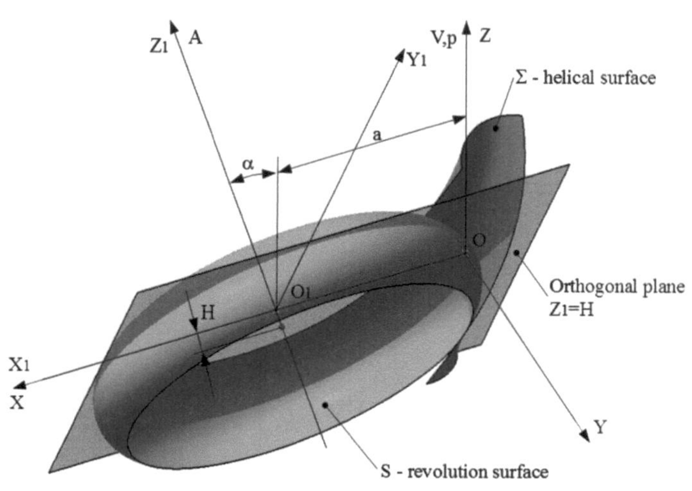

Fig. 1 Relative position of side mill tool's axis regarding the helical surface; reference systems

For the in-plane crossing sections, $Z_1 = H$ (H—variable), it is defined the intersection curve between the plane and the Σ surface, the in-plane curve Σ_H:

$$C_{\Sigma_H} \quad \begin{vmatrix} X_1 = X_1(u); \\ Y_1 = Y_1(u); \\ Z_1 = H. \end{vmatrix} \tag{2}$$

The last equation of system (2) represents, in fact, a geometrical locus onto the helical surface, which, from algebraic point of view, is a dependency between the u and v variable parameters, on type:

$$v = v(u). \tag{3}$$

The enwrapping of the relative trajectories of curves Σ_H, for H arbitrary variable, regarding the reference system of the tool's peripheral surface, determines a geometrical locus which represents the side mill tool's peripheral surface, reciprocally enveloping with the Σ helical surface. The generation process kinematics assumes to make the movements assembly presented in Fig. 2:

- helical movement of the Σ surface (of the blank) with axis $\overrightarrow{V}$ and helical parameter p (the assembly of movements *I* and *II*);
- rotation movement of the side mill, the rotation around the axis $\overrightarrow{A}$, as cutting movement.

The relative generating trajectories are defined as trajectories of points belong to the helical surface regarding the reference system of the future side mill in the movement assembly *I*, *II* and *III*.

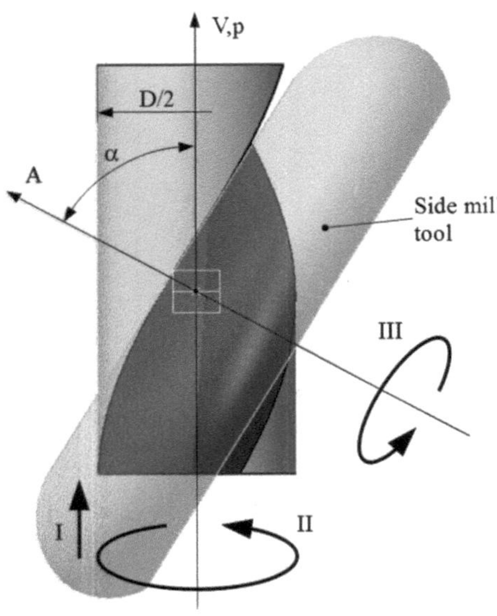

Fig. 2 Helical surface generation kinematics

The characteristic curve of the Σ surface, in the 3 movements assembly does not depend from this component of moving in which the surface is self-generated. In this way, the characteristic curve will depend only by the rotation around the $\vec{A}$ axis.

Consequently, it is possible to state that: the enwrapping of a helical surface, in the rotation movement around a fixed axis is the enwrapping of the relative trajectories family generated in the rotation movement of the curves which represent the in-plane sections of the helical surface, perpendicular to the rotation axis.

In this way, the relative trajectories family, described by the curves on type (2), for H arbitrary variable, is described by the equations:

$$(C_{\Sigma_H})_{\varphi} \left| \begin{array}{l} X_1 = X_1(u)\cos\varphi - Y_1(u)\sin\varphi; \\ Y_1 = X_1(u)\sin\varphi + Y_1(u)\cos\varphi; \\ Z_1 = H, \end{array} \right. \tag{4}$$

with H arbitrary variable.

The enwrapping of the $(C_{\Sigma_H})_{\varphi}$ relative trajectories family, for H variable, represents the revolution surface, which is the side mill's primary peripheral surface.

The enwrapping condition

The normal, in the current point of the Σ_H curve, see (2), has the parametrical equations:

$$N_{\Sigma_H} \left| \begin{array}{l} X_1 = X_1(u) + \lambda\, \dot{Y}_{1_u}; \\ Y_1 = Y_1(u) - \lambda\, \dot{X}_{1_u}, \end{array} \right. \tag{5}$$

with λ variable scalar parameter.

In the rotation movement of the Σ_H curve around the $\vec{A}$ axis, the normal (5) describes a straight lines family, with φ–family's parameter:

$$(N_{\Sigma_H})_{\varphi} \left| \begin{array}{l} X = \left[X_1(u) + \lambda\, \dot{Y}_{1_u}\right]\cos\varphi - \left[Y_1(u) - \lambda\, \dot{X}_{1_u}\right]\sin\varphi; \\ Y = \left[X_1(u) + \lambda\, \dot{Y}_{1_u}\right]\sin\varphi + \left[Y_1(u) - \lambda\, \dot{X}_{1_u}\right]\cos\varphi, \end{array} \right. \tag{6}$$

representing the family of normals to the Σ_H curves.

For $\lambda = 0$, the Eq. (6) represents the generating relative trajectories family (4).

The assembly of these points, for H variable, represents a geometrical locus onto the Σ surface, the characteristic curve, common for S and Σ.

The imposed condition is reduced to condition that the $(N_{\Sigma})_{\varphi}$ family intersecting the $\vec{A}$ rotation axis, in the $Z_1 = H$ plane, hence, to pass through P point, with coordinates:

$$P\left|\begin{array}{l}X_1=0;\\ Y_1=0;\\ Z_1=H.\end{array}\right. \tag{7}$$

The (6) and (7) conditions can be reduced to form:

$$X_1(u)\,\dot{X}_{1u}+Y_1(u)\,\dot{Y}_{1u}=0, \tag{8}$$

for H variable.

For a variation of H between limits constructively defined by Σ surface's form and dimensions, let this H_{min} and H_{max}, see Fig. 1, the (4) and (8) equations assembly represents the coordinates of points from the characteristic curve, onto Σ, in principle in a matrix form:

$$C_\Sigma=\begin{pmatrix}X_{11} & Y_{11} & H_1\\ \vdots & \vdots & \vdots\\ X_{1i} & Y_{1i} & H_i\\ X_{1n} & Y_{1n} & H_n\end{pmatrix}, \tag{9}$$

where, X_{1i}, Y_{1i}, are coordinates of the current point onto the characteristic curve.

By revolving the (9) characteristic curve around the $\vec{A}$ axis, see Fig. 1, the side mill primary peripheral surface is determined:

$$\begin{pmatrix}X_1\\ Y_1\\ Z_1\end{pmatrix}=\begin{pmatrix}\cos\theta & -\sin\theta & 0\\ \sin\theta & \cos\theta & 0\\ 0 & 0 & 1\end{pmatrix}\cdot\begin{pmatrix}X_i\\ Y_i\\ H_i\end{pmatrix}_{i=1,2,\ldots,n}, \tag{10}$$

with θ angular parameter, in the rotation movement around the $\vec{A}$ axis.

The axial section of the side mill is determined in the (R, H) reference system, from (10), using equations:

$$S_A\left|\begin{array}{l}R=\sqrt{X_{1i}^2+Y_{1i}^2};\\ H=H_{ix},\end{array}\right. \tag{11}$$

with i = 1,…, n, regarding the form and dimensions of the generated helical surface, see Fig. 3.

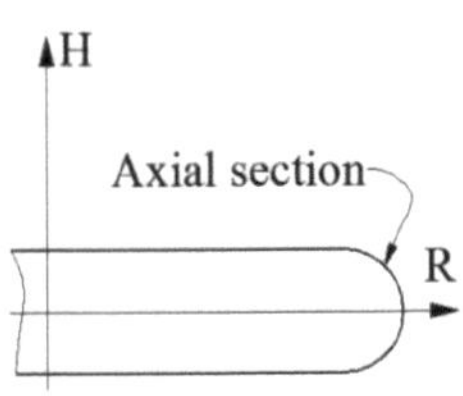

Fig. 3 Side mill tool's axial section

3 Applications

3.1 Side Mill for the Helical Flute of a Helical Drill

The helical flute of a drill is a composite cylindrical helical surface with constant pitch, see Fig. 4. This surface is a ruled helical surface, with generatrix $\overline{AB}$ and a helical surface with curved generatrix, the $\widehat{BC}$ arc, see Fig. 4.

The *XYZ* reference system is defined, with *Z* axis overlapped to the dill's axis. In this reference system the generatrix of the helical flute (*AB* and *BC* zones) are defined.

- The *AB* generatrix

In the *XYZ* reference system, the *AB* generatrix has equations:

$$\left| \begin{array}{l} X = -u \sin \chi; \\ Y = -\frac{d_0}{2}; \\ Z = u \cos \chi, \end{array} \right. \tag{12}$$

with *u* variable parameter, between limits:

$$u_{\min} = 0; \ u_{\max} = \sqrt{\frac{D^2}{4} - \frac{d_0^2}{4}} \cdot \frac{1}{\sin \chi}, \tag{13}$$

where d_0 is the drill's core diameter (constructive value).

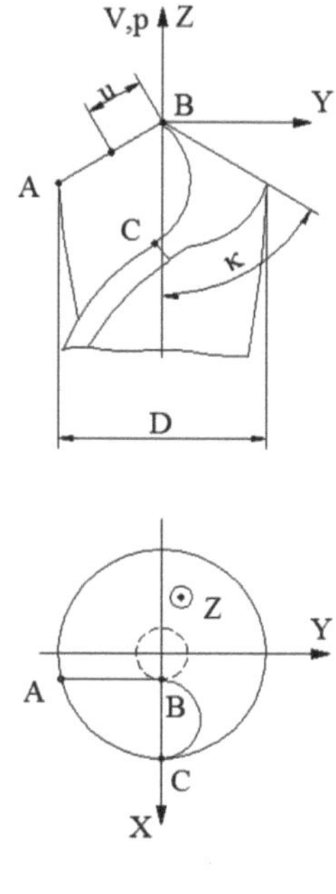

Fig. 4 Composed helical surface of the flute

So, the helical surface with axis $\vec{V}$ (Z) and p helical parameter has equations:

$$\Sigma_{AB}\left|\begin{array}{l} X=-u\sin\chi\cdot\cos v+\frac{d_0}{2}\sin v;\\ Y=-u\sin\chi\cdot\sin v-\frac{d_0}{2}\cos v;\\ Z=u\cos\chi+pv.\end{array}\right. \tag{14}$$

The Σ_{AB} helical surface is a cylindrical helical surface with constant pitch, with p helical parameter, see Fig. 2,

$$p=\frac{D}{2\cdot\tan\omega}, \tag{15}$$

where D is the drill's diameter.

The specific enwrapping condition, see (8), assumes to know the surface's equations in the side mill's reference system, see Fig. 1 and transformation.

From (14), results the parametrical equations of the ruled helical surface in the $X_1Y_1Z_1$ reference system, for the AB zone:

$$\Sigma_{AB}\left|\begin{array}{l} X_1=-u\sin\chi\cdot\cos v+\frac{d_0}{2}\sin v-a;\\ Y_1=\left[-u\sin\chi\cdot\sin v-\frac{d_0}{2}\cos v\right]\cos\alpha+\left[u\cos\chi+pv\right]\sin\alpha;\\ Z_1=-\left[-u\sin\chi\cdot\sin v-\frac{d_0}{2}\cos v\right]\sin\alpha+\left[u\cos\chi+pv\right]\cos\alpha.\end{array}\right. \tag{16}$$

The C_{Σ_H} curve is determined from the intersection condition between the Σ surface with plane

$$Z_1=H. \tag{17}$$

From (17) results:

$$u=\frac{H-\frac{d_0}{2}\cos v\cdot\sin\alpha-pv\cdot\cos\alpha}{\sin\chi\cdot\sin v\cdot\sin\alpha+\cos\chi\cdot\cos\alpha} \tag{18}$$

From (16) the partial derivatives are defined:

$$\begin{aligned} \dot{X}_{1u}&=-\frac{du}{dv}\sin\chi\cos v+u\sin\chi\sin v-\frac{d_0}{2}\cos v\\ \dot{Y}_{1u}&=\left[-\frac{du}{dv}\sin\chi\sin v-u\sin\chi\cos v+\frac{d_0}{2}\sin v\right]\cos\alpha\\ &\quad+\left[\frac{du}{dv}\cos\chi+p\right]\sin\alpha, \end{aligned} \tag{19}$$

$$\frac{du}{dv}=\frac{A(u,v)}{B^2(u,v)}, \tag{20}$$

where:

$$\begin{aligned} A(u,v) &= \left[\frac{d_0}{2}\sin v\cdot\sin\alpha - p\cos\alpha\right]\cdot[\sin\chi\cdot\sin v\cdot\sin\alpha + \cos\chi\cdot\cos\alpha] \\ &\quad - \left[H - \frac{d_0}{2}\cos v\cdot\sin\alpha - pv\cdot\cos\alpha\right]\sin\chi\cos v\sin\alpha; \\ B(u,v) &= \sin v\cdot\sin\chi\cdot\sin\alpha + \cos\chi\cdot\cos\alpha. \end{aligned} \tag{21}$$

A solution for determining the enwrapping condition is proposed. The values $X_1(u)$, $Y_1(u)$, $\dot{X}_{1u}$ and $\dot{Y}_{1u}$ are defined from (16) to (21).

– The *BC* generatrix

The flute for chip removing corresponds with the circle arc *BC*, see Fig. 5.

It is proposed a solution for generation of the helical drill's flute, which avoid the singular points and the discontinuities onto the axial profile of the future side mill, see Fig. 4.

The generating profile of the flute for chip removing is composed from a circle arc with R_0 radius, in a plane tangent to the cylinder which represent the drill's core and, at the same time, tangent to the rectilinear profile of the cutting edge, Fig. 4.

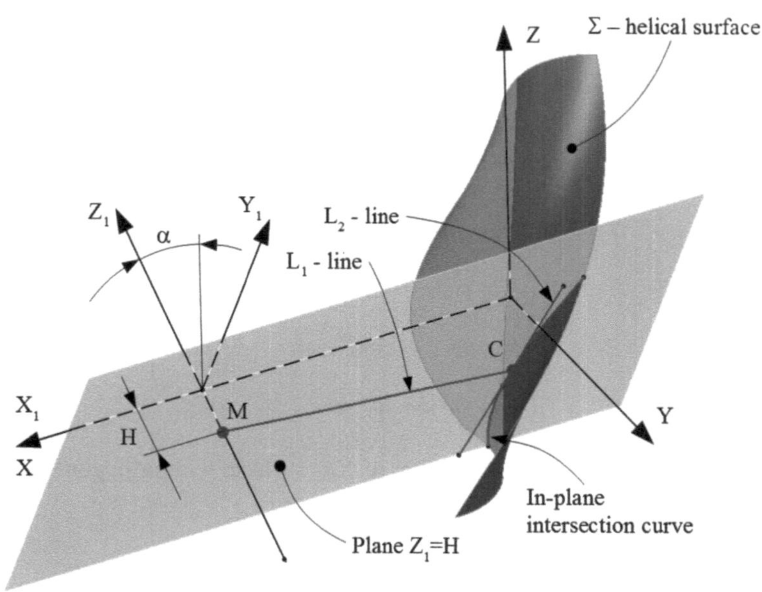

Fig. 5 Geometric elements generated in CATIA

The BC circle's arc equations:

$$\begin{aligned} X &= R_0 \cos\kappa - R_0 \cos\theta; \\ Y &= -\frac{d_0}{2}; \\ Z &= -R_0 \sin\kappa + R_0 \sin\theta; \end{aligned} \tag{22}$$

with θ variable parameter.

The upper limit of the θ parameter is:

$$\theta_{\max} = \text{arc}\cos\left[\frac{1}{R_0}\left(R_0 \cos\kappa - \sqrt{\frac{D^2}{4} - \frac{d_0^2}{4}}\right)\right], \tag{23}$$

with κ working angle of the main cutting edge.

R_O is constructive value.

In this way, the equations of the helical flute Σ_{BC} have become:

$$\Sigma_{BC}\left|\begin{aligned} X &= [R_0 \cos\kappa - R_0 \cos\theta] \cos v + \frac{d_0}{2} \sin v; \\ Y &= [R_0 \cos\kappa - R_0 \cos\theta] \sin v - \frac{d_0}{2} \cos v; \\ Z &= - R_0 \sin\kappa + R_0 \cos\theta + pv. \end{aligned}\right. \tag{24}$$

In the reference system of the side mill, Fig. 1, the Eq. (24) become:

$$\Sigma_{BC}\left|\begin{aligned} X_1 &= R_0 (\cos\kappa - \cos\theta) \cos v + \frac{d_0}{2} \sin v - a; \\ Y_1 &= \left[R_0 (\cos\kappa - \cos\theta) \sin v - \frac{d_0}{2} \cos v\right] \cos\alpha \\ &\quad + (-R_0 \sin\kappa + R_0 \cos\theta + pv) \sin\alpha; \\ Z_1 &= -\left[R_0 (\cos\kappa - \cos\theta) \sin v - \frac{d_2}{2} \cos v\right] \sin\kappa \\ &\quad + (-R_0 \sin\kappa + R_0 \sin\theta + pv) \cos\alpha. \end{aligned}\right. \tag{25}$$

The equations of the crossing section of Σ_{BC} surface, with orthogonal planes of the future side mill, are deduced, Fig. 1:

$$Z_1 = H\,, (H\text{—arbitrary variable}), \tag{26}$$

$$-\left[R_0\left(\cos\kappa-\cos\theta\right)\,\sin v-\frac{d_0}{2}\cos v\right]\,\sin\alpha \\ +\left(-R_0\,\sin\kappa+R_0\,\sin\theta+pv\right)\,\sin\alpha=H. \tag{27}$$

Results:

$$\theta=\operatorname{arc}\cos\left[\frac{H+R_0\,\cos\kappa\,\sin v\,\sin\alpha+R_0\,\sin\kappa\,\cos\alpha}{R_0\left(\sin v\,\sin\alpha+\cos\alpha\right)} \\ -\frac{\frac{d_0}{2}\sin v\,\sin\alpha+pv\,\cos\alpha}{R_0\left(\sin v\,\sin\alpha+\cos\alpha\right)}\right]. \tag{28}$$

The derivatives are defined as:

$$\frac{d\theta}{dv}=\frac{1}{R_0^2\left(\sin v\,\sin\alpha+\cos\alpha\right)^2}\cdot\left[\left(R_0\,\cos\kappa\,\cos v\,\sin\alpha-\frac{d_0}{2}\cos v\,\sin\alpha \\ -\left(-p\,\cos\alpha\right)\left(\sin v\,\sin\alpha+\cos\alpha\right)R_0-R_0\left(H+R_0\,\cos\alpha\,\sin v\,\sin\alpha \\ +R_0\,\sin\kappa\,\cos\alpha-\frac{d_0}{2}\sin v\,\sin\alpha-pv\,\cos\alpha\right)\cos v\,\sin\alpha\right)\right]. \tag{29}$$

Also, see (8), the partial derivatives are defined:

$$\dot{X}_{1v}=R_0\sin\theta\frac{d\theta}{dv}\sin v-R_0(\cos\kappa-\cos\theta)\sin v-\frac{d_0}{2}\cos v; \\ \dot{Y}_{1v}=\left[(R_0\cos\kappa-\cos\theta)\cos v+R_0\sin\theta\frac{d\theta}{dv}\sin v+\frac{d_0}{2}\sin v\right]\cos\alpha \\ +(-R_0\sin v+p)\sin\alpha. \tag{30}$$

The Eqs. (25), (29) and (30) allow to write the condition (8) and to determine the shape of side mill which generates the *BC* arc from the drill's profile.

For a particular case, the values are: $D = 20$ mm, $d_O = 2.5$ mm, $\chi = 60°$, $a = 150$ mm, $p = 17.320$ mm, $\alpha = 60°$, $R_O = 20$ mm.

4 Graphical Method Developed in CATIA

The proposed graphical method supposes the following algorithm.

Two reference systems are generated in the *Generative Shape Design* module, see Fig. 5. The two reference systems are: *XYZ* as reference system joined with the Σ surface, with *Z* axis overlapped to the axis of helical surface and $X_1Y_1Z_1$ reference system joined with the future side mill tool. This last reference system has the Z_1 axis overlapped to the axis of the side mill. The distance between the origins of the reference systems is denoted with a and is measured on the common normal between the axes Z and Z_1. The value of this distance is selected from constructive reasons.

The angle between the tool's axis and the axis of the helical surface can be established as angular parameter and can be calculated as function of the helical parameter and the diameter of directrix helix of the Σ surface:

$$\alpha = \arctan\left(\frac{p_E}{\pi \cdot D}\right), \tag{31}$$

where p_E is the helix pitch and D is the external diameter of the helical surface.

In the XYZ reference system, the cylindrical helix with constant pitch is defined. This helix is the directrix curve of helical surface.

In the XY plane of this reference system, the composed profile is constructed. This profile defined the frontal section of the helical drill's flute.

Using the *SWEEP* command, the flute's helical surface is generated, having as directrix the cylindrical helix with constant pitch and as generatrix the flute's composite profile. The option for the swept surfaces is "with pulling direction", the guide curve is the helix and the pulling direction is Z axis.

A length type parameter, H, is defined. This parameter will control the position of the plane $Z_1 = H$.

Is constructed a plane parallel with X_1Y_1, at distance H from this. Changing the value of the H parameter is easy to change the position of plane where is studied the contact between the helical surface and the revolution one. Intersecting this plane with the helical surface is obtained an in-plane curve, Σ_H, see Fig. 5.

It is generated the point M as intersection between the plane $Z_1 = H$ and Z_1 axis. In the plane $Z_1 = H$ is drawn a skew composed by two line segments. The first segment, L_1 line from Fig. 5, is constricted to has an end onto the $\sum_H$ curve and the other end in the intersection point between $Z_1 = H$ plane and the Z_1 axis.

The second segment, line L_2 from Fig. 5, is constricted to have the midpoint onto the end of first segment, which is onto the in-plane curve. One more restriction is imposed, the tangency between the L_2 line and the curve $\sum_H$. After this, a last perpendicularity restriction between the two segments is applied. It is determined the intersection between the helical surface and the L_1 line.

In this way, is determined a point onto the $\sum$ surface, the C point, see Fig. 5. The normal to $\sum$ surface, which pass through this point, intersect the tool's axis. According to the "generating trajectories" theorem, this point belongs to the characteristic curve.

This intersection point is retained with command *ISOLATE* and a new point onto the characteristic curve is determined, for a new value of the H parameter.

In this way, is possible to determine a discrete form of the characteristic curve, establishing enough points on it.

The characteristic curve's coordinates as so as the axial section of the side mill are presented in Table 1.

The axial section of the side mill is presented in Fig. 6.

Table 1 Coordinates of the characteristic curve and axial section

Nr. crt.	X_1 (mm)	Y_1 (mm)	Z_1 (mm)	R (mm)	H (mm)
AB segment					
1	−25.441	−0.992	3.113	25.460	3.113
2	−28.866	−4.470	2.000	29.210	2.000
3	−30.403	−5.149	1.000	30.836	1.000
4	−31.512	−5.406	0.000	31.972	0.000
5	−32.355	−5.716	−1.000	32.856	−1.000
6	−33.002	−5.844	−2.000	33.515	−2.000
7	−33.491	−5.922	−3.000	34.011	−3.000
8	−33.821	−5.957	−3.920	34.342	−3.920
BC arc					
1	−33.831	−5.954	−3.950	34.351	−3.950
2	−33.850	−5.933	−4.000	34.366	−4.000
3	−34.122	−5.512	−5.000	34.564	−5.000
4	−34.222	−5.095	−6.000	34.599	−6.000
5	−34.159	−4.672	−7.000	34.477	−7.000
6	−33.929	−4.236	−8.000	34.192	−8.000
7	−33.521	−3.774	−9.000	33.733	−9.000
8	−32.902	−3.263	−10.000	33.063	−10.000
9	−32.010	−2.645	−11.000	32.119	−11.000
10	−31.419	−2.244	−11.500	31.499	−11.500
11	−31.073	−1.994	−11.750	31.137	−11.750

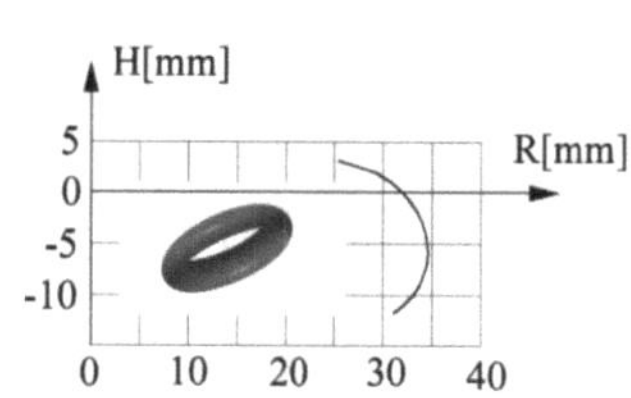

Fig. 6 Axial section of the side mill

5 Conclusions

The relative generation trajectories method as complementary new method for study of geometrical process of surfaces enwrapping, applied for determining of active profile for side mill generating a helical flute with constant pitch is an original solution for profiling tools which generate by enwrapping.

Based on the proposed method was developed an graphical algorithm in CATIA design environment for determining the characteristic curve at contact between a cylindrical helical surface with constant pitch and a revolution surface, with known axis position.

The graphical method is prove as rigorous and easy to apply, allowing to avoid determining errors which can emerge due to inexact positioning of the side mill axis.

The graphical algorithm allows knowing the coordinates of points from characteristic curve in the reference system of the side mill primary peripheral surface.

Acknowledgements This work was supported by a grant of the Romanian National Authority for Scientific Research and Innovation, CNCS–UEFISCDI, project number PN-II-RU-TE-2014-4-0031.

References

1. Litvin FL (1984) Theory of gearing. Reference publication, 1212, NASA, Scientific and Technical Information Division, Washington, DC
2. Oancea N (2004) Generarea suprafeţelor prin înfăşurare (Surfaces generation by winding), vol I–III. Galaţi University Press, ISBN 973-627-170-4, ISBN 973-627-107-2 (vol I), ISBN 973-627-176-6 (vol II), ISBN 973-627-239-7 (vol III)
3. Baicu I (2002) Cercetări privind utilizarea modelării 3D pentru algoritmizarea profilării sculelor aşchietoare (Research regarding the 3D modelling in cutting tools design algorithms), PhD Thesis, "Dunărea de Jos" University of Galaţi
4. Ivanov V, Nankov G, Kirov V (1999) CAD orientated mathematical model for determination of profile helical surfaces. Int J Mach Tools Manuf 38(8):1001–1015
5. Ivanov V, Nankov G (1998) Profiling of rotation tools for forming of helical surfaces. Intern J Mach Tools Manuf 38(M9):1125–1148
6. Berbinschi S et al (2012) 3D graphical method for profiling tools that generate helical surfaces. Int J Ad Manuf Technol 60:505–512. doi:10.1007/s00170-011-3637-3
7. Li G, Sun J, Li J (2014) Process modeling of end mill groove machining based on Boolean method. Int J Adv Manuf Technol 75:959-966. doi:10.1007/s00170-14-6187-7
8. Baroiu N, Teodor V, Oancea N (2015) A new form of plane trajectories theorem. Generation with rotary cutters. Bulletin of the Polytechnic Institute of Jassy, Tome LXI (LXV), Machine Construction, pp. 27–36, ISBN 1011-2855

Mechanistic Approach for the Evaluation of Machine Tools Quasi-Static Capability

Károly Szipka, Theodoros Laspas and Andreas Archenti

Abstract One of the greatest challenges in the manufacturing industry is to increase the understanding of the error sources and their effect on machine tool capability. This challenge is raised by the complexity of machining systems and the high requirements on accuracy. In this paper, a mechanistic evaluation approach is developed to handle the complexity and to describe the underlying mechanisms of the machine tools capability under quasi-static condition. The capability in this case is affected by the geometric errors of the multi-axis system and the quasi-static deflections due to process loads. In the assessment of these sources a mechanistic model is introduced. The model is composed of two parts, combining direct and indirect measurements. The direct measurement modelling method was applied to predict the effects of individual axis geometric errors on the functional point of machine tools. First, the direct measurement is employed to allow measuring each single machine tool axis motion error individually. The computational in the direct measurement model calculates the deviations from a given toolpath in the work space. Then, indirect measurements are used to determine the static stiffness and its variation in the workspace of machine tools. A case study demonstrates the applicability of the proposed approach, where laser interferometry was implemented as direct and loaded double ball bar as indirect measurement. The methodology was investigated on a three and a five axis machine tool and the results demonstrate the potential of the approach.

Keywords Accuracy · Static stiffness · Machine tool

K. Szipka (✉) · T. Laspas · A. Archenti
Department of Production Engineering, KTH Royal Institute of Technology, Brinellvägen 68, 10044 Stockholm, Sweden
e-mail: szipka@kth.se

V. Majstorovic and Z. Jakovljevic (eds.), *Proceedings of 5th International Conference on Advanced Manufacturing Engineering and Technologies*, Lecture Notes in Mechanical Engineering, DOI 10.1007/978-3-319-56430-2_16

1 Introduction

Our industrial society has a fundamental interest in the machine tool market which is strongly connected to the manufacturing industry. In the last 20 years the world's machine tool production and consumption have shown a growing tendency. In this growing market, increasing accuracy to fulfill tighter tolerances of manufactured components and higher flexibility of production systems have a fundamental significance. Proper instrumentation and robust evaluation methods are necessary to evaluate and initiate improvement of machine tools performance.

Accurate production can be accomplished through a controlled and deterministic manufacturing process. In respect of the complexity of machining systems (machine tool, process and their interaction), to increase the accuracy of machined part dimensions, a higher understanding of the sources of the errors is required. Several test methods and international standards have been introduced to raise awareness in the evaluation of the machine tool capability. Mechanistic approaches are focusing on the understanding of the underlying behavior utilizing measurement information and physical rules.

The purpose of this study is to introduce a novel methodology for the prediction of the machine tool errors under quasi-static condition, including geometric errors and the deviations which were induced by an arbitrary applied quasi-static force. The proposed research goal is reached through the combination of two models, which are extended for orthogonal five-axis machine tools. The established methodology consists of a direct, and an indirect measurement modelling approach. During such model building, emphasis has to be given to avoid redundancy and different type of dependencies.

The core of this paper subordinated to the quasi-static description and evaluation of machine tool accuracy. In Fig. 1, the scope of the study, the model and its border and the most important limitations can be seen. However, the short- or long-term

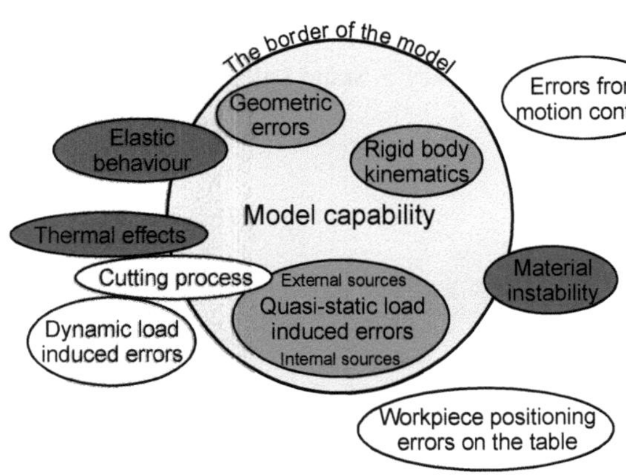

Fig. 1 The scope and the limitations of the developed model

change in the deep grey colored parameters can have an effect on the measurement results (and finally on the model output), thus they were eliminated in the computational model.

The aim of the proposed methodology is to predict machine tool errors under load and quasi-static condition for a given tool trajectory by:

- identifying and mapping the distribution of static stiffness in machine tools workspace, and
- identifying the kinematic behavior of the machine tools.

The first point is realized in the indirect measurement model, where a circular test device called Loaded Double Ball Bar (LDBB) [1] facilitates the investigation of the stiffness of machine tools by measuring the deviations in different load levels, positions, directions and feed rates. This device was used to establish the Elastically Linked System concept to associate the machining system capability to the machined part accuracy level [2]. The direct measurement model is coupled with laser interferometer measurements of error components and predict their effects in the volumetric error level using the homogeneous transformation matrix theory.

2 State of the Art

2.1 Various Sources of Error in Machine Tools

In the manufacturing process various error sources effect the difference between the nominal and the actual positions of the kinematic structure's elements. The most common categorization of these errors are the following: kinematic and geometric errors, thermo-mechanical errors, loads, dynamic forces and motion control [3]. In this approach the error sources are categorised as quasi-static condition related, dynamic, motion control and thermo-mechanical errors.

Thermal deformation can distort the machine structural loop after heat generation. The source of the heat can be an electric motor (machine spindle), the cutting process itself or different type of friction in bearings. Temperature changes in the environment provide different type of affects, which also have to be considered during quasi-static investigation.

Dynamic errors coming from the vibrations of the machine tools, which can increase the machine tool wear and decrease the surface quality of the workpiece. Dynamic behaviour of the machines can be characterized in three main quantities: dynamic stiffness, mass and damping. Although dynamic errors are often traceable their prediction and the compensation are the most difficult of all of the error sources.

2.2 *Errors Under Quasi-Static Condition*

Quasi-static error sources include kinematic errors, and quasi-static load induced errors. Geometric errors are caused by the unwanted motions of the structural components of the machine, such as of guideway carriages, cross-slides and worktables. Geometric imperfections and misalignments lead to these error motions. These imperfections can be derived for instance from the deviations in surface straightness and roughness of the guideway or the preload of bearing. However, temperature changes and other error sources make it difficult to measure geometric errors separately, it is possible to investigate the effects on the kinematic structure of the machine tool.

In case of quasi-static errors, the non-rigid body behavior has to be considered. Location errors and component errors change due to internal or external forces. The weight of the workpiece and the moving carriages can have a significant influence on the machine's accuracy due to the finite stiffness of the structural loop [4].

2.3 *Parameters of Machine Tool Geometric Errors*

The component errors of a machine tool can be classified into four groups according to their characteristics, measurement instrumentation and implementation [5]: linear displacement errors, angular errors, straightness-parallelism orthogonality errors (or so called location errors) and spindle thermal drift. The linear displacement errors and angular errors together are usually called component errors. The static and dynamic errors of the spindle can be significant error sources. Investigations showed that the spindle drift characteristics are more complex than the other error components in the machine tool. In this study the static and dynamic errors due to the spindle imperfections are neglected.

2.3.1 Component Errors

A linear axis of the machine is designed to travel precisely along a reference straight line and stop at the predefined position. From practical point of view, there is deviation between the actual path and the reference straight line at the guideways. The component errors of the linear axis are the linear positioning, the two straightness errors, and the three angular errors (yaw, pitch and roll). The component errors in case of a rotational axis are the two radial errors, the axial error, the two tilt errors and the radial positioning error. The component errors can be seen in Fig. 2 with the notation of ISO 230-1 [6].

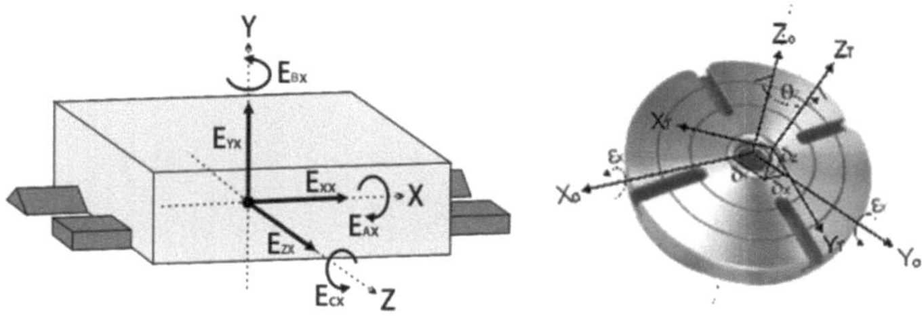

Fig. 2 Representation of the six component errors of a linear (*left*) and a rotary (*right*) axis

2.3.2 Location Errors

However, in literature several different approaches exist, the most common definition for location errors are the following. The linear axis location errors represent orientations of its reference straight line in the reference coordinate system [7]. In case of a rotary axis, location errors are defined analogously, representing the position and orientation of the axis average line of a rotary axis, i.e. the straight line representing the mean location and orientation of its axis of rotation [8]. Typical location errors are the squareness, parallelism and orthogonality error. In Table 1 relevant component and location errors are summarized, for a three- or five-axis machine tool.

For a five-axis machine tool 23 location errors can be differentiated. Since all zero positions of linear and rotary axes can be set to zero when checking the geometric accuracy, 5 local errors can be neglected. The local errors connected to

Table 1 The component and the location errors of a three- or five-axis machine in case of A and C rotary axis

Linear errors				Rotational /Angular errors		
Translational axis	Positioning	Straightness		Roll	Pitch	Yaw
X axis	^{E}XX	E_{YX}	^{E}ZX	^{E}AX	^{E}BX	^{E}CX
Y axis	E_{YY}	E_{XY}	^{E}ZY	^{E}BY	^{E}AY	^{E}BY
Z axis	^{E}ZZ	^{E}XZ	^{E}YZ	^{E}CZ	^{E}BZ	^{E}AZ
Rotational axis	Axial	Radial		Ang. Pos.	Tilt	
A axis	^{E}XA	^{E}ZA	^{E}YA	^{E}AA	^{E}BA	^{E}CA
C axis	^{E}ZC	^{E}XC	^{E}YC	^{E}CC	^{E}AC	^{E}BC

Location Errors E_{AOY} squareness error of Y to Z
E_{COX} squareness error of X to Y
E_{BOX} squareness error of X to Z
E_{BOA} squareness error of A to Z
E_{COA} squareness error of A to Y
E_{YOC} Y offset error from A to C
E_{AOC} parallelism error of C to Z in the reference ZY plane
E_{BOC} parallelism error of C to Z in the reference ZX plane

the spindle are simplified, which means 4 local errors less to be considered. With choosing the Z axis as the primary, the final 8 errors are reached.

2.4 Stiffness of Machine Tools

The stiffness of the machines is generally understood as the ability of a structure to resist deformation due to applied load. The structural stiffness gives the capacity of a structure to resist deformations induced by applied loads. This capacity can be described by its stiffness matrix. The static and dynamic stiffness are defined differently and the stiffness values for the same structure regarding the two definitions will differ.

The scope of this research considers the quasi-static behaviour of machines, thus the loading process has to be slow (quasi-static condition). For periodic loading it means 1 or 0.5 Hz or lower frequency, according to the damping of the system to avoid transient movements. In general, it can be stated that the weakest link, the component where the stiffness is the lowest in the series of the stiffness chain, will dominate the stiffness response of the whole machine. This means a high level of sensitivity and a significance to investigate the stiffness response of machine tools.

3 The Proposed Methodology

3.1 Case Study

A case study including a three- and a five-axis machine tool (Table 2) demonstrates the applicability of the proposed approach. Measurements were implemented on the three axis machine tool, however a model was developed to be able to handle five-axis machine tools as well.

Table 2 The specification of the investigated machine tools

	"Five-axis"	"Three-axis"	"Five-axis"	"Three-axis"
	Effective travel range		Resolutions	
X axis (longitudinal)	1000 (mm)	1000 (mm)	0.001 (mm)	0.001 (mm)
Y axis (transverse)	1100 (mm)	510 (mm)	0.001 (mm)	0.001 (mm)
Z axis (vertical)	750 (mm)	561 (mm)	0.001 (mm)	0.001 (mm)
A axis	230 (°)		16 (")	
C axis	360 (°)		9 (")	

3.2 Direct Measurement Model

Investigations by constructing models for identifying the effect of geometric or kinematic errors on motion accuracy of various types multi-axis machine tools have been made from several different approach. An analytical quadratic model was introduced by Ferreira and Liu [9] where the coefficients of the model were obtained by touch trigger probe measurement. Slocum [10] and Donmez [11] introduced two methodologies with several similarities. Both of them were carried out with case studies for describing geometric and thermal errors.

After the rigid body assumption, homogeneous transformation matrices (HTM) were used to predict and compensate the identified errors. Lin and Shen [12] established a model for five-axis machine tool using the matrix summation approach. Soons, Theuws and Schllekens [13] presented a general methodology and an application in two case study which accounts for errors due to inaccuracies in the geometry, finite stiffness, and thermal deformation of the machine's components. Special statistical techniques were applied to the calibration data to obtain an empirical model for each of the errors. Suh et al. [14] focused on rotary axes in their work. They established a model for the calibration of the rotary table including a geometric error model and an error compensation algorithm with experimental apparatus for CNC controllers.

As HTM theory including rotary axes is well discussed in the state of art, the focus in this paper will be on the data integration and the outcomes of the model.

3.2.1 Direct Measurement of Geometric Errors

A direct measurement method is required to characterize quasi-static positioning capability and measure the individual position dependent and independent geometric error components (e.g., linear positioning, straightness, angular or squareness errors). Laser interferometer systems are dedicated for this task. Most of the instruments enable to measure in low uncertainty and high precision compared to other methods. Since all of the approaches follow the principle of the bottom-up modelling are sensitive to the uncertainty of the contributing components, the uncertainty requirements of the measurements have high priority.

In the case study, measurements were executed on a three axis machine to characterize the geometric errors in Table 1. The laser interferometer and precision leveller (in case of roll) measurement results can be seen on (Fig. 3).

3.2.2 Geometric Error Model Development

In order to predict the functional point (defined in ISO 230-1 [6]) with respect to the nominal position, the first step is to model the kinematic structure, to express the motion of the rigid axes and the linking joints. The local and global coordinate

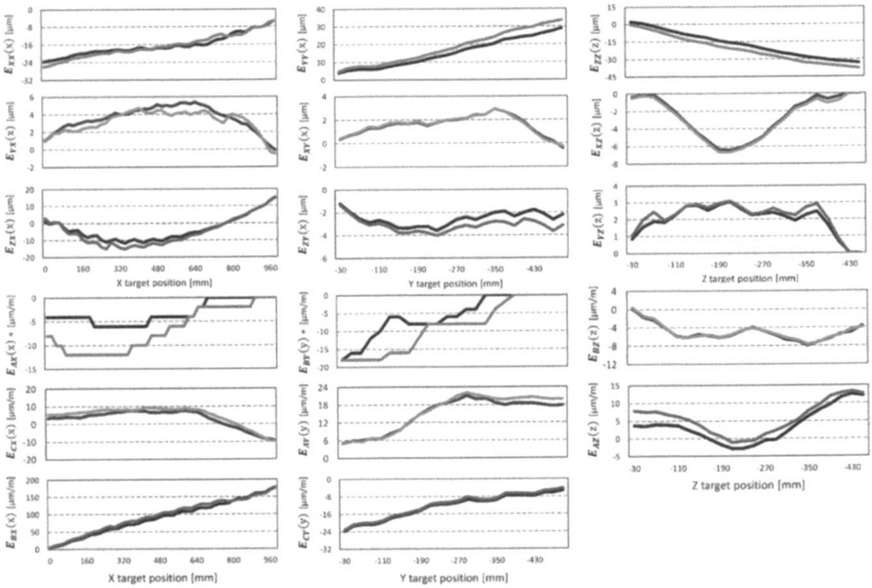

Fig. 3 The measured component errors in different directions (*light* and *dark color* marks the different directions on the axis)

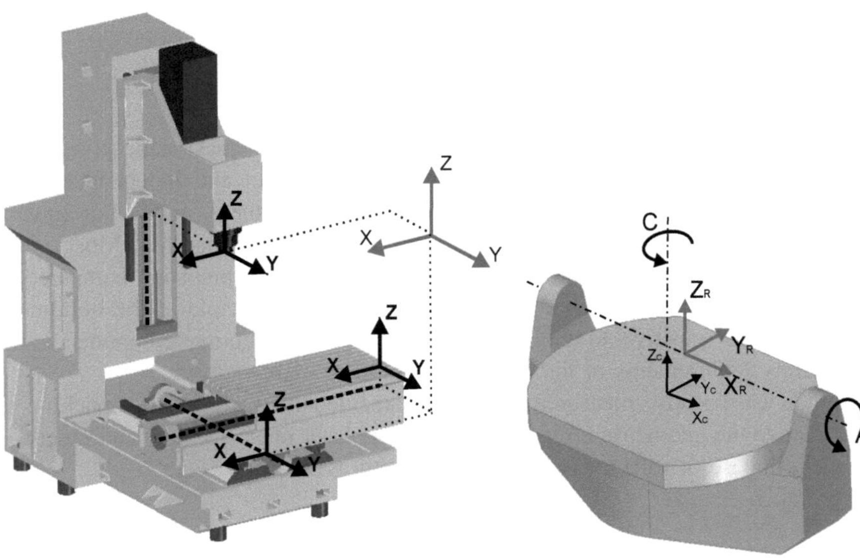

Fig. 4 The assignment of the coordinate frames for the three-axis machine tool. The *red* coordinate system corresponds to the reference coordinate frame. The *black* ones correspond to the local coordinate frames of the X, Y and Z axis (color online)

frames characterize the relative orientation and position of the axes. These have to be chosen to help the integration of the measurement data to the model. The global coordinate frame is defined to meet the corresponding 0 coordinate for each axes for a three axis machine. For a five axes one the global coordinate frame is assigned on the rotational axis, which is closer to the machine bed. For the chosen machine tools the assignments can be seen on the Fig. 4. The model is composed to be applicable for orthogonal axis machine tool with maximum three translational and two rotational axes.

The position independent location errors, and the position dependent component errors for each axis give the considerable error parameters (Table 1), which are measured separately. The measured error parameters are coupled into HTMs to express the relative spatial position and the coordinate transformation of the rigid body frames. All the HTMs can be derived from the sequential multiplication of the three basic translational and three basic rotational HTM:

$$\mathrm{ERR(axis)} = \begin{bmatrix} C\varepsilon_Y \cdot C\varepsilon_Z & -C\varepsilon_Y \cdot S\varepsilon_Z & S\varepsilon_Y & \delta_X \\ S\varepsilon_X \cdot S\varepsilon_Y \cdot C\varepsilon_Z + C\varepsilon_X \cdot S\varepsilon_Z & C\varepsilon_X \cdot C\varepsilon_Z - S\varepsilon_X \cdot S\varepsilon_Y \cdot S\varepsilon_Z & -S\varepsilon_X \cdot C\varepsilon_Y & \delta_Y \\ -C\varepsilon_X \cdot S\varepsilon_Y \cdot C\varepsilon_Z + S\varepsilon_X \cdot S\varepsilon_Z & S\varepsilon_X \cdot C\varepsilon_Z + C\varepsilon_X \cdot S\varepsilon_Y \cdot S\varepsilon_Z & C\varepsilon_X \cdot C\varepsilon_Y & \delta_Z \\ 0 & 0 & 0 & 1 \end{bmatrix} \tag{1}$$

Where δ expresses translation and ε rotation along/around the corresponding axis. The operator S = sin and C = cos. For a certain translational or rotational axis the error matrix can be derived with the substitution of the dimension of the movement and the application of the small angle approximation. The aim of these HTMs is to express the proper spatial effect of the error sources in the workspace.

At the end of the process the position and the orientation of the functional point is determined with respect to the table. The relation between the functional point and the measurement points is a sensitive part of the model as we measure in one zone of the workspace and derive conclusion in another region. To avoid redundancies, the level effect of the angular and tilt errors due to the offset between the functional point, the measurement point was compensated.

3.2.3 Measurement Data Integration to the Model

The error motion of components needs to be known in the trajectory of the functional point to be able to predict its effect on the workpiece geometrical accuracy. Thus the measurement of the displacement errors should be carried out at the toolpath, where the manufacturing will be executed. The uncertainty of the predicted errors relates to the distance between the functional point and the measurement point.

Both linear and rotary axes are sensitive to this deviation, due to angular and tilt errors. For linear axes ISO 230-1 [6] gives a guideline on the effect of angular

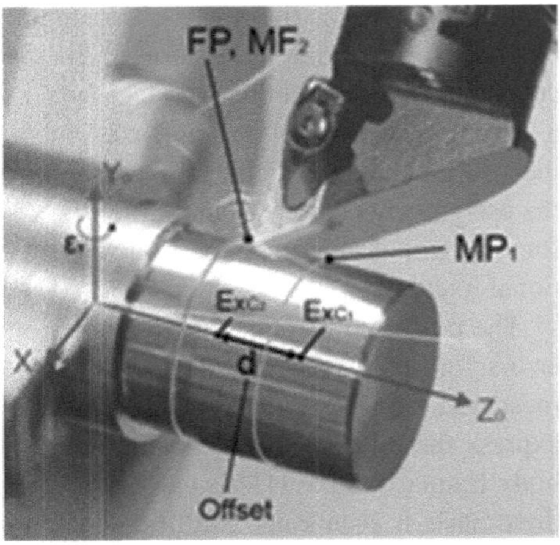

Fig. 5 The level effect of the tilt error of C axis in Y direction (εY) on measuring the radial error of C axis in X direction (ExC) due the offset between the functional point (FP) and the measurement point (MP)

errors. In case of a rotary axis similar effects can be observed due the deviation between measurement and the functional point. Figure 5 shows the level effect of the tilt error of C axis in Y direction on the measured radial error of C axis in X direction due the offset between the functional point and the measurement point. If the measured points are well documented, thus the distances between the functional points and the measurement points are known, the effects of the tilt errors can be corrected from the radial errors. In this research, measurements were taken on the three-axis machine, where separated measurement points were defined for each axis of motion.

3.2.4 The Predicted Geometric Errors

After each error source was integrated to the model and the HTMs were established to express the proper special transformation, the error model is able to predict the superposed effect of the geometrical inaccuracies of each axis.

In Fig. 6 the predicted geometric errors of the AFM R-1000 three-axis machine tool can be seen in case of a circular toolpath. After a sensitivity analysis it can be stated, that the pitch error of X axis has a dominant, around 70% contribution in the predicted errors.

3.3 The Indirect Measurement Model

The proposed research goal in the indirect measurement model is the description of the effect of quasi-static forces, which is related to the static stiffness of the machine tool. In this approach, a computational model is developed to analyze forces and deflections, based on an indirect measurement schema, which allows measuring

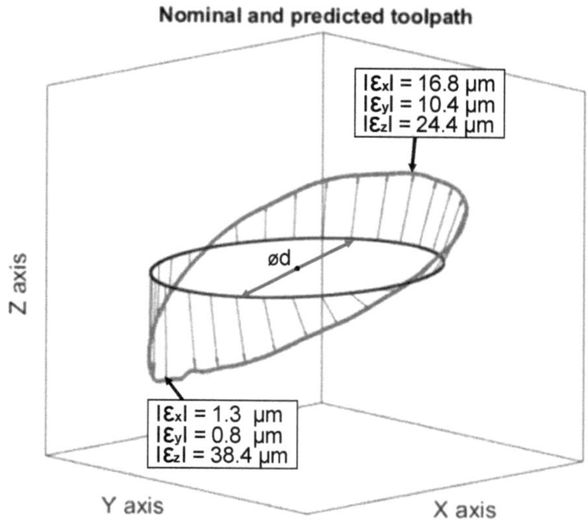

Fig. 6 The predicted geometric errors around a circular toolpath of the AFM R-1000 (Ød = 300 mm)

superposed errors of parallel motion of more machine axes at the same time. Investigations were executed in one certain plane of the machine tools workvolume.

3.3.1 The Loaded Double Ball Bar Measurement Process

In the indirect measurement modelling an indirect test method schema was planned with the application of LDBB to measure machine tool deflections in loaded and quasi-static conditions. Measurements were taken in one plane of the work volume (called workplane). The workplane was investigated in four different positions, which were in premeditated intersections (see Fig. 7).

The linearity of the system is important to be considered for an accurate description of the deformations. Repeated measurements on different load levels need to indicate linear load-deflection characteristic to accurately apply the model.

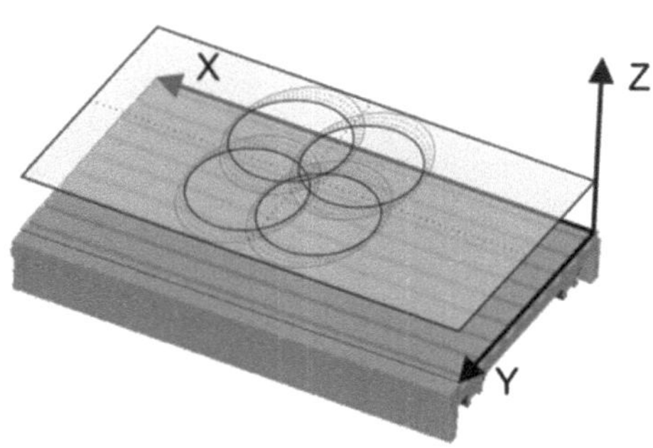

Fig. 7 The proposed composition of the different measurement positions on the machine tool table

3.3.2 Stiffness Prediction for Machine Tools

The computational model of the top-down approach is applied to describe the variation of the position and direction dependent static stiffness of the machine tool. Figure 8 shows the measurement instrument and setup applied for the characterisation. The loaded double ball bar (LDBB) enables the measurement of machine tool deflections under quasi static and (adjustable) loaded conditions.

Regarding the changing relative position and/or orientation between the mechanical components during motion, the resultant deflections will differ in case of different direction (e_f) of the force and different positions (p) at the workspace. Consequently, the static stiffness of the machine, *K(p, ef)* varies as the function of the position and the direction of the applied force:

$$K(\mathrm{p}, \mathrm{e_f}) = \begin{bmatrix} k_{xx}(\mathrm{p}, \mathrm{sgn}(\mathrm{e_{f,x}})) & \dots & \dots \\ \dots & k_{yy}(\mathrm{p}, \mathrm{sgn}(\mathrm{e_{f,y}})) & \dots \\ \dots & \dots & k_{zz}(\mathrm{p}, \mathrm{sgn}(\mathrm{e_{f,z}})) \end{bmatrix} \quad (2)$$

It has to be noted that certain deflections are the result of internal, assembly loads of the machine, including the effect of the own weight of the machine and the preloading of components. However these parameters are included in the presented model through the measurements, their change can affect the result.

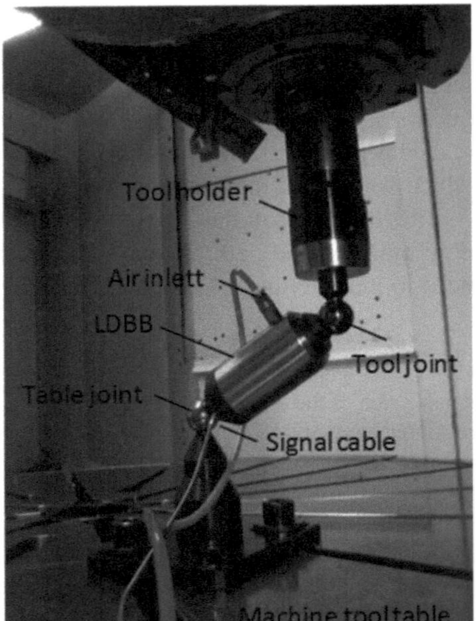

Fig. 8 The loaded double ball bar system [1]

The stiffness needs to be characterized in the machine coordinate system. For this:

- the directional stiffness values have to be predicted from the measured local point stiffness in the spherical coordinate system of the measurements
- the spherical coordinates have to be transformed to the corresponding Cartesian coordinate system of the machine tool

During the top-down modelling, the whole system is decomposed into sub-systems. The deviation profile around the circular measurement trajectory of the LDBB due to the quasi-static force is related to the variation of the static stiffness response of a machine tool in its workspace.

With the LDBB superposed error of parallel motion of more machine axes and their deflections are measured at the same time. In this case the aim of this computational model is to decompose the indirect measurement data and use them as inputs and calculate the static stiffness at different points of the workspace.

3.3.3 Static Stiffness Variation

The output of the computational model is the measured stiffness in the measurement points. The results show the variation of the point stiffness in the workspace. At the intersections of the circular toolpaths, the same points of the investigated workplane have different stiffness values. This is the result of the differences in the stiffness response of the system the direction of the applied forces. This confirms the matrix structure in Eq. 2. The result deviates between 8.4 and 10.8 N/μm. The positional dependency of the static stiffness in certain direction can be investigated in Table 3.

The characteristic of the result shows in the investigated plane of the workspace where the point stiffness of the machine is definitely higher. From the static stiffness point of view, this region is more capable to execute accurate manufacturing because of the higher reduction of the unwanted relative displacement of the tool and the table in different directions. The results also enhance the fact that is stated in ISO 230-1 [6] about the characterization of static compliance, that the standardized measurements in one position of the workspace can give only a brief knowledge about the state of the machine tool.

Table 3 The variation of stiffness depending on the position

	Pos. 1 (%)	Pos. 2 (%)	Pos. 3 (%)
ky+			
Pos. 1	–		
Pos. 2	1.4	–	
Pos. 3	17.3	15.7	–
Pos. 4	10.7	9.1	−5.7
ky-			
Pos. 1	–		
Pos. 2	1.2	–	
Pos. 3	13.4	12.1	–
Pos. 4	8.8	7.5	−4.0
kx+			
Pos. 1	–		
Pos. 2	0.2	–	
Pos. 3	2.8	2.6	–
Pos. 4	−0.7	−0.9	−3.5
kx−			
Pos. 1	–		
Pos. 2	−1.2	–	
Pos. 3	1.9	3.1	–
Pos. 4	−0.2	0.9	−2.1

4 Conclusions

The outcome of this study is a novel evaluation method for the prediction of machine tool errors under quasi-static load. The method is able to take into consideration the effect of the variation of stiffness and geometric errors in the workspace of machine tools. The methodology was established through the combination of the direct and the indirect measurement modelling approach, which is required in case of the combination of direct and indirect measurements.

The direct measurement modelling approach was applied to predict the aggregated effects of the geometric errors in the workspace of a machine tool. The robustness of the results depends on the accuracy of the measurement data, since with the measured error components the uncertainties are aggregated as well. The deeper investigation of the contributing uncertainty factors is advisable for this approach. The sensitivity analysis of the results can highlight the seriousness of each error sources. In the case study, the pitch error of X axis had a significant effect on the results, a correction or compensation for this error can be the most effective intervention from the geometric error point of view.

The indirect measurement modelling method was applied to predict the variation of the static stiffness in the workspace of a machine tool, offering a more accurate characterization of the quasi-static state of a machine tool. During the decomposition the most critical part is the separation of the kinematic errors from the deflections. In this case the evaluation of the composed methodology can be done through experimental cutting, by comparing the part dimensional inaccuracies during finishing and roughing. For the finish the geometric errors will dominate, while for roughing the deflections of the tool and the machine, so the stiffness response of the system dominates. However dynamic effects and the stiffness of the tool itself have to be taken into consideration.

Acknowledgements The authors wish to thank Dr Mikael Hendlind for research contribution on kinematic modelling and M.Sc. Jonny Gustafson for his contribution in the laser interferometer measurements. Centre for design and management of manufacturing systems (DMMS) at the Department of Production Engineering at KTH Royal Institute of Technology is acknowledged for financial support.

References

1. Archenti A, Nicolescu M, Casterman G, Hjelm S (2012) A new method for circular testing of machine tools under loaded condition, 5th CIRP Conference on High Performance Cutting, vol 1, pp 575–580
2. Archenti A (2014) Prediction of machined part accuracy from machining system capability, CIRP Ann Manuf Technol 63: 505–508
3. Schwenke H, Knapp W, Haitjema H, Weckenmann A, Schmitt R, Delbressine F (2008) Geometric error measurement and compensation of machines—an update. CIRP Ann Manuf Technol 57(2):660–675
4. Schellekens P, Vermeulen H, Vermeulen M, Wetzels S, Pril W (1993) Design for precision, current status and trends. Ann CIRP 42(2):557–586
5. Donmez MA, Blomquist DS, Hocken RJ, Lui CR, Barash MM (1986) A general methodology for machine-tool accuracy enhancement by error compensation. Precision Eng J Am Soc Precision Eng 8(4):187–196
6. International Organization for Standardization, ISO 230, Test code for machine tools–part 1: geometric accuracy of machines operating under no-load or quasi-static conditions, 2012
7. Ibaraki S, Knapp W (2012) Indirect measurement of volumetric accuracy for three-axis and five-axis machine tools: a review. Int J Autom Technol 6(2):110–125
8. International Organization for Standardization, ISO 230-7, Test code for machine tools–part 7: geometric accuracy of axes of rotation, 2013
9. Ferreira Placid M, Richard Liu C, University Purdue, Lafayette West (1991) An analytical quadratic model for the geometric error of a machine tool. J Manuf Syst 5(1):51–64
10. Slocum AH (1992) Precision machine-design—macromachine design philosophy and its applicability to the design of micromachines. IEEE micro electro mechanical systems: an investigation of micro structures, sensors, actuators, machines and robots, pp 37-42
11. Donmez MA (1986) A general methodology for machine-tool accuracy enhancement by error compensation. Precision Eng J Am Soc Precision Eng 8(4):187–196
12. Lin Y, Shen Y (2003) Modeling of five-axis machine tool metrology models using the matrix summation approach. Int J Adv Manuf Technol 21:243–248
13. Soons JA, Theuws FC, Schellekens PH (1992) Modeling the errors of multi-axis machines: a general methodology. Precision Eng 14(1):5–19
14. Suh S, Lee E, Jung S (1998) Error modeling and measurement for the rotary table of five-axis machine tools. J Adv Manuf Technol 14:63–656

Assessment of Assembly Process Complexity and Modularity in Mass Customized Manufacturing

Slavomir Bednar, Jan Modrak and Zuzana Soltysova

Abstract Consumers' market is made up of huge number of products as a direct response to the ever increasing demands of customers. This situation is closely related to constantly evolving technical and technological industries. The rapidly increasing demand on products and variety of functional properties enabled a series of fast-growing and emerging processes and operations all over the industry. Unfortunately for companies, such variety results with a complexity of both, products and related processes. This paper deals with the external-product and internal-process complexity in the view of its usability within a mass customized manufacturing. A novel method to identify and measure product and process complexity of an assembly station is proposed on the basis of the number of product and process variants. Subsequently, the aggregated complexity ratios can provide a complexity map of an entire production line. The novel complexity assessment method is applied and verified on the case automotive manufacturing plant. As a result of this application, the newly-proposed complexity assessment methodology is highly useful especially in customized automotive production as an effective tool for assessing the level of complexity of a selected department, or when introducing a new installation. The newly-proposed methodology can be also adopted by a company to decide on what part or sub-module can be produced by external supplied—transfer of complexity—in order to decrease own process complexity and to keep the extent of variety offered to customers. Finally, based on the case application and verifications, the complexity assessment methodology proposed in this paper is highly applicable in the mass customized manufacturing.

Keywords Complexity · Variety · Mass customization · Automotive · Assembly

S. Bednar (✉) · J. Modrak · Z. Soltysova
Faculty of Manufacturing Technologies with a Seat in Presov,
Technical University of Kosice, Bayerova 1, 080 01 Presov, Slovakia
e-mail: slavomir.bednar@tuke.sk

V. Majstorovic and Z. Jakovljevic (eds.), *Proceedings of 5th International Conference on Advanced Manufacturing Engineering and Technologies*,
Lecture Notes in Mechanical Engineering, DOI 10.1007/978-3-319-56430-2_17

1 Introduction

Today's consumer industry is made up of a huge amount of products and their variations, which are in a direct response to the ever increasing demands of customers, not least closely related to the constantly evolving technical and technological equipment of companies. These rapidly increasing demands on products and their functional properties resulted in fast-growing and emerging processes and operations all over the mass-customized (MC) manufacturing industry. In mass-customized manufacturing, a system has to be aware of any product variant to be produced on the same line [1]. It of course influences not only production but also a management and the entire firm and its functions all across the chain. Applying just a very straight principle of breaking up production process into tasks and keeping products standardized through modular production, the principle has a huge cost saving potential, without any doubts [2].

Unfortunately, MC development brought companies to an "unpredictable" and "complex" behaviour of their internal processes [3]. This depends on the type and nature of the production but it is a valid condition of the MC paradigm. Then, managers of production units must generally focus, among others, on incremental product innovation and separated product and process development. If these partial conditions are satisfied, then MC as a strategy can be operated with minimum costs and very efficiently.

This paper discusses the possibility to measure and decide on the less complex assembly line as a whole on the basis of the product and process variety-induced complexity. The novel method proposes how to easily and effectively assess any product and process structure in terms of mass customized manufacturing knowing that different stages of assembly process provide a variety of sub-products using a variety of processes. The objective of the paper is to prove that the newly-proposed method is applicable in terms of purely MC environment—in automotive environment. The method will be used to benchmark two individual takts with the same level of variety against other. It will be shown that even if the sub-product variety reaches the same values, complexity of individual work-stations is not the same depending also on the component modularity. The rest of the paper is organized as follows: In the next sections we will briefly describe the principles of the product and process complexity, their measurement and setting. Later, the methods will be applied on two individual assembly takts similar by the level of variety, not by the operations performed. Mutual relations of the complexity indices will be further explained in terms of their applicability.

2 Related Work

Complexity-related problems arising in a manufacturing are frequently discussed in connection with mass customization. Moreover, complexity management is considered as an important topic among practitioners and academics alike. At the same time, it is not easy to precisely define complexity in manufacturing as there are various view on the production itself. Tiihonen et al. [4] argue that the complexity of production is directly related to the degree of modularity of the product. Probably, any definition of the manufacturing complexity would be necessarily beholden to different types of products, as there are even many definitions of the term 'product configuration' (see e.g., [5–9]). However, ability to measure the complexity in production using a reliable variety-based metric would allow benchmarking of the concurrent or alternative product and process variety platforms. It is possible to identify several pertinent facets of complexity in this domain. Calinescu et al. [10] provided comprehensive view on the various aspects of manufacturing complexity including product structure complexity. According to Jiao et al. [11], a product structure is defined in terms of module types, while product variants derived from this product structure share the same module types and take on different instances of every module type. Liu et al. [12] defines production structure by tree levels, for example: systems, sub-system, modules, in which the components can be added, modified or deleted on each level. Fredendall and Gabriel [13] pointed out that there are four building elements of the complexity that seem to contribute especially to product variety-induced and process complexity. These are: (1) the number of assembled components on entry to mass customized assembly system, (2) the number of manufactured items, (3) the number of levels in a product assembly structure, and (4) the degree of part commonality. It is evident, that the number of assembled components and the number of manufactured products definitely increases the complexity of scheduling and even material control and directly influences also so called process complexity. But the number of components on entry to the assembly system and the number of levels in the product structure independently contribute to the level of product variety-induced complexity.

Several definitions of manufacturing complexity have been provided so far but the very first definition is associated with the Shannon's information theory [14] related to the amount of information (in bits) in uncertainty of information system. From this approach, it is evident that the fewer processes, machines and/or product configurations, the lower is the overall complexity of the system. Zhu et al. [15] proposed a measure of complexity based on the choices that the operator has to make at the station level.

Kim [16] introduced number of metrics for complexity on the basis of system components, elements and their relations. These measures cover majority of system elements but cannot be extended to other manufacturing domains, except for cell production. Their metrics have been further applied and even developed in terms of mass customization by authors [17, 18].

Kampker et al. [19] categorized the term 'product variety' into two types—internal and external. At the same time they emphasize that "product architecture and technology determine the ratio between the varieties externally offered to the market and internally produced". According to Grussenmeyer and Blecker [20], there is a need for a novel complexity management method to support managers in taking proper new product development decisions for the selection of suitable product and related process design alternative.

3 Proposed Approaches to Quantification of Product Complexity and Process Modularity

This section aims to present a methodology to determine the extent of so called product complexity and process modularity via the number of unique product (initial station sub-components) and process (non-variable) alternatives.

Principally, two common types of complexity can be seen in manufacturing [21, 22]:

- Product variety-induced complexity—as number of unique sub-components on entry to assembly station [23];
- Process complexity.

Process complexity can be viewed from several aspects. Accordingly, approaches to their measurement are divided into three categories: topological, operational and dynamic complexity. The first of them is dealing only with structural characteristics (layout of machines), the second one is focused on inclusion of crucial static properties (takes the composition of parts, their scheduling and technological times of operations, etc., into account). The dynamic complexity aims to investigate aspects of complexity depending on time.

Modularity of processes is considered to be an important indicator of the preparedness for MC. There are a number of approaches to quantify the level of modularity. One of them will be applied for the purpose of our research.

3.1 Assessment of Product Complexity

Our efforts in this section will focus on the measurement and assessment of the product variety options. At the very beginning, combinatorial-based methods to quantify product configurations and variations representing product variety extent have been outlined in our previous works [24–26].

In principle, four types of assembly components may appear in any customized assembly. These components are then combined according to technical requirements of the future product and form at least one product configuration, as can be seen on an example in Fig. 1, where two stable, one voluntary optional, two delimited optional and two compulsory optional components enter the assembly process. Assembly worker then chooses the right set of the components to mount one of the 12 possible scheduled product configurations or a sub-product configuration.

There are several possible approaches to measure product component-based complexity. In this work the Axiomatic design-based method by Suh [27] will be transformed and used.

It is based on the transformation of a bipartite assembly graph as in Fig. 1a into a design systems matrix (DSM) where interactions among initial assembly components and resulting product configurations are noted.

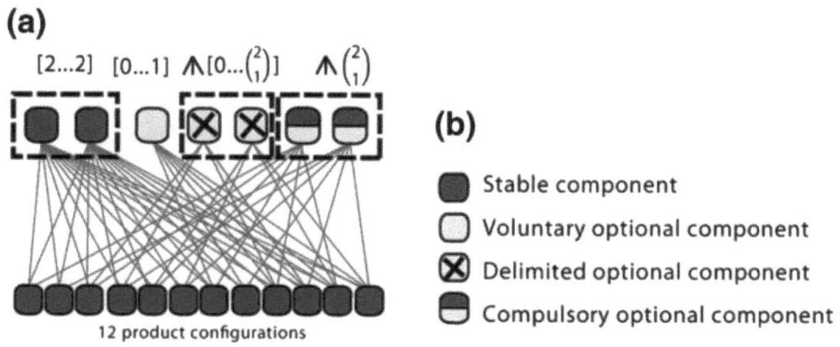

Fig. 1 Assembly structure example with 7 initial assembly components (**a**) explaining four component types in MC (**b**)

Then, the number of interactions per design parameter (DP) are collected and used to obtain so called Systems design complexity (*SDC*) indicator originally derived by Guenov [28], as follow:

$$\mathrm{SDC} = \sum N_j \ln N_j \tag{1}$$

where N_j is the number of interactions per DP.

More details on the method and a simple practical cases from a mass customization environment can be found in our previous works, e.g. [29].

3.2 Assessment of Process Modularity

The principles of modularization in production are based on the groups of operations organized in modules in such manner that the completion of a customized product is performed in production modules, whose number and sequence is individually determined. Process modularity assessment method adopted in this paper aims to benchmark concurrent/alternative production structures already in the design stage of the production layout, or when incorporating new function/physical module into an existing assembly line.

For this purpose, we apply a method of Blackenfelt [30]. The focus is on the maximization of intra-module relations while minimizing inter-module relations. Two following individual sub-metrics have been selected for our purposes:

- Module interdependence (*MI*), as a ratio of the sum of intra-module and intra- and inter-module relations. In case $MI = 1$, system is completely modular with no interactions among modules.
- Average ratio of potential (*ARP*) calculated as ratio of relations inside of modules and the number of potential relations inside them, averaged by the number of modules in the assessed process structure.

Applicability of these sub-metrics is shown in Fig. 2, where synthetic example is used.

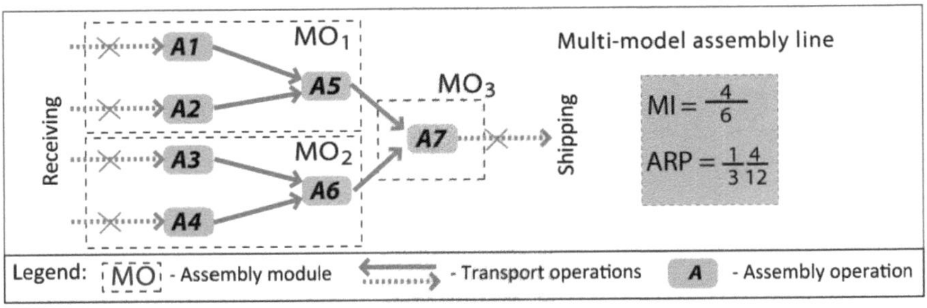

Fig. 2 Calculation example of process modularity

The assessment is based on the principle that the higher are the indices *MI* and *ARP*, the higher is the modularity level.

4 Application of the Proposed Approaches on a Case Study

Subject of the case study is an automotive assembly of a complete SUV car model currently made in Slovakia. For this purpose, we will use the data of an assembly of a single car model. The assembly consists of so called Teams 001 to 036. The teams have equally distributed the assembly operations to finalize the car model or a huge number of variants (8.2×10^{18} car configurations). The purpose of this study is not to verify the summary value of available car model alternatives, but to decompose the this value into product and process variability through complexity and modularity, as each assembly station has to cope with the "popular" product variety through very complex process variety. Next, the proposed metrics, product variety-induced complexity and process modularity will be applied on the case models of assembly Takt A and B to compare their complexity and modularity levels.

First, Case model A—Takt 096 is practically a single assembly work-station with a number of workers performing the pre-defined operations on a car model within a frame of a line takt time. Takt 096 is a set of a very important tasks, connecting the in total 14 motor variants to water-cooling system, water pump and ignition system. In total, 52 individual input sub-components enter the takt, and they are being processed by exactly 40 non-variable, stable processes. Some of the processes are variable and depend on the sub-component to be assembled due to customer selection of a motorization.

Having the process of the Takt 096 analysed, a calculation of a product variety-induced complexity has been provided, as can be seen in Fig. 3 where the product structure with a 52 initial components results with 14 work-shop variants. Then, the bipartite representation of a product structure can be transformed into a SDM of a takt for the calculation of the SDC indicator.

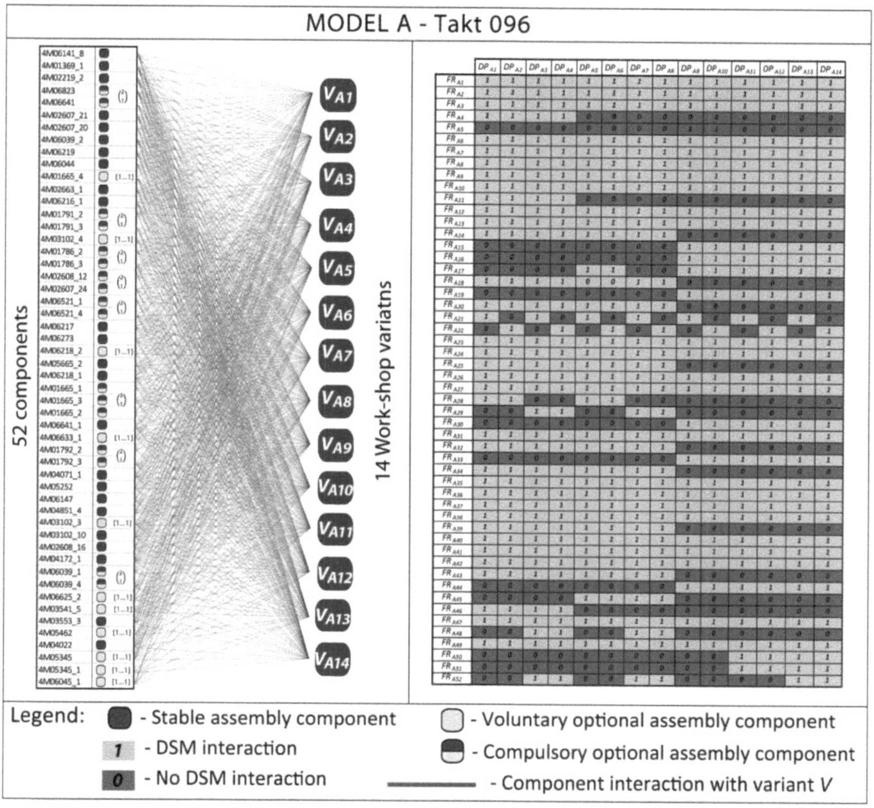

Fig. 3 Assembly structure of Takt 096 with 52 FRs and 14 DPs transformed into a design matrix

Case model B—Takt 041 is again a single assembly work-station with a number of workers performing the operations on a car model and its variant. Takt 041 is a set of a tasks connecting again 14 station variants. There are process such as placing the battery, connecting cables to the battery terminals, connection of selected electronic devices to a central fuse box, night-vision controller and mounting of left and right rear lights. In total, 54 individual input sub-components enter the takt, and they are being processed by exactly 47 non-variable, stable processes. Again, some of the processes are variable and depend on the sub-component. Having the process of the Takt 041 analysed, a calculation of a product variety-induced complexity has been provided, as can be seen in Fig. 4, where the product structure with a 54 initial components results with 14 work-shop variants.

To obtain the adopted value of modularity of the two assembly models, a generic representation of all process had to be constructed to provide a relevant base for modularity assessment. From these models, the values of intra- and inter-relations among the process modules have been counted for every possible station variant. Then, the generic models could be constructed.

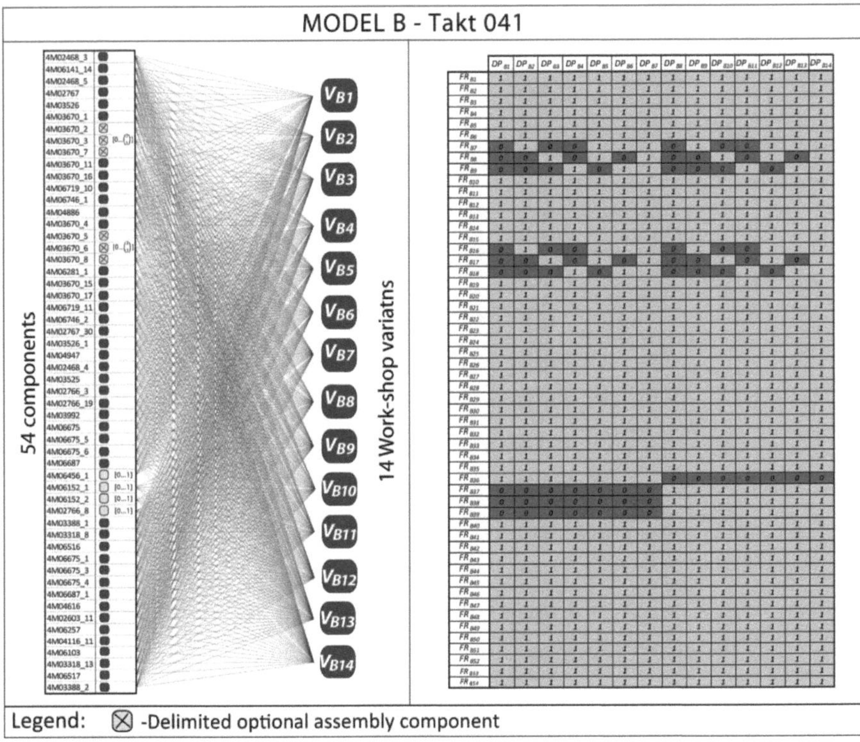

Fig. 4 Assembly structure of Takt 041 with 54 FRs and 14 DPs transformed into a design matrix

The following figure presents a graphical representations of the Assembly Model A and Model B with related values of Modularity metrics *MI* and *ARP* (see Fig. 5).

As can be seen from Fig. 5, customizable assembly modules in automotive flow assembly line do not involve high levels of interactions among work-station teams, rather they work independently on two-three places around the assembled car. The main condition of the multi-model lean line is the takt time. Both models A and B consist of nine individual process o modules.

In these models, there are a number of intra-module relations depending on the selected product variant by customer.

To obtain the values of modularity indices, process maps of each work-station variant had to be analysed, subsequently, related inter- and intra-relations have been counted for whole model altogether.

In Model A assembly (Fig. 5a) illustrates a work-shop setting of Takt 96 with nine individual process modules. To calculate the MI_A, there are in total 349 relations inside all modules of 14 possible work-shop output variants. The number of all relations, including the ones among modules is 469; $MI_A = 0.744$. The value of ARP_A is 349/3906 = 0.089.

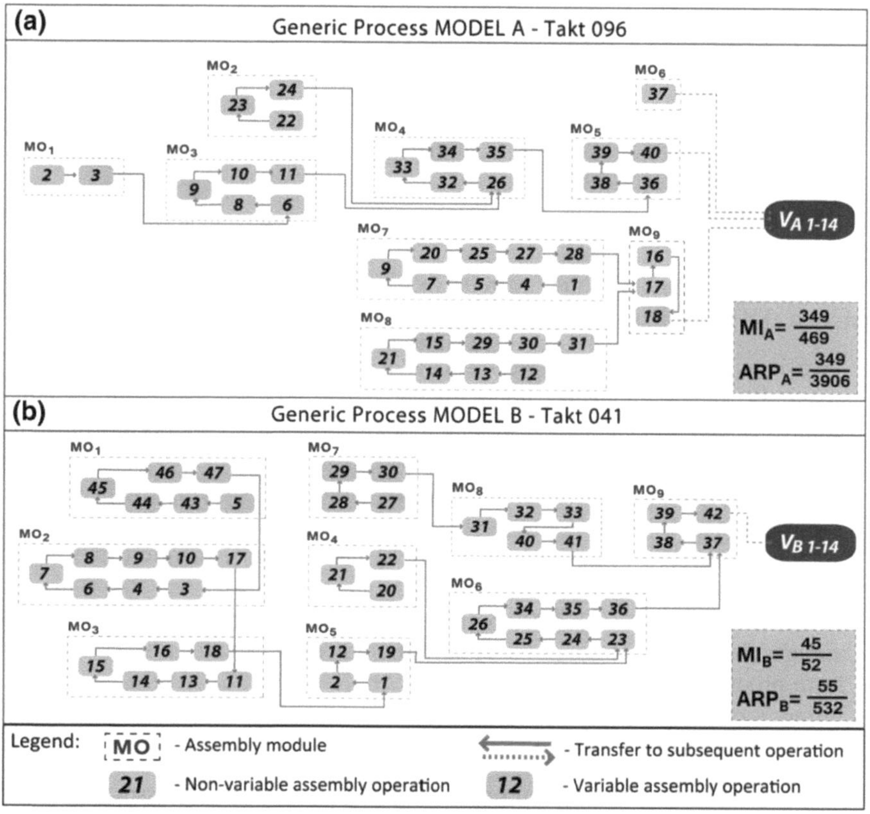

Fig. 5 Generic process models (**a**) and (**b**) and enumeration of modularity indices

Table 1 Numerical values of product and process indices for model A and B

		Model A			Model B		
		Nr. of components	Nr. of variants	Nr. of processes	Nr. of components	Nr. of variants	Nr. of processes
		52	*14*	40	*54*	*14*	47
Variety induced complexity SDC		1788.5			2660.6		
		52	*14*	*40*	54	*14*	*47*
Process modularity	*MI*	0.744			0.865		
	ARP	0.089			0.103		

In Model B assembly (Fig. 5b) illustrates a work-shop setting of Takt nr. 41, again with nine individual process modules. To calculate the MI_B, there are in total 495 relations inside all modules of 14 possible work-shop output variants. The number of all relations, including the ones among modules is 572; $MI_B = 0.86$. The value of ARP_B is 55/532 = 0.1034 (Table 1).

Having higher values of the modularity metrics *MI* and *ARP* with the Model B knowing that both models provide the same levels product variety (14 both), the calculation method demonstrated higher modularity level here than in the case of Model A.

5 Conclusions

To conclude the analysis of the two relatively complex assembly stations (Takt 096 and Takt 041), one may state:

- Two independent assembly work-shops/takts with a similar number of initial assembly sub-components, non-variable stable process and with identical number of takt variants may have different levels of variety-induced complexity and modularity;
- From a product structural perspective, Model B with 684 possible sub-component interactions has higher interconnection of sub-components; only a small number of components vary in 14 takt variants, therefore the higher $SDC_B = 2660.6$ nats;
- Model A consists of only 500 possible sub-component interactions and 14 takt variants have big changes in the composition of each variant; it is less complex from product structural perspective;
- In terms of modularity, Model B demonstrates the higher level of modularity, which is also visible from the SDM in Fig. 5.

To recapitulate the product and process oriented complexity assessment method for assembly operations, the concept does not require a lot of efforts to collect the necessary product and process (structural) data. This is the main requirement of the practitioners or engineers who usually seek to have an easily applicable and not time-consuming method to help them to make a proper decision. This way, production managers may use the method to approximate the desired variety-induce complexity (external view) towards the degree of modularity (company view on complexity). The method itself offers objective evaluation of the process and product with a high level of repeatability.

The limitation of the research is in the structural/static viewpoint of the concept. Our current and future research will focus on assembly, in which operations are performed within a specific time frame—as operational complexity.

References

1. Fettermann DC, Soares Echveste ME (2014) New product development for mass customization: a systematic review. Prod Manufact Res 2(1):266–290
2. Hankammer S, Steiner F (2015) The sustainability potential of mass customization through product service systems in the consumer electronics industry. In: Procedia CIRP 30, pp 504–509. ISSN 2212-8271. doi:http://dx.doi.org/10.1016/j.procir.2015.03.007

3. Alan SB (2009) From simple rules, complex behavior. Mech Eng 131:22–27
4. Tiihonen J, Soininen T, Männistö T, Sulonen R (1996) State-of-the-practice in product configuration—a survey of 10 cases in the Finnish industry. Knowledge intensive CAD. Springer, US, pp 95–114
5. Klein R, Buchheit M, Nutt W (1994) Configuration as model construction: the constructive problem solving approach. In: Proceedings of artificial intelligence in design, pp 201–218
6. Mittal S, Frayman F (1989) Towards a generic model of configuration tasks. In: Proceedings of the eleventh IJCAI, pp 20–25
7. Najman O, Stein B (1992) A theoretical framework for configurations. In: Belli F, Radermacher F J (eds) Proceedings of industrial and engineering applications of artificial intelligence and expert systems: 5th international conference, IEA/AIE-92, pp 441–450
8. Aichner T, Coletti P (2013) Customers' online shopping preferences in mass customization. J Dir Data Digital Mark Pract 15(1):20–35
9. Matt DT, Rauch E, Dallasega P (2016) Trends towards distributed manufacturing systems and modern forms for their design. In: Teti R (ed) International conference on intelligent computation in manufacturing engineering (ICME'15)—proceedings, Capri, Italy, vol 33, pp 185–190
10. Calinescu A, Efstathiou J, Schirn J, Bermejo J (1998) Applying and assessing two methods for measuring complexity in manufacturing. J Oper Res Soc, 723–733
11. Jiao J, Ma Q, Tseng MM (2003) Towards high value-added products and services: mass customization and beyond. Technovation 23(10):809–821
12. Liu XF, Kane G, Bambroo M (2006) An intelligent early warning system for software quality improvement and project management. J Syst Softw 79(11):1552–1564
13. Fredendall D, Gabriel TJ (2003) Manufacturing complexity: a quantitative measure. In: POMS conference—proceedings, 4–7 April, Savannah, GA
14. Shannon CEA (1948) Mathematical theory of communication. Bell Syst Tech J 27(3):379–423
15. Zhu X, Koren SJ, Marin SP (2008) Modeling of manufacturing complexity in mixed-model assembly lines. J Manuf Sci Eng 130(5):313–334
16. Kim YS (1999) A system complexity approach for the integration of product development and production system design. Master of Science, Massachusetts Institute of Technology
17. Krus P (2015) Design space configuration for minimizing design information entropy. In: Proceedings of the ICoRD'15—research into design across boundaries: theory, research methodology, aesthetics, human factors and education, 7–9 Jan. Springer India, pp 51–60
18. Chramcov B, Bucki R, Suchanek P (2013) Logistic optimization of the complex manufacturing system with parallel production lines. J Appl Econ Sci (JAES) 3(25):271–285
19. Kampker A, Burggräf P, Swist M, Nowacki C (2014) Assessment and configuration of a product production system. In: Enabling manufacturing competitiveness and economic sustainability. Springer International Publishing, pp 147–152
20. Grussenmeyer R, Blecker T (2013) Requirements for the design of a complexity management method in new product development of integral and modular products. Int J Eng Sci Technol 5(2):132–149
21. Modrak V, Marton D (2012) Modelling and complexity assessment of assembly supply chain systems. In: Procedia Engineering, vol 48, pp 428–435
22. Modrak V, Marton D, Kulpa W, Hricova R (2012) Unraveling complexity in assembly supply chain networks. In: 2012 4th IEEE international symposium on logistics and industrial informatics (LINDI), pp 151–156
23. Wang H (2010) Product variety induced complexity and its impact on mixed-model assembly systems and supply chains. Doctoral dissertation, General Motors
24. Modrak V, Marton D, Bednar S (2015) The influence of mass customization strategy on configuration complexity of assembly systems. In: Procedia CIRP, vol 33, pp 538–543
25. Modrak V, Marton D, Bednar S (2014) The impact of customized variety on configuration complexity of assembly process. Appl Mech Mater 474:135–140

26. Bednar S, Modrak V (2014) Mass customization and its impact on assembly process' complexity. Int J Qual Res 8(3):417–430
27. Suh NP (2005) Complexity in engineering. CIRP Ann Manufact Technol 54(2):46–63
28. Guenov MD (2002) Complexity and cost effectiveness measures for systems design. In: Manufacturing complexity network conference, Cambridge, UK. ISBN 1-902546-24-5
29. Modrak V, Bednar S (2015) Using axiomatic design and entropy to measure complexity in mass customization. In: Procedia CIRP, vol 34, pp 87–92
30. Blackenfelt M (2001) Managing complexity by product modularisation. Doctoral dissertation, Maskinkonstruktion. doi:http://kth.diva-portal.org/smash/get/diva2:8851/FULLTEXT01.pdf

A Holistic Approach for Developing and Commissioning Data Driven CPPS Functionality in Manufacturing Systems

Hajo Wiemer, Arvid Hellmich and Steffen Ihlenfeldt

Abstract The manufacturing of high-tech materials such as forming of composites often requires complex process chains with a multitude of parameters and parameter interactions. Hence, the manufacturing processes themselves and especially monitoring and controlling those processes becomes increasingly complex. Currently developed Cyber-Physical Production Systems (CPPS) shall comprise data acquisition by sensors, connected actuators, communication functionality as well as data analysis based on mathematical models and autonomous process control. Hence, it becomes possible to detect the current process state and to adjust process parameters in real-time in an optimal way accordingly.

Keywords Cyber-physical production systems · Process data analysis · Process chain · Parameter adaptation

1 Motivation

Production systems such as cutting machine tools or presses did not change in their basic set-up in the last decades. However, functionality was added to increase productivity, flexibility and efficiency. Examples are automatic controller setup [1], hardware in the loop simulation [2], and monitoring for the tool, the production machine and the inherent control loops [3, 4]. With the booming influence of digital services and connection to the World Wide Web, the term "Cyber-Physical System (CPS)" gains influence in the field of production as well. Especially the challenges of Industry 4.0 are only achievable with intelligent Cyber-Physical Production

H. Wiemer (✉) · S. Ihlenfeldt
Institute for Machine Tools and Control Engineering,
Technische Universitaet Dresden (TUD), Dresden, Germany
e-mail: hajo.wiemer@tu-dresden.de

A. Hellmich · S. Ihlenfeldt
Fraunhofer Institute for Machine Tools and Forming Technology (IWU),
Dresden, Germany

V. Majstorovic and Z. Jakovljevic (eds.), *Proceedings of 5th International Conference on Advanced Manufacturing Engineering and Technologies*,
Lecture Notes in Mechanical Engineering, DOI 10.1007/978-3-319-56430-2_18

Systems (CPPS). Apart from new ways of communication as well as extended sensors and sensor networks, CPPS comprise modelling of plants or plant components. Furthermore, new concepts for topology, interfaces and interrelations between components, production systems and to humans were investigated [5].

The paper gives an insight to this emerging kind of production systems in Sect. 2 and subsequently focusses on an application example, especially following questions of data acquisition, data processing and deduction of suitable reactions and adaptations (Sect. 3.2). Furthermore, approaches and levels of process parameter adaptation are discussed in Sect. 3.3, followed by a conclusion and an outlook.

2 Introductions to Cyber-Physical Production Systems

2.1 Cyber-Physical Systems

The terms CPS, CPPS, Industry 4.0 and their interconnection is subsequently discussed based on Fig. 1. Many components of production systems will appear as combination of CPS in the future. They are characterized by a combination of real ("physical") objects or processes and incorporate a virtual, information and data driven ("cyber") counterpart. Following the statement of [6], "their operations are integrated, monitored, and/or controlled by a computational core". In the field of production systems, components, like sensors, actuators or frames, will most likely retain their core functionality and will be extended by an administrative layer with self-description and basic communication possibilities. As Fig. 1 shows, the component layer can be subdivided into three domains.

1. Engineering and/or electronic design focused CPS, e.g. are press frames with integrated sensors (e.g. for force measurement) and the possibility to transmit measured data to a superordinate layer. Another example is an integrated sensor node for temperature and acceleration, based on single core computers (such as Raspberry Pi) with data pre-processing and communication functionality.
2. An example for a mechatronic focused CPS is an integrated active dynamic vibration absorber with sensors, actuators and, again, the possibility to communicate.
3. A third CPS- domain on the component layer is computational science aspect with the Internet of Things (IoT), which is the main enabler (in hard- and software) for putting CPS of the other domains in action. Data storage as well as simulation and running mathematical models of the production system and its components can also be seen in this module.

From the author's perspective, CPS, serving as additional sensors and digital model representations are of primary importance for production systems in the future. Furthermore, [6] states, that the "components are networked at every scale",

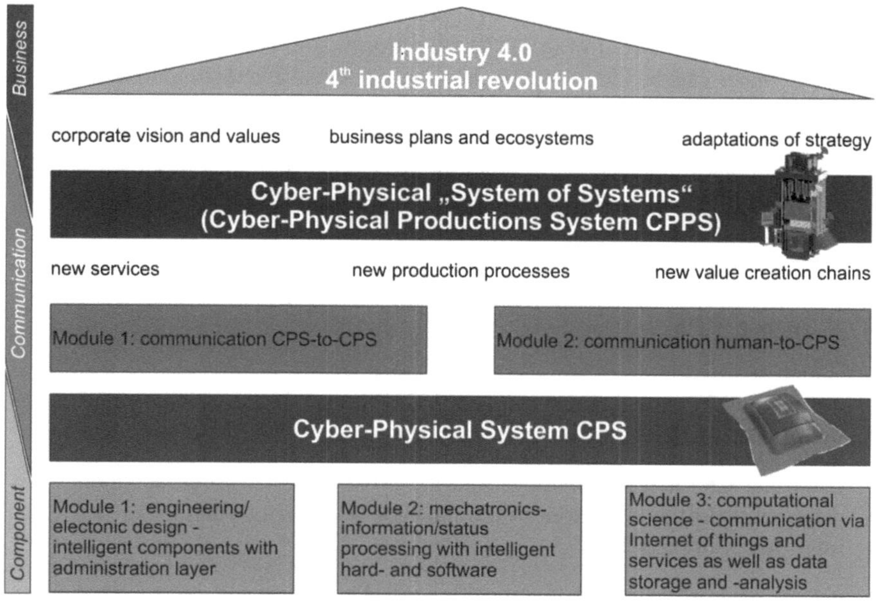

Fig. 1 Schematic illustration of CPS and CPPS (based on [7])

which is represented by the communication layer in Fig. 1. Hereby two basic fields, the communication CPS-to-CPS (based on their specific functionality) and CPS-to-human (e.g. via operation panel, smart glasses, phones/tablets) are to be separated [7].

The emerging possibilities by utilizing CPS and their interconnection lead to new options of services, provided by the production systems. One example is the fairly easy measurement of not yet traced values (e.g. by the controller system). These are forces, temperatures and accelerations. New production processes can be enabled or boundaries of established production processes can be shifted. One Example is given in [8], where the product quality is stabilized in deep drawing processes by increased monitoring and adaptation of parameters, e.g. of the spacers in the tool. Also, value creation by production systems is affected. Not only the manufactured product but also the acquired data is of value for the customer. Especially for branches like automotive- or aviation industry this is a big advantage due to the obligation for documentation.

2.2 *Cyber-Physical Production Systems*

The combination of various CPS from the presented domains forms a "System of Systems" in the general case, which can be specified into a Cyber-Physical Production System for the given field. Taken the CPPS as basis, various new or changed business approaches make an appearance. Corporate visions and values are shifted more and more towards virtual, information and data driven aspects. Business plans can be extended by new ecosystems, where functionality of the production system can be sold and enabled successively, fitted to the customer's needs. In that course, company strategies will have to be changed and adapted to the extended possibilities, degrees of freedom, challenges and customer requirements.

2.3 *Methodology to Transform Production Systems into CPPS*

In order to qualify production systems into CPPS, various extensions are necessary. As visible in Fig. 2:

- controller data and information from the drive system with all actuators,
- data from additional sensors of the machine and the tool and
- input material properties

have to be acquired and put together (Fig. 2). Drive data can be accessed by the machine control through the connected field bus systems [9]. Further transmission of controller and drive system data can be done, e.g. by the OPC unified architecture, which is increasingly established as main enabler for data transmission in connection with Industry 4.0 [10]. Data is collected and further processed by a general production data management tool, described in Sect. 3.2. In order to induce changes to the production system, the data loop needs to be fed back to the controller. In the considered case, no additional actuators are attached to the production system. Hence, only parameters like press speed, controller settings,

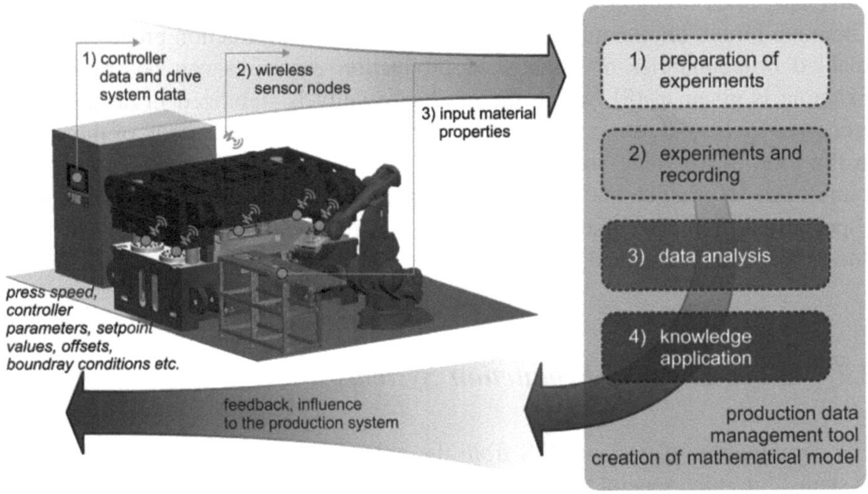

Fig. 2 Production system with extended sensors, process data and the feedback loop

setpoint values, offsets and boundary conditions are transferred from the mathematical model to the controller of the production system (Fig. 2).

3 Application Example Suspension Turret

The course of action to qualify a production system as CPPS is presented, based on the example "suspension turret". It is derived from the collaborative research centre (CRC) 639 "Textile-reinforced composite components for function-integrating multi-material design in complex lightweight applications" (German Research Foundation DFG). The research is focused on the development of textile-reinforced fiber composite components (mainly glass fibre and polypropylene) comprising the necessary process for manufacturing and processing [10]. Therefore, knowledge concerning material- and part properties as well as inter dependencies in manufacturing was generated. It was exploited for a targeted parametrization/choice of input material properties and processing parameters with the goal of a reproducible process. The necessary experiments were both, extensive and complex. To achieve an efficient procedure under the given circumstances, a suitable methodology had to be developed. As displayed in Fig. 3, it integrates a graphical process description and the mathematical modelling of the process-intrinsic interdependencies between

- input material properties,
- the parametrization of the production system
- the environmental influence parameters as well as
- the desired part quality characteristics [12, 13].

Therewith, the mathematical model as a key feature of CPPS is suitable to predetermine the situationally optimal process parameters for the production system. Data driven methodologies to derive the necessary process information are presented for the example of manufacturing suspension turrets.

3.1 Investigated Process

The investigated process chain is shown in Fig. 4. It comprises preforming, sensor placement, consolidation and quality inspection. An insight into the chosen

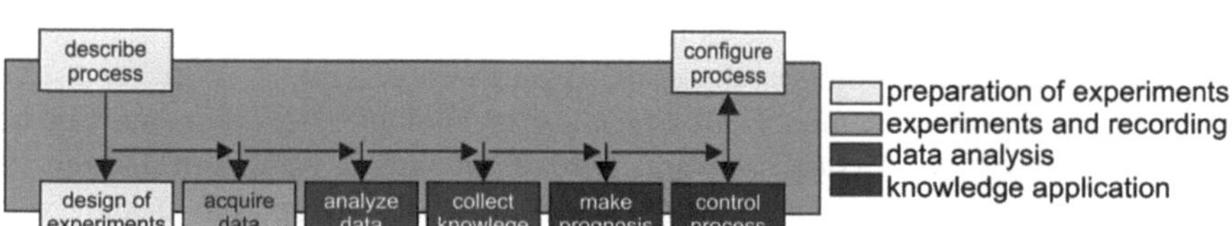

Fig. 3 Methodical approach to the integrated analysis of complex processes [11–13]

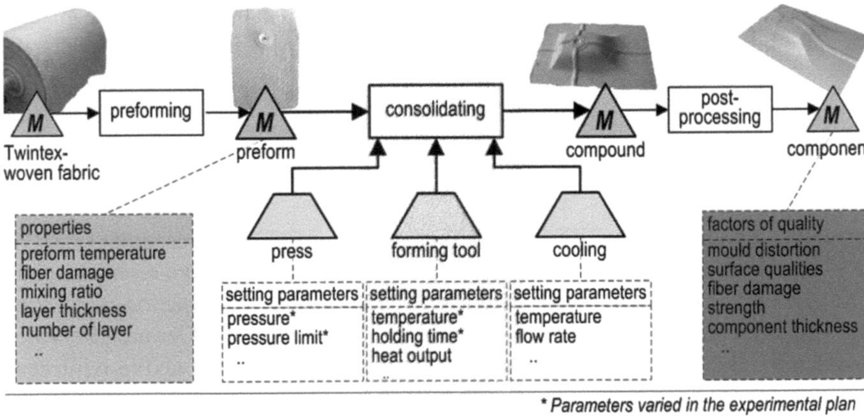

Fig. 4 Graphical model of the process chain “suspension turret” [14]

approach is subsequently given for the consolidation process, comprising the following steps:

- At first the preform is inserted into a variotherm forming tool.
- Subsequently the press machine is closed and the heating process, utilizing electric heating rods, is started. Temperature is measured by internal sensors.
- If the melting temperature of the thermoplastic is reached it engulfs the glass fibre.
- After a defined retention time, tool and part are cooled down (water), the press is opened and the part is removed.
- Afterwards, various quality attributes of the suspension turret are inspected.

3.2 Creation of the Mathematical Model

The purpose of the mathematical model as CPS is to determine suitable process parameters. Furthermore, controllability to the consolidation process was included. For the example, algorithms and prototypic applications were developed based on the introduced holistic data management concept. Subsequently the commercially available IT system “Detact” was developed out of the groundwork in cooperation with the Symate GmbH. Following the methodology, presented in Fig. 3, the process with all relevant parameters has to be described at first.

For the consolidation process these are given in Table 1.

Based on the process model, all necessary technological experiments can be planned by using the Detact software. Out of the broad variety of Design of Experiment (DoE)—algorithms, offered by Detact, the D-optimal design was chosen. The main advantage of the D-optimal design is the possibility to include

Table 1 Lineup of relevant parameters for suspension turret manufacturing

Properties of the preform	Process parameters	Quality characteristics
Temperature	Pressure	Component thickness
Mixing ratio	Temperature	Part strength
Number of layers	Holding time	Surface quality
Layer thickness	Cycle time	
Pattern design of fiber orientation	Maximum energy consumption	

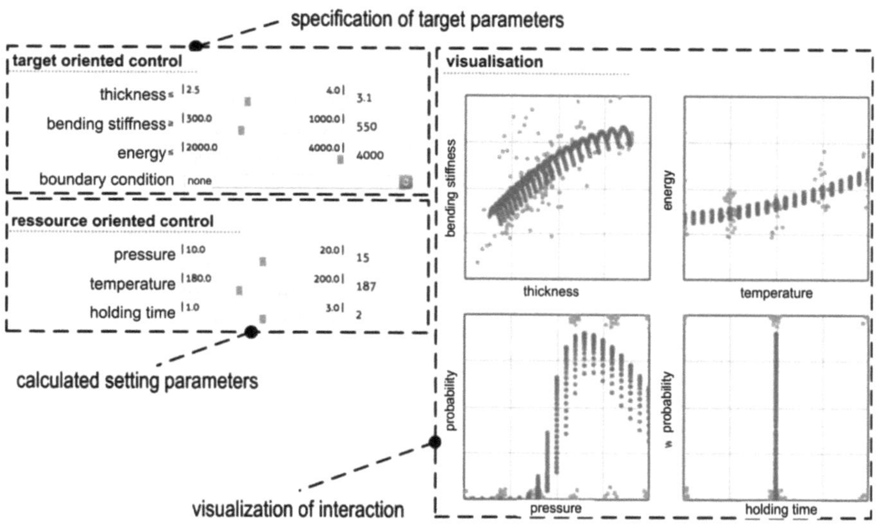

Fig. 5 Mathematical model for process immanent interdependencies (Screenshot Detact)

continuous variation in test values (e.g. temperatures) as well as discrete states such as included sensor (yes/no) or type of material into the DoE.

The significant parameters temperature, pressure, pressure limit and holding time were varied systematically. Interdependencies between the influence parameters and the quality characteristics are captured by a mathematical model as displayed in Fig. 5.

In the left section, target values, such as component thickness and part-strength can be set by sliders. Adequate process parameters for the desired case are automatically computed by the model and can be transferred to the machine control. Depending on the constellation of parameters, the process can also be destabilized, which can be indicated by the model as well. The graphs to the right give insights to the model output from various perspectives. The model also estimates a probability to achieve the desired part quality with the given parameter set. If this is to low, input parameter changes are necessary. The mathematical model is fast enough to be utilized for feedback control. Hence, the basis for an intelligent process control in terms of a CPPS is laid.

3.3 Implementation into the Machine Control

The mathematical model computes sufficient values for parameter adaptation via Detact software. Today, typical controllers are able to change/adapt the values automatically. For this matter, the model needs to be connected with the controller of the press machine. Currently, a basic interconnection between model and machine control is realized. An actual process parameter vector is cyclically written to a specific memory area. The machine controller has access to this area and reads the calculated parameters in advance to every stroke and adapts the controller parameters.

However, the specific requirements of the process, the acquired data in connection with the model also reveal, which frequency of adaptation is necessary and suitable. If the part quality varies a lot and strongly depends on the material and machine properties, a quasi-continuous loop, referred to as "*in-process control*", has to be set up. The frequency of parameter update then is in the range of typical controller sample times, which are 0.5–3 ms [9]. Drive controllers are sampled approx. 4 times faster. Hence, if the focus lies on an in-process control, the model must be able to provide new (setpoint) values in the sample time of the controller. This is a high requirement and limits the complexity of viable mathematical operations and requires fast data transmission.

For smaller dependencies of the part quality, machine parameters shall be updated after each manufactured part. Hence, the sampling interval of the control loop corresponds with the press speed. This "*part to part control*" does not require real-time bus communication, but still emerges challenges, e.g. in measuring the input material properties and part quality. Measurement principles have to be chosen, which are suitable to provide results for every stroke.

The third and less complex type of influencing the controller is a "*lot based control*". Many processes are stable, if the parameters are adjusted according to the characteristics of the input material lot. Hence, parameter changes can be reduced to the rate in which the lot is changes. Hereby, complex mathematical operations can be utilized, based on a broad variety of input material parameters. The characterization of the input material can furthermore be carried out offline in advance to the manufacturing process.

4 Conclusion and Outlook

Increased data acquisition, data transmission and intelligent (sub-) components enhance state of the art production systems to Cyber-Physical Production Systems. Their core components and enabler technologies are introduced together with vivid examples. The mathematical modelling of production processes is one of the key features of CPPS. For the manufacturing of suspension turrets the necessary improvement of a production system is discussed. The focus was laid on a viable possibility to model the manufacturing process along with the influencing parameters. An example for a suitable production data management tool ant its functionality is presented. Furthermore, a distinction between three options to influence the process via the machine control was made.

Based on the findings in paper, several research attempts appear. One of them is to develop feasible sensor nodes, which are easy to be applied and commissioned to existing production systems. Furthermore, the production data management tool can be refined for various production processes and a consistent connection with the controller system. A currently built plant for the manufacturing of cups from cardboard will form the basis for a new generation of process data management with included sensor system and controller connection via OPC UA.

References

1. Mink F, Bähr A, Beineke S (2009) Self-commissioning feedforward control for industrial servo drive. In: 8th International symposium on advanced electromechanical motion systems & electric drives joint symposium, electromotion, pp 1–6
2. Weck M et al (2003) Coupled simulation of control loop and structural dynamics. Ann Ger Acad Soc Prod Eng X(2):105–110
3. Bombińskia S et al (2016) Sensor signal segmentation for tool condition monitoring. In: Procedia of the CIRP, vol 46, pp 155–160
4. Schönherr R, Schlegel H, Drossel W-G (2012) Assessing the performance of servo drive feedback controls. In: Proceedings of the 13th mechatronics forum, international conference, vol 1, Linz, Austria
5. Monostori L et al (2016) Cyber-physical systems in manufacturing. Ann CIRP 65(2):621–641
6. Gill H (2008) A continuing vision: cyber-physical systems. In: Fourth annual Carnegie Mellon conference on the electricity industry future energy systems: efficiency, security, control
7. Roth A (eds) (2016) Einführung und Umsetzung von Industrie 4.0 Grundlagen, Vorgehensmodellund Use Cases aus der Praxis. Springer
8. Spindler K et al (2015) The Audi toolshop—taking the next step into digital dimension. In: 5th international conference on accuracy in forming technology, pp 15–32
9. Hellmich A et al (2016) Drive data acquisition for controller internal monitoring functions. In: XXVII CIRP sponsored conference on supervising and diagnostics of machining systems, Karpacz, Poland
10. Hammerstingl V, Reinhart G (2015) Unified plug & produce architecture for automatic integration of field devices in industrial environments. In: Proceedings of the IEEE international conference on industrial technology, pp 1956–1963
11. Modler N et al (2016) Novel hybrid yarn textile thermoplastic composites for function-integrating multi-material lightweight design. Adv Eng Mater 18(3):361–368
12. Großmann K, Wiemer H (2010) Reproduzierbare Fertigung in innovativen Prozessketten. Besonderheiten innovativer Prozessketten und methodische Ansätze für ihre Beschreibung. Analyse und Führung (Teil 1), ZWF 10, pp 855–859
13. Großmann K et al (2010) Reproduzierbare Fertigung in innovativen Prozessketten. Konzeption eines Beschreibungs- und Analysetools (Teil 2), ZWF 11, pp 954–995
14. Großmann K, Wiemer H, Großmann KK (2013) Methods for modelling and analysing process chains for supporting the development of new technologies. In: Procedia materials science, materials science engineering, vol 2, pp 34–42

Some Aspects Influencing Production of Porous Structures with Complex Shapes of Cells

Katarina Monkova and Peter Monka

Abstract The article deals with some aspects influencing Direct Metal Laser Sintering (DMLS) technology. The 3D printing method is used in production of metal parts, at which a product is built layer by layer. The main advantage of this method is the ability to produce parts which have a very complex geometry and which are produced by classical conventional methods in very complicated way. To be 3D printing technology included in the industrial production of the real components it is required to achieve high mechanical properties of produced parts. The final properties of produced parts are strongly dependent on each individual laser produced track and layer. Impacts of individual factors on the quality of product are described in the article. There is also presented preliminary study related with the modelling and manufacturing of porous structures. It comes to geometrically defined structure with complex shapes that allows weight reduction with sufficient toughness.

Keywords Porous structure · Complex shape · Additive technology · Influencing aspects

1 Introduction

3D printing method is a manufacturing technology used in the production of metal and plastic parts. The product is built layer by layer. The main advantage of this method is the ability to produce parts which have a very complex geometry and

K. Monkova (✉) · P. Monka
Faculty of Manufacturing Technologies with the Seat in Presov,
Technical University of Kosice, Sturova 31, 080 01 Presov, Slovakia
e-mail: katarina.monkova@tuke.sk

P. Monka
e-mail: peter.pavol.monka@tuke.sk

P. Monka
Regional Technological Institute, University of West Bohemia,
Univerzitni 8, 306 14 Pilsen, Czech Republic

V. Majstorovic and Z. Jakovljevic (eds.), *Proceedings of 5th International Conference on Advanced Manufacturing Engineering and Technologies*,
Lecture Notes in Mechanical Engineering, DOI 10.1007/978-3-319-56430-2_19

which are produced by classical conventional methods in very complicated way. Unfortunately, this technology has a number of disadvantages and limitations.

DMLS is the preferred method of 3D rapid prototyping for complex geometries. It allows users to reduce the time and cost for additive manufacturing by completing multiple prototypes directly from CAD data, allowing for more rigorous testing with different alloys. Metal AM is an 'additive' technology that works by fusing together very fine layers of metal powder using a focused laser beam. This powder fusion process can produce complex geometries which might not have been possible using traditional manufacturing techniques [1].

At DMLS (Direct metal laser sintering) technology the parts are made based on a 3D model. There is no need for jigs and tools, no need to perform complex calculations. The preparation phase of production is accelerated and the risk of errors decreases. Thanks to 3D model it is possible to make various changes and modifications of components in the preparatory stage of production. The part is formed from metal powder. There is no need to store different kinds of intermediate product as it is in conventional machining methods.

2 State of the Art

During the manufacturing of simple components, it is generally preferable and especially cheaper to use the classical conventional methods. The intensity of production increases along with the complexity of the part at the conventional production methods. Likewise, there are greater demands for production times, necessary machinery and tools, for operator qualification etc. In 3D printing it is vice versa, because the production of simple components takes the same time as the production of complex parts. Moreover, it is possible to produce the different components in a single operation at the same machine. The production is extended only in the case, if the material will be applied in thinner layers due to better quality of part [2, 3].

Production speed can be increased by making multiple parts at once. This reduces time of secondary movements. It is also important to place right the components inside a workspace. Building speed is significantly affected by the height and volume of the part, by number and thickness of layers. Furthermore, the great influence has required surface roughness, used production method and the type of manufacturing equipment.

Unfortunately, there are also some disadvantages, for example the fact that the printer for its functionality needs several dozen kilograms of metal powder, even if it is made small part. Thus, large finances will be stored in the material because building powder is very expensive.

With the DMLS method it is possible to produce the components from so-called gradient material. The properties of these materials change in a certain direction according to user requirements. These changes are achieved by mixing powders of different characteristics. For use of this material it is needed to have a machine that will be equipped with two or more nozzles by which the building material is fed. Thus, for example it is possible to produce a bronze form, which the chemical composition will change inward into steel. The produced part will be abrasion-resistant, it will be good heat dissipation compared to a form where is a sharp transition between bronze and steel [4, 5].

The most restrictive parameters of 3D printing are: construction materials, production speed, cost, accuracy. During laser sintering it occurs to change in temperature and state of the material what often causes a deformation and stress of the part, which is very difficult to remove. One solution of this problem is to preheat and reheat the manufacturing process. It is not possible to remove these effects completely, but only to reduce them. Other limitations include the inaccuracy which is caused by the thickness of the built layer.

It is not able to create a perfectly continuous surface through the addition of individual layers. It is replaced by "stair" surface which is formed around the edges of the part. This defect is most evident in the construction of cylindrical bodies, which have a horizontal axis. A partial solution is reducing the thickness of the individual layers [6].

An important part of the preparatory process, which directly affects the construction time (productivity) and quality components, is an orientation. On the curved and inclined surfaces often appears so called staircase effect, that negative effects their dimensional accuracy. Unfortunately, there is no way to completely eliminate the staircase effect. A better result can be achieved by reducing the thickness of the individual layers. At this stage, the designer has to specify how the part will be positioned inside the workspace. The orientation of the component has a great influence on its accuracy, surface quality, construction time, the required number of support structures and also on the production costs [7, 8].

To solve this problem, a number of studies was done. For example, Allen and Dutta [8] have developed a method for automatic calculation of load-bearing structures and a decision method to select the most appropriate orientation.

Frank and Fadel [9] designed an expert system, which takes into account various parameters that affect the production of components. This system works with the specified input parameters and created decision matrix according to which the system recommends the best direction and orientation of the building components in the workspace.

Researchers Hur et al. [10] developed an algorithm to calculate the stair effect. Its task was to quantify procedural errors by the volume of material, which has to be removed or added to the components. In addition, it is able to quantify the total assembly time and the volume of the structure.

Mr. Pham et al. [11] developed a support tool for decision making that helps users determine the proper orientation of the part. There are works with a variety of factors, such as construction time, production costs, the problematic features of the component (hole axis, critical surfaces etc.), overhanging ends, volume aids etc. To the individual criteria, it is intuitively allocated scoring that is multiplied by points assigned to each version of orientation. Orientation, which gets the most scoring is most suitable for the orientation of the components in the workspace.

Masood and Rattanawong [12] developed an algorithm for calculating the size of volumetric error that is caused by cutting the CAD model. Volumetric errors are calculated for different orientations turning around user-specified axes. The best orientation is then determined as the part with a minimum density error.

Usually constant thickness of the layer is kept. The parameters changing during production is rarely. However, when creating a part with a constant layer thickness leads to certain drawbacks. It is likely that the components will miss small features and details. In addition, a constant film thickness can increase assembly time, because continuous application of a minimum layer is required to meet the requirement for surface roughness. If applied in varying thickness, some of the above problems could be reduced.

It still holds a rule that it is needed to follow a thickness which is recommended by the manufacturer of production equipment. Values of individual thicknesses vary considerably.

Special category of components, that are made by DMLS technology, are parts with porous structure. The structure it is possible to consider as a solid phase with cavities. Its mechanical properties depend primary on material, from which they are produced, and on topology including geometry and relative density. These functional porous structures can give to the product extraordinary combination of properties such are high strength, stiffness along with low weight and good absorption of energy. It is a reason why the porous structures have become widespread in lightweight structures and they are applied with benefits in biomedicine, aerospace and automotive industries.

Generally, porous structures are divided into two categories: regular and unregular—also called foams. Hanzl et al. [13, 14] deal with producing one type of regular porous structures—so called gyroid structure.

3 Regular Porous Structures Design

Basic building units of geometrically defined porous structures are cells that are created by simple bodies, e.g. cube, cuboid, cylinder or sphere. These geometries are usually well modelled in CAD applications. A little bit more complex structures have their geometries created by curved surfaces. One of these complex structure is Shoen gyroid structure, at which the position and the shape are described by mathematical formula [15]:

$$\sin\left(\frac{2\pi}{L}x\right)\cos\left(\frac{2\pi}{L}y\right)+\sin\left(\frac{2\pi}{L}y\right)\cos\left(\frac{2\pi}{L}z\right)+\sin\left(\frac{2\pi}{L}z\right)\cos\left(\frac{2\pi}{L}x\right)-t=0 \tag{1}$$

where factor L influences the size of basic cell and parameter t decides about cross section of the strut, what influences the relative density of structure. Next parameter —volume ratio V_r of solid phase expresses how many percent of the cell space is filled by material [16]:

$$V_r = \frac{\text{Volume of solid phase}}{\text{Total volume}} 100\ (\%) \tag{2}$$

The topology of Shoen gyroid shows self-supporting properties, so these structures can be made by additive technology without next supports [17].

To get the unit of gyroid structure for research study software Matlab was used, where the gyroid surfaces with various parameter t were plotted. In next step, the surface was exported in STL format into CAD system Siemens NX 10, where the cells were created (Fig. 1).

The samples for testing were produced by DMLS technology using EOS M290 machine with default technological parameters (laser power 400 W, diameter of laser beam 70 μm). Material of samples was the Maraging Steel MS1. The metal powder was sintered by laser in inert atmosphere of Nitrogen. Material purity 99.97% was declared by the device manufacturer. Within preliminary experiments the set of samples with cubic cell size of 6.5 mm and volume proportions 13.5, 9.5, 7.3, 5.8, 5.0 and 4.2% were printed. They are shown in Fig. 2.

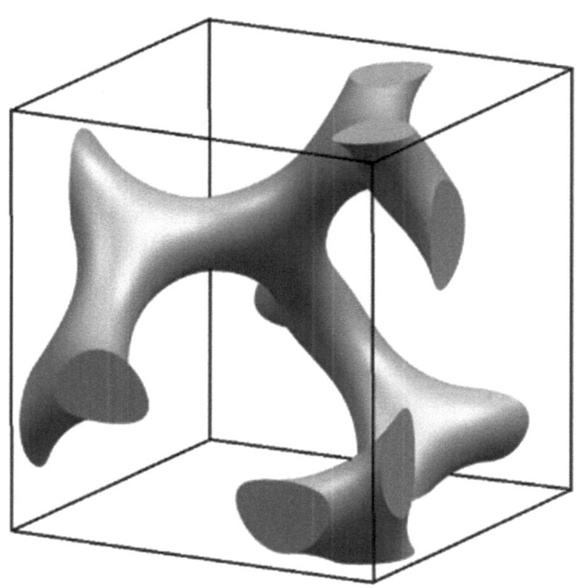

Fig. 1 Basic cell with gyroid structure

Fig. 2 Regular porous structures with various volume proportions [13]

4 Experiments

The goal of test was to load the samples by pressure that after exceeding the yield point of the material causes the irreversible plastic deformation. The test is finished if the set compression is achieved, or if the sample appeared significantly destructive conditions. From pressure diagram are evaluated deformation characteristics, such are yield strength and maximum load capacity. Equipment Zwick/Roell Z250 with force capacity 250 kN was selected for the test. The compression speed of samples was 0.1 mm s^{-1} at the environment temperature of 20 °C. The results of preliminary tests are presented in Fig. 3.

It is clear from the graph displayed above that the relative carrying capacity of samples is slightly rising with volume ration decreasing. It means that the unit of

Fig. 3 Graph with dependency of the force on volume ratio of material at pressure tests

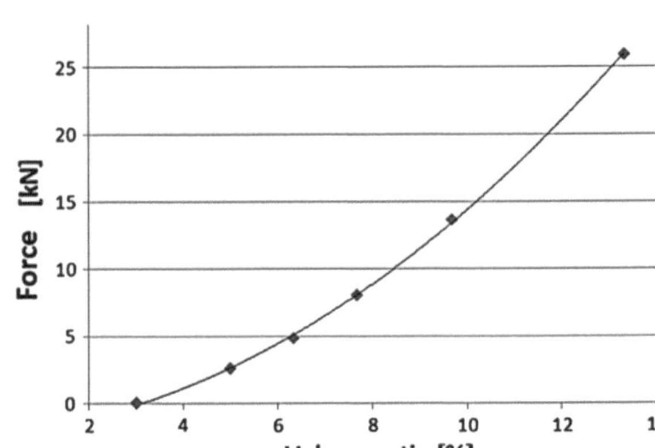

Fig. 4 Deformation of the sample at pressure test [13]

gyroid with lower volume ratio is able to carry a greater relative load to failure then the unit of gyroid structure with higher volume ratio.

However, there are some problems with deformation of samples during the testing. Some porous structures have deformed near the surface, where the sample was separated from a platform, see Fig. 4.

The deformation near the area of separation could be caused by stiffness reduction due to bending of strut structure during the cutting of sample with a band saw.

At margins of samples the self-supporting function was damaged as it is shown in Fig. 5a. To solve this problem, the next one cell to the basic matrix of the sample was added in relevant direction and the sample was detached from support so to be its self-supporting function was preserved (Fig. 5b) [14].

Also, horizontal surfaces should be parallel for pressure tests. The problems occurred at the samples that were deviated by 5° from vertical position; the samples with size of cell of 5 mm were not possible to produce, because after sintering several layers of top cover, the deformations were so great that it was not possible to continue with manufacturing. In Fig. 6 is shown the layer of model, in which the process of printing was discontinued.

Next problems occurred, if the model was oriented so that continuous layers of material were on bottom and top of sample.

Because of high level overheating of applied layer during sintering without adequate heat dissipation and due to insufficient supporting of the area, deformation and scorch have occurred as it is shown in Fig. 7a.

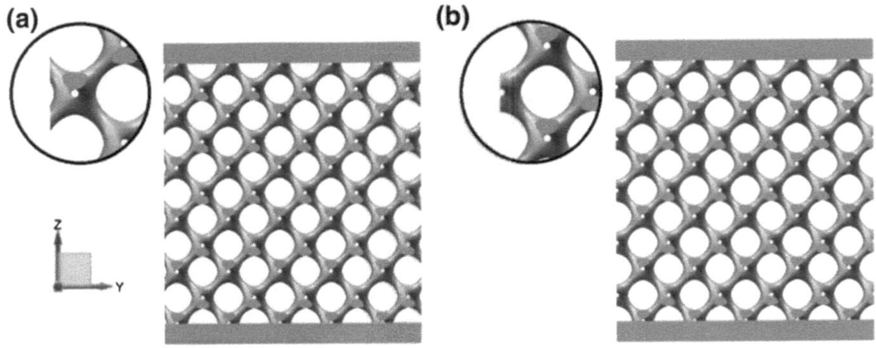

Fig. 5 The margins of the samples **a** without self-supporting function, **b** self-supporting function was preserved [14]

Fig. 6 Printing of the structure with non-parallel surfaces [14]

On the other hand, the preparation of the samples in horizontal position, if the continuous layers of material were on sides of structure (on left and right sides), was smooth and reliable in every manufacturing repetition. Due to the fact that the structure had no imperfections, horizontal position of sintering was evaluated as the most suitable for samples production with rigid plates on the sides of samples, as it is displayed in Fig. 7b.

Fig. 7 The samples produced in **a** vertical, **b** horizontal position [14]

5 Conclusions

Direct metal laser sintering is the technologically mature for industrial production and due to a rising competition between service providers additive manufacturing becomes economically feasible for a growing number of industrial and end-user applications. It is usually used only in special occasions for special type of complex parts or structures that it is not possible to manufacture in other way. However, in the future it will become chipper, so much more producers will be able to use this technology. So, it is very important to know every parameter, which affect the quality of components made by this technology.

Authors in the article show some effects that influence the manufacturing of regular porous structures in spite of the fact that input technological parameters were set the same as the default parameters of the machine. These parameters are: geometry, slope of the sample and its positioning during laser sintering. Realized pressure tests also have indicated that the unit of gyroid with lower volume ratio is able to carry a greater load to failure than the unit of gyroid structure with higher volume ratio.

Acknowledgements The present contribution has been prepared under projects VEGA 1/0614/15 and KEGA 087TUKE-4/2015.

References

1. Ungureanu M, Pop N, Ungureanu N (2016) Innovation and technology transfer for business development. In: Procedia Engineering, vol 149, pp 495–500

2. Majstorovic V et al (2014) CAI model for prismatic parts in digital manufacturing. In: Procedia CIRP, vol 25, pp 27–32
3. Valicek J et al (2016) Mechanism of creating the topography of an abrasive water jet cut surface. Adv Struct Mater 61:111–120
4. Cloots M, Spierings A, Wegener K (2013) Assessing new support minimizing strategies for the additive manufacturing technology SLM. In: Solid freeform fabrication symposium 2013
5. Ferroudji F et al (2014) Large-scale dual axis sun tracking system modeling and static analysis by FEM. Int J Mech Mechatron Eng IJMME/IJENS 14(04):92–97
6. Monkova et al (2014) Inverse processing of undefined complex shape parts from structural high alloyed tool steel. Adv Mech Eng, 1–11
7. Panda A, Jurko J, Pandova I (2016) Monitoring and evaluation of production processes: an analysis of the automotive industry. Springer International Publishing, p 117. ISBN 978-331929442-1
8. Allen S, Dutta, D (1994) On the computation of part orientation using support structure in layered manufacturing. In: Solid freeform fabrication symposium, Austin, pp 59–269
9. Frank D, Fadel G (1995) Expert system-based selection of the preferred direction of build for rapid prototyping processes. J Intell Manuf 6(5):339–345
10. Hur SM, Choi KH, Lee SH, Chang PK (2001) Determination of fabricating orientation and packing in SLS process. J Mater Process Technol 112:236–243
11. Pham DT, Dimov SS, Gault RS (1999) Part orientation in stereolithography. Int J Adv Manuf Technol 15:674–682
12. Masood SH, Rattanawong W (2002) A generic part orientation system based on volumetric error in rapid prototyping. Int J Adv Manuf Technol 19:209–216
13. Hanzl P, Zetek M, Zetkova I Cellular lattice structure produced by selective laser melting and its mechanical properties. In: Katalinic B (ed) Proceedings of the 26th DAAAM international symposium. Published by DAAAM International, Vienna, Austria, pp 0748–0752. ISBN 978-3-902734-07-5. ISSN 1726-9679
14. Hanzl P, Zetková I, Mach J (2017) Optimization of the pressure porous sample and its manufacturability by selective laser melting. Manuf Technol 17(1):34–38. ISSN 1213-2489
15. Paul R, Anand S (2014) Optimization of layered manufacturing process for reducing form errors with minimal support structures. J Manuf Syst. doi:10.1016/j.jmsy.2014.06.014
16. Dobransky J et al (2016) Document optimization of the production and logistics processes based on computer simulation tools. Key Eng Mater 669:532–540
17. Beno P, Konjatic P (2014) Optimization of thin-walled constructions in CAE system ANSYS. TehnickiVjesnik 21(5):1051–1055

Recovery of Metals from Solar Cells by Bioleaching

Mital Chakankar, Cheer Su and Hong Hocheng

Abstract Solar energy is commonly seen as a clean energy source, an alternative to the fossil fuels, as there is no emission of any matter into the environment during operation, endurable operation period, minimum maintenance, robust technique and aesthetic aspects. If one considers the growing use of solar cells, the problem of recycling spent solar cells, the quantities of which will increase rapidly in the coming years, is yet to be solved. Establishing a technology for recycling and reusing the same is thus a necessity. Solar cells contain precious metals including silver, tellurium, and indium along with other heavy metals such as copper, aluminum, lead, arsenic, cadmium, selenium which are hazardous. Limited recycling will result in disposal of these recoverable metals in the waste. There remains the dearth of studies on recycling of old solar cells using microorganisms. In the current study, the recovery of metals from solar cells was investigated using *Thiobacillus ferrooxidans* micro-organism, and its culture supernatant was found to be effective for bioleaching of metals from solar cell powder. B, Mg, Si, Ni and Zn were removed with 100% efficiency within a reasonable time. *T. ferrooxidans* recovered 100% Cr, Mn, and Cu at 60 °C, however, only 41.6 and 13.4% of Al and Te were extracted respectively. These results suggest the higher temperature optimum for metal recovery from solar cells in use of the organism. The possible success in this endeavor will allow reuse of waste solar cells without affecting the environment.

Keywords Metal recovery · Solar cells · Bioleaching · *Thiobacillus ferrooxidans*

M. Chakankar · C. Su · H. Hocheng (✉)
Department of Power Mechanical Engineering, National Tsing Hua University, Hsinchu 30013, Taiwan ROC
e-mail: hochengh@gmail.com

V. Majstorovic and Z. Jakovljevic (eds.), *Proceedings of 5th International Conference on Advanced Manufacturing Engineering and Technologies*, Lecture Notes in Mechanical Engineering, DOI 10.1007/978-3-319-56430-2_20

1 Introduction

Solar cell technology is used eminently worldwide as a source of reliable and cost-effective electricity for industrial, commercial, residential and community applications. The average lifetime of solar cell modules is presumed to be more than 25 years [12]. Tons of mineral materials are used for structural support and transmission of electricity, during the production solar cells, including aluminium, concrete, copper, glass, nickel, steel, and zinc. Moreover, high valued metals are incorporated in solar cells for their special properties like durability and higher levels of light absorption and conversion to electricity that they impart when combined with other mineral materials. These materials include cadmium, gallium, germanium, indium, selenium, and tellurium ([13] cited in [2]). Although the amount utilized is relatively small as compared to other metals, they play a critical role in the performance of thin-film in solar cells.

The increasing importance of solar cells will result in increased demand for these precious metals. The supply of these metals from secondary sources will help to cope with this demand and is likely to increase over time when solar cells are recycled due to breakage or when they reach the end of their useful life of approximately 20 to perhaps 30 years ([14] cited in [2]).

Albeit, solar cells are used as an environmentally friendly way to solve energy problems, yet the problem of recycling spent or damaged solar cells is still to be solved. The quantity of this waste will increase rapidly in the future, hence establishing a technology for their recycling and reuse is obligatory [12]. Also, limited recycling will result in disposal of these recoverable metals in the waste.

Some thermal and chemical methods are reported in the literature for recycling of solar and photovoltaic cells, but they are known to cause the secondary pollution [10, 12, 16]. There remains the dearth of studies of the use of microorganisms for recycling of old solar cells. Therefore, we are proposing a present study to determine the metal content of waste solar cells and application of bioleaching process for removal of these metals.

2 Materials and Methods

2.1 Determination of Metal Content of Solar Cells

The waste solar cells were procured from the local company. The obtained solar cell panels were ground and sieved to obtain a fine powder of solar cell. 10 mL Aqua regia was added to 1 g solar cell powder, in 100 mL beaker and agitated overnight at 150 rpm. Then the contents were heated at 50 °C for 1 h. After acid leaching, the small amount of residue was removed by filtration. The sample was analyzed by inductively coupled plasma optical emission spectrometry (ICP-OES).

2.2 Growth of Micro-organisms and Collection of Cell-Free Supernatant

Thiobacillus ferrooxidans was obtained from the Food Industry Research and Development Institute (FIRDI), Taiwan. The basal 9 K medium was used for growth of *Acidithiobacillus ferrooxidans*. $FeSO_4$ was used as energy source for growth of *A. ferrooxidans* [9]. The flask culture was used as an inoculum (10% v/v) after ten days incubation at 30 °C temperature and 150 rpm shaking speed [6]. The cell-free culture supernatant was collected as described by Hocheng et al. [5, 8].

2.3 Bioleaching of Solar Cells by Using Culture Supernatant

The 1 g of the solar cell powder was mixed separately with the culture supernatants of *T. ferrooxidans* to study the bioleaching of metals. The flasks were then incubated for 72 h in a shaking incubator and aliquots of the samples were taken at 24, 48 and 72 h to analyze metal content by means of ICP-OES.

2.4 Effect of Physicochemical Parameters on Bioleaching of Metals from Solar Cells

Effect of temperature on metal bioleaching from the solar cell was studied. The solar cell powder was added to 100 ml of *T. ferroxidans* culture supernatant in 250 ml flasks separately. For this study, the flasks were incubated at various temperatures (30–50 °C) in a shaker incubator at 150 rpm shaking speed for 72 h. Another study was carried out to find the effect of shaking speed on bioleaching of metals where 1 g of the solar cell powder was added to the culture supernatant (100 ml) in 250 ml flasks separately and were incubated at variable shaking speeds (0–200 rpm) at a certain temperature for 72 h. Effect of concentration of solar cell powder on bioleaching efficiency was studied. The varying concentrations (0.5–5 g) of solar cell powder were mixed with culture supernatant (100 ml) in 250 ml flasks separately and were incubated at an optimum temperature and shaking speed for 72 h. The samples were collected after each experiment and sent for metal content analysis by means of ICP-OES.

2.5 Analytical Methods

After exposure to the leaching solution, the suspension was filtered. It was acidified using nitric acid. Then the clear solution was used for metal analysis. The concentration of metals in the leach liquors was analyzed by Kontron S-35, ICP-OES.

3 Results and Discussion

3.1 Bioleaching of Metals from Solar Cells

Although the solar cells are long lasting, the day is not far when they will turn into one of the waste categories which will remain ever increasing thereafter. So it is the need of time to develop the eco friendly recycling process for the same. Some reports are available on the chemical leaching of waste solar or photovoltaic cells. But till date, there are no reports on the biological recovery of metals from waste solar cells. This study represents the use of various micro-organisms for metal recovery from solar cells. Table 1 shows the chemical composition of solar cell powder. Al, Mg, Si, Zn, Ag and B were found in higher amounts whereas the amount of Co was lowest. One of the high-value metal Te was found to have a concentration of 43.89 μg/g.

Two-step bioleaching process was employed in the present study wherein, solar cell powder was added to the culture supernatant after the growth of microorganisms, in order to avoid the toxic effect of heavy metals. The culture supernatant of *T. ferrooxidans*was employed for the metal recovery from solar cell powder. At the end of 24 h, 35.14% Al, 100% Si and 95.05% Zn were extracted. Increase in the incubation period increased the extraction of Zn (98.31%) and Cu (40.08%). However, there was no notable difference in the recovery of other metals (Data not shown). In their previous work, authors have shown the presence of Fe^{3+}, a strong oxidizing agent in the culture supernatant of *T. ferrooxidans*, which played a key role in biomachining of various metals [4]. It obtains energy from an aerobic oxidation of either iron (II) or reduced sulfur compounds [11]. In an indirect

Table 1 Chemical composition of solar cell powder

Sr. no.	Component	Weight (μg/g)
1	Al	84245.0
2	Mg	821.15
3	Si	516.40
4	Zn	236.65
5	Ag	164.35
6	B	140.65
7	Pb	54.19
8	Te	43.89
9	Cu	10.295
10	Sr	42.68
11	V	9.806
12	Mn	4.446
13	Ni	4.284
14	Cr	3.444
15	Co	0.2166

mechanism, bacteria produce Fe^{3+} by oxidation of Fe^{2+}, which plays a key role for in oxidation and metal extraction from the ore. Gholami et al. [3] also reported the solubilization of heavy metals from spent catalysis due to chemical oxidation by ferric ions. The effects of various process parameters including temperature, shaking speed and pulp density on metal recovery were also studied.

3.2 *Effect of Temperature*

Metal recovery using *T. ferrooxidans* supernatant was found to be dependant on temperature. Different metals were leached at different temperatures (Fig. 1). 100% recovery of Mg, Zn, B, Mn and Cr was obtained within a reasonable time at 60 °C. However, only 41.6 and 13.4% of Al and Te were extracted respectively. These results suggest the higher temperature optimum for metal recovery from solar cells in use of the organism. However, it was also observed that nearly 95% Te was extracted and 20 °C, along with 41.38 and 43.12% of Pb and Sr, respectively. In contrast, it was found that recovery of silver from spent silver oxide-zinc batteries using *T. ferrooxidans* culture supernatant was independent of temperature [7].

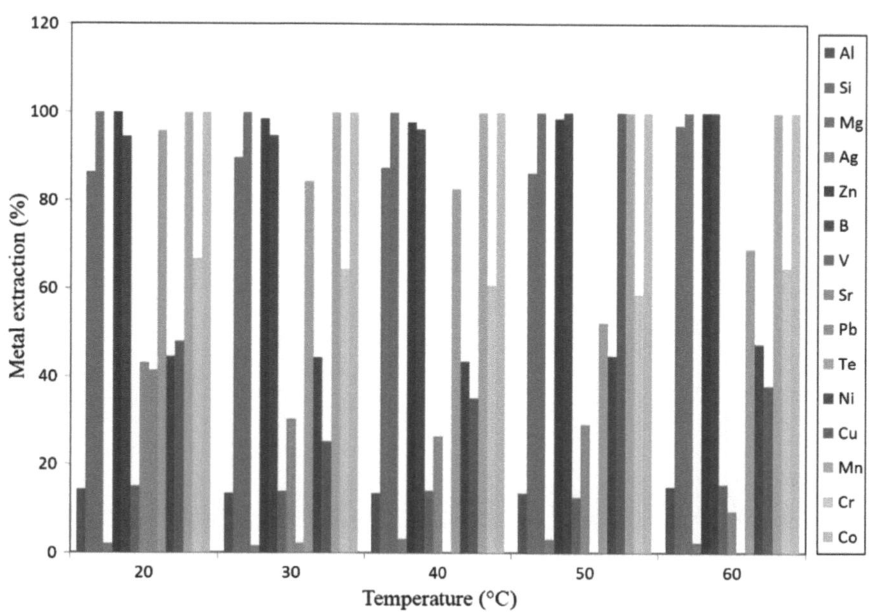

Fig. 1 Effect of temperature on bioleaching of metals from solar cell by *T. ferrooxidans* culture supernatant

3.3 *Effect of Shaking Speed*

Leaching of metals by *T. ferrooxidans* at different shaking speed was very much similar except the extraction of Te and Pb which were recovered differently at different shaking speed. 89.79% Te was recovered at 50 rpm whereas 60.80% Pb was recovered at 100 rpm (Fig. 2).

3.4 *Effect of Pulp Density*

The increase in pulp density had an inhibitory effect on metal recovery by *T. ferrooxidans*. As the pulp density increased metal leaching was decreased, except the recovery of Mn and Co which was 100%, maybe owing to its overall less concentration in the solar cell powder (Fig. 3). Wu and Ting [15] and Amiri et al. [1] also reported the decrease in metal leaching efficiency of the spent medium with an increased pulp density.

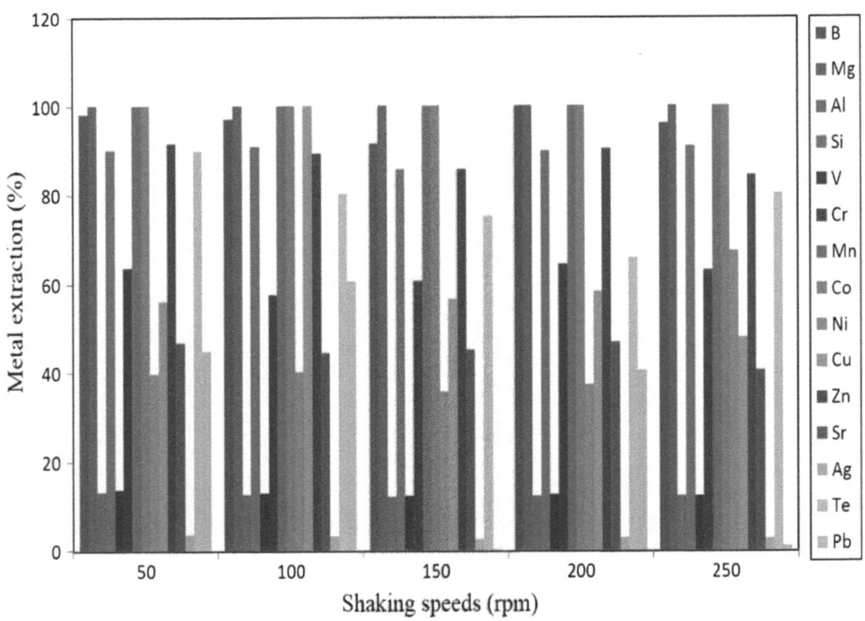

Fig. 2 Effect of shaking speed on bioleaching of metals from solar cell by *T. ferrooxidans* culture supernatant

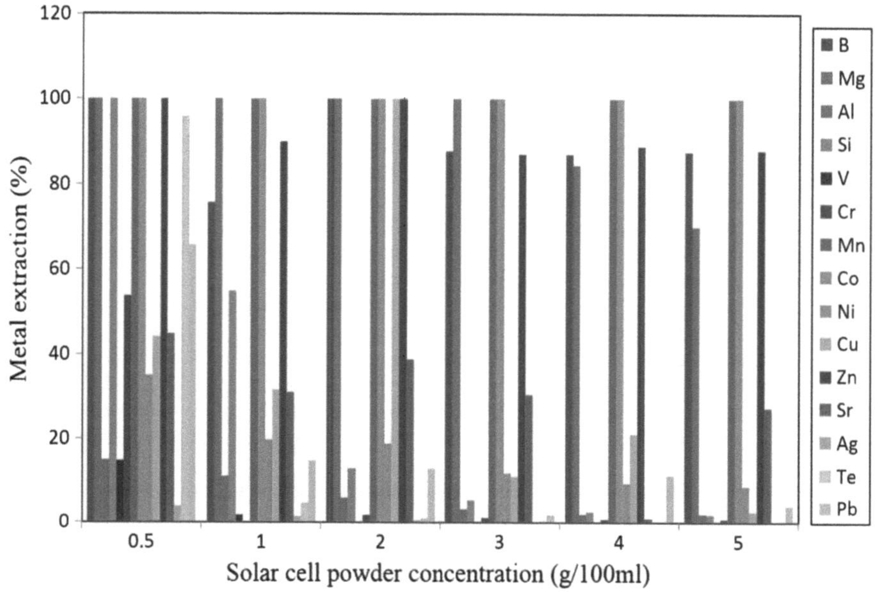

Fig. 3 Effect of increasing pulp density on bioleaching of metals from solar cell by *T. ferrooxidans* culture supernatant

4 Conclusion

The bio-hydrometallurgical approach of recycling waste or damaged solar cells is discussed in this article. Bearing in mind the objective of recovering high-purity materials from the recycling process, bioleaching is the most appropriate and environmentally friendly method. *T. ferrooxidans* shows the potential recovery of heavy metals from waste solar cells along with the recovery of precious metal like Te. This process will allow reuse of waste solar cells without affecting the environment.

References

1. Amiri F, Yaghmaei S, Mousavi SM (2011) Bioleaching of tungsten-rich spent hydrocracking catalyst using *Penicillium simplicissimum*. Bioresour Technol 102:1567–1573
2. Bleiwas DI (2010) Byproduct mineral commodities used for the production of photovoltaic cells. U.S. Geological Survey Circular 1365, 10 p. https://pubs.usgs.gov/circ/1365/. Accessed 04 Jan 2017
3. Gholami R, Borghei S, Mousavi S (2011) Bacterial leaching of a spent MoeCoeNi refinery catalyst using *Acidithiobacillus ferrooxidans* and *Acidithiobacillus thiooxidans*. Hydrometallurgy 106:26–31
4. Hocheng H, Chang J, Jadhav U (2012) Micromachining of various metals by using *Acidithiobacillus ferrooxidans* 13820 culture supernatant experiments. J Cleaner Prod 20:180–185

5. Hocheng H, Hong T, Jadhav U (2014) Microbial leaching of waste solder for recovery of metal. Appl Biochem Biotechnol 173:193–204
6. Hocheng H, Su C, Jadhav UU (2014) Bioleaching of metals from steel slag by *Acidithiobacillus thiooxidans* culture supernatant. Chemosphere 117:652–657
7. Jadhav U, Hocheng H (2013) Extraction of silver from spent silver oxide zinc button cells by using *Acidithiobacillus ferrooxidans* culture supernatant. J Cleaner Prod 44:39–44
8. Jadhav U, Hocheng H (2014) Use of *Aspergillus niger* 34770 culture supernatant for tin metal removal. Corros Sci 85:248–255
9. Jadhav U, Hocheng H, Weng W (2013) Innovative use of biologically produced ferric sulfate for machining of copper metal and study of specific metal removal rate and surface roughness during the process. J Mater Process Technol 213:1509–1515
10. Kang S, Yoo S, Lee J, Boo B, RyuH (2012) Experimental investigations for recycling of silicon and glass from waste photovoltaic modules. Renew Energy 47:152–159
11. Kim S, Bae J, Park H, Cha D (2005) Bioleaching of cadmium and nickel from synthetic sediments by *Acidithiobacillus ferrooxidans*. Environ Geochem Health 27:229–235
12. Klugmann-Radziemska E, Ostrowski P (2010) Chemical treatment of crystalline silicon solar cells as a method of recovering pure silicon from photovoltaic modules. Renew Energy 35:1751–1759
13. Mah O (1998) Fundamentals of photovoltaic materials. National Solar Power Research Institute, Inc., 10 p. http://userwww.sfsu.edu/~ciotola/solar/pv.pdf. Accessed 8 May 8 2009
14. U.S. Department of Energy, Solar Technologies Program (2008) Solar FAQs-Photovoltaics—ALL. U.S. Department of Energy. http://apps1.eere.energy.gov/solar/cfm/faqs/third_level.cfm/name=Photovoltaics/cat=ALL. Accessed 1 Apr 2009
15. Wu HY, Ting YP (2006) Metal extraction from MSW incinerator fly ash Chemical leaching and fungal bioleaching. Enzym Microb Technol 38:839–847
16. Yi YK, Kim HS, Tran T, Hong SK, Kim MJ (2014) Recovering valuable metals from recycled photovoltaic modules. J Air Waste Manage Assoc 64:797–807

Intelligent Wear Identification Based on Sensory Inline Information for a Stamping Process

Johannes Hohmann, Tillmann Schatz and Peter Groche

Abstract In recent years, the use of integrated sensors in stamping tools has strongly increased. Due to this, the opportunity rises to expand monitoring systems from providing simple differentiations between sufficient and insufficient process conditions or the detection of tool failure towards a detailed process understanding based on characteristic parameters describing the current process condition. Therefore, it is necessary to correlate sensor signals with tool and process parameters. However, most of these correlations are not investigated for stamping processes. Due to this, an investigation of correlations is essential to increase the benefits of current process monitoring systems. The aim of this paper is to establish further gradations for the current process condition to increase process stability. Furthermore, process disturbances or tool failures can be detected and upcoming tool changes will be predictable. A first approach is to evaluate correlations between sensor information, like force measurements, and wear mechanisms. This paper presents an approach for the determination of characteristic wear parameters by using the information provided by integrated force sensors. Furthermore, the parameters are established and correlated with wear phenomena on the tool. For the correlation analysis of the measurement and the current state of wear, the active elements of the tool are removed and the surface is measured by a scanning electron microscope (SEM) and a confocal white light microscope to investigate the 3D-surface properties. The experimental investigations are executed with a single pilot shear cutting tool on a high speed stamping press.

Keywords Wear · Wear parameter · Inline measurement · Stamping process

---

J. Hohmann (✉) · T. Schatz · P. Groche
Institute for Production Engineering and Forming Machines,
Technische Universität Darmstadt, Otto-Berndt-Strasse 2,
64287 Darmstadt, Germany
e-mail: Hohmann@ptu.tu-darmstadt.de

V. Majstorovic and Z. Jakovljevic (eds.), *Proceedings of 5th International Conference on Advanced Manufacturing Engineering and Technologies*,
Lecture Notes in Mechanical Engineering, DOI 10.1007/978-3-319-56430-2_21

1 Introduction

Sheet metal stamping is one of the most used manufacturing processes in metal forming industry. Millions of parts are made with high requirements regarding precision and quality every day. While demands on quality and part complexities increase, higher process reliabilities are required. The common approach to reach this requirement is monitoring and optimizing the process to reduce the production cost. Hence, improvements of the monitoring systems lead to a significant benefit in productivity [1].

The use of correlations between process, tool and sensor information result in detectable current process conditions and significantly increased benefits of current monitoring systems. Process disturbances as well as imminent process or tool failures are, therefore, detectable and upcoming tool changes become predictable. Thus, using the acquired process information from the monitoring systems to gain an increased process understanding is essential. Especially correlations between the current tools' wear condition and the inline sensor signals have to be investigated.

Today, knowledge about correlations is the key towards a better process understanding, but yet not investigated sufficiently. Several investigations on stamping processes can be found in scientific literature focusing on the detection and classification of global failures [2–4], the influence of materials [5], the influence of process parameters [6] or the optimization of the process [7, 8] to increase the part quality [9, 10]. These investigations correlate parameters of the semi-finished products or tools before and after the stamping process, but cannot be resumed as condition monitoring. Within these investigations, commonly measured parameters are force, acoustic emission or acceleration. Typically, the sensors are used to detect process deviations or changing part parameters. For the detection of global failures, like a deviating sheet thickness, Zhang et al. [3] pre-processed the acceleration signals by a bispectral analysis to capture the current sheet thickness and decrease the computational load. Another approach to detect a deviating thickness is shown in [4]. The approach correlates the measured acoustic emission with the process force and leads to robust parameters for monitoring the stamping process. The benefit for the process reliability and the productivity depends on the methods used while monitoring the acquired signals. A combination of common methods of monitoring systems like the window technique [11], envelope analysis [12] and limits [13] leads to first classification capabilities of global failures. The main problem for a robust classification is the need of high process understanding and correlations between the current process conditions and the sensor information. As shown in [2–4], global failures are detectable by using common monitoring methods. A comparison of in-process parameters like changing process forces and time-dependent continuous changes of tool conditions like wear mechanisms is not focused in the current state of research. Thus, the aim of the paper is to establish characteristic wear parameters by the provided information of integrated force sensors. Furthermore, the characteristic wear parameters and occurring wear phenomena on the tool are correlated to make the current wear status predictable by using of force signals.

2 Wear Mechanism for Shear Cutting Tools

For shear cutting tools, adhesion, abrasion, surface breakdown and tribochemical wear are the four main wear mechanisms known in literature [14]. Additionally, combinations of different wear types can occur on shear cutting tools [15]. Due to the high accuracy of the tool geometries, increasing abrasion and adhesion mechanisms lead to process deviations or changing tool parameters and finally to insufficient part quality. According to the location of the wear, shell surface, front surface and crater wear are the three types of wear phenomena [16, 17]. Front surface and crater wear mainly occur when processing thin metal sheets under high surface pressure. Shell surface wear arises if the friction between the tool and the sheet metal in the cutting direction during the penetration and retraction of punch is too high [14]. The highest wear occurrence is located at the moving tool elements, commonly the punches [18]. According to the papers cited [15–18], the following investigations are focusing on the correlation of the shell surface wear at the punch and the monitored process forces.

3 Experimental Setup and Procedure

Figure 1 shows the experimental hole stamping tool and the experimental parameters. On the left side of Fig. 1, the structure of the tool is illustrated. During the down stroke the punch (b) penetrates the sheet metal and pushes the material into the die (e). Throughout the upward stroke the stripper (d) removes the remaining

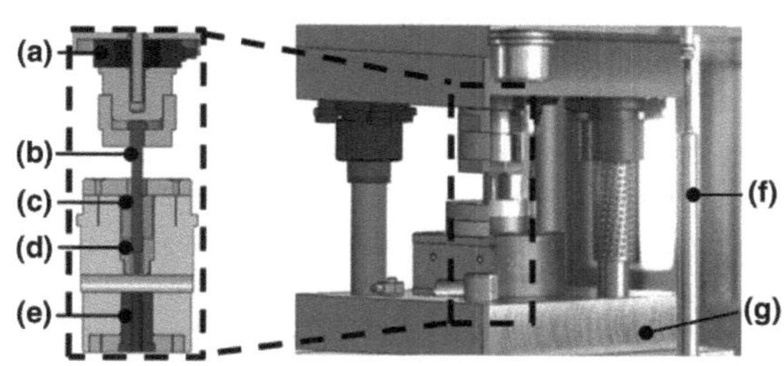

Experimental parameters	
Punch diameter	6 mm
Clearance	0.15 mm
Stroke rate	300 spm
Amplitude	26 mm

Fig. 1 Experimental setup of the hole stamping tool. **a** Force measuring ring Kistler 9051A, **b** punch, **c** guidance, **d** stripper, **e** die, **f** inductive displacement transducer HBM WA/100, **g** bottom tool plate, according to [19]

sheet metal from the punch. During the process the punch is led by the guidance (c). The technical and experimental parameters are shown in the table on the right side of Fig. 1.

The experiments are performed on a high speed press Bruderer BSTA 810. For measuring the process force in the top tool, a load washer Kistler 9051A is integrated (a). In addition to that, the forces are measured by four symmetrically arranged force washers (Kistler 9031B) beneath the bottom tool plate (g). Five Kistler 5067A charge amplifiers are applied for the force measurements. As measuring amplifier a HBM Quantum MX840A is used. The sampling rate is set to 19.2 kHz. All signals are processed by Matlab 2016b. Furthermore, the optical measurements are performed every 5000 strokes with a confocal white light microscope (nanofocus µSurf) to investigate the wear mechanisms on the 3D-surface properties. After 100,000 strokes and in addition to the confocal white light microscope measurement, the shell surfaces of the punches are measured by a scanning electron microscope (SEM) to investigate signs of wear.

4 Characteristic Process Parameter for Wear Identification

As shown in [11–13], commonly the force signals are used to monitor the stamping process. In the present paper, two materials are investigated. An unalloyed steel (DC03–1.0347) and a dual phase steel (DP800–1.943) with a sheet thickness of 2 and 1.5 mm respectively. Experimental investigations up to 100,000 strokes are carried out with 300 strokes per minute (spm). Figure 2 shows the force signals of the measurement in the top (left side) and the bottom (right side) tool after 10,000 strokes with the DC03. The shown signal on the right side is the sum of the four load washers Kistler 9031B in the bottom tool. The signals are divided into punch-, push- and withdraw section (see Fig. 2; I, II and III). The punch-section starts with the penetration of the sheet metal and ends with the cutting impact. In the push-section, the starting point is defined when the material is pushed into the die and ends with reaching the bottom dead center. In the upward stroke the withdraw section starts at the bottom dead center and ends when the remaining sheet metal is stripped off the punch.

The comparison of the measurements in the top and bottom tool shows minor differences in the signals. The acceleration rate after the cutting impact occurs stronger in the bottom tool and merges with increasing number of strokes in the push section (cf. Fig 1a, b). That leads to the assumption, that the information content for monitoring is higher by measuring in the bottom tool. Therefore, the displayed force parameters are the sum of the four integrated sensors in the bottom tool. The maximum in each section (I-III) is calculated using the limit method of common monitoring systems [13]. For the calculation of the start and end points of each section the following mathematical approaches have been used after trimming

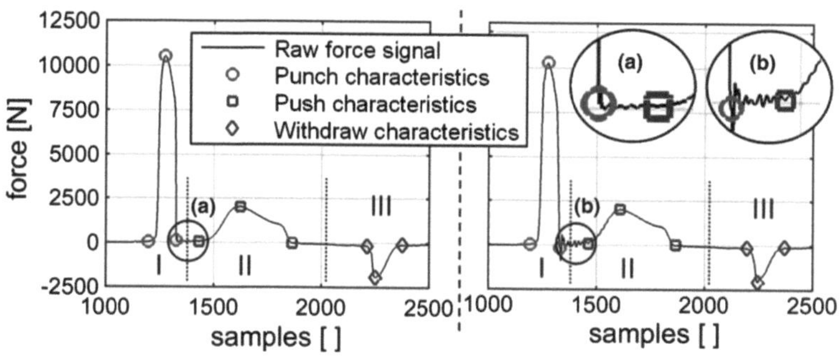

Fig. 2 Force signals of the measurement in the top and bottom tool—DC03, 300 spm, after 10,000 strokes

the raw signals around the respective section. Figure 3 shows the approaches for calculating the characteristic section values with the signal of one Kistler 9031B.

Punch section (I):
The starting point of the punch section is calculated by mirroring the data in the x-axis and multiplying it with a 45° rotational matrix (a). After the transformation, the punching start is determined by transferring the signal from the euclidian-space into the scale-space and calculating the maximum (b) [20]. The endpoint of the punch section is calculated by finding the closest point to the midlevel value of the stroke after the maximum punch signal and finding the first y-value below the starting point's amplitude.

Push section (II):
The possible starting point of the pushing area is the first local minimum of the force after the entire signals maximum (maximum punching force) within the Savitzgy-Golay-Median-Smoothing filtered dataset (c). If this amplitude value is higher or equal to the lower state level (d) of the Savitzgy-Golay-Smoothing dataset, the last amplitude value of the unfiltered dataset equal or lower than the found filtered amplitude value will be set as the push section starting point. Depending on the decay behavior after the cutting impact, a second approach has to be made to calculate the sections starting point, by finding the last amplitude value within the Savitzgy-Golay filtered data set before the push maximum, which is within 3% of the sections global maximum (e). The push section starting point will be equally mapped to the unfiltered data as in the first approach. The endpoint of the pushing section is the first local minimum within the area between the section maximum and the first amplitude value of the Savitzgy-Golay filtered dataset (f), which is equal to or lower than the strokes unfiltered lower state level (d).

Withdraw section (III):
The data is mirrored along the x- and y-axis and afterwards the state levels will be calculated. The init cross (h), final cross (i) and pulse width of the withdraw area is calculated by iterating over the pulse width's tolerance until only one pulse remains

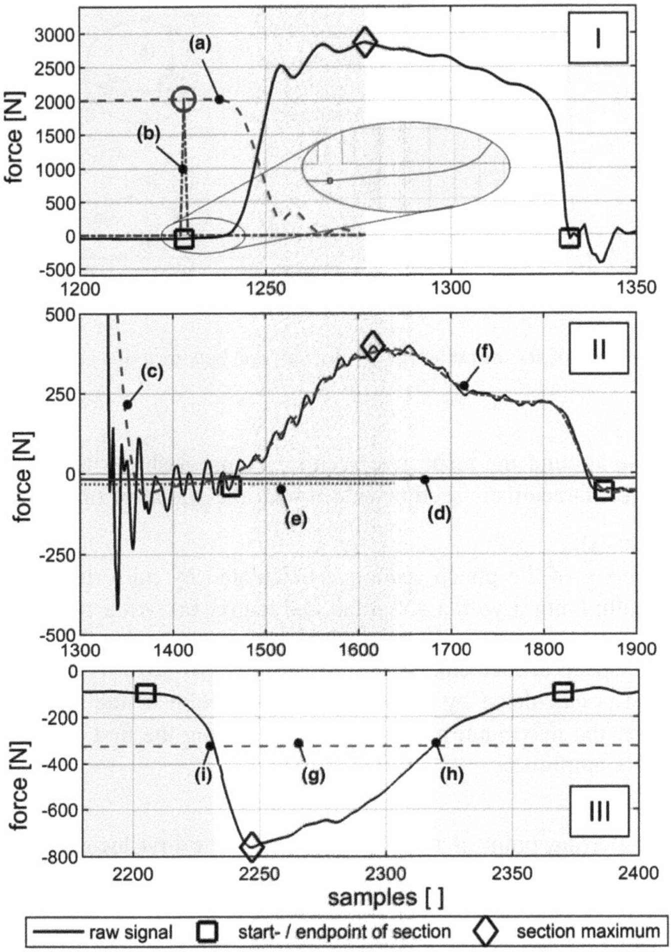

Fig. 3 Calculation of characteristic section values after 40,000 strokes. **a** Transformed signal, **b** transformed signal in scale space, **c** signal filtered with Savitzgy-Golay-Median-Smoothing (Order: 2, Frame length: Pulse width (Punch)/2, Median-Order: Pulse width (Punch)/2, Smoothing-Span: Pulse width (Punch)/2), **d** lower state level, **e** mirrored 3% limit of the section maximum, **f** signal filtered with Savitzgy-Golay (Order: 2, Frame length: Pulse width (Punch)/2), **g** mid reference level, **h** init cross, **i** final cross

within the withdraw area. The crossing points of the mid reference level (g) and the raw signal are the init- and final cross (h, i). The starting point of the withdraw section is the first local minimum after the final cross value. The endpoint of the withdraw section is the first local minimum before the init cross.

Based on the calculated section characteristics showed in Figs. 2 and 3, the maximum and minimum force parameters and the force-time integral parameter (FTI) were determined. Figure 4a shows the results of the measurements of the bottom tool. The maximum punch, push and withdraw forces for the experiments with DC03 fluctuate around a mean value after the first measurement point. The mean value of the punch force is 10,531 N with a deviation of ± 409 N. The push forces deviate with ± 608 N around 2039 N. The deviation of the withdraw force is comparable to the push forces (−2308 N ± 806 N). The results of the experiments with DP800 show the same phenomena of fluctuating maximum forces. The maximum punch force is deviating around 12,745 N with ± 387 N. Until 35,000 strokes there is hardly any increase of the maximum push and the minimum withdraw forces. After 40,000 strokes, the withdraw forces and push forces increase and deviate around −414 and 652 N, respectively.

In Fig. 4b the results of calculated FTI for the push and withdraw section are shown. The test series of DC03 shows, that the FTI of the push and withdraw forces is similar to a rising sawtooth function. The teeth correlate to the changes of the sheet metal coil after 35,000, 60,000 and 95,000 strokes. Due to the storage of the material and no additional process lubrication, the boundary conditions of the first strokes are differing to the inner layers of the coil material. This results in a decrease of the current FTI.

The results for DP800 show, that the FTI of the push and withdraw section increase after 35,000 strokes and start to fluctuate around a mean value. Furthermore, the absolute push forces and absolute withdraw forces as well as the FTI of both forces are significantly lower for the test series with DP800.

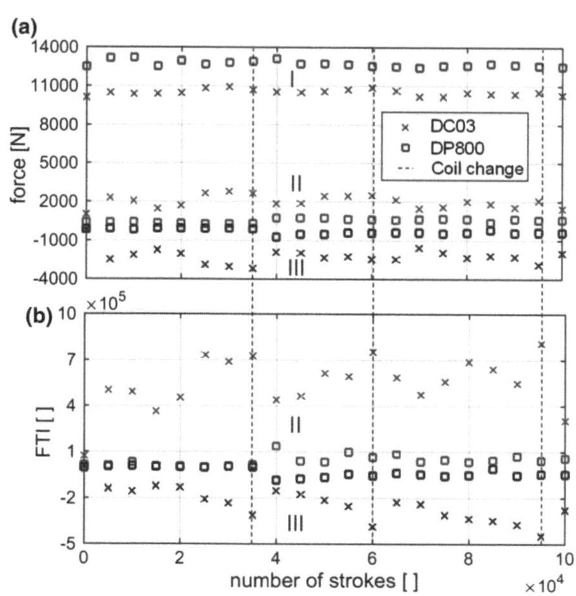

Fig. 4 **a** Maximum and minimum forces of the punch (*I*), push (*II*) and withdraw section (*III*), **b** calculated FTI for the push (*II*) and withdraw section (*III*)

5 Establishing Characteristic Parameters for the Wear Status of the Tool

To establish characteristic parameters of the current wear condition, the surface parameters of the punch have to be correlated with the determined process parameters of each process section. Therefore, at an interval of 5,000 strokes, the punches are removed from the tool for the purpose of performing measurements with the confocal white light microscope. Figure 5 shows the results of the optical measurements in a range between 50,000 and 65,000 strokes with DC03. The shown adhesive wear occurs at the shell surface above the shear cutting edge and moves to the shear cutting edge with increasing stroke number. After the coil change (60,000 strokes), parts of the current adhesive wear decrease and the peaks of the cold welding are flattened. The same phenomena can be observed after the coil changes at 35,000 and 95,000 strokes.

Figure 6 shows the results of the SEM analysis. The upper left picture (a) shows the shell surfaces prior to the experiment with the recognizable milling grooves from production and no signs of wear. In the upper right and lower left (b, c) pictures the punches after 100,000 strokes with DC03 and DP800 are shown.

The result of the SEM analysis indicates heavy adhesive wear on the shell surface and abrasive wear at the shear cutting edge for DC03 (b). In the lower right picture (d), the shell surface was measured by an energy dispersive X-ray spectroscopy (EDX) to confirm adhesive shell surface wear. The result for DP800 (c) shows hardly any wear signs at the punch's shell surface and shear cutting edge. Due to the higher tensile strength and the brittle material behavior, one reason for the non-occurrence of the wear mechanisms is the significant smaller clean-cut section of DP800 (DC03 = $43,5^{+1,2}_{-0,9}\%$; DP800 = $12,9^{+1,7}_{-1,5}\%$). That results in a smaller contact area between the punch surface and the sheet metal.

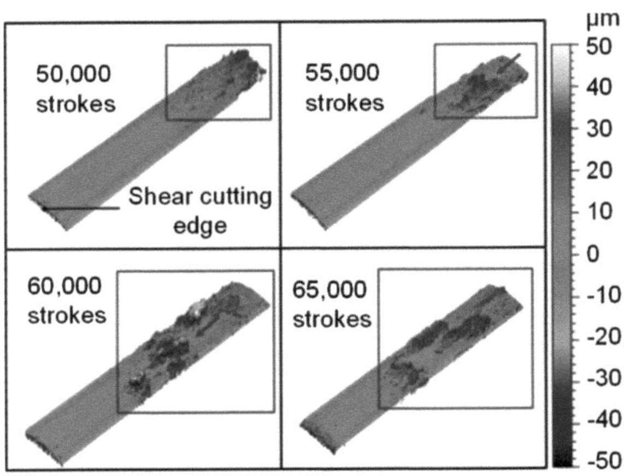

Fig. 5 Optical measurements of the shell surfaces of the punch with the confocal white light microscope in a range between 50 and 65,000 strokes

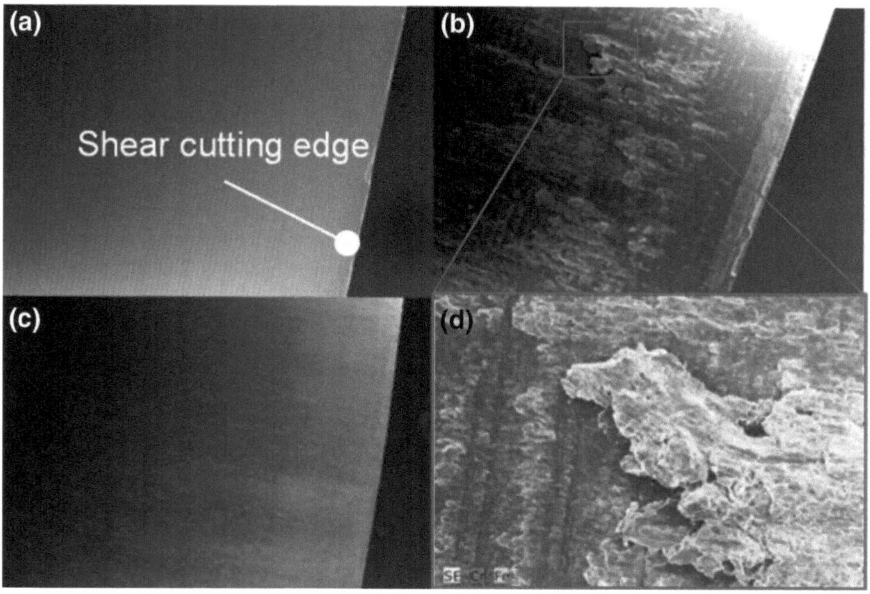

Fig. 6 SEM results of the shell surfaces of the used punches. **a** 0 strokes, **b** test series with DC03 after 100,000 strokes, **c** test series with DP800 after 100.000 strokes, **d** EDX for test series with DC03 after 100,000 strokes

6 Conclusion and Outlook

The investigations show, that observing the maximum process forces is not appropriate to characterize the current process by means of wear conditions of the active parts. The optical measurements indicate, that the adhesive wear of the shell surface is increasing until 100,000 strokes are reached. Within this number of strokes, no significant increase in the punch, push and withdraw force are detectable. The correlations of the optical measurements and the calculated FTI show a dependence on the occurring adhesive wear and the characterized values. With the rise of adhesive wear on the shell surface, the values of the FTI are increasing. Furthermore the position of the wear at the shell surface can be defined by calculating the length between the end point of the punch section and the starting point of the push section. For a decreasing length, the wear migrates in the direction of the cutting edge. These mechanisms can be determined for the test series with DC03. Especially adhesive shell surface wear can be determined by monitoring the FTI. It can be shown, that differing conditions after the coil changes lead to a decrease of the current adhesive wear and the FTI, which is detectable by means of condition monitoring. For materials with high tensile strengths and brittle material behavior, the test series with DP800 shows hardly any wear phenomenon the shell surface and characteristic process parameters for the test series until 100,000 strokes. The lower clean-cut section leads to a decelerated wear occurrence on the shell surface.

To validate the FTI for high tensile strength materials, extended test series with a higher number of strokes have to be performed. Furthermore, the condition monitoring of process characteristics like the position of the maxima and minima, the section lengths and the force-displacement integral of each section could increase the benefits of current monitoring systems and have to be focused within further investigations.

Acknowledgements The results presented in this paper are based on the joint research project "RobIN 4.0". The authors are grateful for the support of the German Federal Ministry of Education and Research (BMBF) within the framework concept "Research for Tomorrow's Production" (funding number 02PJ2700) and managed by the Project Management Agency Karlsruhe (PTKA).

References

1. Klocke F (2001) Prozessüberwachung, Editorial, wt Werkstattstechnik 91, Vol 5, 255–258
2. Bassiuny AM, Li X, Du R (2007) Fault diagnosis of stamping process based on empirical mode decomposition and learning vector quantization. J Machine Tools Manuf 47:2298–2306
3. Zhang GC, Ge M, Tong H, Xu Y, Du R (2002) Bispectral analysis for on-line monitoring of stamping operation. Eng Appl Artif Intell 15:97–104
4. Strache W (2000) Multisensorielle Überwachung des Stanzprozesses. Dissertation, Leibniz Universität Hannover
5. Farzin M, Javani HR, Mashayekhi M, Hambli R (2006) Analysis of blanking process using various damage criteria. J Mater Process Technol 177(1-3):287–290
6. Groche P, Möller N, Hoffmann H, Suh J (2011) Influence of gliding speed and contact pressure on the wear of forming tools. Wear 271:2570–2578
7. Falconnet E, Chambert J, Makich H, Monteil G (2015) Prediction of abrasive punch wear in copper alloy thin sheet blanking. Wear 338–339:144–154
8. Seunghyeon C, Naksoo K (2016) Prediction of tool wear in the blanking process using updated geometry. Wear 352–353:160–170
9. Brecher C, Emonts M, Eckert M, Weinbach M (2014) Double sided irradiation for laser-assisted shearing of ultra high strength steels with process integrated hardening. Phys Procedia 56:1427–1435
10. Ghiotti A, Bruschi S, Regazzo P (2014) Shear surface control in blanking by adaptronic systems. In: 11th international conference on technology of plasticity, Procedia Engineering, vol 81, pp 2512–2517
11. Tan Y, Hahn O, Du F (2005) Process monitoring method with window technique for clinch joining. ISIJ Int 45:723–729
12. Khrebtov P (2011) Neuartiges Verfahren zur Online-Prozessüberwachung und — Fehlerklassifizierung beim Durchsetzfügeverbinden von Blechen. Dissertation, Technische Universität Clausthal
13. Roskam R (1999) In-Prozess-Überwachung von Pressen der Blechverarbeitung. Dissertation, Technische Universität Hannover
14. Lange K (1990) Umformtechnik, Handbuch für Industrie und Wissenschaft, Band 3: Blechbearbeitung. 2. Auflage, Springer, Berlin
15. Loibl D (2003) Standzeit und Teilequalität beim Lochen von Feinblechen mit keramischen Schneidstempeln, Dissertation, Technische Universität München
16. Seidenberg H (1965) Presseneinwirkung auf Werkzeugverschleiß und Grathöhe beim Schneiden von Feinblech im geschlossenen Schnitt. Dissertation, Technische Universität Hannover

17. Fugger B (1984) Untersuchung der Verschleißvorgänge beim Scherschneiden von Feinblech. Dissertation, Universität Hannover
18. Borchert P (1976) Einflüsse der Werkzeuggeometrie und der Maschine beim Schneiden von kaltgewalztem Elektroblech. Dissertation, Technische Universität Hannover
19. Übelacker D, Hohmann J, Groche P (2014) Force requirements in shear cutting of metal-polymer-metal composites. J Adv Mater Res 1018:137–144
20. Liutkus A (2015) Scale-space peak picking. research report inria nancy-grand, France. (https://hal.inria.fr/hal-01103123) 300 spm, after 10,000 strokes

High-Precision Machining of Hard Steel Parts Using Special Abrasive Operations

Wit Grzesik and Krzysztof Żak

Abstract This paper presents the comprehensive approach to the determination of the geometrical and functional characteristics of surface topographies generated on hardened steel parts of about 60 HRC hardness by high-precision cutting and abrasive operations. The initial surfaces were prepared by hard turning (HT) and then finished by precision hard turning (PHT), wheel (WG) and belt (BG) grinding operations. The three surfaces were finished to obtain the Sa roughness parameter of about 0.2 μm after HT and WG operations and 0.04 μm after BG operation. Surface roughness was characterized using the standard 2D and 3D roughness parameters and the fractal and motif parameters. Some important correlations between roughness parameters and surface functional properties are revealed. This study is a part of the research project devoted to the characterization of functional surfaces produced by cutting, abrasive and burnishing operations. The results obtained in the comparative study including hard turning, belt grinding and ball burnishing operations are presented in [1].

Keywords Hard turning · Wheel grinding · Belt grinding · Surface topography · Surface roughness

1 Introduction

It is obviously known that producing high quality machined parts with defined functionality and reliability is strongly integrated with advanced surface metrology. According to ISO Standards, the functional surfaces produced by different

W. Grzesik (✉) · K. Żak
Department of Manufacturing Engineering and Production Automation, Faculty of Mechanical Engineering, Opole University of Technology, 321, 45-271, Opole, Poland
e-mail: w.grzesik@po.opole.pl

K. Żak
e-mail: k.zak@po.opole.pl

V. Majstorovic and Z. Jakovljevic (eds.), *Proceedings of 5th International Conference on Advanced Manufacturing Engineering and Technologies*, Lecture Notes in Mechanical Engineering, DOI 10.1007/978-3-319-56430-2_22

machining operations on such parts as gears, shafts, geared shafts, die and moulds components and hydraulic components can be characterized using the standardized surface areal roughness parameters including *S*-parameters and *V*-parameters sets [2, 3]. For this purpose special machining technologies such as precision machining with $Rz = 2.5–4$ μm and high-precision machining with $Rz < 1$ μm performed on hardened steels (45–60 HRC) using CBN cutting tools are typically used in automotive, hydraulic and die and mold industry sectors [4]. The main advantages achieved concern higher flexibility, producing complete parts in one set-up, lesser ecological impact and higher productivity [5]. One of the technological barriers for partial or full replacement of grinding operations by hard machining or ball burnishing operations is that the characteristics and functionalities of the machined topographies produced are different [6, 7]. This type of a dissimilarity between the hard turned and ground surface topographies was documented even for comparable values of the *Ra* or *Rz* roughness parameters. The comparison of 2D and 3D roughness parameters measured for surfaces generated by precision hard turning and belt grinding operations including bearing area parameters was presented in [7]. Moreover, the topographies machined initially by CBN turning and further modified by superfinishing and ball burnishing operations were compared in terms of possible improvements of functional properties [8]. As mentioned above, this study is focused on the multi-parameter characterization of representative surface topographies produced by high-precision hard turning, wheel and belt grinding operations using both standardized 2D and 3D roughness parameters and fractal dimension and motif parameters. It should be noticed that previous investigations [1, 8–10] assumed the same value of the *Ra*(*Sa*) or *Rz*(*Sz*) roughness parameters.

2 Experimental Details

2.1 Characterization of Workpiece Material and Machining Conditions

Rings made of a 41Cr4 (57 ± 1 HRC) steel were initially turned to $Sa = 0.4$ μm average roughness and subsequently CBN turned with a lower feed of 0.06 mm/rev and wheel and belt ground in order to obtain surfaces with the *Sa* roughness lower than 0.2 μm. The machine tools were Okuma Genos L200E-M CNC precision turning center, conventional grinding machine and special belt grinding head mounted on a conventional lathe (Fig. 1a–c).

Machining conditions for the three machining operations employed were as follows:

1. High-precision hard turning (PHT) (Fig. 1a) using TNGA 160,408 S01030 chamfered CBN insert, cutting speed $v_c = 150$ m/min, feed rate

Fig. 1 Three machining operations performed: CBN hard turning (**a**), wheel grinding (**b**) and belt grinding (**c**)

$f = 0.06$ mm/rev, depth of cut $a_p = 0.05$ mm. All specimens were initially turned using the same cutting tool keeping the feed rate of 0.1 mm/rev and depth of cut of 0.15 mm.

2. Cylindrical grinding (Fig. 1b) using electro-corundum (Al_2O_3) (WG), 350 × 25 × 127 32A grinding wheel with grinding speed of $v_c = 11.9$ m/s, in-feed of $a_e = 0.025$ mm, cross-feed of $f_a = 3.5$ mm/rev.
3. Belt grinding (BG) shown in Fig. 1c using abrasive belts with 30 and 9 μm grains, rotation speed of the workpiece of 900 rev/min, belt feed of 0.06 mm/rev, oscillation frequency of 12 Hz, oscillation amplitude of ±0.5 mm and roller pressure of 2 bars. Finishing process takes 9 s with supplying oil mist produced by a MQL system.

All machining trials were repeated, each for three-times, and the average values of surface roughness parameters were determined.

2.2 Measurements of Surface Roughness Parameters

The surface profiles and topographies produced on the specimens after HT, WG and BG operations were recorded using a 3D contact profilometer with a diamond stylus radius of 2 ± 0.5 μm. 3D roughness parameters were determined and topographies of machined surfaces were visualized using a Digital Surf, Mountains® Map package. The surface topographies were characterization using three groups of parameters including: (a) standardized 3D surface roughness parameters: height, amplitude, horizontal, hybrid and functional [11], (b) non-standardized parameters such as fractal dimension and vectorized valley networks, and (c) standardized motif parameters [2, 3].

3 Experimental Results and Discussion

3.1 Characterization of Machined Surfaces

Figures 2 and 3 present some typical surface profiles (Fig. 2) and topographies (Fig. 3) generated in finishing hard turning (PHT), wheel (WG) and belt (BG) grinding operations. In general, the maximum height *Rz* lower than 1 μm was the surface quality criterion for distinguishing precision machining operations performed [7].

It can be noted that for hard turned and wheel ground surfaces (Fig. 2) the *Ra* values are comparable but the measured values of maximum surface height *Rz* are equal to 1.25 and 2.05 μm respectively. As shown in Fig. 2, *Ra* parameter increases successively from 0.04 μm for two-passes belt grinding to 0.21 μm for hard turning and grinding operations.

In this study the 3D surface roughness parameters were computed as the averages values from about 200 surface profiles which cover the scanned area of 2.5 mm × 2.5 mm. As a result, the characterization of the turned and ground surfaces can be based optionally on the *Rz* and *Sz* parameters as in [5].

Characteristic shapes of the autocorrelation function (AACF) presented in Fig. 4a–c suggest the strong anisotropy (low isotropy) of the three relevant

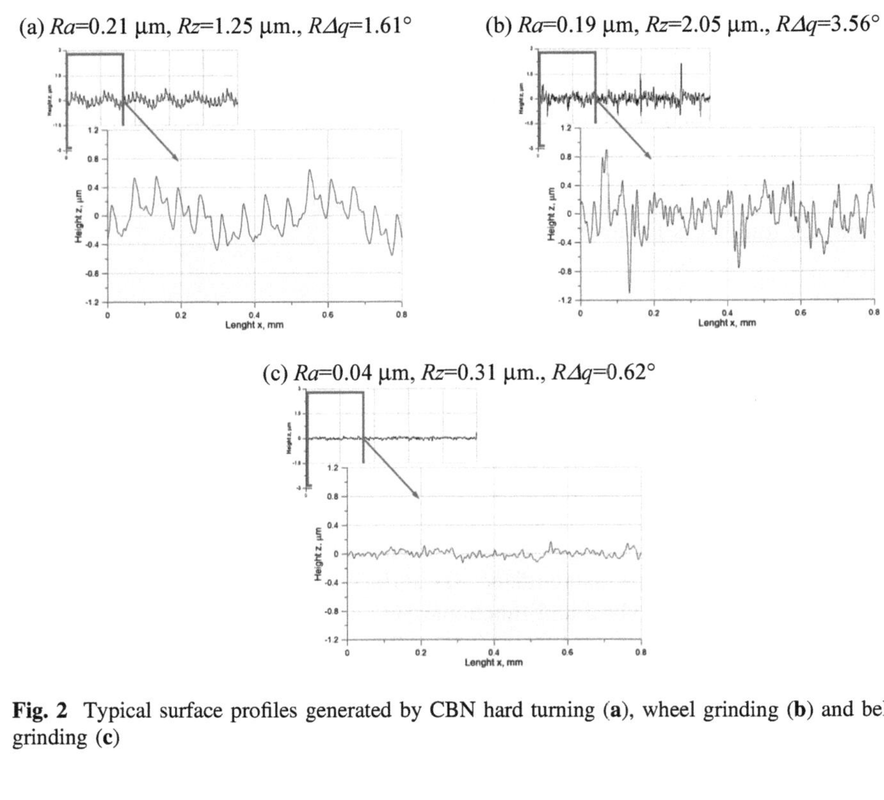

Fig. 2 Typical surface profiles generated by CBN hard turning (**a**), wheel grinding (**b**) and belt grinding (**c**)

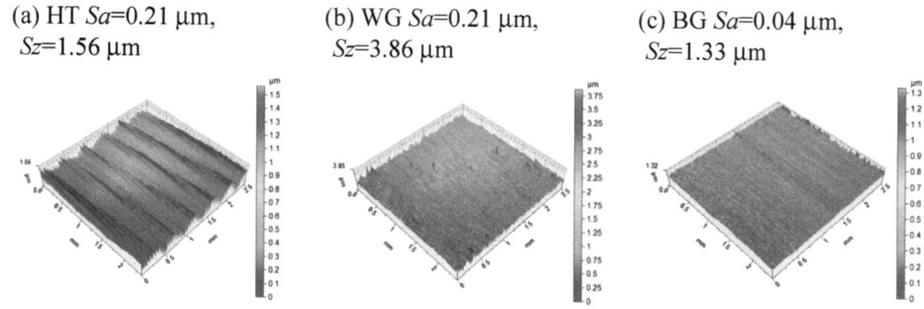

Fig. 3 Three surface topographies generated by CBN hard turning (**a**), wheel grinding (**b**) and belt grinding (**c**)

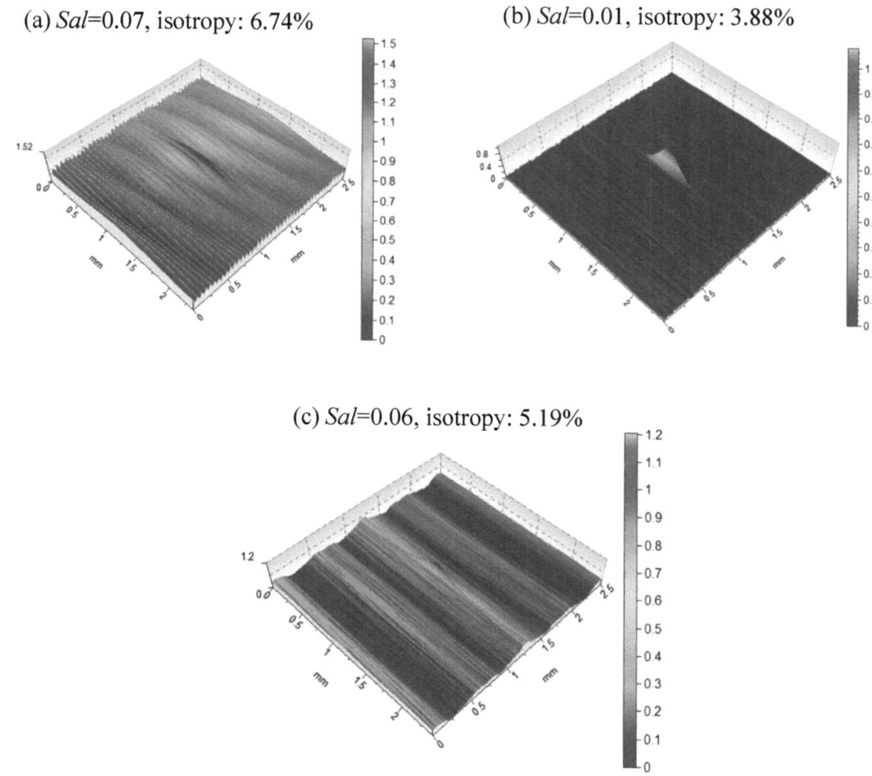

Fig. 4 Distributions of the autocorrelation function for surfaces generated by CBN hard turning (**a**), wheel grinding (**b**) and belt grinding (**c**)

machined surfaces shown in Fig. 3a–c. In particular, the turned surface is periodic-anisotropic (Fig. 4a) but both ground surfaces represent mixed (between anisotropic and random) structures (Fig. 4b, c).

This surface classification is based on the fact that the values of the fastest decay autocorrelation length (*Sal*) are equal to 0.07 for hard turned, 0.01 for wheel ground and 0.06 for belt ground surfaces respectively. A larger value of *Sal* = 0.06 for the belt ground surface coincides with the surface profile dominated by low spatial frequency components [12].

3.2 Characterization of Function Related Parameters

The shapes of 3D BAC's and associated ADF curves obtained for the compared surface topographies are shown in Fig. 5. It was observed that hard turned surfaces have positive skew *Ssk* = 0.24 but both wheel and belt ground surfaces have negative skew equal to (−0.31) and (−0.07) respectively. Moreover, Fig. 5b shows that surfaces produced by hard turning and grinding are distinguished by diametrically different ADF curves which correspond with various bearing and contact properties. The better bearing properties for the ground surface with a negative skewness of *Ssk* = −0.31 corresponds with the case when sharp irregularities produced by hard turning were removed by ceramic wheel (BAC #WG in Fig. 5a). Moreover, values of the areal material ratio *Smr*(*c*), the inverse areal material ratio *Sdc*(*mr*) and the peak extreme height *Sxp* are specified in Fig. 5a.

Figure 6 shows the network of vectorized micro-valleys generated on the machined surfaces produced. This technique can be useful for assessing the fluid retention capability of the machined surface. The maximum depth of valleys ranges between 0.35 and 2.0 μm and the valley width is equal to about 0.5 μm for turned

(a) HT-*Sdc*=0.66 μm, *Sxp*=0.44 μm; WG-*Sdc*=1.97 μm, *Sxp*=0.60 μm; BG-*Sdc*=0.47 μm, *Sxp*=0.12 μm

(b) HT-*Ssk*=0.24, *Sku*=2.56; WG-*Ssk*=-0.31, Sku=5.41; BG-*Ssk*=-0.07, *Sku*=5.31

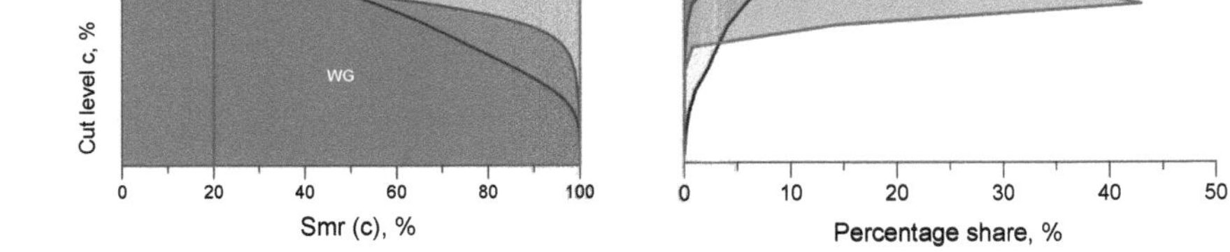

Fig. 5 3D BAC shapes (**a**) and ADF distributions (**b**) for turned (*HT*), wheel ground (*WG*) and belt ground (*BG*) surfaces

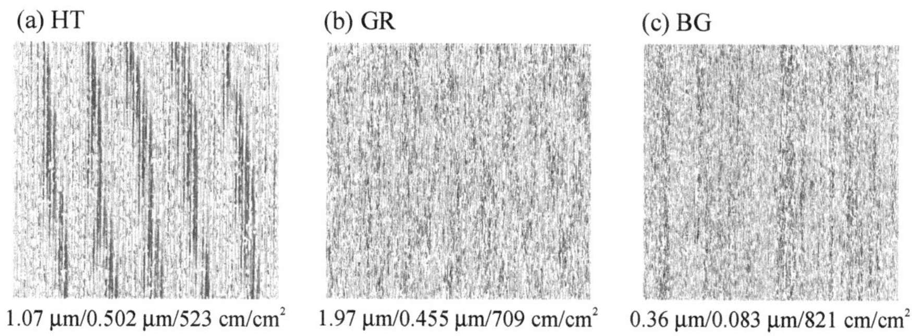

Fig. 6 Vectorized micro-valley networks for turned (**a**), wheel ground (**b**) and belt ground (**c**) surfaces. Three values denote the average depth, width and density of micro-valleys

and wheel ground surfaces and 0.08 μm for belt ground surfaces. The average density of valleys varies between 500 and 800 cm/cm^2 respectively. This specific representation of surfaces confirms evidently that finishing abrasive operations produce smooth surfaces with a larger number of deeper valleys (Fig. 6b, c) which causes the retention capability increases. These data corresponds well with the distributions of the volume functional parameter (*Vmp* and *Vvv*) shown in Fig. 7. In this study the functional analysis of the bearing areal ratio curves is based on the four volume parameters including the peak material volume (*Vmp*), the core material volume (*Vmc*), the core void volume (*Vvc*) and the valley void volume (*Vvv*) parameters [2].

The values of volume parameters obtained for the three machined surfaces are equal to (in order HT/BG/WG): *Vmp* = 0.0125/0.015/0.0026 μm^3/μm^2; *Vmc* = 0.254/0.224/0.05 μm^3/μm^2; *Vvc* = 0.342/0.292/0.066 μm^3/μm^2; *Vvv* = 0.02175/0.038/0.0072 μm^3/μm^2. As mentioned above, higher value of *Vvv* = 0.038 μm^3/μm^2

Fig. 7 Functional volumetric parameters for different finishing operations

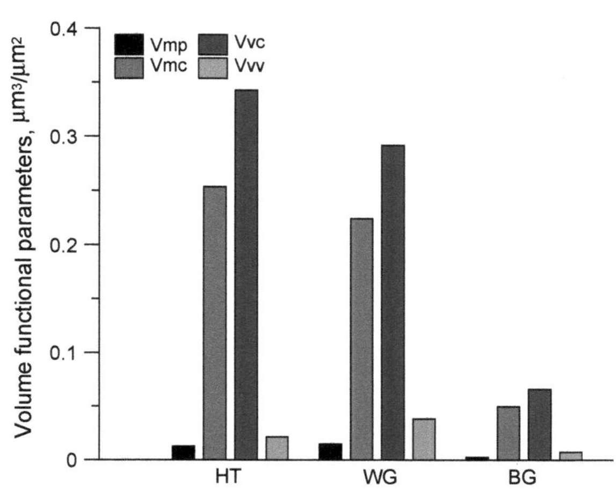

indicates better fluid retention ability of WG surfaces in comparison to turned and very smooth surfaces produced by belt grinding.

The function related parameters considered in this study include three areal (*V*) material ratio parameters-the reduced core (*Sk*), peak (*Spk*) and valley (*Svk*) height and their ratios-*Spk*/*Sk*, *Svk*/*Sk*, *Spk*/*Svk*. They are shown and compared in Figs. 8 and 9. The ratio of *Spk*/*Sk* is recommended [3] to distinguish between two surfaces with indistinguishable roughness average *Sa*.

Figure 8 shows that surfaces produced by turning and wheel grinding operations have comparable values of the reduced core height Sk of about 0.70/0.60 μm and further surface smoothing causes that the Sk parameter is reduced to 0.13 μm. In addition, the reduced peak (*Spk*) height and reduced valley (*Svk*) height differ

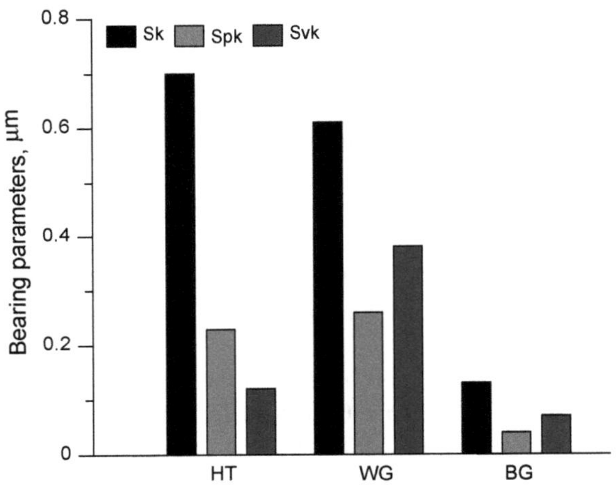

Fig. 8 Areal bearing parameters for different finishing operations

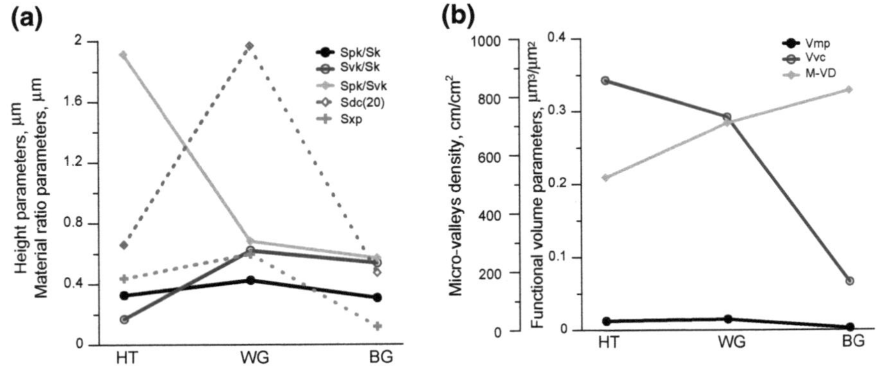

Fig. 9 Functional relationships between selected 3D *V*-parameters

substantially depending on the machining operations employed (Fig. 8). The lowest *Spk* of 0.04 μm corresponds to belt ground surfaces and the highest *Svk* of 0.38 μm to wheel ground surfaces. The comparison of areal bearing parameters suggests that belt ground surfaces have the highest wear resistance and, in turn, the wheel ground surfaces have the best fluid retention capability.

Figure 9a shows that both turned and ground surfaces are characterized by comparable *Spk*/*Sk* values of about 0.3–0.4. In addition, the highest *Svk*/*Sk* ratio of about 0.6 determined for wheel ground surfaces corresponds with better fluid retention abilities (in contrast for turned surface this ratio is about 0.2). Moreover, as shown in Fig. 9b, the ratio of *Spk*/*Sk* correlates well also with the *Vmp* volume parameter, whereas the ratio of *Svk*/*Sk* with the *Vvc* volume parameter and in a lower scale with micro-valleys density. Figure 7a presents other possible relationships between the ratio of *Spk*/*Svk* and *Sdc* and *Sxp* material ratio parameters.

3.3 Characterization of Spatial and Hybrid Parameters

The standard set of 3D roughness parameters specifies four spatial parameters, three of which are texture parameters. The density of summits within the scanned area is equal to *Sds* = 3001.6 1/mm^2 for BG surfaces and 1605.5 1/mm^2 for WG and 1440.7 for HT surfaces. The relatively small texture aspect ratio *Str* = 0.04–0.07 is characteristic for surfaces with stronger directionality of lays (anisotropy). In general, values of Str parameter less than 0.1 are characteristic for highly anisotropic surfaces [11]. The texture direction *Std* close to 90° obtained for all three surfaces indicates that the dominant surface lay is perpendicular to the measurement direction. The corresponding values of *Sal* parameter are specified in Fig. 4.

Additional geometrical differences in the surface topographies analyzed can be emphasized by comparing values of three 3D hybrid parameters. Very low slopes *Sdq* of about 1°–3° are characteristic for very smooth surfaces produced by precision turning and belt grinding operations [12]. The values of the average summit curvature *Ssc* were equal to 0.005 and 0.007 μm^{-1} for the turned and belt ground surfaces and about 0.018 μm^{-1} for the wheel ground surface. In general they are typical for machined surfaces (0.004–0.03 μm^{-1} given in [11]). The *Sdr* parameter (the developed interfacial area ratio) of 0.17% is higher for wheel ground surfaces (Fig. 3b). On the other hand, for turned and belt ground surfaces (Fig. 3a, c) the values of 0.04 and 0.01% were recorded.

3.4 Surface Characterization Based on Motifs and Fractals

In surface engineering [3] the motif analysis is performed on the unfiltered surface profile which is divided into a series of windows [11, 13]. As a result, such

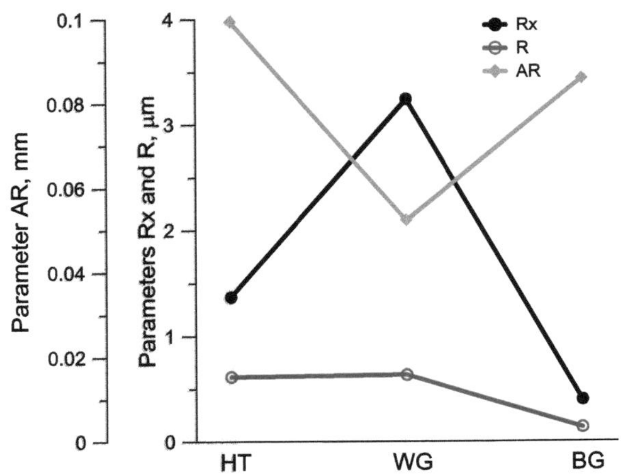

Fig. 10 Examples of the motif graphs for hard turned (*HT*), wheel ground (*WG*) and belt ground (*BG*) surfaces

parameters as the mean depth of roughness motif *R*, the mean spacing of roughness motif *AR* and the largest motif height *Rx* are estimated.

Figure 10 shows that wheel ground surfaces contains distinctly deeper pits (Rx = 3.24 µm) than hard-turned and belt ground surfaces (Rx = 1.37 and 0.40 µm respectively). This fact is in accordance with appropriate volume bearing parameters (Fig. 8). In addition, Fig. 11 shows that the *Rx* motif parameter correlates with the *Rz* parameter stronger rather than Sz because motifs are based on 2D analysis. Moreover, the *R* motif parameter of 0.14–0.64 µm is also independent of the machining operations used and coincides with the *Rz* changes.

The values of fractal dimension *Sfd* determined by means of the method of enclosing boxes are equal to 2.37, 2.41 and 2.54 for turned, wheel ground and belt ground surfaces respectively. In this study, functional relationships between fractal

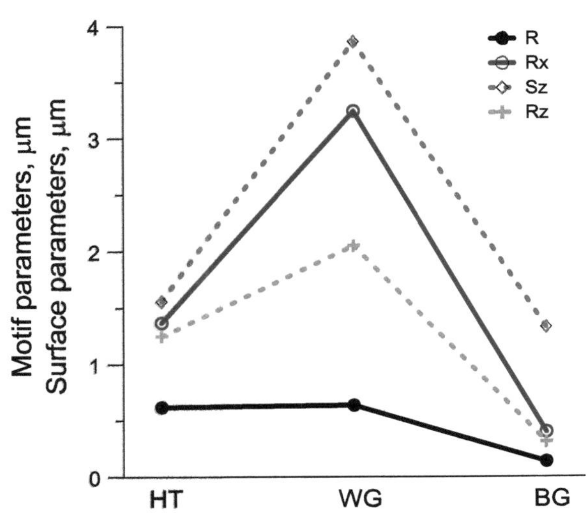

Fig. 11 Functional relationships between *Sz*(*Rz*) and *Rx*(*R*) motif parameters

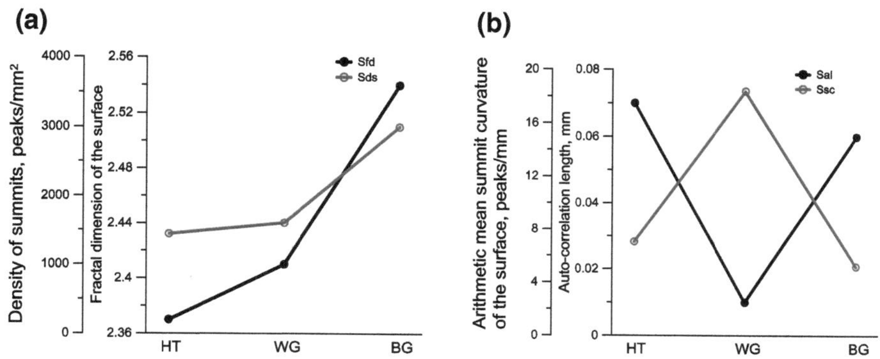

Sfd: HT-2.37, WG-2.41, BG-2.54
Sds: HT-1441 $1/mm^2$, WG-1606 $1/mm^2$, BG-3002 $1/mm^2$

Fig. 12 Functional relationships between selected 3D *S*-parameters and fractal dimension

dimension *Sfd* and *Sal*, *Ssc* and *Sds* spatial and hybrid parameters were examined and the results are shown in Fig. 12. It can be observed in Fig. 12a that the *Sfd* is strongly correlated with the density of summits (*Sds*) in such a way that the highest *Sfd* = 2.54 corresponds with the maximum value of *Sds* = 3001.6 $1/mm^2$ determined for the belt ground surface. At the second order, this correlations can be related to the arithmetic summit curvature (*Ssc*) and the autocorrelation length Sal parameter which characterize the uniformity of the texture (Fig. 12b).

4 Conclusions

This experimental study were performed to assist both design and manufacturing engineers to answer the fundamental question how high–precision machining operations can produce the surface topographies with such desired functional properties as resistant to wear, fluid retention ability, resistant to contact loads, etc.

The following specific comments are formulated based on the measured data and analyses carried out.

1. The multi-parameter comparison of hard turned, wheel ground and belt ground surfaces confirms that surfaces characterized by comparable *Sa* (*Ra*) and *Sz* (*Rz*) roughness parameters have distinctly different spatial features and functional properties.
2. The distributions of the PSD (APSD) function and vectorial maps of micro-valleys suggest that the hard turned surfaces are periodic-anisotropic (Sal = 0.07, Str = 0.07). On the other hand, the structures of wheel (Sal = 0.01, Str = 0.04) and belt ground (Sal = 0.06, Str = 0.05) surfaces are random anisotropic.

3. The shapes of areal bearing curves and appropriate functional parameters for wheel ground surfaces are decisive in obtaining enhanced fluid retention abilities. This is due to a larger negative *Ssk* value of about −0.3 and higher *Vvv* volume of 0.038 $\mu m^3/\mu m^2$.
4. Hard turned and wheel ground surfaces have comparable *Vmp* and *Spk* parameters and, as a result, comparable tribological properties. The improved wear resistance of the belt ground surface is due to producing surfaces with minimum $Vmp = 0.0026\ \mu m^3/\mu m^2$ and $Spk = 0.04\ \mu m$ values.
5. It was reasoned that surface functionality depends on the technological conditions used and can be predicted based not only on the selected functional 3D parameters but can be assisted by motif parameters and fractal dimension.

References

1. Grzesik W, Rech J, Żak K (2015) High-precision finishing hard steel surfaces using cutting, abrasive and burnishing operations. Proc Manuf 1:619–627
2. Jiang XJ, Whitehouse DJ (2012) Technological shifts in surface metrology. CIRP Ann Manuf Technol 61(2):815–836
3. Leach R (2013) Characterization of areal surface texture. Springer, Berlin
4. Klocke F (2011) Manufacturing processes 1. Cutting. Springer, Berlin
5. Davim JP (2011) Machining of hard materials. Springer, London
6. Klocke F, Brinksmeier E, Weinert K (2005) Capability profile of hard cutting and grinding processes. CIRP Ann Manuf Technol 54(2):557–580
7. Grzesik W, Rech J, Wanat T (2007) Surface finish on hardened bearing steel parts produced by superhard and abrasive tools. Int J Mach Tools Manuf 47(2):255–262
8. Grzesik W, Żak K (2012) Modification of surface finish produced by hard turning using superfinishing and burnishing operations. J Mater Process Technol 212(1):315–322
9. Grzesik W, Żak K, Kiszka P (2014) Comparison of surface textures generated in hard turning and grinding operations. Proc CIRP 13:84–89
10. Grzesik W, Żak K (2015) Possibilities of the generation of hardened steel parts with defined topographic characteristics of the machined surfaces. J Manuf Sci Eng 137(1): 014502-1-014502-5. doi:10.1115/1.4028895
11. Griffiths B (2001) Manufacturing surface technology. Surface integrity and functional performance. Penton Press, London
12. Hashimoto F, Guo YB, Warren AW (2008) Surface integrity difference between hard turned and ground surfaces and its impact on fatigue life. Ann CIRP 55(1):81–84
13. Michigan Metrology 3D surface roughness and wear measurements, analysis and inspection. Available at www.michmet.com

Process and Microstructure in Materials-Affected Manufacturing

Steven Y. Liang and Zhipeng Pan

Abstract The fundamental understanding of manufacturing processes has been long focused on the geometric, mechanic, and thermal aspects leading to the product shape and finish. However, the effects of process mechanics attributing to material microstructural properties and constitutive characteristics are essential but not yet well understood due to the intricacy of multiple scale process-materials interaction physics. Further, the effects of materials mechanics on the process behaviors, in the context of stress and heat generations, carries significant practical relevance but has not been fully addressed in science. This is to state that manufacturing processes, such as metal forging, polymer compression modeling, 3-D printing, et al., commonly involve a significant amount of mechanical, thermal, and even chemical loadings that interact strongly with part material microstructural evolutions, which in turn determine the performance and functionality beyond just the shape and finish of the end products. On the other hand, the materials microstructure in terms of grain size, texture, phase field, etc. can also change the stress and heat generation mechanics of the manufacturing process. The scope of this paper is to present the "materials-affected manufacturing" connotation in exploring how process mechanics and materials mechanics interact retroactively with each other, and based upon this connotation better predictions of force, temperature, residual stress, and final part properties and functionalities can be possible. The materials-affected manufacturing analysis methodology involves an iterative blending scheme in combining microstructural synthesis and material homogenization analysis to allow for the interactive effects of materials dynamics and processing mechanics to be considered simultaneously. This paper discusses the basic formulation, computational configuration, and experimental validation in the example cases of machining operations with material recrystallization, grain size variation, recrystallization, texture, and phase field in consideration. Explicit calculation of material microstructure evolution path is provided as functions of process parameters and materials attributes. To factor the material microstructure

---

S.Y. Liang (✉) · Z. Pan
Woodruff School of Mechanical Engineering, Georgia Institute of Technology, Atlanta, GA 30332, USA
e-mail: steven.liang@me.gatech.edu

V. Majstorovic and Z. Jakovljevic (eds.), *Proceedings of 5th International Conference on Advanced Manufacturing Engineering and Technologies*,
Lecture Notes in Mechanical Engineering, DOI 10.1007/978-3-319-56430-2_23

states into the thermo-mechanical coupling process, the material microstructure terms are introduced into the traditional material constitutive model with hardened steels and titanium alloys as examples. Results show that residual stresses and machining forcescan be better modeled and predicted in the materials-affected manufacturing analysis platform.

Keywords Machining · Microstructure · Flow stress · Force · Residual stress

1 Introduction

The development of high precision machining technology enables the complicated shape control, high geometrical accuracy and good surface integrity of the end product [1–3]. The geometrical shape control is achieved by the precise machine tool path planning and error compensation. Appropriate design of the final workpiece material mechanical and microstructural property is required for good surface integrity. The machining process conditions could significantly influence the resultant surface integrity of the final workpiece material. The service functionality of the precision machining end product strongly depends on workpiece material properties. The main consideration in terms of the final workpiece properties includes mechanical attributes (residual stress profile, yield stress, surface hardness), microstructure states (grain structure and orientation, phase composition). The workpiece material properties in the machining process are directly influenced by the process conditions.

Appropriate selection of the machining parameters could help to improve the functionality performance of the end product [4]. For the hard to machine material, such as titanium, nickel based alloys and hardened steels, the high precision machining still faces considerable challenges [5–7]. Significant microstructural evolution has been observed in the machining process [8–10]. The material mechanical properties are strongly dependent on the microstructural states. Also workpiece surface corrosion resistance, microhardness are also influenced by the microstructural attributes. The machining induced residual stress profile is a critical factor for the workpiece corrosion resistance. For biomedical or aerospace industry, the grain refinement would be desirable for the strengthening.

The thermo-mechanical loading introduced from high speed machining will unfavorably affect the workpiece material properties, such as augmented grain size, reduced surface hardness, and tensile residual stress profile [11, 12]. Most of the current research work only focuses on the thermo-mechanical coupling process, where the microstructural evolution effect is largely ignored. However, obvious microstructural change has been observed in the machining process, especially for multiphase material, such as titanium alloys and nickel based alloys. Therefore, it is important to understand the thermo-mechanical-microstructural coupling effects.

The combined effect of server plastic deformation, large strain, high strain rate and high temperature in the primary shear zone and workpiece/tool interface would promote the microstructure evolution such as dislocation density change, grain size evolution and multiphase material phase transformation. The early work on the microstructure change in the machining process is reported by Xu et al. [13] in the grinding of ceramics. The different material microstructure effect on the material removal mechanism is investigated. The white layer is observed in the hard turning of hardened steel by Chou et al. [14], which results from workpiece material phase transformation effect. Similarly, the extensive grain refinement and strain induced martensite phase transformation is observed by Ghosh et al. [9] in surfaced turning of 304L stainless steel. The grain refinement and uniform nanocrystalline structure also is found in the chip in turning of copper by Swaminathan et al. [15].

The investigation of machining induced microstructure change would not only benefit the machining process optimization to achieve machining end product with good service functionality, the machining tool selection and improvement could also be obtained. However, the microstructural level investigation of machining is still at its debut stage, where most of the research focuses on the experimental investigation. The aim of the current paper is to summarize the current existing research work on the machining induced microstructure change and discuss a computational frame work for the machining induced microstructure evolution investigation. The case studies are proposed in hardened steels and titanium alloys.

2 Microstructure Evolution Modeling

2.1 *Phase Transformation and Dynamic Recrystallization of Ti–6Al–4V*

Titanium alloys could be divided into three types based on the crystal structure, α alloys, β alloys and $\alpha + \beta$ alloys [16]. The α alloys have α stabilizer such as aluminum and tin with a *hcp* structure at the room temperature. High strength, toughness are the main characters of α alloys. The β alloys are in the state of bcc phase which contains large amount of β isomorphous additions, such as vanadium, niobium and tantalum. The low strength characterizes the basic mechanical property of the β alloys. For the $\alpha + \beta$ alloys, more than one α stabilizers together with β stabilizers exist. The adjustment of the microstructural states could control the mechanical properties of $\alpha + \beta$ alloys. So various heat treatment method could control the strength and fracture toughness of the material in a wide range.

Ti–6Al–4V is a typical $\alpha + \beta$ alloy, which contains 6 wt% α phase stabilizing aluminum and 4 wt% β phase stabilizing vanadium. The equilibrium state microstructure contains the hcp structured α phase with scattered distribution of β phase at the room temperature, as shown in Fig. 1 [17]. The microstructure property may vary depending on the prior heat treatment. Basic microstructural types in *Ti–6Al–4V* includes grain boundary allotriomorph, primary α, Widmanstatten and martensitic. The cooling rate could greatly influence β precipitates distribution and morphology.

The mechanical properties of the *Ti–6Al–4V* material are dependent on the microstructural states. The dominating factor that influences the mechanical

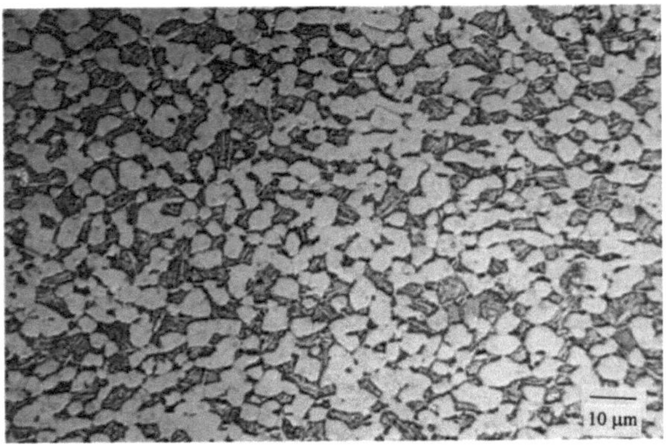

Fig. 1 Typical microstructure of *Ti–6Al–4V* alloy [17]

properties is the α phase colony size. The yield strength, fracture toughness and ductility could be greatly improved by reducing the colony size of α phase.

The JMAK model has been widely used to describe the dynamic recrystallization process of crystalline material by considering the strain, strain rate and temperature [18]. The basics of *JMAK* model is the calculation of the recrystallized volume fraction of the material as a function of time. The grain size is obtained from the grain growth rate and nucleation. The dynamic recrystallization is defined with the Avrami equation as

$$X_{drex} = 1 - \exp\left[-\beta_d\left(\frac{\varepsilon - a_{10}\varepsilon_p}{\varepsilon_{0.5}}\right)^{k_d}\right] \tag{1}$$

where ε is the strain, ε_p is the peak strain, X_{drex} is the volume fraction of dynamically recrystallized material. $\varepsilon_{0.5}$ is the strain for $X_{drex} = 0.5$ and it is given by

$$\varepsilon_{0.5} = a_5 d_0^{h_5} \varepsilon^{n_5} \dot{\varepsilon}^{m_5} \exp(Q_{act} m_5/RT) + c_5 \tag{2}$$

where R is the gas constant, d_0 is the initial diameter of the grain, a_5, h_5, n_5, m_5, c_5 are material constants which could be determined by experiments and regression analysis, Q_{act} is the activation energy. A critical strain at which the dynamic recrystallization would occur is defined as ε_p = 0.8 ε_p. The peak stain ε_p is denoted as

$$\varepsilon_p = a_1 d_0^{h_1} \dot{\varepsilon}^{m_1} \exp(Q_{act} m_1/RT) + c_1 \tag{3}$$

where a_1, h_1, m_1, c_1 are material constants. The grain size after recrystallization is given by

$$d_{drex} = a_8 d_0^{h_8} \varepsilon^{n_8} \dot{\varepsilon}^{m_8} \exp(Q_{act} m_8/RT) + c_8 \tag{4}$$

where a_8, h_8, n_8, m_8, c_8 are the material constants. The average grain size is calculated with a mixture rule as

Table 1 JMAK parameters of *Ti–6Al–4V* material

Peak strain	a_1	h_1	m_1	$Q_{act}m_1$	c_1		a_2		
	0.0064	0	0.0801	30,579 J/mol	0		0.8		
DRx kinematics	a_5	h_5	n_5	m_5	$Q_{act}m_5$	c_5	β_d	k_d	a_{10}
	0.022	0	0	0.11146	26,430 J/mol	0	0.9339	0.5994	0.0311
DRx grain size	a_8	h_8	n_8	m_8	$Q_{act}m_8$	c_8			
	150	0	0	−0.03	−6540	0			

$$d = d_0(1 - X_{drex}) + d_{drex}X_{drex} \tag{5}$$

The initial average grain size is characterized as $d_0 = 15$ μm. The *JMAK* parameters of *Ti–6Al–4V* are listed in Table 1.

The microstructure modelling of *Ti–6Al–4V* consists of two phases, α phase and β phase. The initial microstructure of *Ti–6Al–4V* is bimodal, mainly composed α grains with low concentration of β. In the thermal heating process, α destabilizes and the transformation from α to β starts above the β transformation temperature according to the phase transformation curves. Also, in the cooling down process, the β phase starts to slowly decompose into α phase. Therefore, two different α phase need to be distinguished, Widmanstatten and grain boundary [19]. In the current study, to simplify the model, it is assumed that the material only consists of primary α and β phase.

For the heating process where the phase transformation from α to β takes place, a simplified Avrami model [20] is used as,

$$\vartheta_v = 1 - e^{\left\{A\left(\frac{T-T_s}{T_e-T_s}\right)^D\right\}} \tag{6}$$

where T is the temperature, T_s = 600 °C is the phase transformation starting temperature, T_e = 980 °C is the temperature when the process ends, A_s and D_s are material constants to be determined. The calculation of A_s and D_s could be conducted through an experimental curves of the phase transformation. In the current work, A_s and D_s are selected as −1.86 and 4.35 from a previous study [21].

In the cooling down process, the β to $\alpha + \beta$ transformation is characterized by the *TTT* curve, as shown in Fig. 2. As for the α to β transformation in cooling, the coefficient is used as a mean value of a set of data from literature, which could be described by the function of time as

$$\xi = 1 - e^{-bT^n} \tag{7}$$

where b is the material constant and n = 1.32 is the Avrami number. The dynamic recrystallization of the grain growth and phase transformation model are implemented in the finite element code for microstructural evolution simulation.

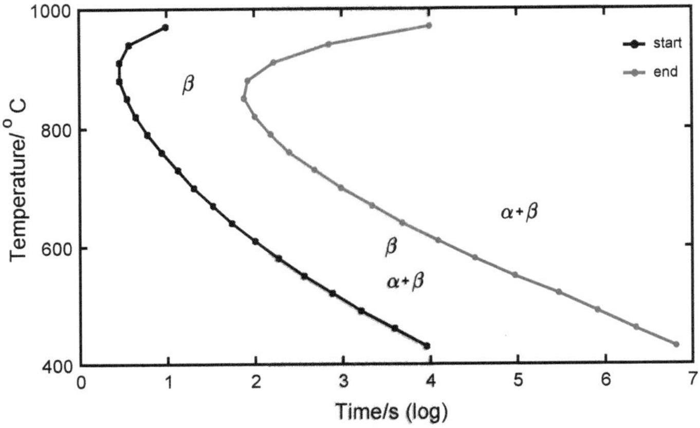

Fig. 2 The TTT curve of *Ti–6Al–4V*

2.2 *Dynamic Recrystallization of 4130 Steel*

AISI 4130 steel is a widely used hardened steel alloys because of its strong hardness and large yield strength. The grain structure of AISI 4130 steel is shown in Fig. 3[22]. The application of AISI 4130 steel includes bearing rings, transmission gears and crankshaft. However, those superior properties of AISI 4130 makes it hard to machining, which imposes great limitation on the material removal rate. The challenges in the machining of AISI 4130 steels comes from the large machining forces, bad surface quality, server tool wear and large dimensional distortion. A predictive force model scheme is proposed by Ji et al. [23] for the machining force optimization for orthogonal turning of AISI 4130. The effect of microstructure

Fig. 3 Typical microstructure of AISI 4130 steel alloys [22]

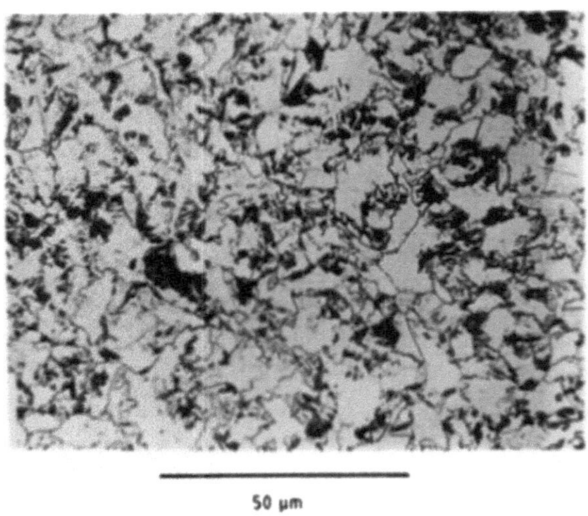

on machining force in turning of Al–Si alloys is investigated by Grum et al. [24]. Hodgson et al. [25] provides the models to predict Xdrex, the recrystallized volume fraction, under static, dynamic and post dynamic recrystallization. X_{drex} is basically an exponential function of $t/t_{0.5}$. t is the time and $t_{0.5}$ is the time of 50% softening for all steels. $t_{0.5}$ is a function of inverse temperature. So based on the time and temperature during turning, the grain size drop could be calculated. Sun et al. [26] adopted empirical relationship to create a theoretical model to describe recrystallization kinetics. Later, Sajadi et al. [27] built the relationship between peak stress, temperature and strain rate. And the mean hot deformation activation energy Q_{act} of AISI 4130 was determined to calculate Zener-Hollomon parameter. The theory from these work is able to predict mechanical and thermal parameters but is not combined with classic machining theory. Current models [28–30] are able to predict forces in different materials including AISI 4130 but none of them take grain size change into consideration. With the similar approach for the *Ti–6Al–4V*, the average grain size could also be calculated in the machining process.

For the AISI 4130 steel, which is a C–Mn steel, the Sellar's model [31] could be used in the form of,

$$d_{drex} = \varepsilon^{-0.5} d_0^{0.4} \exp\left(\frac{45{,}000}{RT}\right) \tag{8}$$

where ε is the plastic strain, d_0 is the initial average grain size. The recrystallized volume fraction X_{drex} could be calculated as,

$$X_{drex} = 1 - \exp\left[-0.693\left(\frac{t}{t_{0.5}}\right)^{1.5}\right] \tag{9}$$

where $t_{0.5}$ is the time when half of the material recrystallizes, defined as,

$$t_{0.5} = k_{md} Z^{n_{md}} \exp\left(\frac{Q_{md}}{RT}\right) \tag{10}$$

where k_{md}, n_{md} and Q_{md} are material constants, selected as 2.5×10^{-6}, −0.8, and 230 kJ/mol respectively; R is the universal gas constant, T is the temperature, Z is the Zener-Hollomon parameter defined as,

$$Z = \dot{\varepsilon} \exp\left(\frac{Q_{def}}{RT}\right) \tag{11}$$

where $\dot{\varepsilon}$ is the plastic strain rate, Q_{def} is the material constant, selected to be 300 kJ/mol for all C–Mn alloy steels [31]. The *JMAK* parameters are taken from Hodgson's [25] experimental measurement, which are listed in Table 2.

Table 2 The JMAK parameters for AISI 4130 steel

k_{md}	n_{md}	Q_{md}	Q_{def}
2.5×10^{-6}	−0.8	230 kJ/mol	300 kJ/mol

3 Microstructure Sensitive Flow Stress Model

The *JC* constitutive material flow stress model has been widely used in the machining process [32], which could be denoted as

$$\sigma = (A + B\varepsilon^n)\left[1 + C\ln\left(\frac{\dot{\varepsilon}}{\dot{\varepsilon}_0}\right)\right]\left\{1 - \left[\frac{T - T_0}{T_m - T_0}\right]^m\right\} \tag{12}$$

where *A*, *B*, *C*, *m*, *n* are materials constants, ε is the equivalent plastic strain, $\dot{\varepsilon}$ is equivalent strain rate, $\dot{\varepsilon}_0$ is the reference strain rate, typically taken as 1 s^{-1}, *T* is the material temperature, T_m is the material melting temperature and T_r is the room temperature. Since the *JC* model is purely based on experiment and data fitting, a lot of modified *JC* models have been developed to more accurately capture the material flow stress from the physical side of material deformation process. To account for the temperature dependent flow softening at high temperature, Calamaz et al. [33] suggested a strain and temperature tangent term. A later self-consistent model (SCM) is proposed by Zhang et al. [34] to account for the phase transformation in the dual phase *Ti–6Al–4V*. However, the SCM does not explicitly calculate the phase transformation and is only based on an iterative fitting method.

The flow stress of the dual-phase *Ti–6Al–4V* depends strongly on its microstructure. Due to increased temperature in the machining process, significant microstructure evolution could occur [35]. The grain morphology and volume fraction of different phase can vary significantly depending on the machining condition. The hexagonal α phase has much stronger yield stress than the BCC β phase. The room temperature yield stress of *Ti–6Al–4V* can vary from 850 MPa to 1100 MPa in different heat treatment conditions. So, a reasonable flow stress model for *Ti–6Al–4V* should include the initial volume fraction of the two phases and the phase transformation in the machining process.

In the current study, since the volume fraction of each phase has been calculated. With the flow stress for each phase σ_α and σ_β calculated from Eq. (6), the flow stress of the dual phase material can be easily obtained from the mixture rule as,

$$\sigma_{\alpha+\beta} = \eta\sigma_\alpha + (1 - \eta)\sigma_\beta \tag{13}$$

where η is the volume fraction of the α phase. In the current study, we assume that only the initial strength *A* is different in *JC* model for different phases. This is a reasonable assumption because the biggest difference between the α and β phase is the initial yield strength. The material initial yield stress *A* is a strong function of the grain size, which could be described by the Hall-Petch equation, as

$$A = A_{hp} + K_{hp} D^{-0.5} \tag{14}$$

where A_{hp} and K_{hp} are the Hall-Petch parameters. For the *Ti–6Al–4V* material, the *A* values are obtained by a linear regression method from the experimental flow stress data at different volume fraction of β phase provided in Zhang's [34] paper, as listed in Table 3 [36].

Also, the modified *JC* model parameters of AISI 4130 are listed in Table 4. The modified *JC* model is implemented as the user subroutine in the finite element code. A detailed implementation of the grain growth and phase transformation is shown in Fig. 4.

Table 3 Modified *JC* model parameters for α and β phases of *Ti–6Al–4V*

Phase	A_{hp}	K_{hp}	B	n	C	m	T_m (°)	$\dot{\varepsilon}_0$
α	517.31	201.68	683.10	0.47	0.035	1	1668	1
β	296.55	100.84	314.55	0.47	0.035	1	1668	1

Table 4 Modified *JC* model parameters for AISI 4130 alloy steel

A_{hp}	K_{hp}	B	C	m	n	T_m (°)	$\dot{\varepsilon}_0$
574.58	112	750	0.008	1	0.25	1432	1

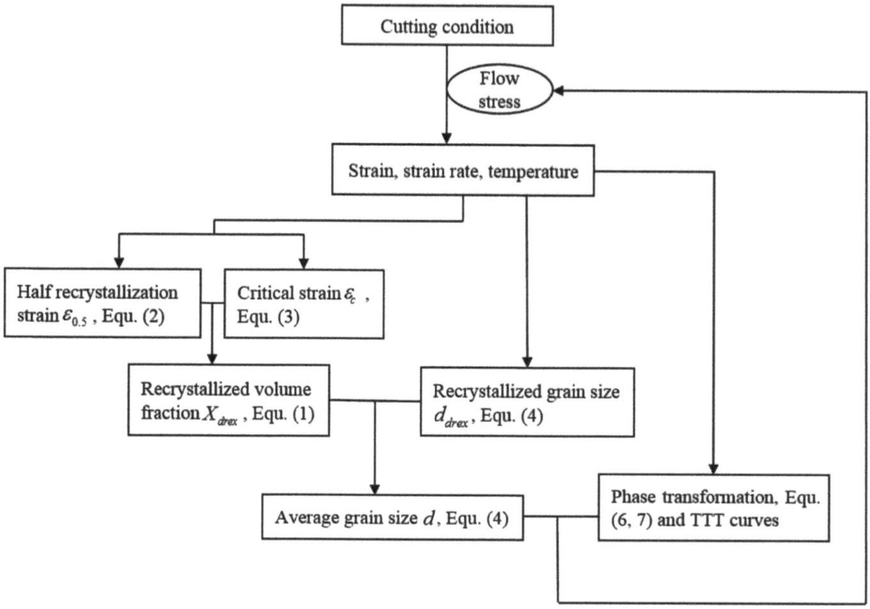

Fig. 4 Schematic illustration of the grain growth and phase transformation implementation

4 Phase Transformation and Grain Size Prediction

The grain size evolution and phase transformation in the orthogonal turning process are predicted with the above proposed method. The initial material average grain size is 15 μm, the α phase volume fraction is 95%. The cutting insert with a tool edge radius of 5 μm is used. The rake angle in orthogonal turning is 5°. A cutting speed at 55 m/min is used, as reported in a previous research. The predicted average grain size and α phase volume fraction are plotted in Fig. 4. The machined surface has slight grain refinement, shown in Fig. 5a. Also, significant amount of β phase generated in both the chip and machined workpiece surface, as indicated in Fig. 5b.

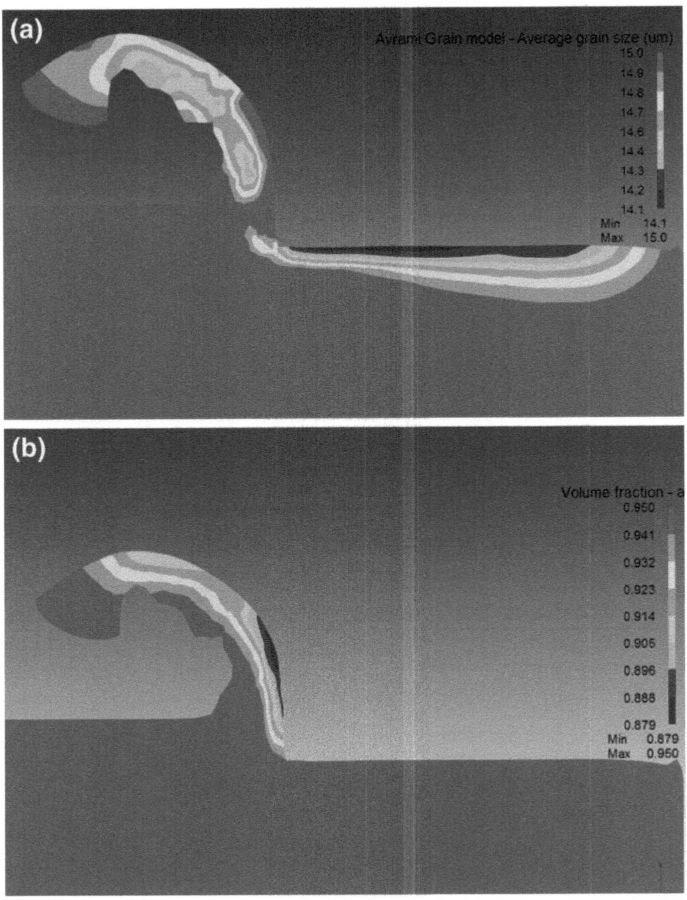

Fig. 5 The predicted grain size (**a**) and volume fraction of α phase (**b**) at cutting speed of 55 m/min, depth of cut 0.076 mm

5 The Force Prediction

For the machining of *Ti–6Al–4V*, four sets of different orthogonal turning conditions are used here to validate the proposed model. The cutting insert edge radius is measured to be 13 μm. A cutting speed is selected as 0.5 m/s. Tool rake angle is 8°. The width of cut is fixed at 3.8 mm. The predicted force and measurement data are plotted in Fig. 6 for comparison at two different depth of cut. The model with a grain size evolution resolves a better prediction compared with the traditional *JC* flow stress model. By varying the rake angle from 8° to 15°, with a constant cutting speed, the predicted forces are plotted in Fig. 7. A better prediction is also observed in Fig. 7.

Similarly, the application of the microstructure sensitive flow stress model is implemented in the hard turning of AISI 4130 steel for further validation. Five machining experiments of AISI 4130 are used for the force model validation. A thin wall cylindrical workpiece is used. The wall thickness is measured to be 4.775 mm.

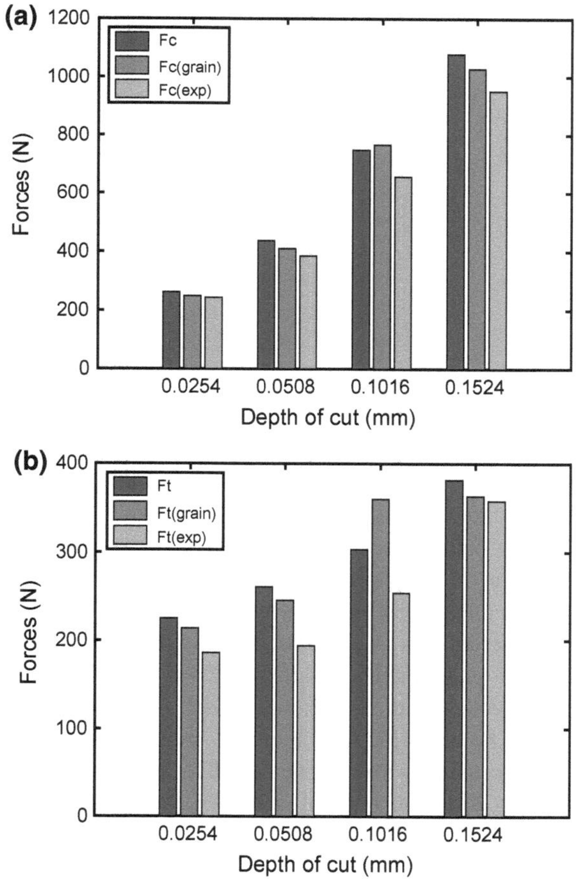

Fig. 6 Cutting force F_c (**a**) and ploughing force F_t (**b**) with a rake angle of 8° at different depth of cut

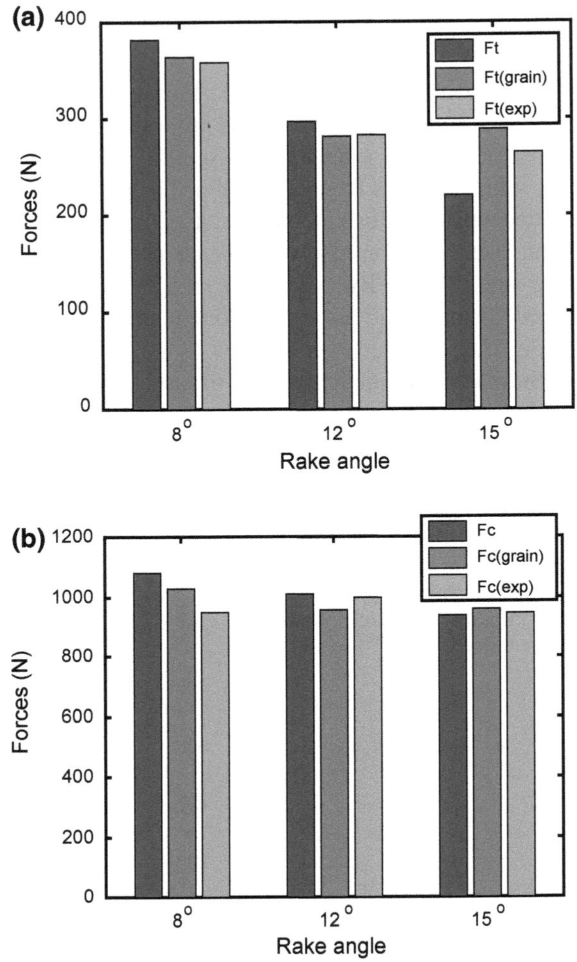

Fig. 7 Cutting force F_c (**a**) and ploughing force F_t (**b**) at a depth of cut 0.153 mm with different rake angles

A Sandvik tungsten carbide tool is mounted to a tool holder to achieve a 5° rake angle and 11° relief angle.

The cutting speed is fixed at 1.049 m/s, the machining forces are plotted as a function of different feed rates, as shown in Fig. 8. To show the microstructure effects on the machining force prediction. The predicted force without a grain size consideration is also imposed in Fig. 8.

The grain model obtains a closer approximation to the measurement data, as compared with the traditional model. A general trend is found that, both the cutting force *Fc* and *Ft* will increase monotonically with the increasing feed rate. Additionally, to investigate the effect of cutting speed on the machining forces, the turning feed rate is fixed at 0.0508 mm/rev by varying the cutting speed. The *Fc* and *Ft* are plotted as a function of cutting speed, as shown in Fig. 9.

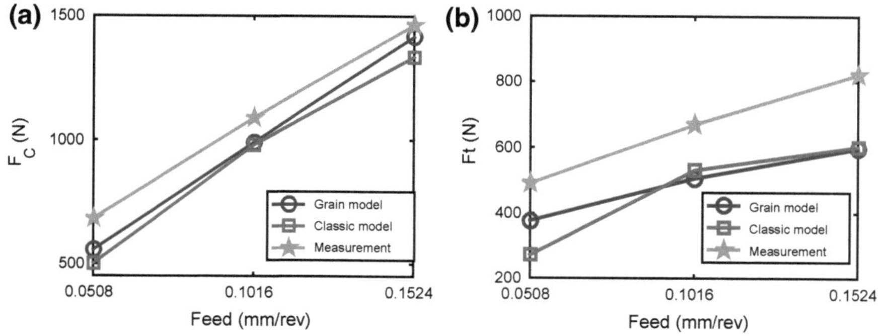

Fig. 8 When cutting speed is fixed at 1.049 m/s, the cutting forces F_C (**a**) and F_t (**b**) as a function of feed rate

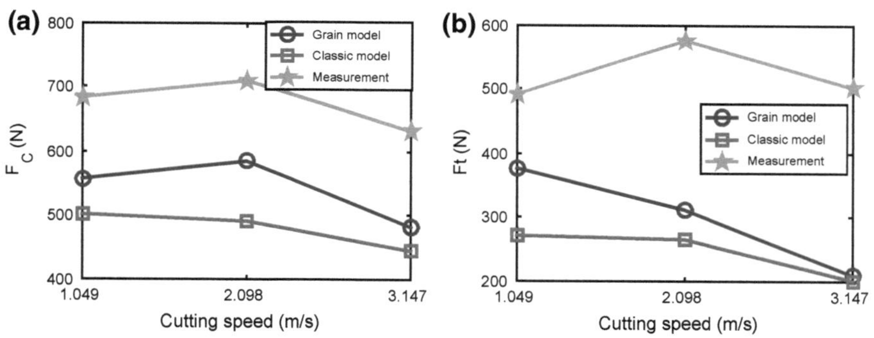

Fig. 9 When feed rate is fixed at 0.0508 mm/rev, the cutting forces F_C (**a**) and F_t (**b**) as a function of cutting speed

Both the *Ft* and *Fc* follows the similar trend, as when the cutting speed increase, the force first increases and then decreases. Also, the grain model gives a better prediction compared with the classic model.

6 Residual Stress Prediction

By implementing the proposed microstructure sensitive flow stress model, the residual stress on the machined workpiece surface could be predicted with an analytical model. The residual stress prediction is first applied for the *Ti–6Al–4V* material. With a constant feed rate of 0.1 mm/rev, depth of cut 0.1 mm and cutting speed at 26.4 m/min, the residual stress is plotted as a function of distance from the machined surface into the workpiece, as shown in Fig. 10. Since a two-dimensional stress distribution assumption is used, in which the stress in the workpiece axial direction is negligible. The largest magnitude of stress value is found to be on the

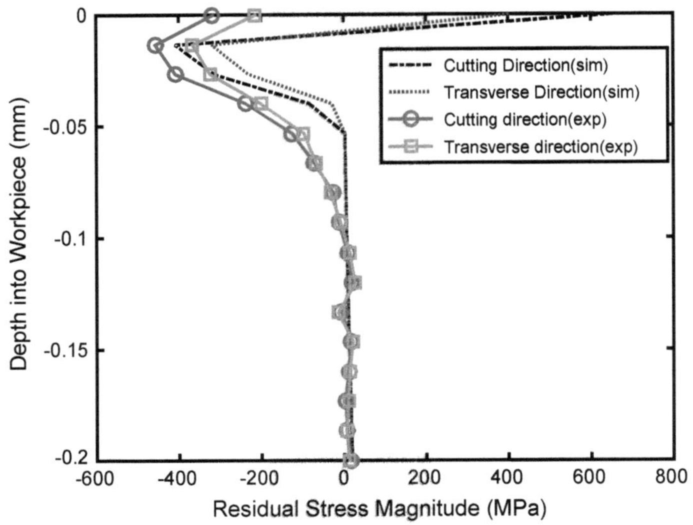

Fig. 10 Prediction and measured residual stress at feed rate 100 μm/rev, cutting speed 26.4 m/min, depth of cut of 100 μm

machined surface. With the increasing depth into the workpiece, the tensile residual stress first decrease and change to compressive at a certain depth. After that, the compressive residual stress reaches its peak value and then decreases to zero. When the depth is around 0.1 mm, the magnitude of the residual stress is around zero. So in the current machining condition, the residual stress affected depth is around 100 μm. A good agreement is found between the prediction and experimental measurement from Ratchev et al. [37]. However, large discrepancy is found on the surface, where the prediction shows tensile residual stress, but the experimental measurements show the compressive stress. Those errors could be from the oxidation on the machined surface.

For the residual stress in the cutting direction σ_{xx}, with the increasing depth into workpiece, the tensile residual stress changes to compressive. After the peak value of compressive residual stress occurs, the compressive residual stress gradually reduces to zero. A good agreement between the measurement data and prediction is found in both σ_{xx} and σ_{yy} (Fig. 11).

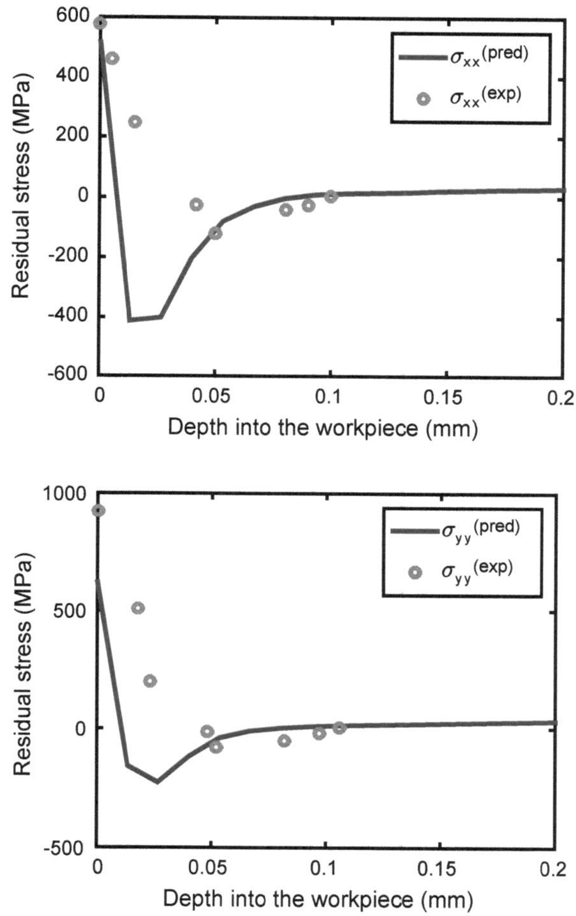

Fig. 11 Comparison of predicted residual stress between experimental measurement at cutting speed of 1.049 m/s, depth of cut = 0.0508 mm, width of cut = 4.775 mm

7 Conclusion

A materials-affected manufacturing computational framework for the material dynamic recrystallization and phase transformation in the machining process is proposed in the current work. The *JMAK* model is used for the explicit grain size evolution calculation by assuming an isothermal condition. With the temperature history input, phase composition of different phases is calculated from the TTT curve and Avrami equations. A modified *JC* flow stress model is developed by considering the grain size and phase volume fraction effects. The proposed model is applied in the case study of *Ti–6Al–4V* and AISI 4130 steel alloys for the machining forces and residual stresses predictions. Experimental data are provided for the model validation. Better force and residual stress predictions are obtained compared with the traditional model. The proposed framework could provide a machining process optimization scheme at a microstructural level.

References

1. Lim EM, Menq CH (1997) Integrated planning for precision machining of complex surfaces. Part 1: cutting-path and feedrate optimization. Int J Mach Tools Manuf 37(1):61–75
2. Cheung CF, Lee WB (2000) A multi-spectrum analysis of surface roughness formation in ultra-precision machining. Precision Engineering 24(1):77–87
3. Picard YN et al (2003) Focused ion beam-shaped microtools for ultra-precision machining of cylindrical components. Precis Eng 27(1):59–69
4. Ulutan D, Ozel T (2011) Machining induced surface integrity in titanium and nickel alloys: a review. Int J Mach Tools Manuf 51(3):250–280
5. Pan Z et al (2016) Analytical model for force prediction in laser-assisted milling of IN718. Int J Adv Manuf Technol. doi:10.1007/s00170-016-9629-6
6. Lei S, Liu W (2002) High-speed machining of titanium alloys using the driven rotary tool. Int J Mach Tools Manuf 42(6):653–661
7. Pan Z et al (2017) Modeling of Ti–6Al–4V machining force considering material microstructure evolution. Int J Adv Manuf Technol. doi:10.1007/s00170-016-9964-7
8. Fergani O et al (2016) Microstructure texture prediction in machining processes. Procedia CIRP 46:595–598
9. Ghosh S, Kain V (2010) Microstructural changes in AISI 304L stainless steel due to surface machining: effect on its susceptibility to chloride stress corrosion cracking. J Nucl Mater 403 (1):62–67
10. Barbacki A, Kawalec M, Hamrol A (2003) Turning and grinding as a source of micro structural changes in the surface layer of hardened steel. J Mater Process Technol 133(1):21–25
11. Lu C (2008) Study on prediction of surface quality in machining process. J Mater Process Technol 205(1):439–450
12. Zhipeng Pan AT, Shih DS, Garmestani H, Liang SY (2017) The Effects of Dynamic Evolution of Microstructure on Machining Forces. In: Proceedings of the institution of mechanical engineers, Part B: Journal of Engineering Manufacture
13. Xu HH, Jahanmir S (1994) Effect of microstructure on abrasive machining of advanced ceramics. In: A collection of papers presented at the 96th annual meeting and the 1994 fall meetings of the materials and equipment/Whitewares/Refractory Ceramics/Basic Science: Ceramic Engineering and Science Proceedings, Vol. 16, Issue 1. 1995. Wiley Online Library
14. Chou YK, Evans CJ (1999) White layers and thermal modeling of hard turned surfaces. Int J Mach Tools Manuf 39(12):1863–1881
15. Swaminathan S et al (2007) Severe plastic deformation of copper by machining: microstructure refinement and nanostructure evolution with strain. Scripta Mater 56 (12):1047–1050
16. Filip R et al (2003) The effect of microstructure on the mechanical properties of two-phase titanium alloys. J Mater Process Technol 133(1):84–89
17. Gil F et al (2001) Formation of α-Widmanstätten structure: effects of grain size and cooling rate on the Widmanstätten morphologies and on the mechanical properties in Ti–6Al–4V alloy. J Alloy Compd 329(1):142–152
18. Kugler G, Turk R (2004) Modeling the dynamic recrystallization under multi-stage hot deformation. Acta Mater 52(15):4659–4668
19. Suárez A et al (2004) Modeling of phase transformations of Ti–6Al–4V during laser metal deposition. Physics Procedia 12:666–673
20. Malinov S et al (2001) Differential scanning calorimetry study and computer modeling of $\beta \Rightarrow \alpha$ phase transformation in a Ti–6Al–4V alloy. Metall Mat Trans A 32(4):879–887
21. Ducato A, Fratini L, Micari F (2013) Prediction of phase transformation of Ti–6Al–4V titanium alloy during hot-forging processes using a numerical model. In: Proceedings of the institution of mechanical engineers, Part L: Journal of Materials Design and Applications, p. 1464420713477344

22. Jiles D (1988) The effect of compressive plastic deformation on the magnetic properties of AISI 4130 steels with various microstructures. J Phys D Appl Phys 21(7):1196
23. Ji X et al (2014) Modeling the effects of minimum quantity lubrication on machining force, temperature, and residual stress. Mach Sci Technol 18(4):547–564
24. Grum J, Kisin M (2003) Influence of microstructure on surface integrity in turning—Part II: the influence of a microstructure of the workpiece material on cutting forces. Int J Mach Tools Manuf 43:1545–1551
25. Gibbs RK Hodgson PD (1992) A mathematical model to predict the mechanical properties of the hot rolled C–Mn and microalloyed steels. ISIJ Int 32(12):10
26. Hawbolt EB, Sun WP (1997) Comparison between Static and Metadynamic Recrystallization an application to the hot rolling of steels. ISIJ Int 37(10):1000–1009
27. Sajadi SV, Ketabchi M, Nourani MR (2010) Hot deformation characteristics of 34CrMo4 steel. J Iron Steel Res Int 17(12):65–69
28. Xia Ji XZ, Li Beizhi, Liang Steven Y (2014) Modeling the effects of minimum quantity lubrication on machining force, temperature, and residual stress. Mach Sci Technol Int J 18 (4):547–564
29. Srinivasa YV, Shunmugam MS (2013) Mechanistic model for prediction of cutting forces in micro end-milling and experimental comparison. Int J Mach Tools Manuf 67:18–27
30. Kim DH, Lee CM (2014) A study of cutting force and preheating-temperature prediction for laser-assisted milling of Inconel 718 and AISI 1045 steel. Int J Heat Mass Transf 71:264–274
31. Roucoules C, Hodgson P (1995) Post-dynamic recrystallisation after multiple peak dynamic recrystallisation in C–Mn steels. Mater Sci Technol 11(6):548–556
32. Johnson GR, Cook WH (1983) A constitutive model and data for metals subjected to large strains, high strain rates and high temperatures. In: Proceedings of the 7th international symposium on ballistics, The Hague, The Netherlands
33. Calamaz M, Coupard D, Girot F (2008) A new material model for 2D numerical simulation of serrated chip formation when machining titanium alloy Ti–6Al–4V. Int J Mach Tools Manuf 48(3):275–288
34. Zhang X, Shivpuri R, Srivastava A (2014) Role of phase transformation in chip segmentation during high speed machining of dual phase titanium alloys. J Mater Process Technol 214 (12):3048–3066
35. Shivpuri R et al (2002) Microstructure-mechanics interactions in modeling chip segmentation during titanium machining. CIRP Ann Manuf Technol 51(1):71–74
36. Pan Z et al (2016) Prediction of machining-induced phase transformation and grain growth of Ti–6Al–4V alloy. Int J Adv Manuf Technol 87(1):859–866
37. Ratchev S et al (2011) Mathematical modelling and integration of micro-scale residual stresses into axisymmetric FE models of Ti–6Al–4V alloy in turning. CIRP J Manuf Sci Technol 4(1):80–89

Structural FEM Analysis of Thermal Sprayed Coatings Under Conditions of Contact Pressure and High Temperature

Piotr Jabłoński and Piotr Czajka

Abstract State of art in the field of wear resistance indicate that, coatings produced with thermal spray methods, can significantly contribute to increase wear rate under conditions of wide range of sliding speeds and varying temperature values. In the beginning of this paper mechanical properties, manufacturing technology and possible applications of thermal sprayed coatings were reviewed. Mechanical properties for selected MMC (metal-matrix composites) materials applied for such coatings were introduced. Determining both stress and deformations of HVOF anti-wear coatings was necessary in order to estimate possible use of these coatings in application on machine parts exposed to high contact pressure and increased temperature. As well, relationship between different materials participation and mechanical properties of produced coatings were mentioned. Analysis method, model and boundary conditions were stated as method of experiment. Numerous structural FEM analysis were performed in order to determine stress and deformations in model samples under conditions of contact pressure and thermal conditions. High temperature was applied in prepared simulations to verify stress and deformation of samples caused by difference in base material and coating coefficient of thermal expansion. 3D structural simulations were conducted, in order to verify stress in different sections. Finite elements size and shape were optimized for obtaining possible short time of both mesh generation and solving of numerical problems. Contact pressure conditions were assumed as same for all simulations, for increasing result comparison possibilities. Analyses were performed for different geometry of samples, differing in coating thickness and surface curvature. Simulations of coatings on both round and flat surfaces were conducted in order to designate influence of surface curvature on coating stress under conditions of contact pressure and high temperature. Results of performed simulations were presented, including detailed information about stress in various direction,

P. Jabłoński (✉) · P. Czajka
Poznan University of Technology, Institute of Mechanical Technology,
pl. Marii Curie-Skłodowskiej 5, 60-965 Poznan, Poland
e-mail: piotr.le.jablonski@doctorate.put.poznan.pl

P. Czajka
e-mail: piotr.ad.czajka@doctorate.put.poznan.pl

V. Majstorovic and Z. Jakovljevic (eds.), *Proceedings of 5th International Conference on Advanced Manufacturing Engineering and Technologies*,
Lecture Notes in Mechanical Engineering, DOI 10.1007/978-3-319-56430-2_24

deformations and shear stress in area between base material and coating. Correlations between results and geometrical parameters of model were presented, especially including curvature and thickens of coating. Discussion of obtained results and recorded correlations was presented in last paragraphs of this paper. Authors find obtained results as useful initial data in further research on field of thermal sprayed coating application on machine parts exposed to high contact pressures.

Keywords FEM · Coefficient of thermal expansion · Thermal spray · Coatings

1 Introduction

Surface and coatings technology has been developing with increasing rate through last years. This fact is not surprising at all, as surface engineering opens new ways for shaping physiochemical properties of numerous object of everyday use. Various known methods of coatings technology are proven to be capable of significant change in surface properties of processed object, while keeping substrate unchanged. This fact is great advantage of technical coatings, as very often properties of substrate material differ from requirements for object's surface properties.

Main fields of applications for thermal sprayed coatings are faced towards improving surface properties of treated object. Especially, resistance to various kinds of wear, chemical corrosion and heat influence can be significantly increased [1]. As well, other applications of these techniques are known, like creating heat barriers, parts regeneration and for aesthetic causes. Wide range of possible processed materials, makes thermal spray technology very promising and all-purpose method for manufacturing technical coatings. However, work described in following paper was focused on tribological applications of these.

Numerous materials and methods of deposition can be adopted for thermal spray coating fabrication. However, among described techniques, high velocity oxy-fuel (HVOF) deposition method presents some interesting capabilities. Firstly, this method allows fabrication of coating of low porosity. This fact is very important, while considering tribological properties. Also possibility of achieving relatively high adhesion between coating and substrate material is important in case of high contact pressure applications [2]. Among all processed coating materials, Cr_3C_2–NiCr compound can be proposed as promising material for tribological applications. Especially, for applications involving increased temperatures. There are several research reports among state of art, stating that this material presents excellent wear resistance in high temperatures, reaching 600 °C thanks to its mechanical properties, Cr_3C_2–NiCr composite can be successfully used on machine parts exposed to abrasive wear under conditions of increased temperature [3].

HFOV thermal sprayed coating performance is significantly influenced by residual stresses which develop during deposition process. These stresses can affect life and behavior of the coating layer, therefore must be taken into consideration [4–6]. The phenomenon depend largely on thermal conditions of deposition process. According to the state of art [4, 5], residual stresses that develop in the deposit are combination of quenching and cooling stresses. Quenching stresses arise during deposition process as the result of rapid cooling of sprayed particles from their melting temperature to the temperature of substrate after they reach its surface [4, 5]. As stated in [4, 8], quenching stresses have always a tensile nature. When deposition process is finished and secondary cooling is started, the temperature of coating system drops from the temperature that was reached during spraying process to room temperature. The cooling stresses are then induced by mismatch in thermal shrinkage of substrate and coating caused by the difference in CTE between substrate and coating material [4, 8]. It is stated in [4, 7], that if CTE of coating is smaller than that of the substrate a compressive stress would develop in coating whereas a tensile stress would develop when CTE of substrate is smaller than that of the coating. Depending on the sign of quenching and cooling stresses values, residual stresses in coating could be of tensile or compressive nature [6–8]. According to [4, 8], compressive residual stresses are generally more favorable than tensile ones. Their presence may not harm the coating system, and what is more they can even contribute to improvement of adhesion bonding and fatigue strength and also inhibit formation of cracks in coating layers [8]. However, if generated, compressive residual stresses exist in relatively large values, they may lead to buckling or even delamination of the coating [7]. Tensile nature of residual stresses is usually more harmful to coating/substrate system, because it lead to such problems as initiation and propagation of cracks which result in interface delamination, loss of adhesion and fatigue failure [4, 7, 8].

Because of relatively high thickness of these coatings and different properties of substrate and coating materials (different in values of both coefficients of thermal expansion-CTE and Young's modulus E), characteristics of stresses occurring within coatings is significantly hard to describe and understand. Therefore aim of conducted research described in following paper was to develop FEM simulations, which would help in better understanding of stress distribution in HVOF thermal sprayed coatings under harsh conditions of high contact pressure and thermal load. Following paper covers whole process of preparation and processing of FEA simulations for described coatings. Use of mathematical models and simplifications, as well as discretization of model geometry of coating were presented. Obtained results were stated and interpreted. Conclusions and final comprehension were aimed to describe and understand most significant correlations between geometrical parameters of coating and maximum magnitude of stress. These data are expected to be essential for further development of HVOF coatings in the field of high temperature and high contact pressure resistance.

2 FEM Structural Simulations Preparations

Conducted FEM structural simulations, were aimed to determine stress distribution:

- in contact zone (between indenter and coat),
- across coating thickness,
- in the interface of substrate and coating,
- in substrate.

Simulations assumed constant force load applied by indenter and constant temperature condition. Before conducting simulations, several steps were made, such as geometry definition, modelling of loads and temperature, meshing and boundary conditions determination.

2.1 *Geometry for Simulation*

Point contact between flat surface and spherical indenter was assumed. As well, additional simulations involving contact between hemispherical coated surface and hemispherical indenter also were proposed for simulation plan. Simulations were conducted with various coating thickness, and in case of coating on hemispherical surface-various curvature of surface. Dimensions and shape of hemispherical indenter were constant during all conducted simulations. Overview of geometrical model prepared for simulation is presented in Fig. 1.

Geometrical CAD model was parameterized in order to conduct parametric simulations with various values of R and t (surface curvature radius and coating thickness, respectively). Geometrical model was applied into FEM environment as a multi-body solid geometry.

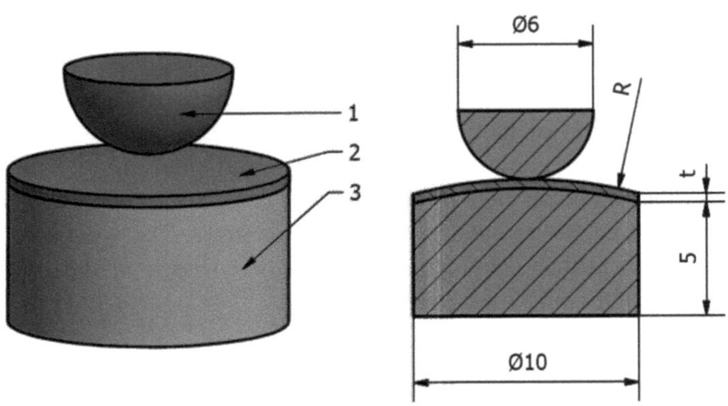

Fig. 1 Layout of geometry model for simulation. *1* indenter, *2* coating, *3* substrate. Variable dimensions: R surface curvature, t coating thickness

2.2 Loads and Temperature

To simulate contact pressure and conditions of increased temperature, two kind of external loads were applied to developed model. To induce Hertzian stress, axial force was applied onto flat surface of hemispherical indenter (see Fig. 1). Distribution of load force on flat surface of indenter was uniform. Magnitude of force was equal to 100 N. During all simulations, magnitude, direction and distribution of load was constant.

In order to simulate thermal stress that occurs because of thermal deformation, constant temperature condition was adopted. As reference temperature, 22 °C was assumed. Uniform, constant temperature of 500 °C was applied in FEM test environment to geometry of both base material and coating. Thermal state was assumed as stable, so no heat flux is considered during simulation, and temperature stays constant; there is no heat flow between model and environment, as well there is heat exchange between model elements.

2.3 Boundary Conditions

Boundary conditions were introduced in order to remove unneeded degrees of freedom of interacting bodies. Firstly, the lowest flat surface of model was constrained with frictionless. Coating model, was connected with base material via contact relation. Definition of contact between base material and coating model was determined as contact with infinite friction. This kind of contact definition allowed to fully constrain coating model and simulate adhesion forces between coating and base material. In order to introduce contact between indenter and coating, frictional contact was adopted. Magnitude of coefficient of friction was 0.6 [9]. Finally, upper indenter model was constrained. In order to avoid unexpected motion of this body, only one degree of freedom was allowed-axial movement of indenter. Because of symmetry of model, loads and boundary conditions, whole model was assumed to be axisymmetric by main vertical axis. Thanks to that assumption, despite the fact that base element was supported with frictionless constraint, whole system didn't expect rigid body motion.

2.4 Meshing and Mathematical Model

Finite element mesh for all bodies was assumed to consist of structured hexahedrons. Smooth change of element size was ensured. Size of mesh for all bodies was set as 0.2 mm. Mesh needed to finer in critical area. There were three regions selected for increase of mesh resolution: central region of coating model and regions near to contact area of base material and indenter. In these areas, element

Fig. 2 Mesh overview

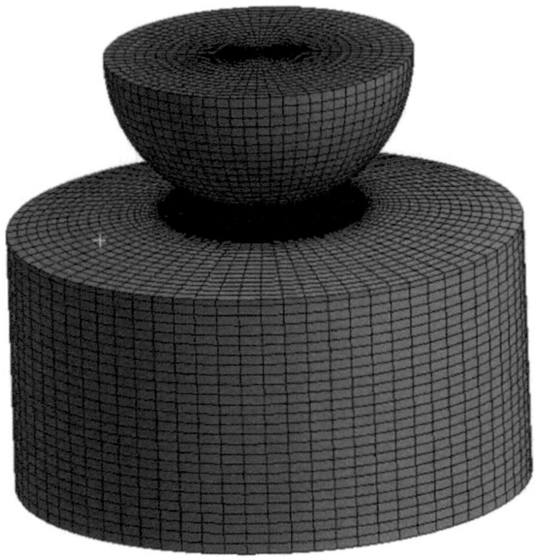

Fig. 3 Mesh modification

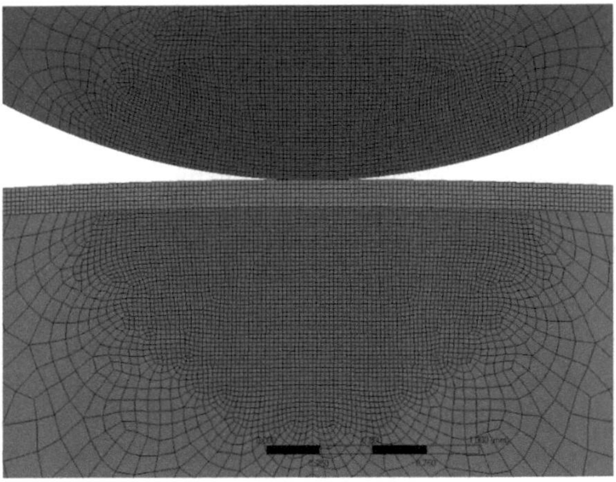

size was reduced to 0.05 mm. This mesh modification was applied because of the fact, that in these regions stresses and mesh deformations were expected to be the greatest. Also, too coarse mesh in area of contact can result in high numerical errors and mesh penetration. Figure 2 presents overview of developed 3D mesh. Figure 3 shows area of mesh modification.

2.5 Simulation Characteristics

Simulation was conducted in static structural module of Ansys Workbench 16 FEM environment. Because of expected non-linearity caused by contact constraints, all conducted analysis were using iterative Newton-Raphson Method solver. As well, force convergence was checked along iterations. Especially, contact problem was carefully defined. Augmented Lagrange equations were selected as method of contact formulation. Iterative model of contact assumed contact stiffens correction at each simulation iteration. Maximum number of iteration for each simulation was set to 200. In order to ensure proper contact processing and increase precision of calculations, penetration tolerance was set to 0.001 mm.

2.6 Material Properties

Two type of materials were implemented in conducted simulation. As material for substrate and indenter, structural steel was adopted. For coating, Cr_3C_2–NiCr (80% Cr_3C_2) composite was used. Because of the fact, that mechanical properties of this material are strongly influenced by processing parameters and spray method, only averaged properties can be stated [10]. Table 1 presents selected mechanical properties of adopted materials.

2.7 Simulation Plan

Simulations were conducted with two varying parameters where R stood for hemisphere curvature radius, and t presented thickness of coating. For each of 10 values of curvature radius of coating and one flat case, simulations were conducted for 5 different coating thickness, resulting totally in 55 simulations. Table 2 presents simulation plan.

3 Simulation Results

According to initial parameters and conditions presented in previous paragraphs, numerous simulations were conducted. Detailed schedule of simulations and range of geometrical parameters were presented in Table 2 (see Sect. 2.7). Following

Table 1 The properties of substrate and coating material

Material	Young modulus [GPa]	CTE [10^{-6} K^{-1}]	Density [kg/m^3]	Hardness
1.0038 Steel	200	12	7780	120 (HB)
Cr_3C_2–NiCr	260	10.5	6800	800 (HV microhardness)

Table 2 Simulation plan

t-coating thickness [mm]	R-surface curvature
0.2; 0.3; 0.4; 0.5; 0.6	None (flat); 10; 20; 30; 40; 50; 60; 70; 80; 90; 100

Totally: 55 simulations (5 for flat surface and 50 for curved with various curvature)

paragraphs presents results of these simulations. Results overview especially focus on distribution of HMH (Huber-Misses-Hencky yield theorem) equivalent tensile stress. Magnitude of maximum equivalent tensile stress was selected as main determinant of complex stress state for this problem, because of complicated, multidirectional stress distribution. However, similar approach can be found among recent contributions [11]. To present stress occurring in transition zone between coating and base material, magnitude of shear stress on surface between coating and base material was also evaluated for better complexity of considerations.

First simulations were conducted only with force load. Respectively, following simulations were conducted only under conditions of increased temperature and as complex state- with both force load and temperature applied. For these initial simulations, geometrical parameters t (coating thickness) and R (surface curvature) were constant, $t = 0.2$ mm and $R = 10$ mm.

In case of complex state, simulations with varying geometrical parameters were executed according to simulation plan.

3.1 Contact Pressure

Simulations of case with only force load applied were conducted. Force magnitude was 100 N (see Sect. 2.2 for detailed information about loads). For this case two simulations were conducted for flat and curved surface. Radius of curvature was set to 10 mm. Thickness of coating was constant and was equal to 0.2 mm.

Results of this simulation were presented on following pictures. Figure 4 presents overview of stress distribution in axial cross-section of model for curved case. Figure 5a includes close-up of stress distribution for flat surface. Analogous view of stress distribution for case with curved surface is presented in Fig. 5b. Distribution of shear stress, tangent to surface separating coating from base material, is shown in Fig. 6. Please note that this view is coaxial with direction of force load, and shows top view of base material. Figure 6 shows only half of surface, as whole problem is axisymmetric.

Detailed information about magnitudes of recorded stress, results discussion an interpretation will be included in following paragraphs.

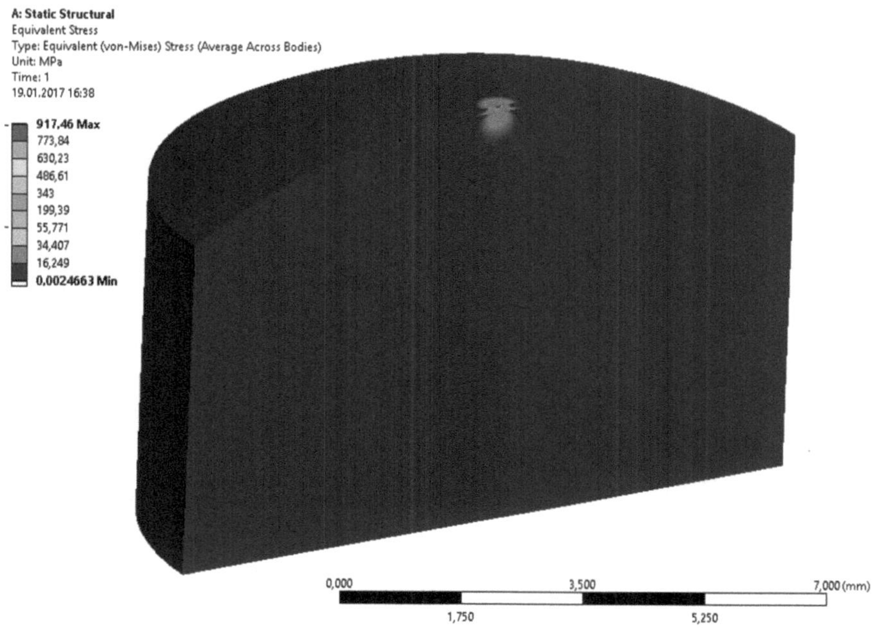

Fig. 4 HMH stress distribution in cross section of curved coating model (R = 10 mm, t = 0.2 mm)

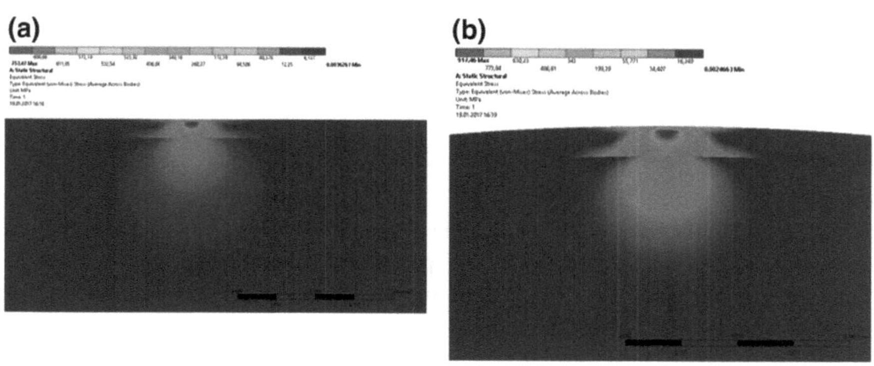

Fig. 5 Detailed view of HMH stress distribution in coating close to contact area: **a** flat surface (t = 0.2 mm), **b** curved surface with R = 10 mm, t = 0.2 mm

3.2 *Thermal Stress*

Thermal stress was defined as another, next to axial force, load for described problem. Following simulations were conducted only with thermal condition, which was 500 °C. Reference temperature was 22 °C, resulting in 478 °C increase

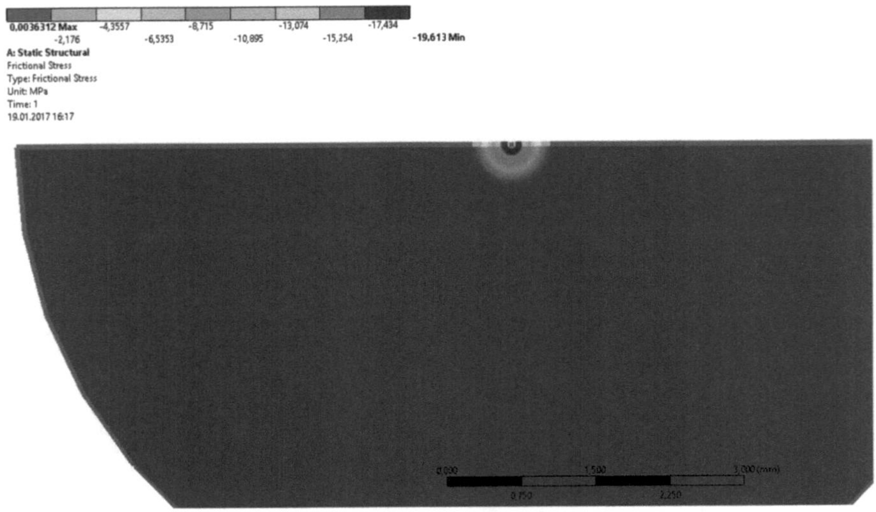

Fig. 6 Aerial distribution of shear stress on transition surface between coating and base material. Note that presented view is parallel to direction of force load (R = 10 mm, t = 0.2 mm)

in temperature. As coating material has lower coefficient of thermal expansion than base material (see Sect. 2.6), significant thermal stress is expected to occur. For this case, sample distribution of HMH stress for curved state is presented in Fig. 7. Surface shear stress for surface between coating and base material is shown in Fig. 8. Curvature of model was set to 10 mm, as in previous case.

3.3 Complex State

Finally, simulations of complex state, considering both force and thermal load were conducted. In this case, simulations were performed according to main simulations plan (see Table 2, Sect. 2.7). Force load and thermal conditions were analogous to previous simulations (Sects. 3.1 and 3.2). Following figures present equivalent reduced stress (HMH) distributions for sample cases. Figure 9 presents overview of stress distribution in complete model including contact and thermal stress. Figure 10a, b presents stress distribution across substrate and coating respectively for flat and curved surface. Aerial distribution of shear stress on transition surface between coating and base material is shown Fig. 11.

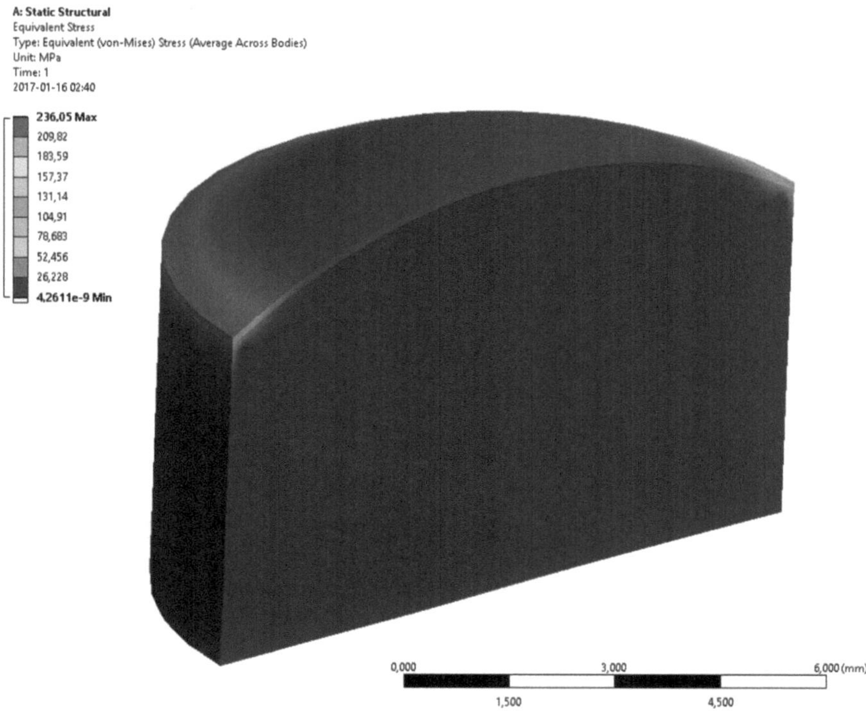

Fig. 7 Distribution of equivalent reduced stress in axial cross section for curved model (R = 10 mm, t = 0.2 mm)

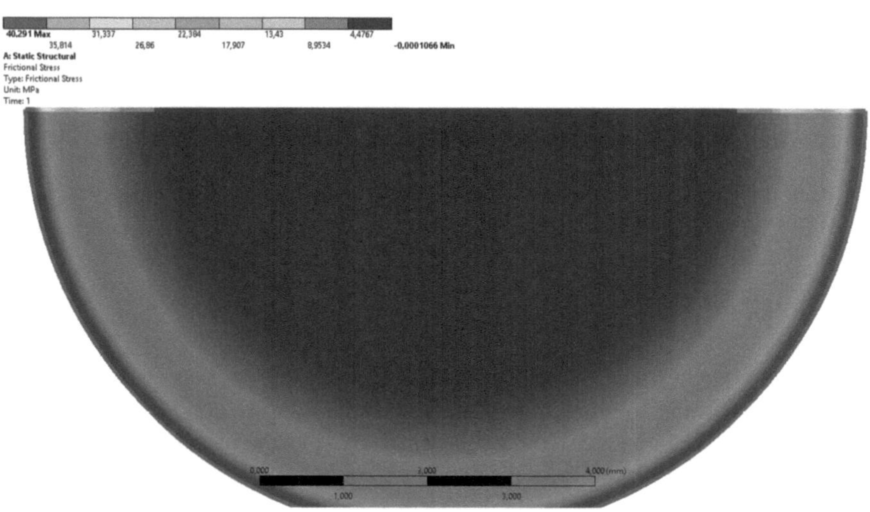

Fig. 8 Aerial distribution of shear stress on transition surface between coating and base material. Note that presented view is parallel to main axis of model (R = 10 mm, t = 0.2 mm)

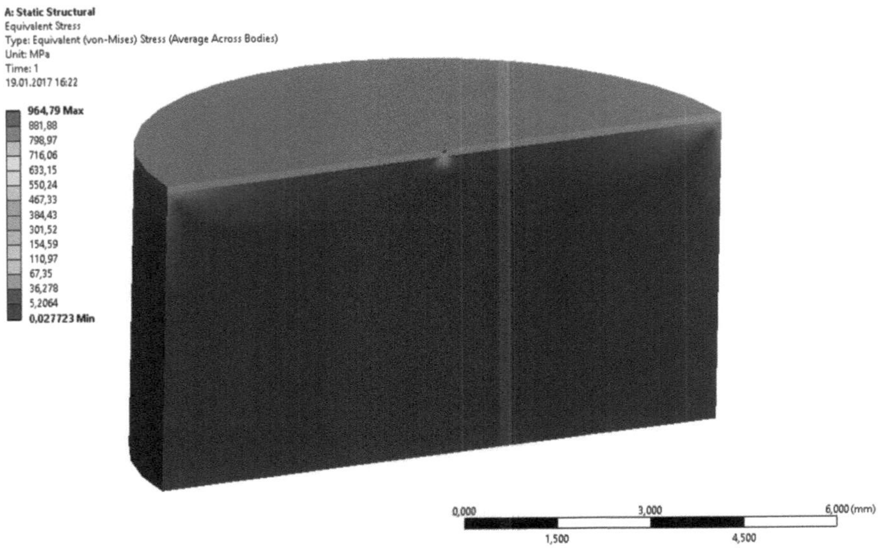

Fig. 9 Stress distribution for complex load state and flat surface (t = 0.2 mm)

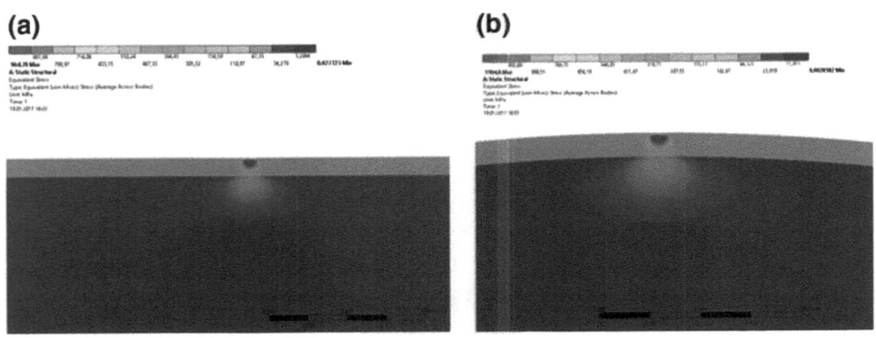

Fig. 10 Detail view of HMH stress distribution in flat (**a**) and curved (**b**) model under contact and thermal stress (t = 0.2 mm)

3.4 Correlations

For complex load, which sample results were presented in Sect. 3.3, numerous simulation according to simulation plan (see Table 2) were conducted. Based on obtained data, relations between geometrical parameters of coating (thickness, curvature) and results were formulated. As determinants of stress intensity within coating, maximum value of HMH equivalent stress were proposed. As well, shear stress maximum value for each simulation was taken into account.

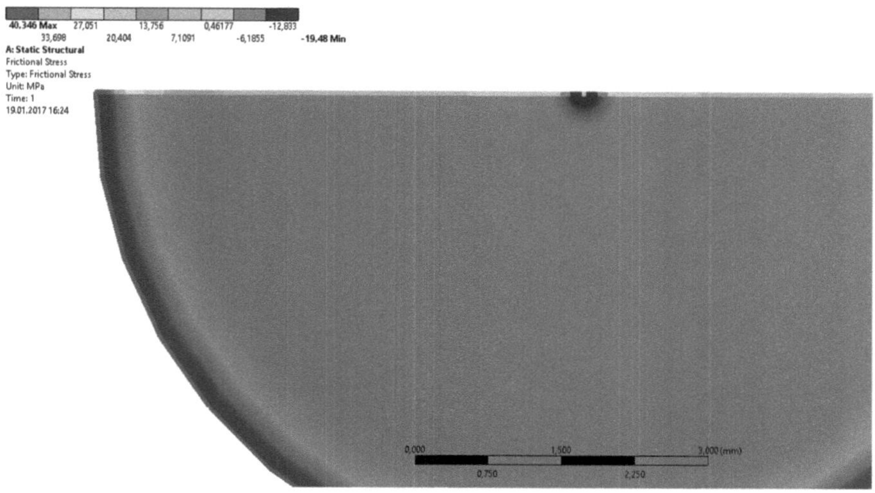

Fig. 11 Aerial distribution of shear stress on transition surface between coating and base material for complex case. Note that presented view is parallel to main axis of model (R = 10 mm, t = 0.2 mm)

Graph presented in Fig. 12 includes relationship of maximum magnitude of HMH equivalent stress in coating and surface curvature. Highest values of reduced equivalent stress reached 1.14 MPa for the lowest coating thickness. Magnitude of stress decreases, and after reaching minimum of 0.87 MPa, starts to increase. Local minimum for this trend can be noticed at surface curvature radius 70 mm.

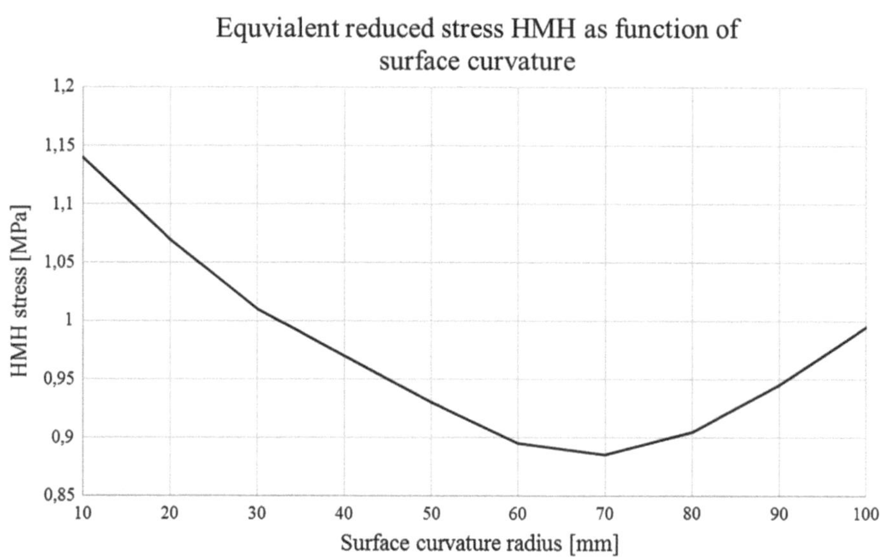

Fig. 12 Graph presenting recorded HMH stress depending on surface curvature radius

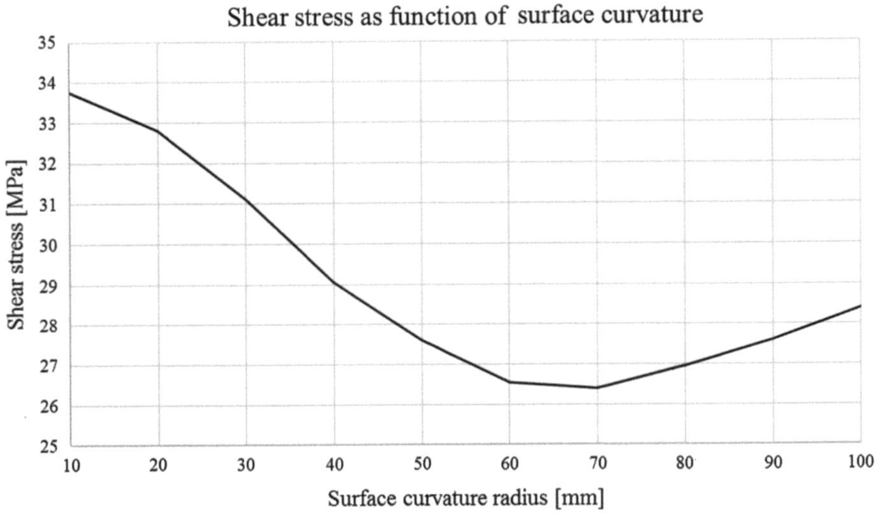

Fig. 13 Graph presenting shear stress magnitude as function of surface curvature radius

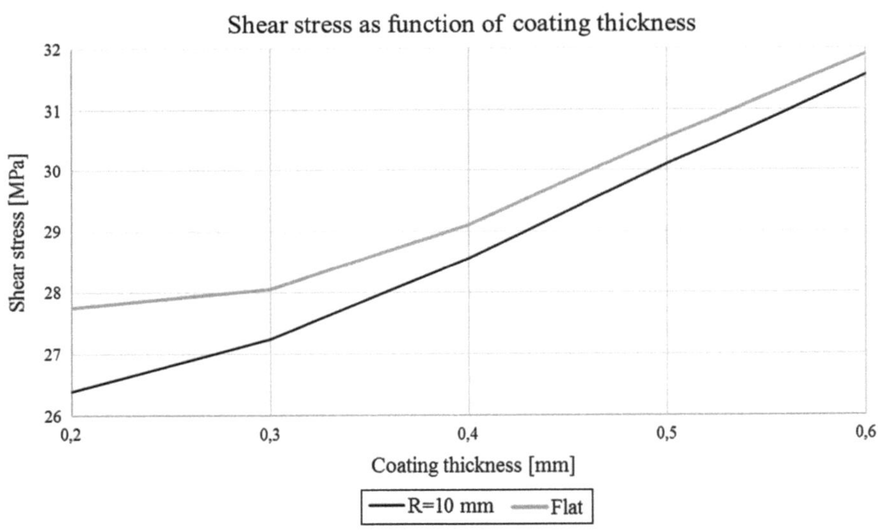

Fig. 14 Summary graph presenting variations of shear stress depending on coating thickness for flat and hemispherical surface

Analogous trends can be observed for change of shear stress with increasing coating curvature radius. Local minimum of stress magnitude value is reached with curvature radius equal to 70 mm. Trends observed in Figs. 12 and 13 are convergent.

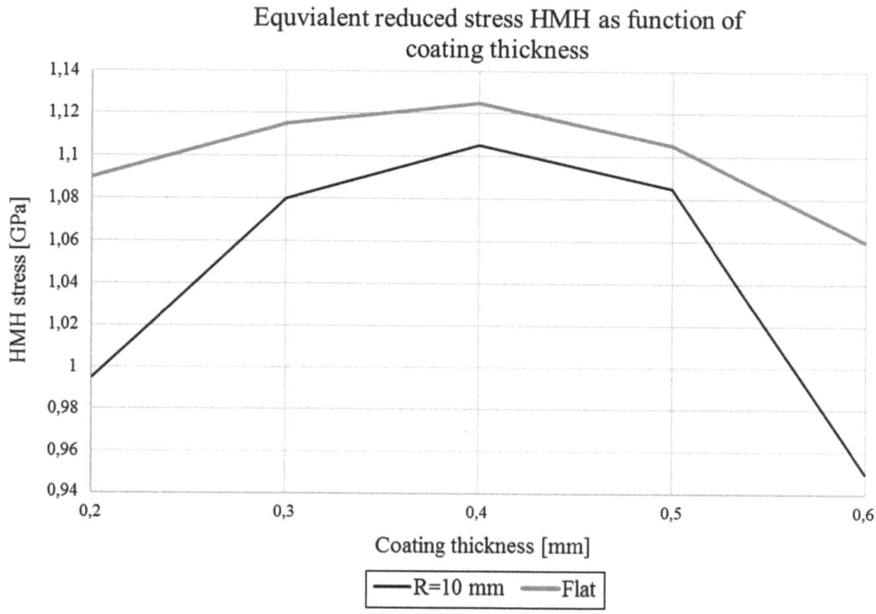

Fig. 15 Summary graph presenting variations of HMH equivalent stress depending on coating thickness for flat and hemispherical surface

Figures 14 and 15 present respectively influence of coating thickness on shear stress magnitude in the substrate-coating interface and reduced equivalent HMH stress. Shear stress presents constant monotonous increase with increasing coating thickness for both flat and curved coating. Values of shear stress are higher in case of flat substrate surface than for hemispherical surface.

Correlation between HMH reduced equivalent stress and coating thickness presents noticeable local maximum at thickness of 0.4 mm. Values of HMH stress in case of flat substrate were higher than in case of hemispherical coating surface. However, for mid-range of coating thickness magnitude, values of HMH reduced equivalent stress were most convergent for both flat can hemispherical geometry.

4 Conclusions

Conducted simulations allowed to clearly designate correlations between input geometrical parameters and recoded stress magnitudes. Apart from magnitude scale, recorded correlations bring complex and interesting data. All presented simulation results show non-linear relations. Moreover, local extrema can be observed on several graphs (see Figs. 12, 13, 14 and 15). Knowledge about presence of minima and extrema can be used during design process of coating and durability estimation.

In case of shear stress and equivalent reduced HMH stress, it can be observed that increase of surface curvature effected in local decrease of maximum stress value, however over specific value of coating surface curvature (70 mm), maximum recorded stress magnitude increased. This correlation are analogues for both HMH and shear stress, despite significant difference in value scale.

As predicted, increase of coating thickness caused monotonous increase of maximum shear stress value. As well, this correlation stays valid for both curved and flat coating. However, values of shear stress in case of flat surface were higher than in case of hemispherical surface. Although, along with increase of coating thickness, difference between stress magnitude for both cases (flat, curved) is decreasing. All of these relation are convergent with state of art on field of coating thermal stresses.

Obtained data and trends can be applied as an information source for further research. Especially, material model for simulations can be improved. Presented simulation model assumes linearity of material properties, both for substrate and coating. However, because of possible non-linearity connected with temperature change and high range of recorded stress value, linear models of material properties can cause some calculation errors. Especially, value of coefficient of thermal expansion, according to state of art, changes with increasing temperature. Another development can be made on field of geometry, as tests with more complex shape of coating and indenter can be helpful in estimating stress distribution in thermal sprayed coatings. It must be mention that the model which was used to obtain presented data, did not examine influence of residual stresses generated during deposition process on performance of proposed coating system under contact pressure and thermal load.

References

1. Fauchais PL, Heberlein JVR, Boulos MI (2014) Thermal spray fundamentals from powder to part. Springer, New York
2. Valentinelli L, Valente T, Casadei F, Fedrizzi L (2004) Mechanical and tribocorrosion properties of HVOF sprayed WC–Co coatings. Int J Corros Processes Corros Control 39:301–307. doi:10.1179/174327804X13884
3. Wenchao Z, Liu LB, Mentging Z, Ligang Z (2015) A comparison between WC–10Co–4Cr and Cr_3C_2–25NiCr coatings sprayed on H13 steel by HVOF. Trans Nonferrous Met Soc China 25:989–1005. doi:10.1016/S1003-6326(15)64011-0
4. Arif AFM, Al-Athel KS, Mostaghimi J (2016) Residual stresses in thermal spray coating. Compr Mater Finish 3:56–70. doi:10.1016/B978-0-12-803581-8.09199-2
5. Stokes J, Looney L (2009) FEA of residual stress during HVOF thermal spraying. J Mater Eng Perform 18:21–25. doi:10.1007/s11665-008-9262-0
6. Stokes J, Looney L (2004) Residual stress in HVOF thermally sprayed thick deposits. Surf Coat Technol 177–178:18–23. doi:10.1016/j.surfcoat.2003.06.003

7. Oladijo OP, Sacks N, Cornish LA, Venter AM (2012) Effect of substrate on the 3 body abrasion wear of HVOF WC-17 wt.% Co coatings. Int J Refract Metal Hard Mater 35:288–294. doi:10.1016/j.ijrmhm.2012.06.011
8. Jalali Azizpour M, Norouzi S, Mohammadi Majd H, Sajedipour D, Mohammadi Sadr R, Derakhshan Mehr M, Shoabi SA, Mohammadi R (2011) Development trend in investigation of residual stresses in WC-Co coating by HVOF thermal spraying. Int Sch Sci Res Innov 6:473–476. doi:scholar.waset.org/1999/11791
9. Zhang Y, Li J, Huang J, Ding C (1998) Mechanical and tribological properties of plasma-sprayed Cr_3C_2–NiCr, WC–Co, and Cr_2O_3 coatings. J Therm Spray Technol 7: 242–246. doi:10.1361/105996398770350981
10. Li J, Ding C (2001) Determining microhardness and elastic modulus of plasma-sprayed Cr_3C_2–NiCr coatings using Knoop indentation testing. 135:229–237. doi:10.1016/S0257-8972(00)01066-5
11. Jackson RL, Green I (2005) A finite element study of elasto-plastic hemispherical contact against a rigid flat. J Tribol 127:343–354. doi:10.1115/1.1866166

Intelligent Dual Curve-Driven Tool Path Optimization and Virtual CMM Inspection for Sculptured Surface CNC Machining

N.A. Fountas, S. Živković, R. Benhadj-Djilali, C.I. Stergiou, V.D. Majstorovic and N.M. Vaxevanidis

Abstract This paper investigates the profitability of a dual-curve driven surface finish tool path under the concept of optimizing crucial machining parameters such as toroidal end-mill diameter, lead angle and tilt angle. Surface machining error as well as tool path time are treated as optimization objectives under a multi-criteria sense, whilst a central composite design is conducted to obtain experimental outputs for examination and, finally, fit a full quadratic model considered as the fitness function for process optimization by means of a genetic algorithm. A benchmark sculptured surface given as a second-order parametric equation was tested and simulated using a cutting-edge manufacturing modeling software and best parameters recommended by the genetic algorithm were implemented for validation. Further assessment involves the virtual inspection to selected profile sections on the part. It was shown that the approach can produce dual-curve driven tool trajectories capable of eliminating sharp scallop heights, maximizing machining strip widths as well as maintaining smoothness quality and machining efficiency.

N.A. Fountas · R. Benhadj-Djilali
Faculty of Science, Engineering and Computing, Kingston University, Roehampton Vale Campus, Friars Avenue, Kingston upon Thames, London SW15 3DW, UK

S. Živković
Coordinate Metrology Lab, Military Technical Institute Belgrade, Ratka Resanovića 1, 11030 Belgrade, Serbia

N.M. Vaxevanidis (✉)
Laboratory of Manufacturing Processes and Machine Tools (LMProMaT), Department of Mechanical Engineering Educators, School of Pedagogical and Technological Education (ASPETE), ASPETE Campus, N. Heraklion, GR 14121 Athens, Greece
e-mail: vaxev@aspete.gr

C.I. Stergiou
Department of Mechanical Engineering, Piraeus University of Applied Sciences, GR 122 44 Egaleo, Greece

V.D. Majstorovic
Faculty of Mechanical Engineering, University of Belgrade, Kraljice Marije 16, 11120 Belgrade, Serbia

V. Majstorovic and Z. Jakovljevic (eds.), *Proceedings of 5th International Conference on Advanced Manufacturing Engineering and Technologies*, Lecture Notes in Mechanical Engineering, DOI 10.1007/978-3-319-56430-2_25

Keywords Dual-curve driven tool paths · Toroidal end-mills · Multi-criteria optimization · Sculptured surface machining · Genetic algorithms

1 Introduction

A large number of multi-axis surface finish tool paths are now available to machine sculptured surfaces such that reduced cycle times as well as high quality criteria are met. Multi-axis surface finish tool paths are accompanied with a number of machining parameters ought to be determined with special emphasis to inclination angles for the two rotational axes found in 5-axis CNC machining. Even though a resemblance exists among various swept-cut style tool paths in terms of parameter specification through the manufacturing software interface, the controllability of the trajectory flow may be different and remains unknown to the public [1]. Such tool paths available to commercially available CAM software systems are iso-scallop [2], iso-planar [3] as well as iso-parametric [4].

To ensure consistency in cutting performance and capture profitability, experimental investigations ought to be conducted using a given tool path to be applied on a free-form surface, prior to its implementation. To that extend, several researchers have examined a variety of tool paths for complex surfaces in order to provide insight concerning their settings aiming at optimizing criteria such as scallop height, tool path length, surface deviation using the theoretical CAD model as a reference, etc. [5], as well as physical machinability indicators such as roughness [6]. Related research concerning experimental investigations for tool path strategies also involves the optimal formulation of various tool paths using artificial techniques. Ülker et al. [7] implemented an immune approach to produce properly spaced cutter locations given a cut tolerance, for both u and v directions referring to the 3D parametric space. Agrawal et al. [8] implemented a genetic algorithm to formulate master cutter paths that will smoothly unite the tool trajectories of different limiting contours.

The cutting tool type is the most important factor to which corresponded surface finish tool path parameters are refer to. Hence, machining surface creation is largely depended on the tool type and its contact to that surface [9]. In recent years toroidal as well as barrel type end mills have been attracting much attention whilst a large number of research works report their benefits against other tool types like ball end mills or flat-end mills. Benefits of torus end mills include larger effective radii towards path trajectories [10], gouge avoidance owing to the absence of sharp edges when machining with normal to surface, as well as attaining larger machining strip widths [11]. Hendriko in [12] developed an extended analytical boundary method to determine the machining geometry during free-form milling. The algorithm was developed by considering inclination angles. The proposed method was successfully implemented to calculate the scallop for two model parts with different surface profiles. The accuracy was verified by comparing the scallop height calculated using the proposed method with those measured using a cutting-edge machining simulation environment. For these experiments, a toroidal cutter was applied. Investigation for the usage of toroidal end-mills is mainly conducted for tool path planning in terms of tool positioning such that a curvature matching between the tool's geometry and the given sculptured surface exists in every cutter contact point. This assumption spans a noticeable number of research works for tool positioning, such as those presented in Roman et al. [13] dealing with the "rolling-ball" method applied to 3½½-axis machining, as well as in Warkentin et al. [14] where a number of discrete tool positioning methods for 5-axis surface tool paths are assessed using a toroidal cutter.

The main thrust of this paper is to examine a particular multi-axis surface finish tool path based on two drive curves for its cutting style definition, aiming at attaining the lowest surface deviation and the lowest machining time. Note that values for machining parameters ought to produce a tool path under the preset cut tolerance for the successful interpolation. First, the tool path's control is described whilst a number of machining experiments are established using a benchmark sculptured surface, to obtain results for machining deviation error and tool path time. From these experiments a fitness model under the principles of least squares approach is developed; the representation of which a genetic algorithm utilized to sequentially evaluate it so as to arrive at optimal solutions for machining parameters; cutting tool diameter as well as 5-axis inclination angles (lead and tilt). The experiments are conducted and analyzed using a machining simulation environment [15].

2 Intelligent Tool Path Parameter Selection for Dual-Curve Driven Trajectories

2.1 *Dual-Curve Driven Multi-axis Surface Tool Paths*

According to the philosophy underpinning the control of dual-curve driven tool surface paths, the largest possible machining strip widths can be attained by positioning the cutting tool given its geometry, radial cut engagement (stepover), and tool inclination angles. The first drive-curve is responsible for indicating a set of cutter contact points for the formulation of the primary cutting path. The secondary cutting path comes to be as the outcome of the first drive curve with the narrowest strip width among the above maximum machining strip width at each cutter contact point. Thereby, the tool path strategy repositions the tool utilizing the pair of cutter contact points along the two drive curves, resulting in uniformly planned trajectories, free of gouging and large surface deviation error (Fig. 1). By taking into account Fig. 1, for a given surface *S*, the primary drive-curve dc_i^1 is discretized to $cc_{i,j}$ sampling cutter contact points, with $cc_{i,j} = (j = 1, 2, \ldots, n)$, considering the cut tolerance and the discretization algorithm of CAM software for the strategy. For every $\mathrm{cc_{i,j}}$ point on $\mathrm{dc_i^1}$ the corresponding cutter contact point $\mathrm{cc_{i+1,j}}$ may be obtained, as well as the related machining strip width $\mathrm{msw_{i,j}}$ at $\mathrm{cc_{i,j}}$. In other words,

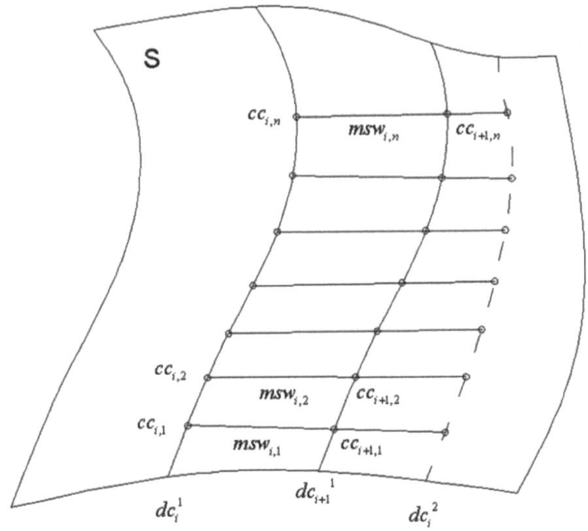

Fig. 1 General representation of a dual-curve driven tool path mechanism

$msw_{i,j}$ is directly related to the path interval or the step over distance of S. The next drive-curve dc_{i+1}^1 is the result of dc_i^1 and $msw_{i,j}$ whereas the curve linking $cc_{i,j}$ and $cc_{i+1,j}$ is finally the secondary drive-curve dc_i^2 for the tool path.

An important function of this tool path control is also the sequential alternation between primary and secondary drive curves that leads to scallop elimination between subsequent step over passes, smoothing this way the adjacent regions. Further details on this particular tool path generation as well as new developments on the basis of dual-driven 5-axis surface CNC planning can be found in [16].

3 Experimental Results

The sculptured surface selected for the analysis was a benchmark $S(u, v)$ second-order parametric surface described by the following equations. The surface was of an open-form contour, commonly found in mold/die industry and it has been used by many researchers [16, 17].

$$S(u,v) = \begin{bmatrix} -94.4 + 88.9v + 5.6v^2 \\ -131.3u + 28.1u^2 \\ a_1 + a_2 \end{bmatrix} \tag{1}$$

with,

$$a_1 = 5.9(u^2v^2 + u^2v) - 3.9v^2u + 76.2u^2 \text{ and } a_2 = 6.7v^2 - 27.3uv - 50.8u + 25v + 12.1.$$

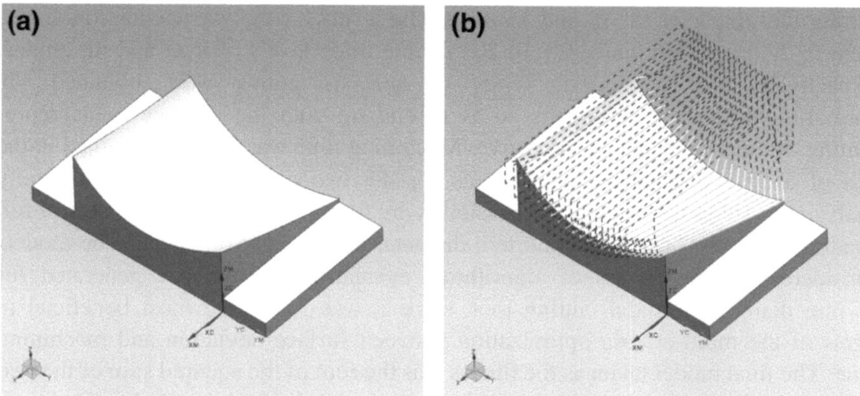

Fig. 2 Benchmark parametric surface: **a** 3D CAD benchmark model; **b** dual-curve driven surface finish tool path for machining experiments

Figure 2a shows the CAD model of the benchmark sculptured surface modeled using the above equations; whilst Fig. 2b depicts the dual-curve driven surface finish tool path that was investigated.

3.1 *Fitness Function Formulation and Genetic Algorithm Parameters*

The proposed tool path optimization scheme suggests the implementation of a genetic algorithm as an efficient heuristic technique capable of converging to an optimum solution for the objectives involved to the problem. Prior to its involvement the search space representing the problem's domain should be formulated via the fitness function generation. Aiming at building a robust fitness model for the algorithmic evaluations, a number of machining simulation experiments was conducted adopting the central composite design under the response surface methodology (RSM) having two continuous factors (the two inclination angles), one categorical factor (the cutting tool diameter), and two equally weighted responses referring to the machining error and tool path time.

Experiments were simulated in Siemens® NX-11 CAM environment using the aforementioned tool path strategy. Spindle speed was set to 16,000 rpm whilst feed rate was set to 5000 mm/min. Cut tolerance was set to 0.1 mm; large enough so as to obtain noticeable variations that would ease examination efforts. Step over distance was controlled through scallop height equal to cut tolerance; 0.1 mm for the same reasons. Hence different adjacent regions were examined according the cutting diameter. All three toroidal cutters simulated were of corner radius equal to 3 mm. The combination of the independent parameters and their three 'low-middle-high' levels gave a total of 21 experiments; 7 for each torus-type cutting tool diameter (8; 12 and 16 mm). The levels for the two inclination angles were 0, 22.5 and 45°, and −20, 0, 20° for the three levels of lead and tilt angles, respectively. For each test, series of analysis points were obtained via normal-to-surface measurements so as to end up with the average value representing the surface deviation objective. Machining time was directly obtained in the end of each experimental test. Results for the two objectives were statistically analyzed to study the correlation possibility by fitting the data via three sets of full quadratic models, one per cutting tool diameter and one per objective. The models considered for the fitness of algorithmic evaluations were those generated for 16 mm diameter toroidal cutting tool, since it was found the most beneficial in terms of the multi-criteria optimization between surface deviation and machining time. The final model taken as the fitness was the root of the squared sum of the two objectives after their normalization procedure to remove the inherent bias owing to different magnitudes. Table 1 shows the parameters, their levels as well as the experimental results from the machining simulations.

Optimization for simultaneously minimizing the criteria is subjected to the specified ranges of machining parameters. The ranges are determined considering the lower and upper levels of the experimental design presented above.

Table 1 Design of experiments and results for objectives

Run	Lead (°)	Tilt (°)	TorusRc3 (mm)	SurfDev (mm)	TlpTime (min)	SurfDevNorm	TlpTimeNorm	Pareto
1	0	−20	8	0.18624	53.20	0.332951945	0.996254682	53.2
2	45	−20	8	0.062233	44.20	0.111257509	0.827715356	44.2
3	0	20	8	0.179627	53.19	0.321129505	0.996067416	53.19
4	45	20	8	0.041593	44.15	0.074358195	0.826779026	44.15
5	22.5	0	8	0.1211	38.13	0.216497426	0.714044944	38.13
6	22.5	0	8	0.1211	38.13	0.216497426	0.714044944	38.13
7	22.5	0	8	0.1211	38.13	0.216497426	0.714044944	38.13
8	0	−20	12	0.55936	53.25	1	0.997191011	53.253
9	45	−20	12	0.278307	35.06	0.497545409	0.656554307	35.061
10	0	20	12	0.55216	53.20	0.987128146	0.996254682	53.203
11	45	20	12	0.739373	34.47	1.32181958	0.645505618	34.478
12	22.5	0	12	0.11824	27.01	0.211384439	0.505805243	27.01
13	22.5	0	12	0.11824	27.01	0.211384439	0.505805243	27.01
14	22.5	0	12	0.11824	27.01	0.211384439	0.505805243	27.01
15	0	−20	16	0.153747	53.40	0.274862343	1	53.4
16	45	−20	16	0.067887	30.04	0.121365489	0.562546816	30.04
17	0	20	16	0.195833	53.31	0.350101902	0.998314607	53.31
18	45	20	16	0.320147	29.32	0.57234518	0.54906367	29.322
19	22.5	0	16	0.16306	21.31	0.291511728	0.39906367	21.311
20	22.5	0	16	0.16306	21.31	0.291511728	0.39906367	21.311
21	22.5	0	16	0.16306	21.31	0.291511728	0.39906367	21.311

When dealing with engineering optimization problems it is a common technique that heuristics will run several times under a trial-and-error experimentation so as to find the optimal parameter setting referring to intelligent operators. This should be done given the fact that no certain recommendations for best combinations among intelligent parameter settings exists. By employing the genetic algorithm to optimize the fitness function, ten sequential runs were executed using the *Mathworks Matlab® R2016a* optimization toolbox. The parameters for the genetic algorithm implemented to optimize the objective functions were as follows:

- Population size: 50
- Generations: 60
- Scaling function: Rank
- Selection scheme: Stochastic uniform
- Elite count: 0.05 * population size = 2
- Crossover type: Single-point crossover
- Crossover fraction: 0.8

The population type was encoded as double vector and the range for the search was (low: [0, −20], high: [45, 20]). Ranking was applied for scaling fitness values so as to select parents for next generations. Selection assigns higher probability for reproduction to better fitted individuals whose fitness scores are ranked accordingly. Ranking is achieved by considering all fitness scores attained by candidate solutions. Individuals selected for reproduction followed the stochastic uniform scheme where each parent corresponds to a sub-string of the chromosome's binary string of length proportional to its scaled value. According to this scheme, the algorithm allocates a parent from the sub-string through its equidistant movement towards the whole chromosome string. The default value determining the elite count for individuals was 0.05 times the population size which in this case is equal to 50. Hence 2 elite individuals are guaranteed to survive to the next generation.

Crossover type was a single-point. Crossover fraction determines the number of individuals other than elite ones which are produced through crossover operator. Remaining individuals are produced owing to mutation. Since optimization for parameters is restricted to lower and upper ranges mentioned, a constraint dependent mutation function was applied to the module. The outputs selected for evaluation were the best fitness value attained, the best mean fitness and the best individual (final point). This implies that the optimum result obtained by the genetic algorithm corresponding to the Pareto combined criterion for the tool path is expected to be lower than, or at least equal to the minimum Pareto value attained for the experimental results. Mean fitness values of final points the algorithm attained were also assessed and compared against the average of the experimental results as well as the average of predicted ones by applying the full quadratic model.

The correlation coefficient R^2 for the final multi-criteria Pareto objective implemented as the fitness function was found equal to 82.22%. The optimal

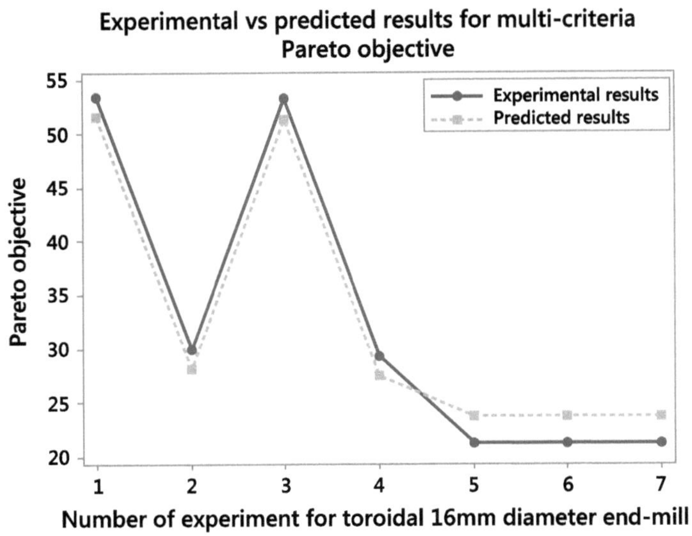

Fig. 3 Experimental versus predicted results for multi-criteria Pareto objective

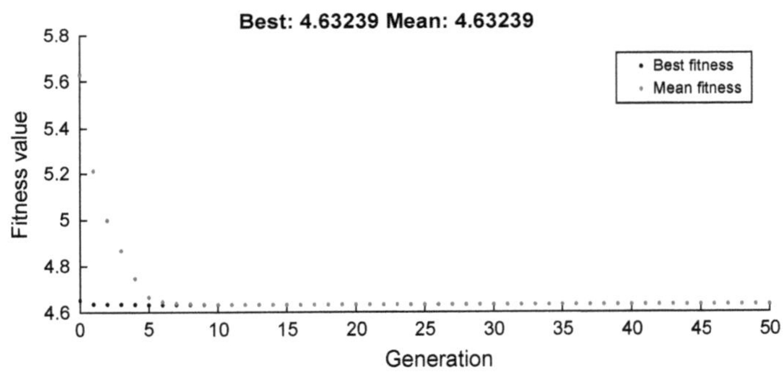

Fig. 4 Fitness function evaluations

machining parameters recommended were the average values from ten series of evaluations whilst they checked whether they fall in the same range as the one restricting the values for parameters selection from the experimental design shown in Table 1. Figure 3 shows the correlation among the normalized experimental multi-criteria Pareto results versus the one corresponding to the prediction capability of the full quadratic multi-criteria Pareto objective.

The convergence diagram of the best fitness score obtained is depicted in Fig. 4. The resulting fitness score is by far lowest than the lowest experimental result for Pareto objective (21.311) as well as the lowest one predicted by the regression Eq. (23.764). To quantitatively illustrate the benefits of the optimization approach, the best value representing the fitness score was expressed as a percentage of the aforementioned lowest experimental results. By taking into account the lowest experimental result, the genetic algorithm reduced the mean for the Pareto objective by 78.28%, whereas considering the lowest predicted result when employing the regression model the gain is 80.52% reduction of the corresponding mean.

3.1.1 Simulation Results

Optimal settings for machining parameters recommended by the genetic algorithm were passed onto Siemens® NX 11 CAM environment to validate the proposed experimental investigation and optimization procedure for this particular tool path. These parameters were: Ø16Rc3 toroidal end-mill; Lead angle = 30.857° and Tilt angle = 19.99°. Figure 5 shows the CAM simulation outputs of the lowest experimental result and the optimum one using the genetic algorithm. Figure 5a refers to the output obtained when using a lead angle equal to 22.5° and tilt angle equal to 0° under the Ø16Rc3 toroidal end-mill. Figure 5b refers to the CAM simulation output when employing the parameters recommended by genetic algorithm for the process. The former tool path came with a total machining time equal to 28 min and 37 s, whereas the latter tool path came with a total machining time equal to 21 min and 31 s. By taking into account the machining strip widths remained as excess material on the part's free form surface, it can be seen that they are more profound in the case of employing parameters from the lowest experimental result's combination than those resulted in the optimized case.

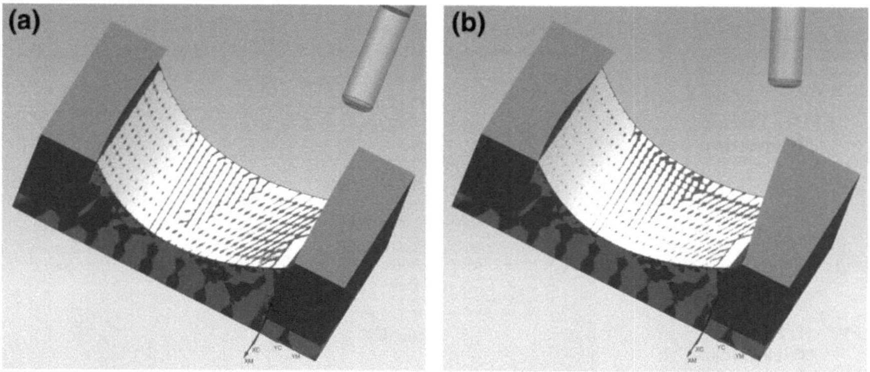

Fig. 5 **a** CAM simulation output using the lowest experimental result parameters; **b** CAM simulation output using the parameters recommended by the genetic algorithm

3.1.2 Virtual CMM Inspection Results

Further validation of the experimental results involves the virtual CMM inspection. For the benchmark part machined using the optimal parameters for the dual-curve driven surface finish tool path, a series of measurements were performed in Decartes coordinate system XYZ [18]. Figure 6 shows details of all steps performed in the virtual inspection.

Measured values were verified using the [+0.1 mm −0.1 mm] tolerance field. Figure 7 shows the surface deviation error between the virtually machined surface and the curved profile of the design surface in four selected sections. The selected sections are referred to the OYZ plane at X = −5 mm, X = −30 mm, X = −60 mm, and X = −90 mm respectively. One may notice that the error distribution in all three resulting profiles falls within the specified tolerance limits, whereas no undercuts or gauges are detected. Moreover it is evident that most of the red dots illustrating the surface machining error vary from [0 +0.05] or even less than +0.05 mm of tolerance as the Y coordinates increase. Regions to where large scallops are to be experienced are the ones distributed among the adjacent tool path trajectories where no sharp edges were finally tracked. Machining strip widths are also found continuous and smooth whilst the peaks in the figure suggest large, yet rounded scallops that join these uniform machining strip widths.

The developed methodology was successfully tested on a real test part. Figure 8 shows an actual coordinate inspection of the manufactured test part.

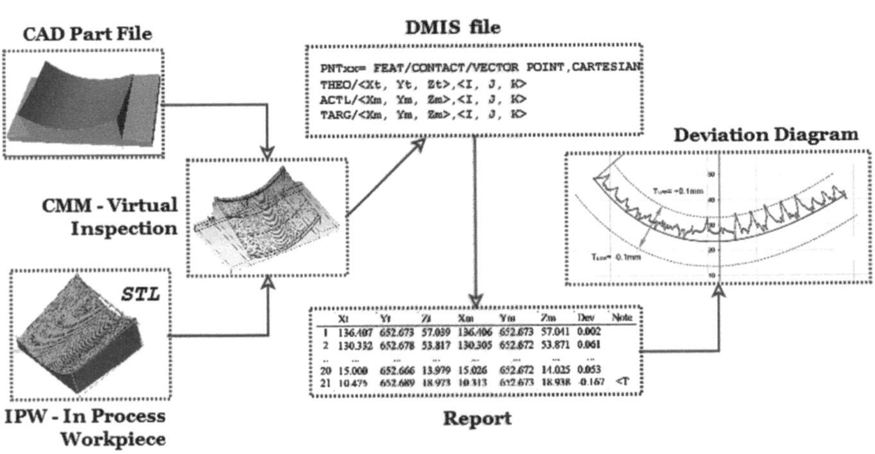

Fig. 6 Virtual inspection steps

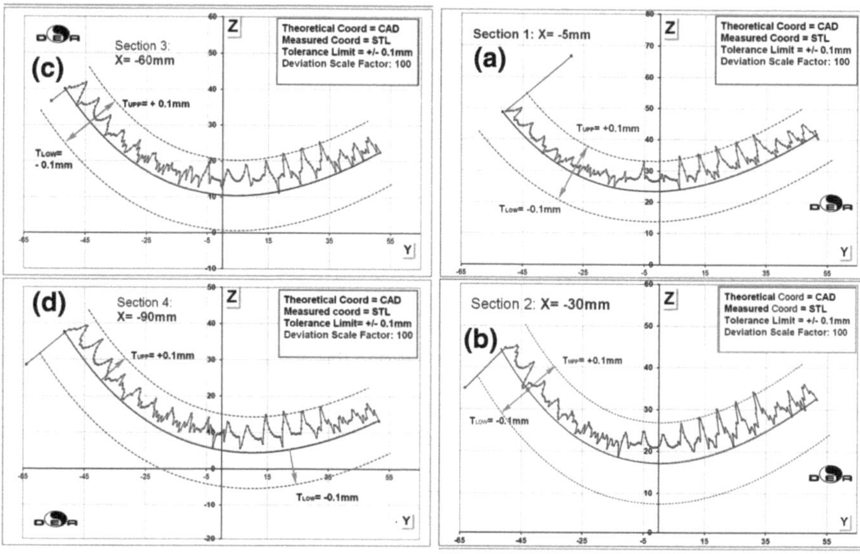

Fig. 7 Measured results of the machining-simulated surface: **a** section X = −5 mm; **b** section X = −30 mm; **c** section X = −60 mm; **d** section X = −90 mm

Fig. 8 Coordinate inspection of the manufactured test part

4 Conclusions

A particular multi-axis surface finish tool path, based on two drive curves for its cutting style definition, was virtually examined. Its efficiency was decided upon the context of finding optimal parameter values for the two inclination angles lead and tilt, whilst machining by employing toroidal end-mills. To characterize the performance of such a tool path, a genetic algorithm was implemented undertaking the multi-criteria minimization between surface machining error and tool path time. A regression model built through an experimental design played the role of a fitness function for the genetic algorithm. Despite the frequently mentioned shortcomings in terms of generalization, the regression model was capable of describing the problem's domain such that first-order continuous profiles could be generated on the tested sculptured surface. Looking further ahead, more tool paths for multi-axis surface finish will be examined so as to enhance shops' practices implementing advanced manufacturing software for sculptured surface CNC machining.

References

1. Lin Z, Fu J, Shen H, Gan W (2014) An accurate surface error optimization for five-axis machining of freeform surfaces. Int J Adv Manuf Technol 71:1175–1185
2. Suresh K, Yang DCH (1994) Constant scallop-height machining of free-form s urfaces. J Eng Ind 116(2):253–259
3. Ding S, Mannan MA, Poo AN, Yang DCH, Han Z (2005) The implementation of adaptive isoplanar tool path generation for the machining of free-form surfaces. Int J Adv Manuf Technol 26(7):852–860
4. Subhajit S, Dey PP (2014) A new iso-parametric machining algorithm for free-form surface. Proc IMechE Part E: J Process Mechanical Engineering 228(3):197–209
5. Zhang XF, Xie J, Xie HF, Li LH (2012) Experimental investigation on various tool path strategies influencing surface quality and form accuracy of CNC milled complex freeform surface. Int J Adv Manuf Technol 59:647–654
6. De Souza AF, Machado A, Beckert SF, Diniz AE (2014) Evaluating the roughness according to the tool path strategy when milling free form surfaces for mold application. Proc CIRP 14:188–193
7. Ülker E, Turanalp ME, Halkaci HS (2009) An artificial immune system approach to CNC tool path generation. J Intell Manuf 20(1):67–77
8. Agrawal RK, Pratihar DK, Choudhury AR (2006) Optimization of CNC isoscallop free form surface machining using a genetic algorithm. Int J Mach Tools Manuf 46(7–8):811–819
9. Bedi S, Ismail F, Mahjoob MJ, Chen Y (1997) Toroidal versus ball nose and flat bottom end mills. Int J Adv Manuf Technol 13(5):326–332
10. Redonnet JM, Djebali S, Segonds S, Senatore J, Rubio W (2013) Study of the effective cutter radius for end milling of free-form surfaces using a torus milling cutter. Comput Aided Des 45(6):951–962
11. Lee YS (1998) Non-isoparametric tool path planning by machining strip evaluation for 5-axis sculptured surface machining. Comput Aided Des 30(7):559–570
12. Hendriko H (2017) Analytical boundary method for obtaining feed scallop of toroidal cutter in multi-axis milling. Key Eng Mater 728:48–53
13. Roman A, Barocio E, Huegel JC, Bedi S (2015) Rolling ball method applied to 3½½-axis machining for tool orientation and positioning and path planning. Adv Mech Eng 7(12):1–12
14. Warkentin A, Ismail F, Bedi S (2000) Comparison between multi-point and other 5-axis tool positioning strategies. Int J Mach Tools Manuf 40:185–208
15. Tunç LT, Ozkirimli OM, Budak E (2016) Machining strategy development and parameter selection in 5-axis milling based on process simulations. Int J Adv Manuf Technol 85: 1483–1500
16. Rufeng X, Zhitong C, Wuyi C, Xianzhen W, Jianjun Z (2010) Dual drive curve tool path planning method for 5-axis NC machining of sculptured surfaces. Chin J Aeronaut 23:486–494
17. Warkentin A, Ismail F, Bedi S (2000) Multi-point tool positioning strategy for 5-axis machining of sculptured surfaces. Comput Aided Geom Des 17(1):83–100
18. Sibalija TV, Zivkovic SP, Fountas NA, Majstorovic VD, Macuzic JZ, Vaxevanidis NM (2016) Virtual optimisation of CAI process parameters for the sculptured surface inspection. Proc CIRP 57:574–579

Advanced Manufacturing and Industrie 4.0 for SME

Detlef Kochan and Ronald Miksche

Abstract Technological development is driven by the need of permanent improvements in terms of productivity, functionality and quality. Large companies have the economical power for the practical realization for further developments such as new strategic goals like Industrie 4.0. This situation is fortunately given in SMEs but at the same time most of the innovations are developed in small companies which leads to the conclusion that especially in the starting phase small companies need special external support. The paper will describe made experience in the field of Advanced Manufacturing including some examples.

Keywords Advanced manufacturing · CNC · CIM · CAD/CAM · PLM · Additive manufacturing · Smart factory · CPPS · Industrie 4.0

1 Introduction

International competition requires innovations along the entire process chain. Product development, design and manufacturing of high-quality products has to be constantly improved in order to allow a realization in shortest times at reasonable cost as well as reduced energy and resources consumption.

The essential applications are realized in the automotive and aerospace industry; machine and plant engineering; consumer, goods, medical equipment and others.

The automotive industry in Germany and throughout Europe is noticed to be the pass finder in designing and implementing innovative and integrated process chains [1]. This particular market constantly sets new standards and represents the reference point for both established and cutting-edge technology. Under the general framework of complete PLM strategies some of the most essential working fields will be focused on.

D. Kochan (✉) · R. Miksche
Zentrum Für Angewandte Forschung Und Technologie e. V. an der HTW Dresden, 120701, 01008 Dresden, Germany
e-mail: detlef.kochan@zaft.htw-dresden.de

V. Majstorovic and Z. Jakovljevic (eds.), *Proceedings of 5th International Conference on Advanced Manufacturing Engineering and Technologies*, Lecture Notes in Mechanical Engineering, DOI 10.1007/978-3-319-56430-2_26

2 Characteristic of the International Situation

The challenges for manufacturing in Europe are summarized in Fig. 1 by Dr. Filos [2].

This figure is based on carefully SWOT analysis of European manufacturing experts, especially Dr. Filos. It is characterized by strength weaknesses, opportunities and threats. The detailed explanation is included in [2] as paper Manufacturing Innovation and Horizon 2020 by Dr. E. Filos.

Re-Industrialization: What are some countries doing?

- **Germany**
  Use the potential of Cyber-physical Systems (Industrie 4.0)
- **UK**
  Launched Advanced Manufacturing supply Chain Initiative focusing on R&D and skills
- **France**
  Has included "Fantasies of the Future" and "Robotics" to its 34 plans for reindustrialization
- **Finland**
  R&D + I support to ICT-enabled and sustainable manufacturing
- **USA**
  Advanced Manufacturing Partnership initiative launched in 2011, with Nat'l Network of Manufacturing Innovation

The main activities in several countries are demonstrated with the following short information. Information concerning Industrie 4.0 will follow in Sect. 3. Concerning the United States, main activities were started in a so-called "Industrial Internet Consortium IIC" as a non-profit organization of ATaT, Cisco, GE, IBM and Intel with

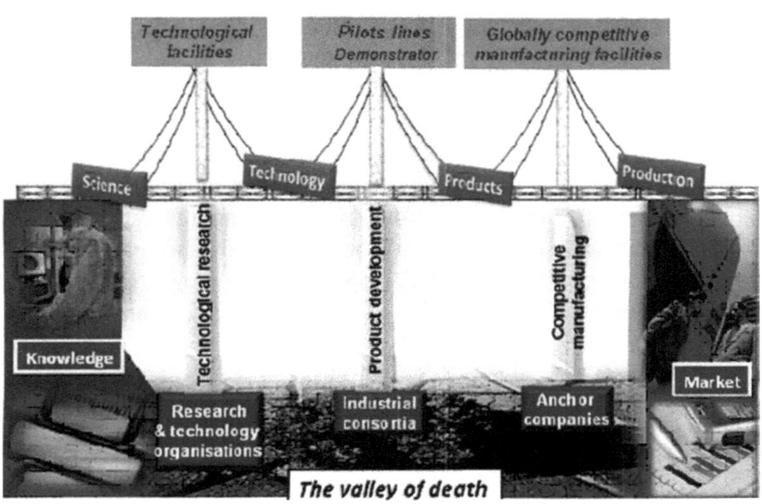

Fig. 1 General situation in Europe

currently about 200 members. Similar activities in China were named *Internet+* or *made in China 2025*. Most interesting in comparison with the German strategy is the "Industrial Value Chain Initiative" of the Japanese industry and government. The activities in South Korea are summarized with the goal "Smart Factory".

3 Industrie 4.0 Key Features and Aspects

Key features of the German future strategy, which can also be summarized under the name 4th industrial revolution, are the following:

- future-oriented project of the German Hightec-Strategy
- developed and proposed by German top-industry, in cooperation with industry-related science and industrial organizations
- main goal: acceleration of manufacturing sector by informatization
- essential aim: "Smart Factory", characterized by flexibility, resource efficiency and ergonomic design
- integrated realization of value added and business processes
- technological basis via Cyber Physical Production Systems (CPPS) and "Internet of things"

A wide area of responsibilities is given through these features. A simplified representation by Henke is showing the activities in a 3-level-model (Fig. 2).

Related to the general developmental trends are the following aspects:

The entire technological developments have to be realized by the so-called "bottom-up" principle. That means that especially for the future-oriented strategy Industry 4.0 the optimization of value-added process chains is basic and start point for all further activities.

Especially in SME, the production level has significant meaning when it comes to aspects of strategic focus of Advanced Manufacturing. It can be assumed that most of

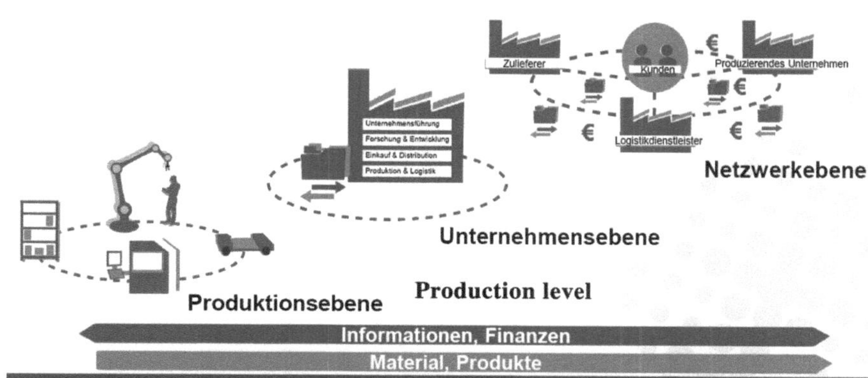

Fig. 2 Digitization on different levels

the producing companies, including suppliers for the automotive industry, have a minimum configuration of automatized, e.g. numeric controlled, machines. Related to that is the assumption that the beginning of the digital age can be dated about 75 years back in time. It is common knowledge that the first numeric controlled machines were designed in the USA in 1950. In Europe, as well as in Germany, these dynamic developments started to take place in 1960. The first specialist book by Simon about NC machines [3] looked from the perspective of so-called "statistics processing machines" at grave impacts on "information-processing machines". The whole development of increased use of freely programmable computers in CN-mode of operation is well-known as well as the functional build of complex processing machines, such as machining centers, flexible production systems and computer-integrated production in the 1990s. Naturally, by developing the first NC-machines in the industrial countries in the 1960s, NC-programming systems were developed and put into practical use. In 1969, the first IFIP conference PROLAMAT (Programming Languages for Machine Tools in Roma) made it possible to exchange experience internationally for the first time. The thereof originated advancement in 3D-CAD/CAM systems is, at the present, an important starting and reference point for further possibilities of "Reverse Engineering and Additive Manufacturing" [4, 5]. The latest developments in PLM (Product Lifecycle Management) are based on this known level of development, which also include several areas of use and fitting simulation programs.

This issue is shown in Fig. 3.

Based on this figure, the important question arises which technological procedures will be the starting point for further industrial development. Figure 4 gives means to answer this question as it shows the statistical distribution of essential processing machines and plants. This overview clearly shows that metal cutting machine tools represent the dominating part. It also illustrates that a significant part is constituted by highly automated machines and plants—procedures such as rotating, milling, drilling or procedures using geometrically defined edges represent the predominating part. In connection to the development into Smart Factory and CPPS (Cyber Physical Production Systems) arises the question if the traditionally

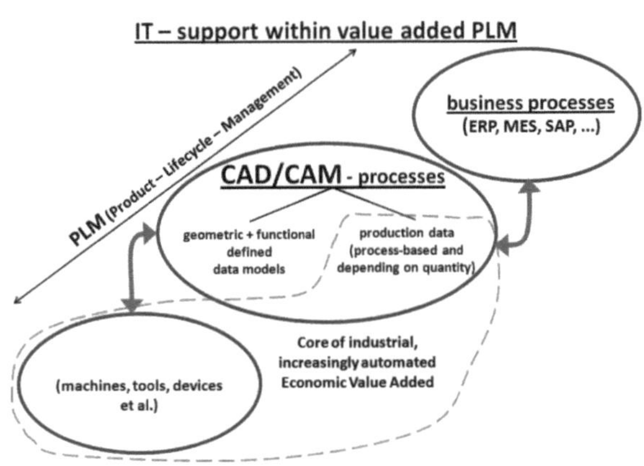

Fig. 3 Characterization of the 3./4. Industrial revolution

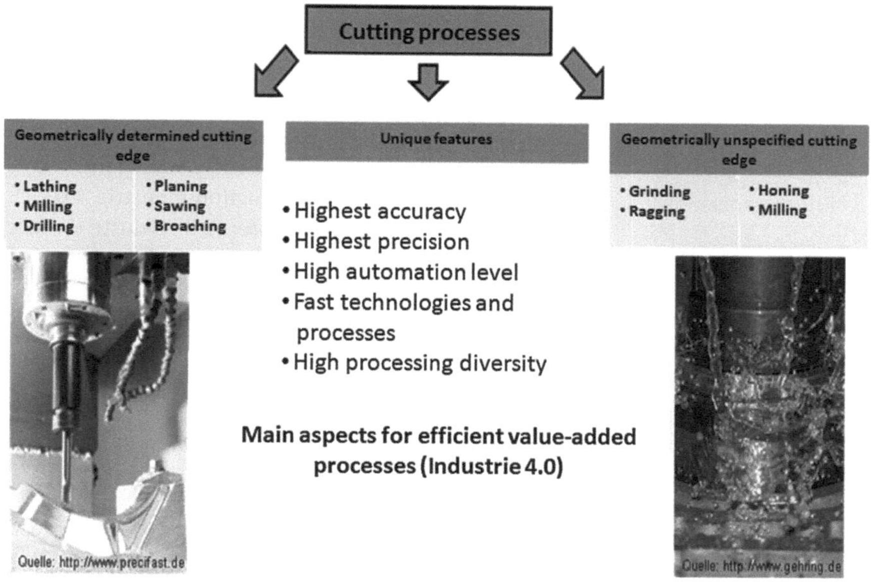

Fig. 4 Characteristics of cutting procedures

known cutting basic procedures will also be an important reference point in the future. This general question will be answered with the help of Fig. 4.

This illustration sheds light on the crucial unique characteristics of cutting procedures. The claim of pinpoint accuracy and precision will continue to be answered primarily with traditionally known cutting procedures with geometrically defined edges and geometrically undefined edges in the upcoming years.

The various fields of application of different machine tools are shown in Fig. 5.

Two aspects of the future development of digital product and process development systems are important from the point of view of design and realization of efficient value chains. On the one hand, it is the aspect of IT realization of complex value chains using CAx and CAD/CAM solutions which are increasingly summarized as PLM tools. This also includes aspects such as simulation application in various areas and the potential of virtual reality. The framework of such complex PLM applications forces to consider the second important aspect. This is the aspect of automation of production of high performance processes (HSC, HPC) and, to a certain extent, also new procedures of material processing such as laser and others. Based on these general aspects, the most important areas of application with the main focus on industrial development will be characterized in the following:

Topics—Main Fields from Manufacturing Innovations and Processes

- Product and process optimization.
  - PLM-Tools for integrated digital process chains
  - IT-Applications for product and process innovations in leading companies

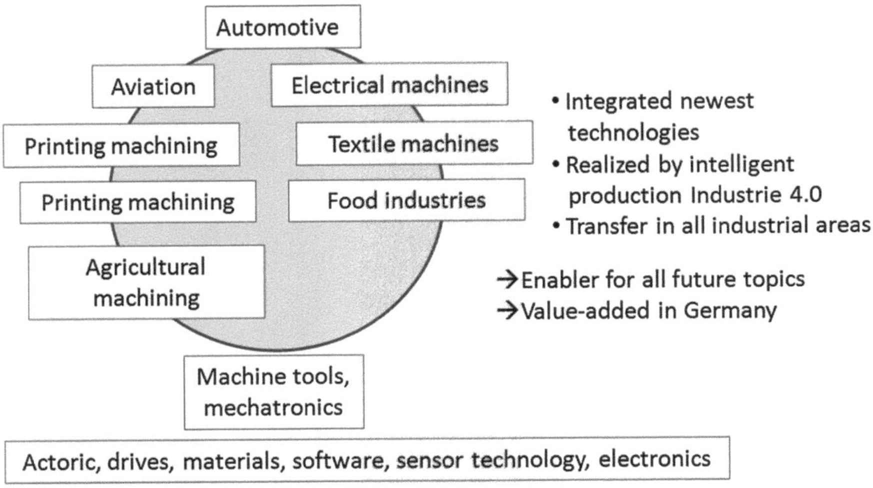

Fig. 5 Machine tools as mother of all other machines

  - Automatic process planning, simulation and integration in product- and production data management.

- Software developments for multi-axis high speed and high precision manufacturing.
- Rapid innovation by Additive Manufacturing and Mass Customization.
- Process control and monitoring for new manufacturing procedures.
- Manufacturing quality control, optical measurement and reverse engineering.
- Sustainable production resource management.
- Knowledge engineering and future-oriented education.

The realization of this scope of tasks can be shown on the example of the tasks of the automotive industry. A fitting example is the German top company BMW. The company produces 30 different vehicles organized in various categories, such as super-luxury class, luxury class, higher middle class, middle class, compact class and small cars. Considering the different parts of each individual type of car, a consequence of the 10–15 different parts of the producer are the 10,000–15,000 challenges for the automotive producer in the complex competition. Assuming that only around 20% of the key components are produced by the OEMS, it is clear that around 80% of the parts have to be produced by suppliers. This range of tasks needs to fulfill the demands of Advanced Manufacturing which includes:

- High-productivity
- Excellent functionality
- High quality
- Economical aspects
- Environmental aspects

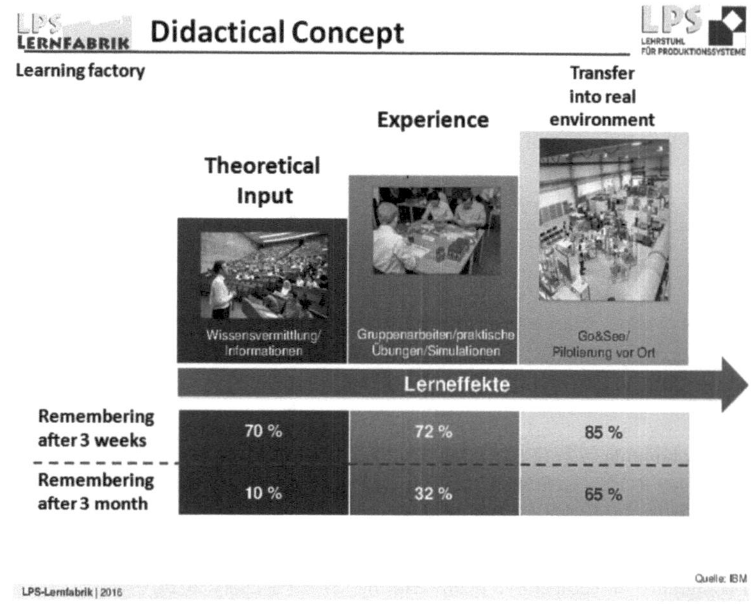

Fig. 6 Didactical concept of the learning factory

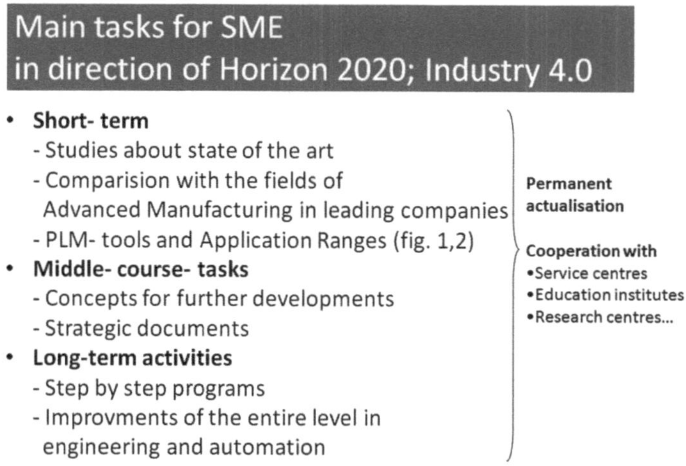

Fig. 7 Main tasks for SME in direction of Horizon 2020; industry 4.0

Concrete examples of the synthetic industry and of transformative procedures will be shown in the presentation using various examples.

For the distribution of advanced knowledge, several universities in Germany start to include learning factories in their institutions (Fig. 6).

The principles of the learning factories are not only of significant importance for the deepened education and training in universities and colleges, but also for specific continuing education of experts of the industry. Such a basis of the usage of advanced knowledge and abilities creates a new scope of tasks for SME, especially for industrial further development (Fig. 7).

References

1. GTI: investing innovation internationalize. Industry clusters in Eastern Germany. www.powerhouse-eastern-germany.de/en
2. Kovacs GL, Kochan D Digital product and process development systems. Springer, Heiderlberg
3. Bundesministerium für bildung und forschung: industrie 4.0 innovationen für die produktion von morgen. www.bmbf.de
4. Kochan D, Miksche R (2013) Session keynote: additive manufacturing for new thermal forming process chains. In: Coma'16 international conference on competitive manufacturing, 27–29 January 2016
5. Miksche R, Kochan D Additive manufacturing for improved Al and Mg-parts. In: Coma'16 international conference on competitive manufacturing, 27–29 January 2016

Industrial Product Life Cycle Stages and Lifecycle Eco-design

Anca Iuga, Vasile Popa and Luminița Popa

Abstract The article addresses the issue of industrial product life cycle, respecting the principles of circular economy, in order to solve the challenges for Romanian industry according with European Union requirements. The Lifecycle Eco-Design Strategies are considered key issues regarding the sustainable development of industrial products. The authors have identified the correlations between the elements of Product Life Cycle Stages and Lifecycle Eco Design Strategies.

Keywords Product life cycle · Eco-design · Circular economy

1 Introduction

The production and consumption of goods and services is the main promoter of Romania' economic development, contributing greatly to improved quality of life, but can be made and a number of its damage. How Romanian population produces and consumes, express the degree of development and environmental issues awareness. In terms of humanity today, carrying out such activities on a planetary scale, it has already generated a high level of pollution and global warming, caused depletion of certain natural resources and seriously threaten the maintenance of biodiversity.

A. Iuga (✉)
Department of Engineering and Industrial Management, Transilvania University, Mihai Viteazul 5, 500036 Brașov, România
e-mail: anca.iuga@unitbv.ro

V. Popa
Regional Departament of Defense Resources Management Studies, National Defence University, Mihai Viteazul 160, 500183 Brașov, România

L. Popa
Department of Automation and Information Technologies, Transilvania University, Mihai Viteazul 5, 500036 Brașov, România

V. Majstorovic and Z. Jakovljevic (eds.), *Proceedings of 5th International Conference on Advanced Manufacturing Engineering and Technologies*, Lecture Notes in Mechanical Engineering, DOI 10.1007/978-3-319-56430-2_27

Usually, the negative impact of production and consumption activities is accentuated by increasing population and per capita consumption.

1.1 Sustainable Consumption and Production

The sustainable consumption and production are focused on more efficient use of natural resources and energy, reducing greenhouse gas and other environmental impacts.

In other words, production and use of products and services means to affect as little as possible the environment.

The sustainable consumption and production objective is to meet the people basic needs about the goods and services while increasing their quality of life and ensuring sufficient resources for future generations.

The sustainable consumption targets our lifestyle, the buying behavior and the way we use and eliminate products and services. The sustainable production focuses on reducing the environmental impact of production processes and on designing better products. The efficient use of resources and energy helps Romania's future competitiveness. The sustainable consumption and production maximize the potential of businesses to transform environmental challenges into economic opportunities while providing consumers a better deal.

Housing, household goods, food and travels are responsible for 70–80% of all environmental impacts casused by pollution and greenhouse emissions. In assessing the state of sustainability in production and personal, household, community or industrial consumption, there are used a number of relevant indicators. If sustainable production concerns economic sectors key (agriculture, energy, industry, transport), the sustainable consumption takes into account the demand for goods and services necessary to meet people basic needs and improving quality of life: food, health, housing, clothing, environment and party leisure [1].

2 Eco-design

Eco-design addresses the first stage of the supply chain: the product development stage. This new concept aims to "eliminate the environmental impact of the product and the production process".

Although the design itself is a "clean" process, it is one that causes most elements of the environmental impact of the product. Once the design is complete and manufacturing technologies have been fixed, remain very little ways to increase their process efficiency and minimizing the emissions. In addition, most advanced recycling technology has to cope with the requirements defined in the design phase.

In general, about 80% of the product's environmental impact can be determined during the design phase: in terms of costs over the life cycle, the situation is much the same.

In such circumstances it is crucial to consider economic and environmental aspects of the first product design stage. Eco-design is to integrate environmental aspects into the design phase, taking into account the entire product life cycle from the purchase of raw materials to product disposal. Particle "eco" means both economy and ecology.

2.1 *The Content of Eco-design Process*

The Eco-design process, in terms of product, is depending on the goals and can be composed as following:

1. Product improvements;
2. Products redesign;
3. New product design;
4. New production system.

2.1.1 Product Improvements

The product improvements means an update of the product and the environmental restriction are given point by point in accordance with manufacturing techniques. In the first phase, it is necessary to require obyain a certain number of product information to quickly identify improvements to be made.

This information can be obtained:

- tallying: fewer parts, the number of materials, number of screws or other fasteners;
- measuring: the consumption of energy (during use), product weight during the disassembly of the main parts and the amount of potentially hazardous substances;
- computing the cost of possible environmental improvements and the benefits of these improvements.

It is intended, for example: reducing the amount of used raw materials, suppressing a toxic surface treatment, eliminating a certain type of coverage, adding an anti-pollution device, creating a system for collecting waste products, reducing the time of assembly and disassembly, reducing the weight of the product or the package, reducing energy consumption and so on.

2.1.2 Product Design

Product redesign is product rethinking. In this case the project requires further study of the existing product and mobilization of an important design team. Ideally, it should not impose any restriction on cost, schedule and quality to facilitate the creativity work.

At this level, the technologies can evolve to minimize the energy consumption in several stages of the life cycle of the product. At the product level, the materials components can be changed with non-toxic materials. The project involves the enterprise (which may have its methods of production and assembly) and suppliers or contractors who must in turn change the patterns, the materials and manufacturing technologies.

2.1.3 New Product Design

New product design is the first level of innovation, which is part of a genuine policy of innovation with a long-term strategic vision. Innovation is a radical issue, since this is a breakthrough innovation (the transition from paper letters to emails or from thermal engines to electric motors).

The innovation could expose an enterprise to enormous risks:

- Product alternatives can produce large investments with uncertain compensation;
- Product alternatives may do unnecessary current investments and oblige the certain suppliers to modify their entire processes.
- The clients may be reluctant to purchase a new product or service if the environmental benefits are not combined with other indispensable criteria (cost, performance, quality, etc.);

A Romanian enterprise needs to adopt a strategy of product development with a very broad and very long term that takes into account future market developments.

2.1.4 New Production System

New production system is even innovation: the design of products and services that require profound changes of the infrastructures and organizations.

A good example of this level of eco-design is food production shifting from production based agriculture to production based industry or changes based on new information technologies (e.g. video conferencing).

For now, most of the enterprises concerned with eco-design rarely exceed the redesign of their products. Eco-innovation targets improvements in the company, product or company client relationship thus: reducing consumption of embedded raw materials; reducing energy and water consumption during usage; fewer toxic materials embedded in product; easy maintainable, repairable and recyclable product; product performance optimization; efficient production processes (e.g., energy consumption, waste generation, emissions to air, water, etc.); reduction/elimination of toxic substances in the product/process; improving the company's innovation capabilities; increased productivity, positive image, good market positioning; reducing costs during usage; less risk of use (e.g. less noise, emissions, hazardous substances); easy of operation, better product performance; technological or functional improvements.

2.1.5 Essentials for Adopting Eco-design Strategies

When a company decides to draft an eco-design project within its policy of protecting the environment, it sets the targets for eco-development and one or more strategies to achieve them.

The company objective is to improve the functions, performance and product life, without increasing ecological potential impacts on the entire life cycle.

These principles are found in a number of very different strategic axes, whose choice will depend on the company's long-term or short goals, its market objectives and type of products.

2.1.6 Eco-design Strategies According with Circular Economy Principles

Targets for improvement are different; the adopted strategies will also be different:

- In pre-production phase: there are using recycled and less energy materials;
- In the manufacturing phase: there are using processes that require fewer materials;
- In the distribution—transport phase: the volume and weight reduction, use of recycled materials for packaging etc.;
- In the operational phase: less energy consumption, increasing the life duration, help for maintenance;
- In the end of life phase: help for disassembly, protect the quality of materials, etc.

Business motivation for eco-innovation is now widely accepted because:

- the green market worth trillions of dollars;
- the retailers require suppliers to meet consumers "green" requirements;
- a "green" reputation leads to financial value;
- small investment in "green" can lead to big savings;
- the new generation of workers cherishes sustainability and demands "green" jobs (96% of young people wants employers to be environmentally responsible).

3 Product Life Cycle Stages and Lifecycle Eco Design Strategies [2, 3]

The main principles to be considered for adoption of Eco Design Strategies (Table 1) are: resource protection (it aims to reduce the amount of resources needed to start product manufacturing); waste prevention. It focuses on reducing the pollution and damage caused by a product in its life cycle, eliminating the cause of environmental impact; optimization of product service.

In the Fig. 1, one of "Lifecycle Eco Design" Strategy named "Reduction of the environmental impact in the user stage", has the highest weighted score (20.00%). Another "Lifecycle Eco Design Strategy" named "New Eco-Concept Development", has the second position of weighted score (14.54%).

In these cases, the managers of Romanian companies which are trying to implement sustainable development of industrial products, respecting the principles of circular economy, need to take in consideration these two very important "Lifecycle Eco Design Strategies" for the next years.

In the Fig. 2, two of "Product Life Cycle Stages" named "Research and development" and "Research design and development" have the highest weighted score (12.72%). Another "Product Life Cycle Stage" named "Market needs", has the second position of weighted score (10.90%).

In these cases, the managers of Romanian companies which are trying to implement sustainable development of industrial products, respecting the principles of circular economy, need to take in consideration these three very important 'Product Life Cycle Stages" for the next years.

To begin a process of eco-design, an enterprise will be motivated by one or more internal and external factors.

It can be said that the important factors motivating an enterprise in a work of eco-design seem to be: the pressure of customers, competition and legislation, reducing costs and improving corporate image indoors. In the Table 2 are represented the stages of the design process and the measures according with ISO/TR **14062.**

The basic tools in cross—designing are checklists that indicate where something must be done and what must be done; tse tools help us to think about environmental issues and does not omit any. Repeated checks lead to improvements.

The Romanian state action to stimulate the emergence and development of circular economy may result in a policy package that should include:

- Reducing and eventually eliminating subsidies for activities that affect environmental quality and redirect them towards the circular economy;
- Increasing eco-taxes to discourage pollution and emission of greenhouse and use the revenues from these taxes to support the green economy;
- Returning the products which are at the end of their life to companies that produced them, by regulations which apply the principle of extended producer responsibility;

Table 1 Product life cycle stages and lifecycle eco design strategies matrix (realized by the authors)

No.	Product life cycle stages	Lifecycle eco design strategies								Total scores	Weighted Score (%)
		New eco-concept development	Selection of low-impact materials	Reduction of material	Optimization of eco-production techniques	Efficient distri-bution system	Reduction of the environmental impact in the user stage	Optimization of initial life-time	Optimization of end-of-life system		
		(1)	(2)	(3)	(4)	(5)	(6)	(7)	(8)	(9)	(10)
1	Market needs	X	X	X	–	X	X	–	X	**6**	**10.90**
2	Research and development	X	X	X	X	–	X	X	X	**7**	**12.72**
3	Idea generation	X	X	X	–	–	X	X	–	**5**	**9.09**
4	Opportunity identification and concept definition	X	X	–	–	X	X	–	–	**4**	**7.27**
5	Research design and development	X	X	X	X	–	X	X	X	**7**	**12.72**
6	Prototype, production	X	X	X	X	–	X	–	–	**5**	**9.09**
7	Distribution and manufacturing	–	X	X	X	X	X	–	–	**5**	**9.09**
8	Marketing	X	–	–	–	X	X	–	–	**3**	**5.45**
9	Sales	X	–	–	–	X	X	X	–	**4**	**7.27**
10	Maintenance/service	–	–	–	–	X	X	X	X	**4**	**7.27**
11	Products feed-back	–	–	–	–	X	X	X	X	**4**	**7.27**
12	Removal and disposal and recycling	–	–	–	–	–	–	–	X	**1**	**1.81**
Total scores		**8**	**7**	**6**	**4**	**7**	**11**	**6**	**6**	**55**	**100**
Weighted Score (%)		**14.54**	**12.72**	**10.90**	**7.27**	**12.72**	**20.00**	**10.90**	**10.90**	**100**	**100**

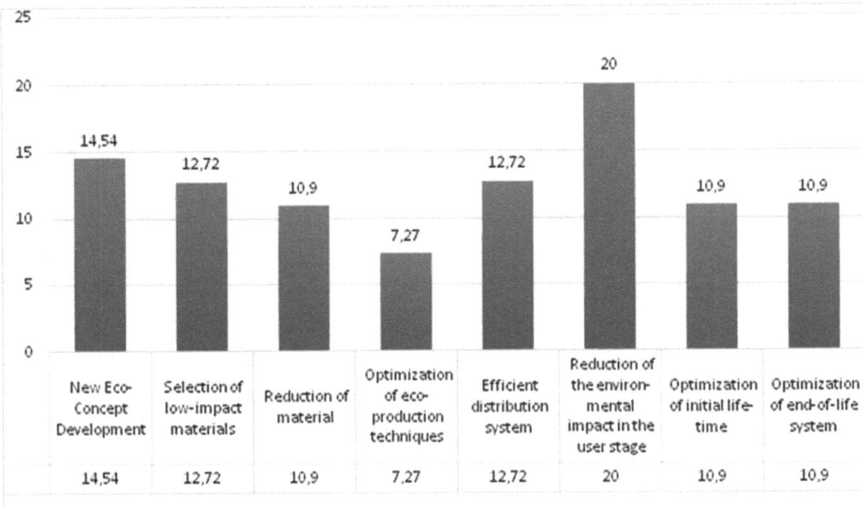

Fig. 1 Lifecycle eco design strategies (figure is realized by the authors)

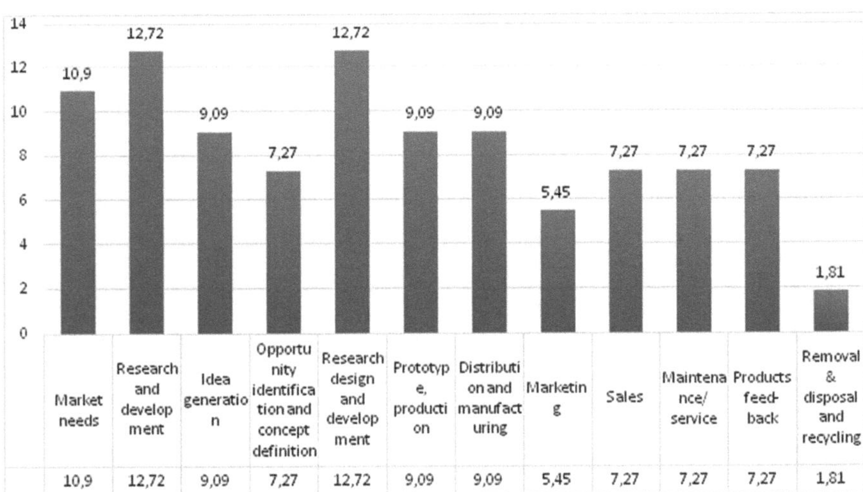

Fig. 2 Product life cycle stages (figure is realized by the authors)

- Eco-labeling for all consumer products made environmentally, to inform consumers and give them the opportunity to choose and encourage consequently the production and sale of such products;
- More rigorous regulation of activities related to natural resource use and rigorous application of the law when deviations from the established norms;
- Financial support of research activities contributing to the objectives of environmental protection and circular economy development;

Table 2 The stages of the design process and the measures according with ISO/TR 14062

Stages	Eco-design activities
Planning	Clarify: what is the product idea? Which are the economic, technical and environmental priorities for our product? It is a new product or a development of an older product? Status quo: what eco-design activities rely? Use the relations with environmental management systems Be aware of the business environment: customers and market needs, legislation, eco-label, niche markets, products competition
Conception	Integrate the eco-design aspects when setting specifications Check feasibility (technological, financial) Use guides and checklists to refine specifications Communicate with your supply chain
Design	Use design tools and databases of eco-design Seek alternatives to problems Lifecycle scenarios developed to better understand the evolution of the product Analyze assembly/disassembly
Testing, prototype	Compare with previous products generation Analyze the objectives achievement
Market launch	Promote to customer groups the environmental excellence of your product Promote additional characteristics: quality and operational cost Customer awareness
Review	Evaluate the success of the product (which arguments really counted for the customer) Identify opportunities for further improvement for the next generation of products What innovations will appear (in the firm and the market)? What our competitors are doing?

- Financial support of training programs for occupations and specific skills of circullar economy;
- Encouraging and promoting environmental projects eligible for attracting European funds for their realization.

4 Conclusions

The Romanian Eco-design strategies (according with circular economy principles) that we evaluate today are very far from the stated objective of the EU on green sustainable development.

This discrepancy is due to the fact that economic decisions often ignore natural and human capital that we have available.

References

1. Robins N, Roberts S (1997) Changing consumption and production patterns: unlocking trade opportunities. International Institute for Environment and Development UN Department of Policy Coordination and Sustainable Development
2. Popa Luminita I, Popa Vasile N (2013) PLM innovation matrix for a complex product development process. Appl Mech Mater
3. Popa L, Popa V (2013) Remanufacturing eco-design applied in industrial products lifecycle management. Appl Mech Mater
4. Sundin E, Elo K, Lee HM (2012) Design for automatic end-of-life processes. Assembly Autom Emerald Indight. http://www.emeraldinsight.com/doi/abs/10.1108/01445151211262447
5. Alting L (1995) Life cycle engineering and design. CIRP Ann Manuf Technol 44:569–580
6. Hauschild M, Jeswiet J, Alting L (2005) From life cycle assessment to sustainable production: status and perspectives. CIRP Ann Manuf Technol 54(2):1–21
7. Zufia J, Arana L (2008) Life cycle assessment to eco-design food products: industrial cooked dish case study. J Cleaner Prod 16:1915–1921
8. Rebitzer G et al (2004) Life cycle assessment Part 1: framework, goal and scope definition, inventory analysis, and applications. Environ Int 30(5):701–720

Study on the Machinability Characteristics of Inconel 718 Super Alloy During Micro-Milling

Branislav Sredanovic, Globocki Lakic, Davorin Kramar and Janez Kopac

Abstract The increase of demands for products miniaturization has led to a need to research and explore the possibilities of micro-machining of special alloys. This paper presents the experimental study on the micro-milling of Inconel 718 super alloy. It is austenitic nickel-chromium based material which is oxidation resistant. Inconel 718 super alloy is used in the aerospace, automotive and energetic device industry. It applies for extreme environments subjected to high pressure and high temperatures. In this study, mentioned super alloy is machined with long neck micro-end-mill with diameter 0.6 mm, which is intended for side and channel milling of high aspect ratio features. The effects of cutting parameters on output machinability parameters were monitored. During experimental research, there are made conclusions about surface roughness, build-up on edges, cutting forces and etc. For purposes of practical application, general indicative data, area of cutting parameter values, and guidelines for micro-machining of Inconel 718 are given.

Keywords Micro-milling · Inconel 718 · Super-alloy · Analyzing

B. Sredanovic (✉) · G. Lakic
Faculty of Mechanical Engineering, University of Banja Luka,
Vojvode Stepe Stepanovica 71, 78000 Banja Luka, Bosnia and Herzegovina
e-mail: sredanovic@gmail.com

G. Lakic
e-mail: gordana.globocki@gmail.com

D. Kramar · J. Kopac
Faculty of Mechanical Engineering, University of Ljubljana, Askerceva 6,
1000 Ljubljana, Slovenia
e-mail: davorin.kramar@fs.uni-lj.si

J. Kopac
e-mail: Janez.Kopac@fs.uni-lj.si

V. Majstorovic and Z. Jakovljevic (eds.), *Proceedings of 5th International Conference on Advanced Manufacturing Engineering and Technologies*,
Lecture Notes in Mechanical Engineering, DOI 10.1007/978-3-319-56430-2_28

1 Introduction

Micro-manufacturing experienced a very strong development in the last decade, along with the development of medicine, automotive industry, aero and space industry, energetics, optics, home devices industry and etc. The best indicator of micro-component industry development is the total value of production at the global level. In 2014, the amounted value is to 13.7 billion US dollars, while for 2018 foresees a jump to 22.5 billion US dollars. The biggest growth of micro-component industry, in amounted value 22.8% is in the medical industry, while the machinery industry predicts a growth of 11.5%. Micro-manufacturing develops in several fields: development of machines, development of manipulation systems, development of tools, development of processes, and etc. The introduction of this developments leads to the application of the whole spectrum of new principles, concepts and resources. Micro-manufacturing are divided into two technology types: microsystem technologies (MST) and micro-engineering technologies (MET). The first refers to production of components at micro-electro-mechanical systems (MEMS) with special processes. The second type relates to the part production with common machining technology (electro discharge machining, laser beam machining, electron beam machining, cutting, forming, casting and etc.).

Micro-engineering technologies can be used for the production of high precision parts, micro-parts and micro-features [1, 2]. With MET, can be processed different materials, but the highest percentage refers to metals and metal alloys. Steels and special alloys, which belong to the hard-to-machine materials group, are used for producing of functional parts subjected to high mechanical and thermal loads [3, 4]. Micro-milling is milling with cutting tools smaller than 1000 μm, or small structures which one or two perpendicular dimensions is up to 1000 μm [1–4].

In this paper, focus is on micro-machining of nickel-chromium Inconel 718 super-alloy, because of its excellent mechanical properties. Also, there are neediness for producing of high aspect ratio micro-parts. Nowadays, super-alloy micro-parts are machined in use of non-conventional processes such are laser beam machining (LBM) and electro-discharge machining (EDM). Using of this processes allows machining of 2D or simple 3D structures. There are great impact of generated heat, problems with machined surface integrity, long machining time, and increasing of production cost. The aim of this research is analysis of cutting parameters influencing and conclusions about the cutting conditions that allow the milling of Inconel 718. The milling process is carried out with flat-end-mill cuter with long-neck and diameter 0.6 mm and minimal quantity lubrication (MQL).

Investigation on solid-tool based machining of Inconel 718 super-alloy has shown difficulties in achieving of acceptable surface integrity and tool wear [5]. In studies [6] and [7], Ucun et al. investigated effect of cutting parameters, coating material and the built-up edge on the surface roughness in micro-milling of Inconel 718, with flat end mill with diameter 760 μm and 1.6 mm long tool neck, and concluded that coated tools are given the best surface roughness performance.

The same authors [8], investigated dry cutting process, minimum quantity lubrication and cryogenic pre-cooling as the cutting conditions in micro-milling of Inconel 718. In papers [9] and [10] is shown similar research on Inconel 718 machining but on macro level cutting process. Rahim et al. [11] used laser assisted micro-milling of Inconel 718. Author used FEM method for modelling of machining process. Kuram et al. [12] investigated the influences of cutting parameters on surface roughness and cutting forces, in micro-milling of Ti6Al4V alloy and Inconel 718, with 0.8 mm diameter flat-end mill cutter. Lu et al. [13] performed comparison between experimental and simulation data and evaluated residual stress on Inconel 718, in using of different cutting conditions. In [14] authors investigated the influencing of cutting condition and tool coats on output parameters in micro-milling of Inconel 718. Lu et al. [15] researched and modeled cutting forces during nickel based alloy micro-milling. Thepsonthi and Ozel [16] conducted experiments and FEM modelling of the micro-milling of Ti6Al4V alloy by cBN coated tool. Jaffery et al. [17] investigated process parameter influence on machinability of Ti6Al4V micro-milling. Biermann et al. [18] investigated of cutting parameter influence on micro-milling of NiTi alloys. Influence of cooling and lubrication techniques are investigated by Vazquez et al. [19]. Similar researches are performed in [20].

2 Experimental Setup

In experiment was used the high speed milling center Sodick MC430L (Fig. 1) with tree axis control, hybrid bearing high speed spindle (maximum spindle revolution of 40.000 min^{-1}), and HSK-25 tool holder mounting. The machine has linear magnetic motors for feed motion with resolution of 100 nm and accelerate up to 1 G. Machine is equipped with touch probe workpiece setting devices, BLUM Micro automatic tool compensation devices, and MQL system.

Tool was two flute flat-end-mill by SECO, TiAlN layer coated. It has diameter is 600 μm, corner radius 0.05 mm, and extra-long neck 8 mm. Helix revolution angle is 7.25° and height 0.8 mm. Tool mounting diameter is 3 mm. Experiment was performed in using of MQL. Workpiece was nickel-chromium based Inconel 718 supper-alloy, with oxidation and high temperature resistance. This super-alloy has tensile strength 1350 MPa and hardness 40 HRc. Used Inconel 718 super-alloy consists: 53% Ni, 19% Cr, 18% Fe, 5.1% Nb, 3% Mo, 1.1% Ti, 1% Co, 0.5% Al, 0.35% Si and 0.08% C. Workpiece dimensions was 50 × 20. Each channel on upper workpiece surface was machined with different combination of cutting parameters.

Measuring chain, with piezoelectric three-component dynamometer KISTLER 9257B, was used to measure three component cutting forces. For the data processing, LabVIEW software with specially designed software framework was used. The framework enables data transmission to software MATLAB. Surface roughness, tool edge wear and channel depth was measured with ALICONA

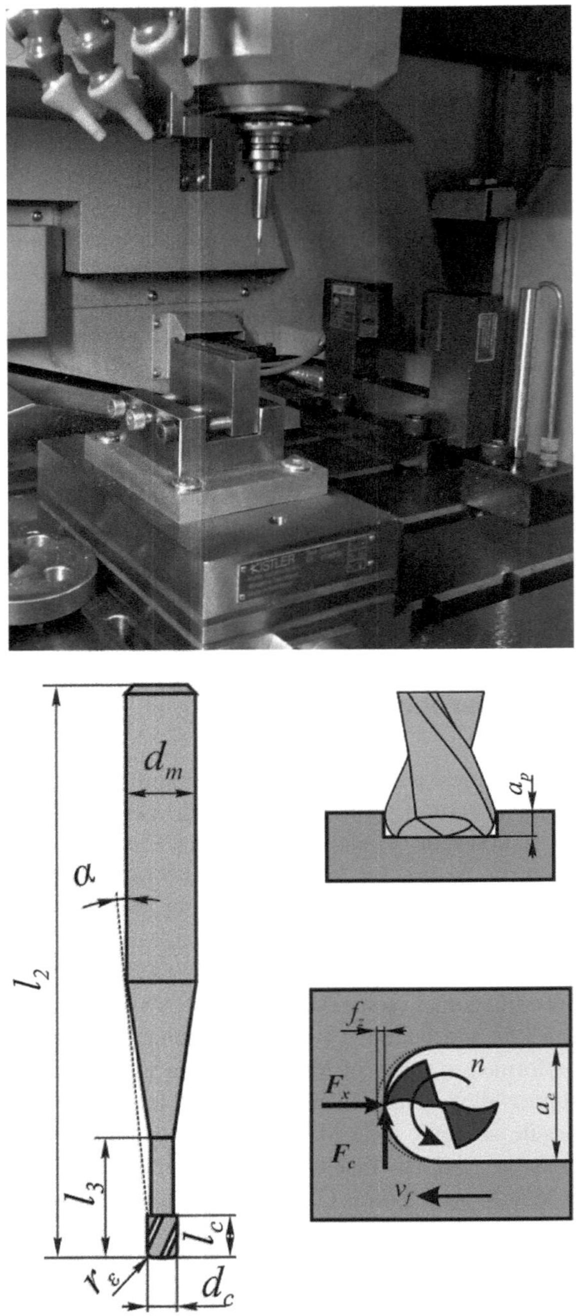

Fig. 1 Experimental setup and parameter description

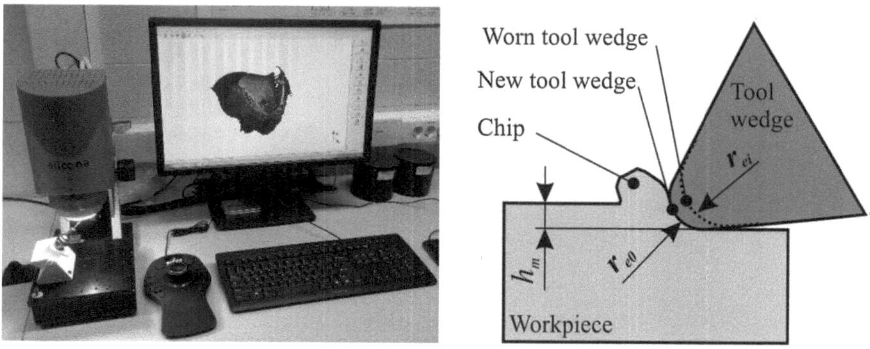

Fig. 2 ALICONA measuring device (*left*) and principle of cutting wedge wear (*right*)

InfiniteFocus. It is optical 3D non-contact measuring device for shape form and surface roughness measurement (Fig. 2). Measurement was performed by photo recording of horizontal layers on every 10 nm, and proceeds the sharpest parts of captured image. After that, software made a complete record of 3D surface relief.

3 Results and Discussion

In this section was shown results of experimental research. Results was shown in connected two parts. First part shows study on cutting forces, channel depths and surface quality for nine different combinations of feed and depth of cutting, and constant cutting speed. In this part, cutting speed is medium value level of second part of study. Second part shows study on tool wear for three different cutting speed, and constant feed and depth of cutting. In this part, feed and depth of cutting are medium value level of first part of study.

Values of cutting force component are shown on Fig. 3. This values represents average value of measured forces signals during milling of channel. Cutting forces were measured with sampling rate 3.6 kHz (about 10 samples per cycles). During measuring amplifier was set to measuring range from −10 to 10 N. In this experimental setup feed cutting force F_f was measured along machined channels, main cutting force F_c was measured along chip separation line. Penetration cutting force F_p was measured, but values was up to 0.5 N for all parameter combinations. Cutting force components were measured for different combination of feed per tooth f_z and different depth of cutting a_p, but cutting speed was constant v_c = 40 m/min. Based on values showed on diagraph can be concluded that cutting forces increase with increasing of feed and depth of cutting. All force components increases with increasing of feed by the linear dependence. In other hand, increasing of depth of cutting leads to non-linearity increasing of cutting force component values. Component F_c has higher values that component F_c, according theoretical assertion.

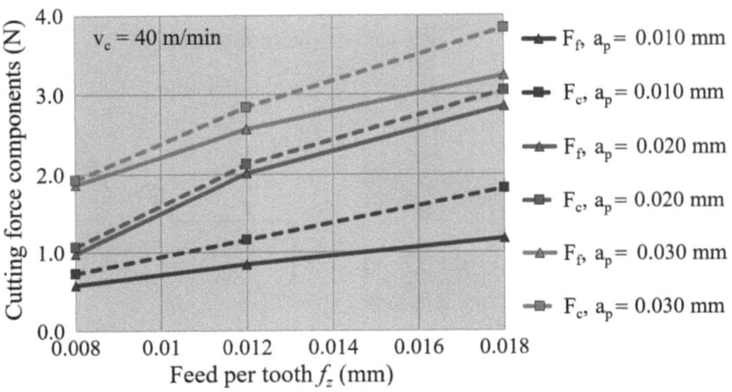

Fig. 3 Values of cutting force components F_f and F_c (for cutting speed v_c = 40 m/min)

Channel depth deviation Δh is difference between theoretical channel depth and channel depth measured after milling. For each channel, depth was measured on four different sections of each channel. For each section, measuring was conducted perpendicularly on channel direction. Measuring data was shown that measured values deviates maximum 5% from average value for all channels measuring. Average value of channel depth deviation for different cutting parameters combination is shown on Fig. 4. In case of channel machining, cutting force component F_f is the main reason for tool deflection which leads to channel depth deviation. Can be concluded that the value channel depth deviations follow the increasing of cutting force component value F_c and F_f, which shown in the previous figure. Values of deviation for feed 0.012 mm and different cutting depths is 0.43, 0.53 and 0.62 μm respectively, which is a difference of 20% approximately. In case of using of micro-mills with long neck and small diameter, can be concluded great sensitivity on increasing of cutting forces, which leads to intensively tool deflections. Tool deflection reduces the workpieces dimensions accuracy. In micro-milling, this can lead to decrease of surface quality also.

Fig. 4 Channel depth deviations for different cutting parameter combinations (v_c = 40 m/min)

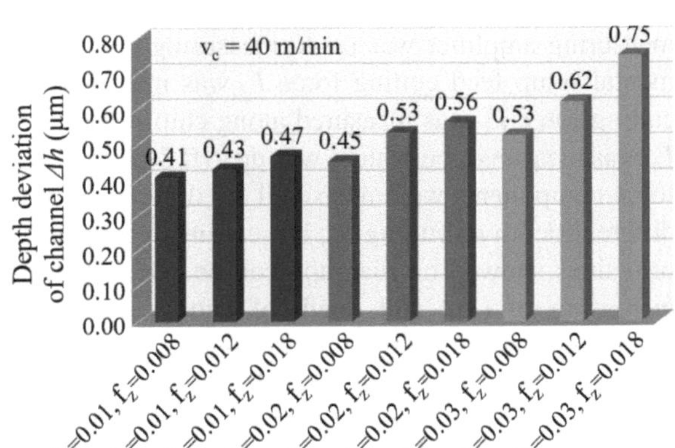

The measuring and evaluation of surface quality parameters in micro-machining technologies is very complicated. In other hand, surface roughness is one of the most imported machinability parameter in micro-technologies. In this paper was used non-contact principle of surface roughness measuring. In using of ALICONA InfiniteFocus device, there was obtained digital scan of machined relief, suitable for software processing (Fig. 5). Digital model of surface allow measuring roughness of machined surface and deviations of dimensions.

Scanned channels is shown on Fig. 6. There is shown scanned channels machined by different cutting parameter combinations, same in cutting forces measuring. Can be concluded that surface quality decrees with increased of feed and depth of cutting.

Fig. 5 Results of software data processing with ALICONA device (magnification 20×)

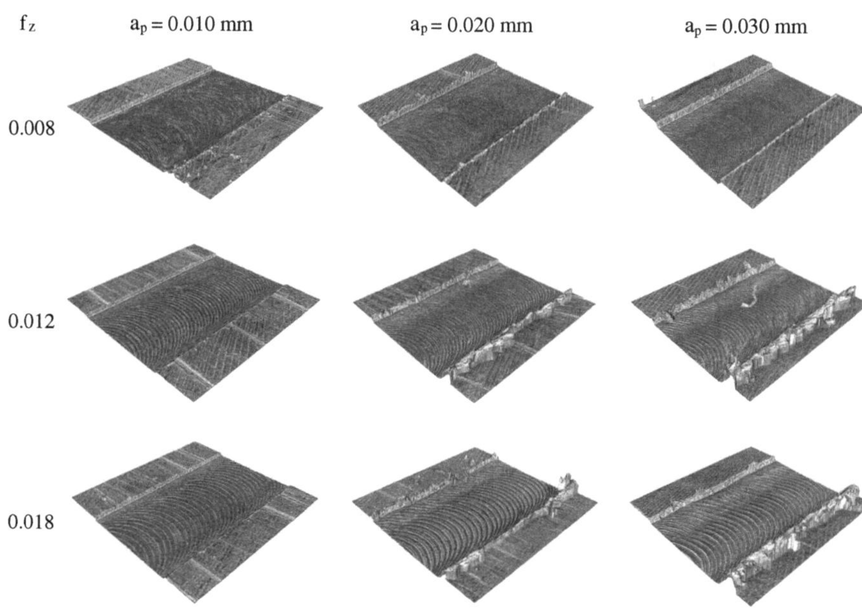

Fig. 6 Scanned channels with ALICONA device (magnification 20×)

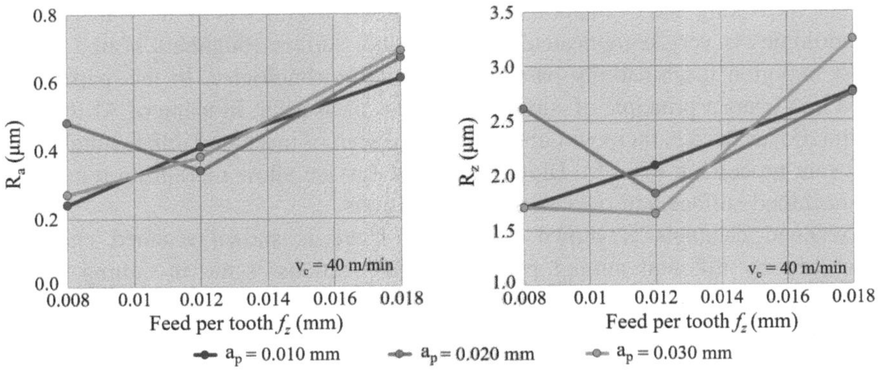

Fig. 7 Surface roughness parameter values for different parameter combinations (v_c = 40 m/min)

In Fig. 7 is shown influence of feed and depth of cutting on surface roughness parameters: average surface roughness (R_a) and mean peak to valley height of roughness profile (R_z). This parameters was measured on four different points in each channel. On each point in channel, reference length was set along the channel. Concluded and shown values shown on graph, are the average values for each channel. In this measuring procedures, maximum error was 8.5%, and maximum relative error was 4.3%, which given on measuring of channel which machined with a_p = 0.030 mm and f_z = 0.012 mm.

Can be concluded that surface roughness increase with increase of cutting parameters, except for low cutting parameter values. High values of surface roughness for lower values of depth of cutting and feed were caused by phenomenon of "size effect". This phenomenon is caused by proportions of the thickness of the affected workpiece material by cutting wedge and cutting wedge radius, because edge is not ideal sharp. In this study, cutting wedge on new tool was evaluated as round shape with average radius 8.5 µm. Combination of feed 0.008 mm and depth of cutting 0.01 mm is given high values of surface roughness that in other combinations. In this case, workpiece material is not cut but it is plugged.

Build up is one of the important quality parameter in micro-machining. Values of it parameter is relative high for tough materials, such as Inconel 718. Measuring procedures for build-up edge (BUE) evaluation was consist measuring on four different sections on each channel, and calculation of average value. Average values BUE for each channel, machined by different cutting parameters, is shown of Fig. 8. Build up on cutting edge is higher on down milling side of channel because of geometric properties of process.

Tool wear in micro-milling is evaluated by cutting tool diameter reduction, or tool wedge radius increasing. Tool wedge radius increasing is important parameter, because it causes size effect phenomenon, which leads worse output cutting parameters, such are surface quality and accuracy. Tool wear measuring was performed for different cutting speeds, but constant feed and depth of cutting,

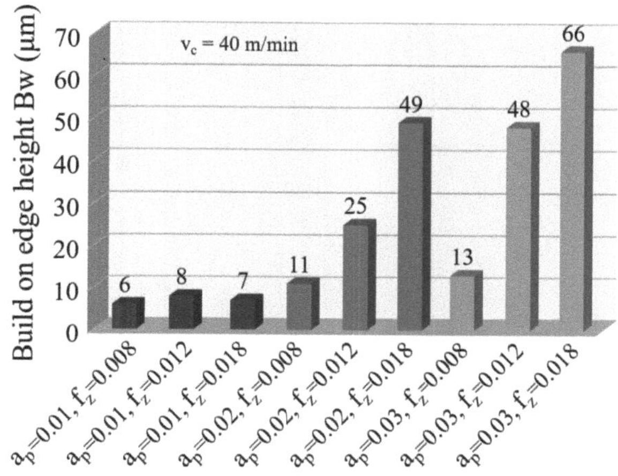

Fig. 8 Build-up on channel edge for different cutting parameter combinations

because cutting speed was defined as most influential parameter in tool wear. Procedure for tool wear measuring was consist machining of new lonest channels, process stopping on each 20 s, fast tool wedge scanning of ALICONA device, and continue of channel machining.

In Fig. 9 is shown increase of tool wedge round radius during micro-milling process with different cutting speed. Evaluation processes was stopped after 4 min. During tool wedge wear evaluation, surface roughness was measured at start and at end of procedure of each channel machining. With increasing of cutting speeds there was rapidly changing of tool wedge wear intensity. For cutting speed 25 and 40 m/min, wedge radius is increases from initial 8.5 μm to double value for short time. But for cutting speed of 55 m/min, for same machining time, it is increases for triple. It is caused by intensively generated heat during chip separation, which is

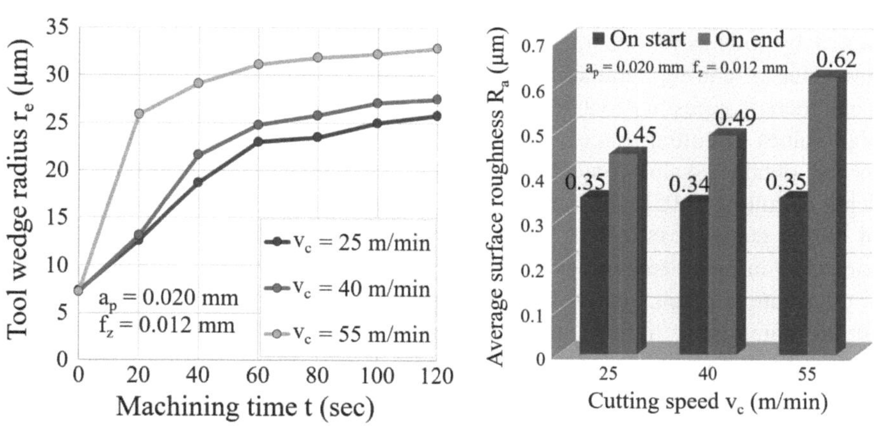

Fig. 9 Tool wedge wear increasing (*left*) and surface roughness on start and end of milling (*right*)

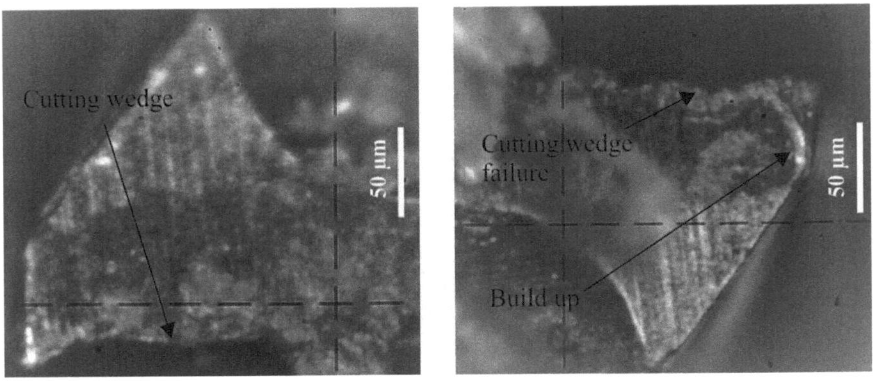

Fig. 10 Build-up on cutting wedge under optical microscope

very pronounced in Inconel 718 machining. Also it is caused by using of MQL techniques and the absence of intensive cooling. As mechanism of tool wedge wear is noticed as adhesive wear and abrasive wear (Fig. 10). This mechanism is caused by high percent of nickel-chromium in this super alloy.

4 Conclusions

In this paper study on machinability of Inconel 718 super alloy during micro-milling was performed. This is done in order to analyze the possibilities micro-machining of it. Study is realized by machining of micro-channels with long neck flat end mill with diameter 0.6 mm, in use of MQL technique of lubrication. For different combinations of cutting parameter there were measured output process parameters such are cutting forces, channel depth deviations, surface quality parameters and tool wear. Can be concluded that Inconel 718 super-alloy can be milled by long neck flat micro end mills. In this case, it is necessary to avoid low-value of cutting parameters because the phenomenon of size effect. In other hand, there is necessary to avoid high value of cutting parameters because it causes high values of cutting forces, surface roughness and tool wear. Based on experimental obtained values, can be concluded that combination of feed 0.012 mm, depth of cutting of 0.02 mm and cutting speed of 40 m/min gives satisfactory value of output cutting parameters, and productivity at least, which is very important parameter in micro-technology.

As future research, analyze of machinability criteria for different combination of cutting parameters, tool geometric parameters, and workpieces material will be performed. Base on experimental data, can be developed machinability database of hard-to-machine materials, models and performed optimization. Aim of future research is development of software application for obtaining of optimal cutting parameters in micro-milling of hard-to-machine materials.

References

1. Dornfeld D, Min S, Takeuchi Y (2006) Recent advanced in mechanical micromachining. Ann CIRP 55(2):745–768
2. Camara MA, Rubio JCC, Abrao AM, Davim JP (2012) State of the art on micromilling of materials, a review. J Mater Sci Technol 28(8):673–685
3. Bissacco G, Hansen HN, De Chiffre L (2006) Size effects on surface generation in micro milling of hardened tool steel. Ann CIRP 55(1):593–596
4. Aramcharoen A, Mativenga PT (2009) Size effect and tool geometry in micromilling of tool steel. Precis Eng 33:402–407
5. Dudzinski D, Devillez A, Moufki A, Larrouquere D, Zerrouki V, Vigneau J (2004) A review of developments towards dry and high speed machining of Inconel 718 alloy. Int J Mach Tool Manuf 44:439–456
6. Ucun I, Aslantas K, Gockce B, Bedir F (2014) Effect of tool coating materials on surface roughness in micromachining of Inconel 718 super alloy. Proc Inst Mech Eng B-J Eng 288 (12):1550–1562
7. Ucun I, Aslantas K, Bedir F (2015) The performance Of DLC-coated and uncoated ultra-fine carbide tools in micromilling of Inconel 718. Precis Eng 41:135–144
8. Ucun I, Aslantas K, Gockce B, Bedir F (2014) The effect of minimum quantity lubrication and cryogenic pre-cooling on cutting performance in the micro milling of Inconel 718. Proc Inst Mech Eng B-J Eng 229(12):2134–2143
9. Pusavec F, Hamdi H, Kopac J, Jawahir IS (2011) Surface integrity in cryogenic machining of nickel based alloy-Inconel 718. J Mater Process Technol 211(4):773–783
10. Wang ZY, Rajurkar KP, Fan J, Lei S, Shin YC, Petrescu G (2003) Hybrid machining of Inconel 718. Int J Mach Tool Manuf 43(13):1391–1396
11. Rahim EA, Warap NM, Mohid Z, Ibrahim R, Rafai N (2015) Numerical analysis of laser preheating for laser assisted micro milling. Appl Mech Mater 773–774:332–336
12. Kuram E, Ozcelik B (2017) Optimization of machining parameters during micro-milling of Ti6Al4V titanium alloy and Inconel 718 materials using Taguchi method. Proc Inst Mech Eng Part B: J Eng Manuf 231(2):228–242
13. Lu X, Lu Y, Wang F, Zhao C (2016) Research on surface residual stress of micro-milling nickel-based superalloy Inconel 718. Int J Nanomanuf 12(1):82–92
14. Lu X, Jia Z, Wang H, Si L, Wang X (2016) Surface roughness prediction model of micro-milling Inconel 718 with consideration of tool wear. Int J Nanomanuf 12(1):93–108
15. Lu X, Jia Z, Wang X, Li G, Ren Z (2015) Three dimensional dynamic cutting forces prediction model during micro-milling nickel based super alloy. Int J Adv Manuf Technol 81:2067–2086
16. Thepsonthi T, Ozel T (2013) Experimental and finite element simulation based investigations on micro-milling Ti-6Al-4V titanium alloy: effects of cBN coating on tool wear. J Mater Process Technol 213:532–542
17. Jaffery SHI, Khan M, Ali L, Mativenga PT (2016) Statistical analysis of process parameters in micromachining of Ti-6Al-4V alloy. Proc Inst Mech Eng Part B: J Eng Manuf 230(6):1017–1034
18. Biermann D, Kahleyss F, Krebs E, Upmeier T (2010) A study on micro-machining technology for the machining of NiTi: five-axis micro-milling and micro deep-hole drilling. J Mater Eng Perform 20(4–5):745–751
19. Vazquez E, Gomar J, Ciurana J, Rodríguez CA (2015) Analyzing effects of cooling and lubrication conditions in micromilling of Ti6Al4V. J Clean Prod 87:906–913
20. Prakash JRS, Rahman M, Kumar AS, Lim SC (2001) Effect of minimal quantities of lubricant in micro-milling. 10th international conference on precision engineering (ICPE) Yokohama, Japan. pp 309–313

A Multi-parameter Experimental and Statistical Analysis of Surface Texture in Turning of a New Aluminum Matrix Steel Particulate Composite

N.M. Vaxevanidis, N.A. Fountas, G.V. Seretis, C.G. Provatidis and D.E. Manolakos

Abstract Metal matrix composites (MMCs) represent a new generation of engineering materials in which a strong reinforcement is incorporated into a metal matrix to improve its properties including specific strength, specific stiffness, wear resistance, corrosion resistance and elastic modulus. Aluminum matrix composites (AMCs), a specific type of MMCs, are rapidly replacing conventional materials in various engineering applications, especially in the aerospace and automobile industries due to their attractive properties. From the literature already published it is evident that the machining of AMCs is an important area of research, but only very few if any studies have been carried out using metal particles reinforced AMCs. A multi-parameter analysis of surface finish imparted by turning to a new L316 stainless steel flake-reinforced aluminum matrix composite is presented. Surface finish is investigated by examining a number of surface texture parameters. Spindle speed as well as feed rate was treated as the independent variables under a constant depth of cut whilst roughness parameters were considered as the responses under an L9 orthogonal array experimental design. ANOVA analysis was also conducted to study the effect of the two cutting variables on the surface texture responses.

N.M. Vaxevanidis (✉) · N.A. Fountas
Laboratory of Manufacturing Processes and Machine Tools (LMProMaT), Department of Mechanical Engineering Educators, School of Pedagogical and Technological Education (ASPETE), ASPETE Campus, N. Heraklion, GR 14121 Athens, Greece
e-mail: vaxev@aspete.gr

G.V. Seretis · C.G. Provatidis
Section of Mechanical Design and Automatic Control, School of Mechanical Engineering, National Technical University of Athens (NTUA), Heroon Polytechniou 9, GR 15780 Zografou, Athens, Greece

D.E. Manolakos
Section of Manufacturing Technology, School of Mechanical Engineering, National Technical University of Athens (NTUA), Heroon Polytechniou 9, GR 15780 Zografou, Athens, Greece

V. Majstorovic and Z. Jakovljevic (eds.), *Proceedings of 5th International Conference on Advanced Manufacturing Engineering and Technologies*, Lecture Notes in Mechanical Engineering, DOI 10.1007/978-3-319-56430-2_29

Keywords Surface texture · Aluminum matrix particulate composite (AMPC) · Stainless steel flakes (SSF) · Turning · Multi-parameter analysis

1 Introduction

Metal matrix composites (MMCs) represent a new generation of engineering materials in which a strong reinforcement is incorporated into a metal matrix to improve its properties including specific strength, specific stiffness, wear resistance, corrosion resistance and elastic modulus [1]. Aluminum matrix composites (AMCs), a specific type of MMCs), are rapidly replacing conventional materials in various engineering applications, especially in the aerospace and automobile industries due to their attractive properties. Since further improvement of aluminum alloys seems to be hardly achieved by conventional technologies, the reinforcement material has to play the key role in maximizing the mechanical performance of the composites. The use of carbide, oxide or diamond-structure additives for fabrication of aluminum matrix particulate composites (AMPCs) have demonstrated positive effects on the mechanical properties of the composites [2]. Up-to now, the most popular reinforcements are silicon carbide (SiC) and alumina (Al_2O_3) whilst aluminum, titanium, and magnesium alloys are commonly used as the matrix phase [3–5]. On the other hand, metal additives have not been widely investigated yet as reinforcements in AMPCs [6].

From the literature already published it is evident that the machining of AMCs is an important area of research [7–9], but only very few—if any—studies have been carried out using metal particles reinforced AMPCs [10]. Metal cutting operations are widespread in manufacturing industry and the prediction and/or the control of relevant machinability parameters always attracts interest. One basic machinability parameter is the surface texture, since it is closely associated with the quality, reliability and functional performance of components. Turning is the primary operation in metalworking industry for producing axisymmetric components. These components, typically, possesses critical features that require specific surface finish and the best possible functional behaviour. Due to inadequate knowledge of the complexity of the process and factors affecting the surface integrity in turning operation, an improper decision may cause high production costs and low machining quality. The proper selection of cutting tools and process parameters for achieving high cutting performance in a turning operation is, therefore, a critical task [11–13].

The present study concerns with a multi-parameter experimental and statistical analysis of surface texture in turning of stainless steel flakes (SSF) reinforced cast aluminum matrix particulate composite (AMPC). For comparison purposes, rods of the matrix material (without reinforcement) were tested with the same machining conditions. Rotational speed, n (rpm) and feed rate, f (mm/rev) were assigned to an L9 Taguchi orthogonal array as independent variables whilst depth of cut was kept constant. The desired response (performance characteristic) was an increased number of surface roughness parameters including amplitude, hybrid, and random ones. The multi-parameter analysis of surface finish was selected since the evaluation of the roughness with one or two only parameter is ambiguous [14]. The correlation of these parameters with the machining conditions was investigated. Then, statistical analysis and ANOVA was applied to quantitatively allow exploration of the effect which each machining input yields on roughness outputs. Regression analysis followed next, aiming at providing a mathematical model capable of correlating the experimental data with low estimation error; a full quadratic prediction model—as a direct outcome of ANOVA analysis—is proposed for this purpose.

2 Surface Roughness Parameters

Surface roughness evaluation is very important for many fundamental phenomena such as friction, contact deformation, heat and electric current conduction and positional accuracy. For this reason surface roughness has been the subject of experimental and theoretical investigations for many decades. The real surface geometry is so complicated that a finite number of parameters cannot provide a full description. If the number of parameters used is increased, a more accurate description can be obtained. This is the main reason for introducing new parameters for surface evaluation [15]. More than one hundred parameters have been proposed in the literature, owing to the fact that a real profile shape is usually very complicated, and the functional requirements for the components are very high in today's industry [14]. Since the estimation of the roughness performed with one parameter is ambiguous, the multi-parameter estimation of roughness is recommended [14–18]. Surface roughness parameters are normally categorized into three groups according to its functionality. These groups are defined as amplitude parameters, spacing parameters, and hybrid parameters [15]. The surface roughness parameters under study were:

- the arithmetic average, R_a.
  Arithmetic average roughness (R_a) is by far the most commonly used parameter in surface finish measurement and for general quality control. Despite its inherent limitations, it is easy to measure and offers a good overall description of the height characteristics of a surface profile.
- the maximum height of the profile; R_t or R_{max}.
  This parameter is very sensitive to the high peaks or deep scratches. R_t is defined as the vertical distance between the highest peak and the lowest valley along the assessment length of the profile.
- the skewness of the amplitude distribution, R_{sk}.
  Skewness parameter (R_{sk}) is typically used to measure the symmetry of the profile about the mean line and is sensitive to deep valleys or high peaks. A symmetrical height distribution, i.e. with as many peaks as valleys, has zero

skewness. Profiles with peaks removed or deep scratches have negative skewness. Profiles with valleys filled in or high peaks have positive skewness.

- the kurtosis of the amplitude distribution, R_{ku}.
  Kurtosis (R_{ku}) typically describes the sharpness of the probability density of the profile. If $R_{ku} < 3$ the distribution curve is said to be platykurtoic and has relatively few high peaks and low valleys. If $R_{ku} > 3$ the distribution curve is said to be leptokurtoic and has relatively many high peaks and low valleys.

3 Experimental

In this study, Al–Mn matrix alloy containing Mg was used as matrix material. The purity of the matrix alloy was 96.6%. Stainless steel 316L flakes (SSF) were used as reinforcing fillers. Al–Mn/SSF composite rods with SSF content equal to 0.25% wt were produced. For comparison, rods of pure matrix aluminum alloy (Al–Mn), with no reinforcement were produced as well. The fabrication of the aluminum/stainless steel flakes composites was reported in [6]; therefore it s not presented here.

Due to their non-uniform roundness, the as-cast rods were pre-machined on a conventional lathe so as to prepare the measuring zones for roughness tests. Figure 1a illustrates one of the two as-cast rods produced whilst in Fig. 1b a typical Ø20 rod prepared for cutting experiments is shown.

Cutting experiments were performed on a Mondiale Gallic 14 conventional lathe equipped with a NC positioning unit. The ISO 6 L1616 P20-grade brazed carbide cutting tool (PVD TiN coated) from Sandvic Coromant® with 0.8 tip radius for continuous turning was employed. Coolant was not used during the tests.

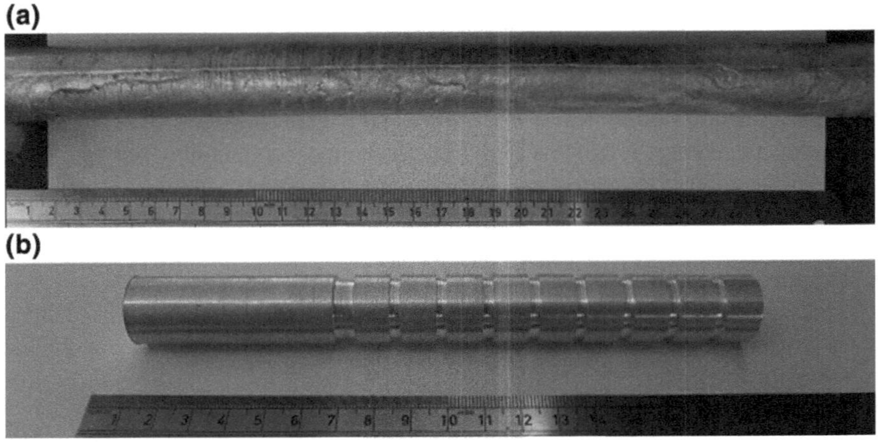

Fig. 1 **a** As-cast Al alloy/SSF rod; **b** rod prepared for cutting experiments

Spindle speed and feed rate were treated as the independent machining parameters whilst roughness parameters including R_a, R_t, R_{sk} and R_{ku} were treated as the responses on an L9 Taguchi orthogonal array design. The depth of cut was kept constant for all turning passes at a = 1.0 mm.

The surface roughness analysis was performed using a Rank Taylor-Hobson® Surtronic 3 profilometer equipped with the Talyprof® software. The cut-off length was selected at 0.8 mm whilst three measurements were conducted on every pass at the longitudinal direction. A typical filtered profile of a turned surface is presented in Fig. 2. Measured average values for all surface roughness parameters under study together with the corresponding cutting variables are tabulated in Table 1 for both the matrix material (Al–Mn alloy) and the Al–SSF composite.

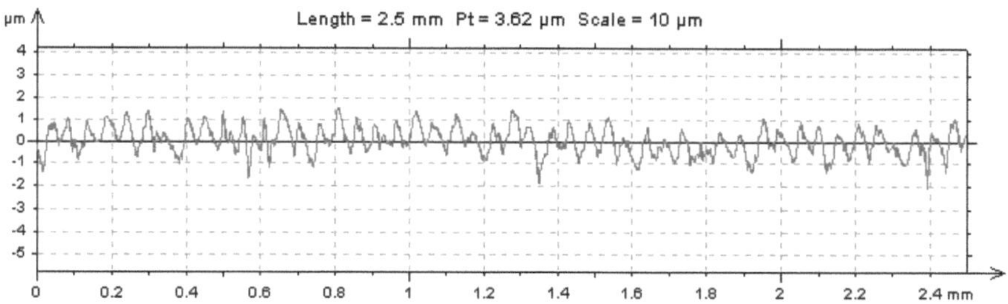

Fig. 2 Roughness profile of a turned surface (n = 200 rpm, f = 0.05 mm/rev, a = 1.0 mm)

Table 1 Measured values of surface roughness parameters for Al–Mn matrix material and Al–SSF composite

Run	Machining parameters		Responses (surface roughness parameters)							
	n (rpm)	f (mm/rev)	R_a (μm)		R_t (μm)		R_{sk}		R_{ku}	
			Al–Mn	Al–SSF	Al–Mn	Al–SSF	Al–Mn	Al–SSF	Al–Mn	Al–SSF
1	200	0.05	0.55	1.86	4.06	12.88	−0.32	0.11	3.22	2.92
2	200	0.27	5.44	8.43	31.33	46.53	−0.15	0.52	2.59	2.79
3	200	0.60	13.32	8.57	66.73	44.10	0.80	0.27	2.81	2.46
4	800	0.05	0.626	0.65	4.82	4.85	−0.01	−0.21	3.48	3.10
5	800	0.27	6.30	8.21	36.5	45.67	0.00	0.04	2.62	2.48
6	800	0.60	7.99	7.77	48.47	47.00	0.067	0.54	2.75	3.67
7	1600	0.05	0.75	0.88	5.30	6.54	−0.26	−0.19	2.84	2.84
8	1600	0.27	9.80	7.96	57.13	50.13	0.11	0.38	2.89	3.33
9	1600	0.60	8.45	8.50	47.03	51.50	0.37	0.44	2.46	3.33

4 Results and Discussion

4.1 Variation of Surface Roughness Parameters

The variation of the surface roughness parameters with the independent machining parameters, namely, spindle speed n, (rpm) and feed rate f, (mm/rev) for both Al–Mn an Al–SSF materials, are shown in Figs. 3 and 4 for R_a, R_t and R_{sk}, R_{ku} respectively. Note that results for materials are depicted in the same experimental graph using a continuous trend line for Al–Mn matrix material and a dashed trend line for Al–SSF composite.

In the case of pure Al–Mn material the main observations are: Ra increases almost linearly when increasing feed rate for a fixed low spindle speed value; i.e. n = 200 rpm. However, for the same range of feeds, R_a tends to be reduced as spindle speed increases. Experimental observations for R_t are quite similar to those for R_a, as it is expected, whereas R_{sk} changes its increasing trend when increasing feed rate, from almost linear (see R_a) to clearly exponential for the lowest spindle speed set. For the middle level set for spindle speed (n = 800 rpm) the measurement range for R_{sk} gets narrower as it happens to the one corresponding to R_a. Then, when it comes to the highest spindle speed level of 1600 rpm, the R_{sk} exponential trend changes from a probability density (convex region on top) to a cumulative probability (convex region on bottom). Kurtosis of the profile (R_{ku}) tends to get lower for middle feeds under the implementation of low and middle spindle speeds (i.e. 200 and 800 rpm respectively) whilst it reaches its lowest value for the highest feed rate level combined with the high spindle speed values (i.e. 1600 rpm).

In the case of Al–SSF composite material the main observations are: When turning under the lowest level for feed rate (i.e. f = 0.05 mm/rev), R_a is kept low for all spindle speed levels, therefore it is deemed as the most influential parameter affecting R_a. Lowest value for R_a was obtained for spindle speed equal to 800 rpm with 0.05 mm/rev feed rate. For the same feed, R_a slightly increases when spindle reaches 1600 rpm. Experimental observations for R_t are quite similar to those for R_a, as it is expected. R_{sk} reaches its highest value when turning the Al–SSF composite using a spindle speed equal to 800 rpm with a feed rate equal to 0.27 mm/rev.

4.2 Chip Formation and Tool Wear

Chip formation involves the plastic deformation of the shear zone in machining. Typically, in metal cutting the shear zone can be analyzed based on the thin-plane model or the thick-plane model. At higher cutting speeds the shear zone approximates to the thin-plane model which allows for easy determination of the shear zone angle. In practice MMCs are machined at relatively high-cutting speeds and

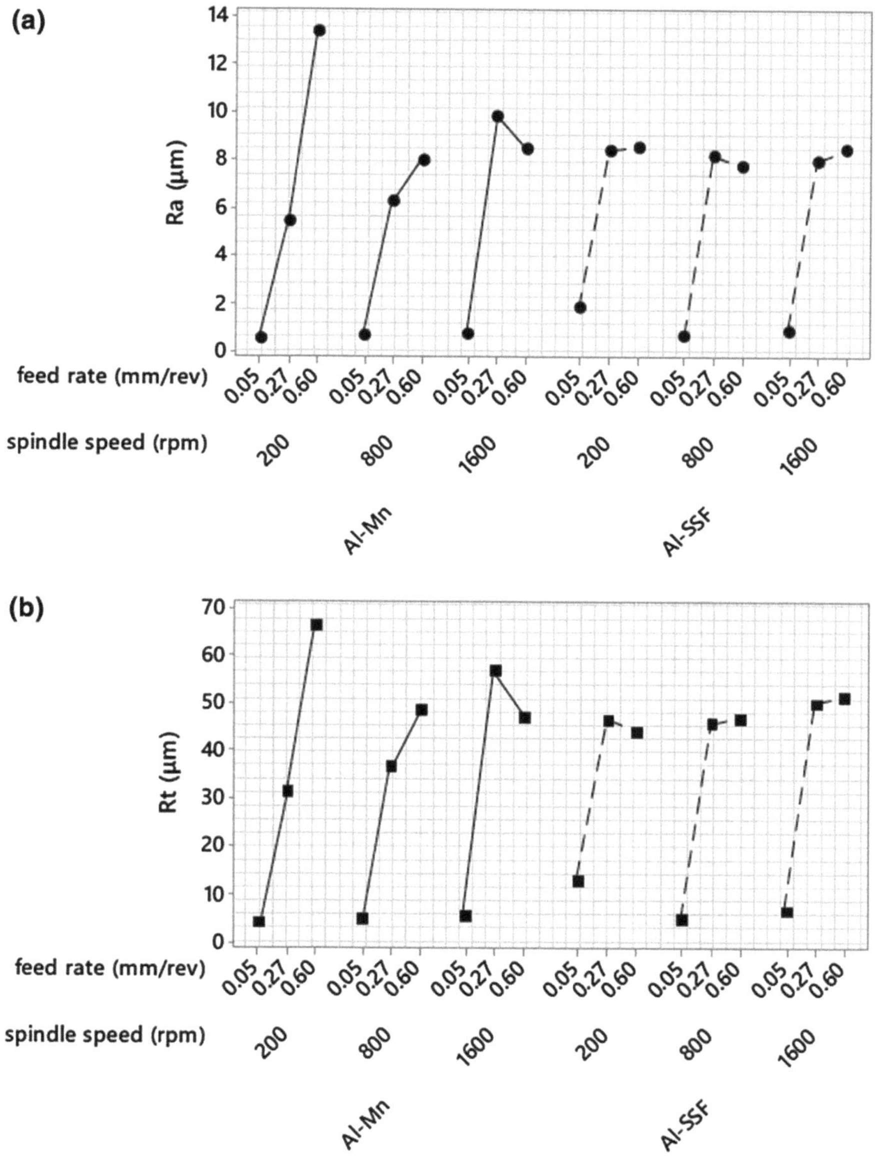

Fig. 3 Roughness parameters versus feed rate and spindle speed for Al–Mn and Al-SSF composite: **a** R_a; **b** R_t (—■— Al–Mn; - - -■- - - Al–SSF composite)

therefore the approximation to a thin-plane model is representative of the cutting process. On the other hand, the formation of chips cannot be only related to the nature of the shear zone but is also related to the material properties, microstructure and instabilities in the cutting process [19].

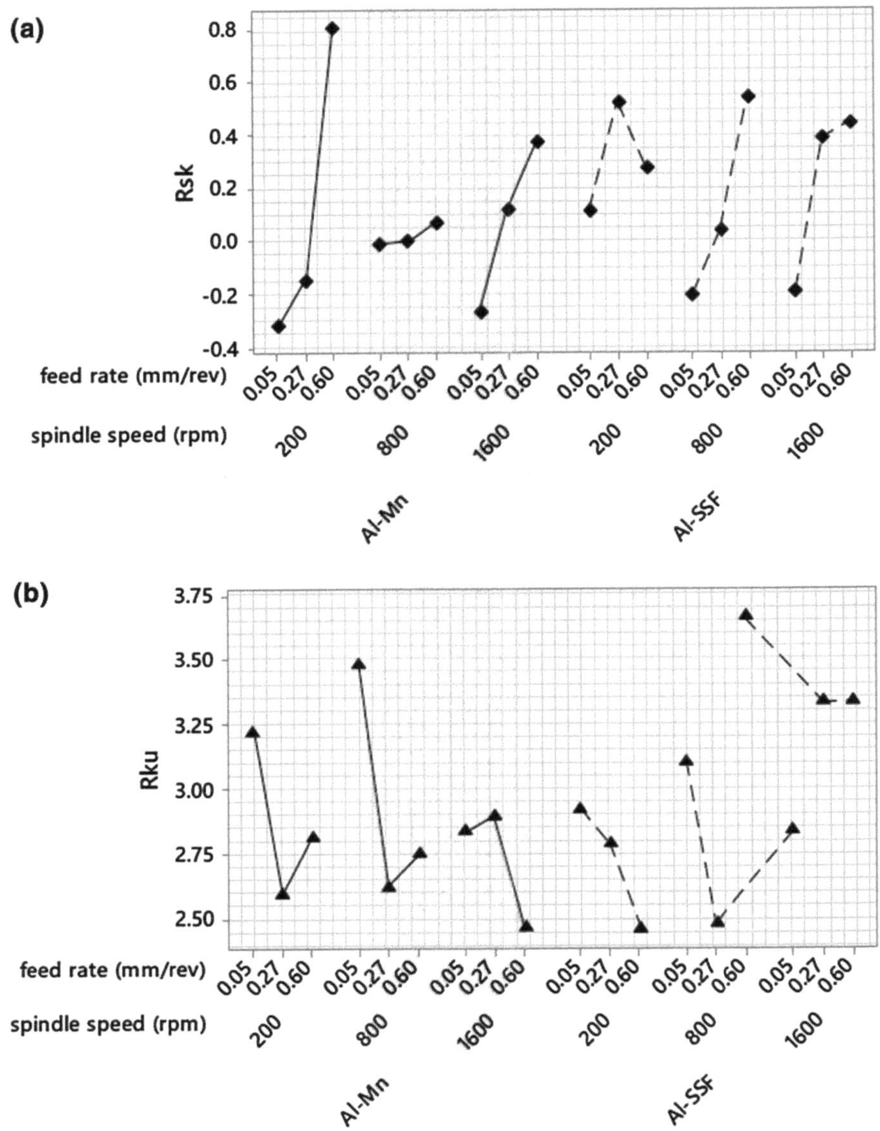

Fig. 4 Roughness parameters versus feed rate and spindle speed for Al–Mn matrix and Al–SSF composite: **a** R_{sk}; **b** R_{ku} (—■— Al–Mn; - - -■- - - Al–SSF composite)

As far as the turning of particulate reinforced MMC is concerned, it is evident, from the available literature, that the reinforcement material, type of reinforcement (particle or whisker), volume fraction of the reinforcement, and matrix properties as well as the distribution of these particles in the matrix are the main factors that affect chip formation and crater wear; thus the overall machinability of these composites.

In general, the reports on the effect of machining parameters on chip formation and tool wear (types and mechanisms) when turning MMPC are rather contradictory. It has been stated that chip formation mechanism of metal matrix composites resembles, if not replicates, the behavior of monolithic materials [20]. When aluminum is the primary matrix material, the occurrence of BUE is expected and has been noted by many authors [3, 9].

During present research, for the case of Al–Mn matrix turning, no special features were observed than those reported in the literature. It should be noted that significant built-up edge was observed mainly when operating under high speeds and low feeds. The film of material producing the built-up edge was radically gauged away by the chip produced without leaving a noticeable tool wear; but a thin film of aluminum which seemed to play the role of an adhering layer; see Fig. 5.

Unlike Al–Mn matrix case, built-up edge was not observed during the turning of the Al–SSF composite owing, probably, to the tendency of stainless steel flakes to remove the built-up edge, leaving only small scratches on the aluminum alloy film. Figure 6 illustrates the physical appearance of chips produced during the nine experimental runs according to Table 2.

The most common types of chips observed were serrated and discontinuous chips combined, sometimes, with sawtooth-type chips. The chip formation mechanism is accompanied by a severe plastic deformation of the shear zone. The addition of the SS flakes in the aluminum alloy reduces the ductility of the material

Fig. 5 Built-up edge appearance on the tool tip of the carbide brazed cutting insert (Al–Mn matrix)

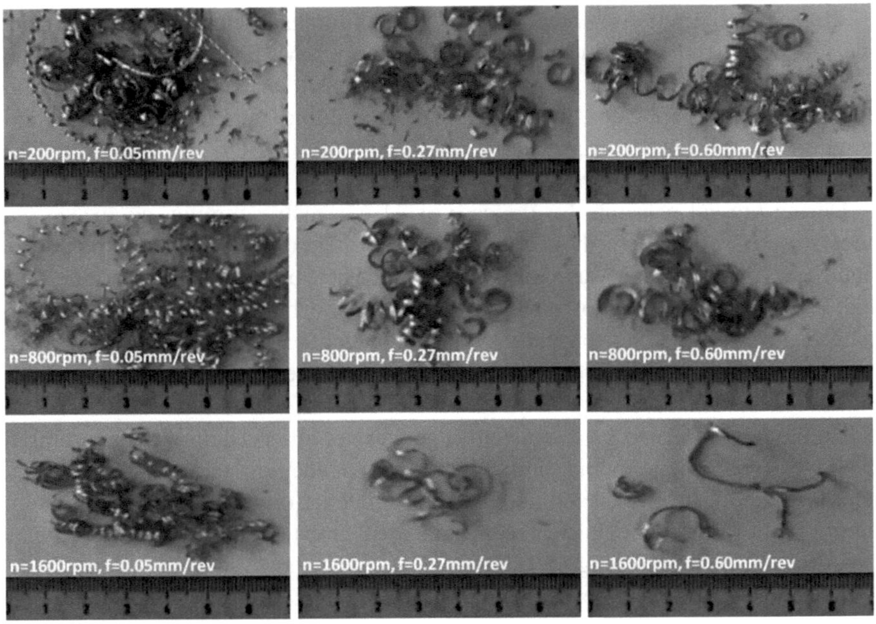

Fig. 6 Physical appearance of chips produced during turning of Al–SSF composite

contributing to the segmental chip. Chip breakability seems to be improved due to the presence of the reinforcement particles in the AMC; see also similar remarks in Refs. [19, 21].

4.3 *Statistical Analysis and Regression*

Statistical analysis was performed in order to examine the influence of cutting conditions upon all four surface roughness parameters. MINITAB® 17 software was employed to perform the analysis and to obtain the necessary outputs for results interpretation. By considering the results in terms of main effects referring to both materials studied and for all four surface roughness parameters; arithmetic roughness average R_a, is minimum at the second level of spindle speed—n (rpm) as well as at the first level of feed rate—f (mm/rev); see Fig. 7a, b. The maximum height of the profile R_t, follows a similar trend in terms of machining parameter effects as the one indicated in R_a.

The R_{sk} parameter in the case of Al–Mn, is favored by employing the second level of spindle speed—n (rpm) as well as at the first level of feed rate—f (mm/rev); see Fig. 8a; thus, these levels represent the optimal conditions according to the experimental design. The same behavior is also observed for the case of Al–SSF turning where R_{sk} is minimized for the second level of spindle speed—n (rpm) and

Table 2 Full quadratic regression equations for predicting roughness parameters of Al–Mn material

Response	Regression model	R^2 (%)
R_a	$= -1.81 - 0.00318 * n + 48.5 * f + 3 * 10^{-5} * n^2 - 39 * f^2 - 0.00725 * n * f$	90.16
R_t	$= -11.6 - 79 * 10^{-2} * n + 271.0 * f + 10^{-5} * n^2 - 237 * f^2 - 0.0316 * n * f$	91.29
R_{sk}	$= -0.290 - 10^{-4} * n + 1.12 * f + 0.79 * f^2 - 6 * 10^{-4} * n * f$	69.32
R_{ku}	$= 3.287 + 4 * 10^{-4} * n - 3.4 * f + 3.8 * f^2 - 3 * 10^{-5} * n * f$	68.39

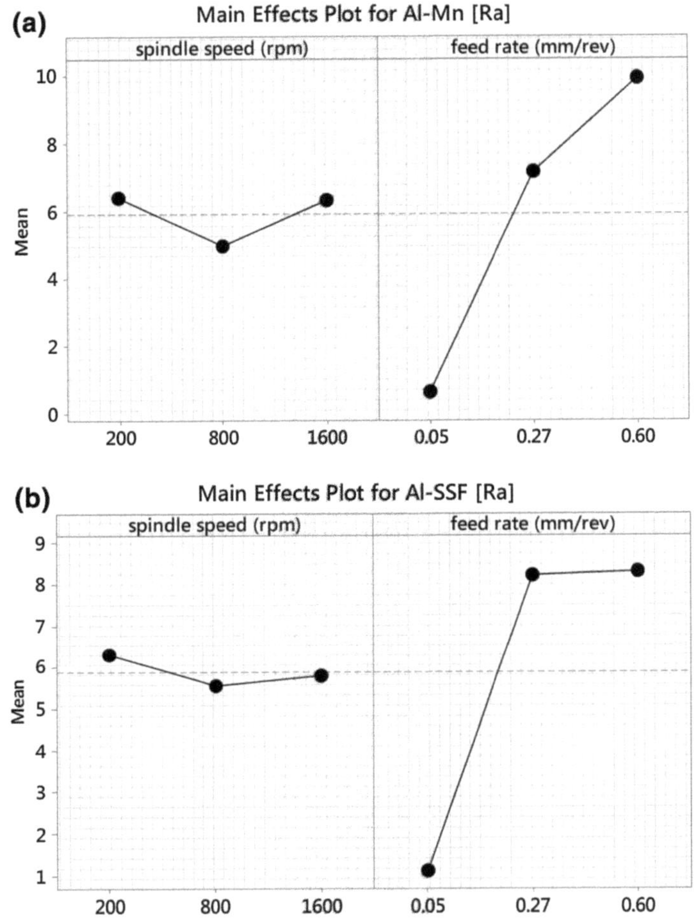

Fig. 7 Main effects plots for mean surface roughness (R_a): **a** Al–Mn matrix alloy; **b** Al–SSF composite

the third one of feed rate—f (mm/rev); see Fig. 8b. Main effects for R_{ku} parameter suggest a different behavior when comparing the results obtained for Al–Mn to those obtained for Al–SSF. For the former material R_{ku} is minimized for the third level of spindle speed—n (rpm) and the third level of feed rate—f (mm/rev); see Fig. 8c, whilst for the latter one a different optimization trend—towards the usage of the first level of spindle speed—n (rpm) and the second level of feed rate—f (mm/rev); see Fig. 8d, was identified.

The same indications may be observed by examining interaction plots for both materials investigated as well as the roughness parameters. Figure 9 depicts the interaction plots for Al–Mn matrix material whilst in Fig. 10 the interactions results obtained for the Al–SSF composite are presented. Optimal values for roughness

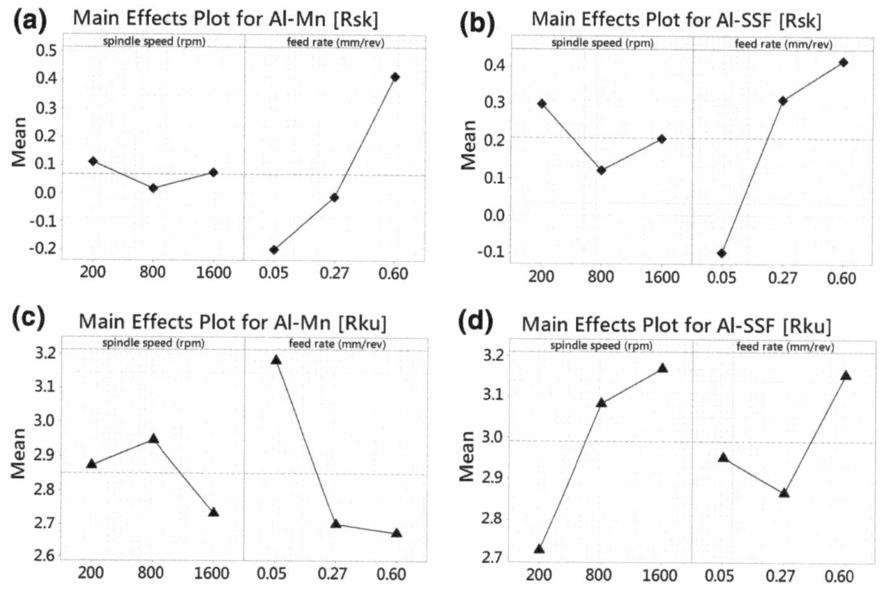

Fig. 8 Main effects plots for Al–Mn and Al–SSF: **a** Al–Mn, R_{sk}; **b** Al–SSF, R_{sk}; **c** Al–Mn, R_{ku}; **d** Al–SSF, R_{ku}

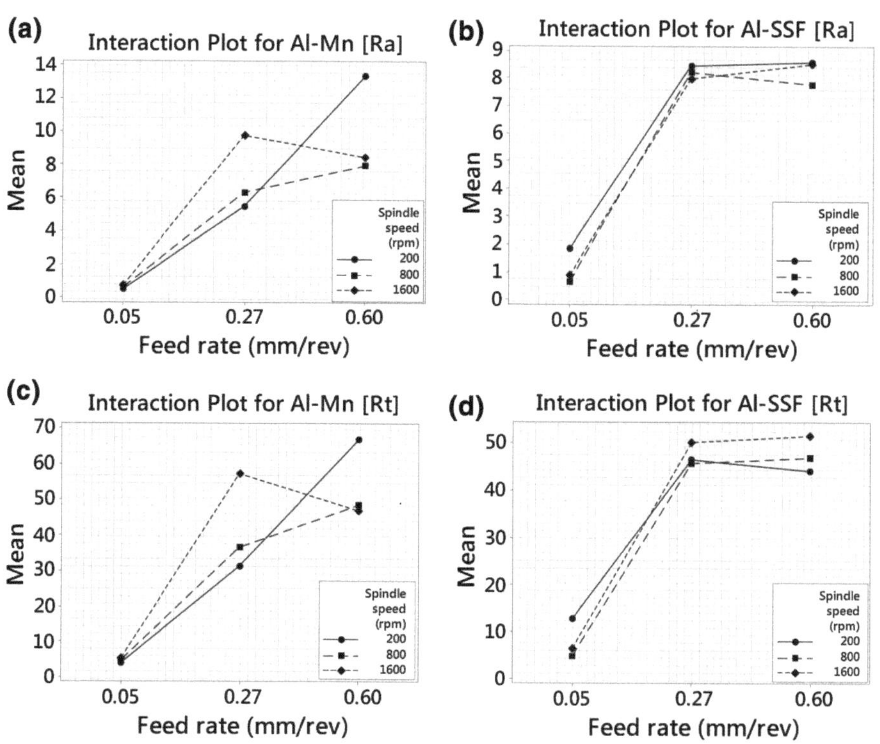

Fig. 9 Interaction plots for test materials: **a** Al–Mn R_a; **b** Al–SSF R_a; **c** Al–Mn, R_t; **d** Al–SSF, R_t

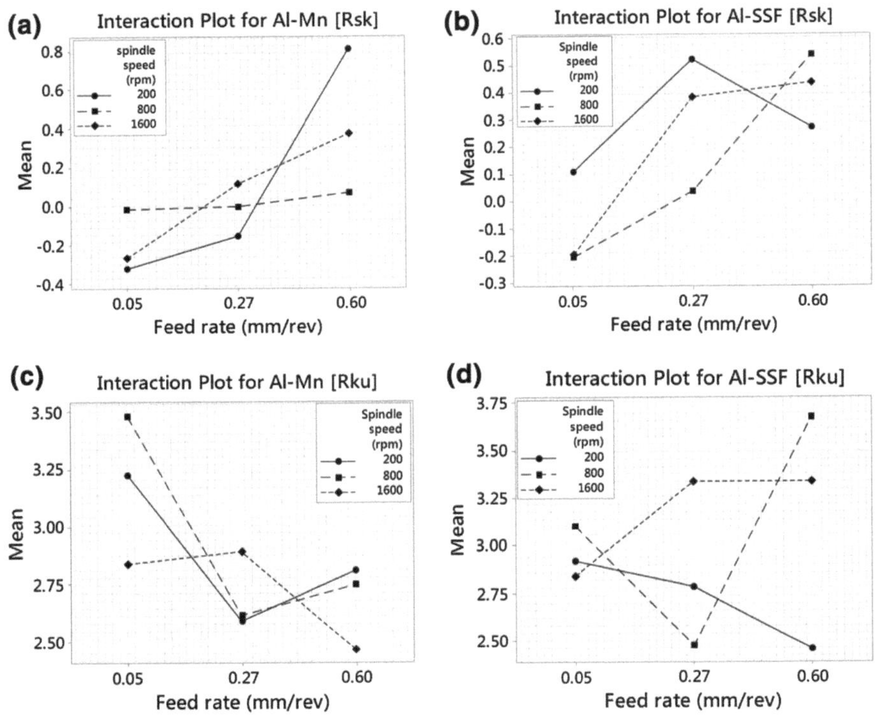

Fig. 10 Interaction plots for test materials: **a** Al–Mn, R_{sk}; **b** Al–SSF, R_{sk}; **c** Al–Mn, R_{ku}; **d** Al–SSF, R_{ku}

parameters are determined for all independent parameter levels capable of minimizing the mean value. These levels are identical to those observed for the main effects plots discussed above.

According to ANOVA analysis, P-values for regression models referring to Al–Mn material were 0.056, 0.080, 0.427 and 0.442 for R_a, R_t, R_{sk} and R_{ku} respectively whilst P-values for regression models referring to Al–SSF composite material were 0.001, 0.001, 0.244 and 0.721 for R_a, R_t, R_{sk} and R_{ku} respectively. From the statistical analysis it was concluded that feed rate holds dominant effect against spindle speed for all roughness parameters. Note that indications for P-values larger than 0.05 are insignificant.

In general, it should be noted that surface quality, tool wear, cutting forces and productivity are the essential outputs in turning. The sought-for optimum cutting conditions should be corresponding to the best surface quality, least tool wear and cutting forces and highest productivity. Since these are competing objectives, it seems that there is no unique combination of cutting conditions that would correspond to the optimal conditions at the same time; see also [5, 14].

Based on *ANOVA* outputs several prediction models of different types can be formulated. These models typically include: (a) the general linear model for first-order; (b) the least squares approach—second order model;—and (c) the exponential one [18]. In this paper only full quadratic models for the two materials and all roughness responses were developed whilst the implementation of exponential models is yet to be investigated. The independent parameters are correlated to the response following the expression in Eq. 1.

$$y = f(x_1, x_2, \ldots x_n) + e \tag{1}$$

where y denotes the response, x_i for i = 1, 2, … n are the independent parameters and e is the statistical error of the measurements that is usually omitted in f. Thereby, the quadratic model is of the form:

$$y = b_0 + \sum_{i=1}^{n} b_i x_i + \sum_{i=1}^{n} b_{ii} x_i^2 + \sum_i \sum_j b_{ij} x_i x_j + e \tag{2}$$

where b_i, b_{ii} and b_{ij} are the linear, the quadratic and the interaction terms respectively, whilst x_i, x_j are the coded variables corresponding to spindle n (rpm) and federate f (mm/rev). The final equation relating the response to independent parameters is extracted by specifying the coefficients after conducting the ANOVA analysis, with the method of least squares. Hence, the second order models for roughness responses given the problem investigated, will be in the form:

$$Rx = b_0 + b_1 n + b_2 f + b_3 n^2 + b_4 f^2 + b_5 nf \tag{3}$$

Final full quadratic regression models developed following the aforementioned concept, are presented in Tables 2 and 3, for Al–Mn matrix material and Al–SSF composite, respectively.

Regression equations are accompanied with their coefficient of determination, (R^2). Coefficient of determination is the result of the division between sum of squares of regression (SS-Regression) and the total sum of squares (SS-Total), hence; R^2 = SS-regression/SS-Total. The higher the value of R^2 the better the determination occurs for the coefficients of regression equations. By examining R^2 values in Tables 2 and 3, it is indicated that R_a and R_t parameters are highly correlated to their corresponding coefficients and independent parameter values; whilst the opposite is revealed for R_{sk} and R_{ku} parameters and for both materials. R_{sk} and R_{ku} parameters remain uncorrelated and their models cannot be employed.

Table 3 Full quadratic regression equations for predicting roughness parameters of Al–SSF composite

Response	Regression model	R^2 (%)
R_a	$= -0.107 - 0.0027 * n + 49.68 * f + 10^{-6} * n^2 - 57.97 * f^2 + 10^{-2} * n * f$	99.75
R_t	$= 0.85 - 0.014 * n + 268.5 * f + 6 * 10^{-5} * n^2 - 324.7 * f^2 + 0.016 * n * f$	99.54
R_{sk}	$= 0.114 - 7 * 10^{-4} * n + 2.3 * f - 2.8 * f^2 + 6 * 10^{-3} * n * f$	80.46
R_{ku}	$= 2.8 + 6 * 10^{-4} * n - 2.1 * f + 2.3 * f^2 + 10^{-3} * n * f$	49.32

5 Conclusions

A multi-parameter experimental and statistical analysis of surface texture in turning of a new aluminum matrix steel particulate composite (AL–SSF) was presented. Spindle speed and feed rate were treated as the independent machining parameters whilst roughness parameters including R_a, R_t, R_{sk} and R_{ku} were treated as the responses on an L9 Taguchi orthogonal array design. The correlation of these responses was successfully achieved for R_a and R_t parameters; whereas it was shown that R_{sk} as well as R_{ku} remain uncorrelated according to experimental results. The machinability behaviour of Al–Mn/SSF composite under the longitudinal turning process was examined. Looking further ahead high speed machining experiments considering cutting speed; feed rate and depth of cut to measure both roughness and cutting force, are to be conducted to further investigate the machinability of the new composite. Further research will also examine prediction capabilities of other types of regression models such as power law with proper logarithmic transformations using the experimental results, as well as exponential.

References

1. Kok M (2005) Production and mechanical properties of Al_2O_3 particle-reinforced 2024 aluminium alloy composites. J Mater Process Technol 161(3):381–387
2. Bains PS, Sidhu SS, Payal HS (2016) Fabrication and machining of metal matrix composites: A review. Mater Manuf Processes 31(5):553–573
3. Tomac N, Tannessen K, Rasch, FO (1992) Machinability of particulate aluminium matrix composites. CIRP Ann-Manuf Technol 41(1):55–58
4. Kainer KU (ed.) (2006) Metal matrix composites: custom-made materials for automotive and aerospace engineering. Wiley, New York
5. Aramesh M, Shi B, Nassef AO, Attia H, Balazinski M, Kishawy HA (2013) Meta-modeling optimization of the cutting process during turning titanium metal matrix composites (Ti-MMCs) Procedia CIRP 8:576–581
6. Seretis GV, Kouzilos G, Polyzou AK, Manolakos DE (2016) Provatidis, effect of stainless steel flakes content on mechanical properties and microstructure of cast aluminum matrix composites, submitted to Materials Research Express, Sept 2016
7. Kannan S, Kishawy HA (2008) Tribological aspects of machining aluminium metal matrix composites. J Mater Process Technol 198(1):399–406
8. Kumar GV, Rao CSP, Selvaraj N (2011) Mechanical and tribological behavior of particulate reinforced aluminum metal matrix composites–a review. J Miner Mater Charact Eng 10(01):59
9. Gururaja S, Ramulu M, Pedersen W (2013). Machining of MMCs: a review. Mach Sci Technol 17(1):41–73
10. Basavarajappa S, Chandramohan G, Rao KN, Radhakrishanan R, Krishnaraj V (2006) Turning of particulate metal matrix composites—review and discussion. Proc Inst Mech Eng Part B: J Eng Manuf 220(7):1189–1204
11. Vaxevanidis NM, Galanis NI, Petropoulos GP, Karalis N, Vasilakakos P, Sideris J (2010) Surface roughness analysis in high speed-dry turning of a tool steel. In: ASME 2010 10th biennial conference on engineering systems design and analysis, ESDA 2010, pp 551–557

12. Vaxevanidis NM, Fountas NA, Galanis NI, Bounas I, Sideris J (2011) Multiparametric analysis of surface roughness in high-speed turning of a high-alloyed tool steel. In: Proceedings of 7th BalkanTrib international conference, pp 351–358
13. Aramesh M, Shi B, Nassef,AO, Attia H, Balazinski M, Kishawy HA (2013) Meta-modeling optimization of the cutting process during turning titanium metal matrix composites (Ti-MMCs) Procedia CIRP 8:576–581
14. Petropoulos GP, Vaxevanidis NM, Pantazaras CN, Antoniadis AT (2006) Multi-parameter identification and control of turned surface textures corresponding to various cutting factors—new typology charts. Int J Adv Manuf Technol 29:118–128
15. Gadelmawla ES, Koura MM, Maksoud TMA, Elewa IM, Soliman HH (2002) Roughness parameters. J Mater Process Technol 123(1):133–145
16. Grzesik W, Brol S (2003) Hybrid approach to surface roughness evaluation in multistage machining processes. J Mater Process Technol 134(2):265–272
17. Petropoulos G, Pandazaras C, Vaxevanidis NM, Ntziantzias I, Korlos A (2007) Selecting subsets of mutually unrelated ISO 13565-2: 1997 surface roughness parameters in turning operations. Int J Comput Mater Sci Surf Eng 1(1):114–128
18. Petropoulos G, Vaxevanidis NM, Pandazaras C (2004) Modeling of surface finish in electro-discharge machining based upon statistical multi-parameter analysis. J Mater Process Technol 155:1247–1251
19. Shin YC, Dandekar C (2012) Mechanics and modeling of chip formation in machining of MMC. In: Machining of metal matrix composites. Springer, London, pp 1–49
20. Hung NP, Yeo SH, Lee KK, Ng KJ (1998) Chip formation in machining particle-reinforced metal matrix composites. Mater Manuf Processes 13:85–100
21. Pramanik A, Zhang LC, Arsecularatne JA (2008) Machining of metal matrix composites: effect of ceramic particles on residual stress, surface roughness and chip formation. Int J Mach Tool Manuf 48(15):1613–1625

Influence of the Determination of FLC's and FLSC's and Their Application for Deep Drawing Process with Additional Force Transmission

B.-A. Behrens, A. Bouguecha, C. Bonk, D. Rosenbusch, N. Grbic and M. Vucetic

Abstract This contribution deals with the experimental and numerical analysis of the material fracture behavior of steel sheet material HCT 600 X + Z (1.0941) in thickness s_0 = 1.0 mm using the Nakajima test. Firstly, the Nakajima test is carried out, whereby the major and minor strains are measured with the optical measuring system ARAMIS. Here, two different estimation methods for determination of the strains are applied and sensitivity of estimated results related to different strain gauge lengths was analyzed. Hereby, the forming limit curve (FLC) is determined experimentally. Subsequently, the FEA of a Nakajima test was carried out and compared with corresponding experimental results. The flow behavior of HCT 600 is modelled using a planar anisotropic material model based on the Hill's 1948 criterion, which was validated in previous work. Using FLC the simulation-based determination of the forming limit stress curve (FLSC) is carried out. Furthermore, two deep drawing processes, conventional and process with activation of additional force transmission are carried out producing the rectangle cups. Here, the larger process window is achieved. Within the numerical investigation of the material fracture behavior the FLC and FLSC are applied by FEA of both deep drawing processes. Finally, the assessment of the performed material modelling is presented.

B.-A. Behrens · A. Bouguecha · C. Bonk · D. Rosenbusch · N. Grbic (✉) · M. Vucetic
Institute of Forming Technology and Machines, Leibniz Universität Hannover,
An der Universität 2, 30823 Garbsen, Germany
e-mail: grbic@ifum.uni-hannover.de

B.-A. Behrens
e-mail: behrens@ifum.uni-hannover.de

A. Bouguecha
e-mail: bouguecha@ifum.uni-hannover.de

C. Bonk
e-mail: bonk@ifum.uni-hannover.de

D. Rosenbusch
e-mail: rosenbusch@ifum.uni-hannover.de

M. Vucetic
e-mail: vucetic@ifum.uni-hannover.de

V. Majstorovic and Z. Jakovljevic (eds.), *Proceedings of 5th International Conference on Advanced Manufacturing Engineering and Technologies*,
Lecture Notes in Mechanical Engineering, DOI 10.1007/978-3-319-56430-2_30

Keywords Forming limit curve · Forming limit stress curve · Deep drawing

1 Introduction

In order to meet requirements by automotive industry regarding the reduction of vehicles mass in the car body, further development of existing production processes for producing the structural parts made of sheet metal, are required. Here, the continuous innovation of deep drawing and bending processes are crucial with the aim to develop processes with a high level of productivity, minimum number of production stages, maximum utilization of the material, low revision need, aspired energy balance and high shape or dimensional accuracy [1]. For this purpose a new tool system for deep drawing as well as for stretch-bending with additional mechanical force transmission by means of a counter punch is designed at the Institute of Forming Technology and Machines (IFUM). For a failure-free production of structural parts regarding the wrinkling and fracture initiation by necking or cracks, the prior simulation-based process design is of major importance. Beside the hardening behavior, the material fracture behavior has to be considered in the simulation. Here, the development of material models for an accurate modelling of material behavior is crucial [2].

2 Material Modelling

Depending on the specific material characteristics of sheet metal, different material models can be used for FEA. The most common of them are the material model by von Mises [3] developed for modeling the isotropic material behavior and the material model by Hill [4], which can be used for modeling the anisotropic material behavior.

According to the results of previous characterization tests the flow criterion by Hill [4] with isotropic hardening rule is used in Abaqus for modelling the material behavior of the steel metal HCT 600 X + Z. For this purpose, the HCT 600 was investigated by quasi-static uniaxial tensile tests according to DIN EN ISO 6892. Thus, the shape of the yield locus based on six material-specific coefficients, calculated from anisotropic coefficients (r-values), is determined (Fig. 1a). To improve the extrapolation of the flow curve determined by tensile test, the steel alloy was investigated in a hydraulic bulge test according to ISO 16808:2014 (Fig. 1b) [5]. In order to approximate the experimental determined flow curves with a minimum of deviation a combined extrapolation approach based on Swift and Hockett-Sher by is used (Fig. 1c) [6].

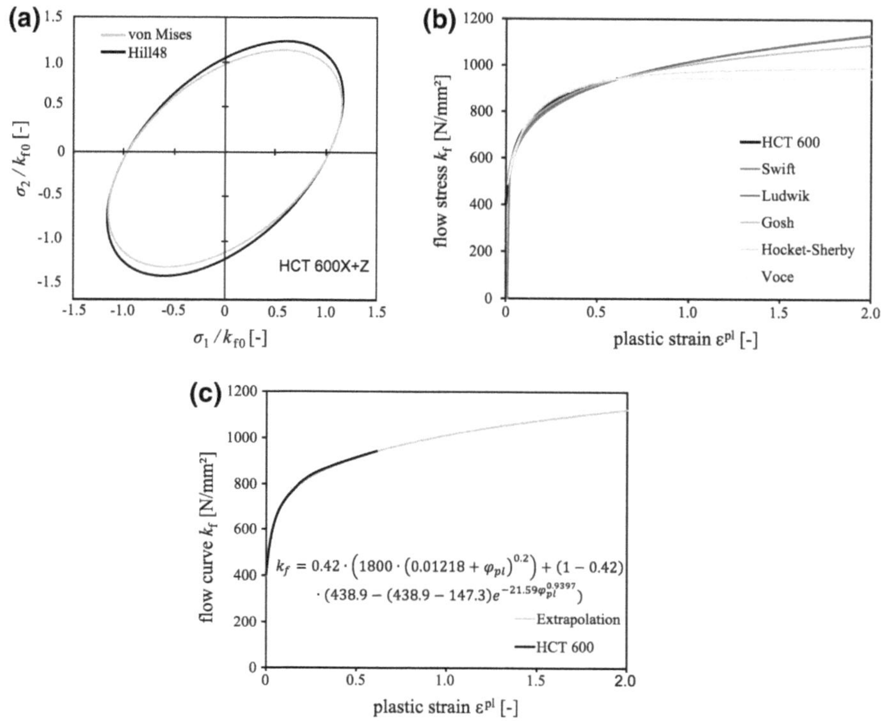

Fig. 1 Material modelling by Hill48 for HCT 600. **a** Yield locus. **b** Extrapolation approaches. **c** Extrapolation based on Swift and Hockett-Sherby

3 Damage Modelling

Various damage models are used to describe the forming limits of sheet metals in FEA. It can be distinguished between strain-based and stress-based damage models [7]. Furthermore, a distinction is made between models which predict ductile failure, like fracture initiation by necking, and models for the prediction of cracks. Due to the fact, that the material failure in deep drawing processes occurs primary by necking, the strain-based model related to the forming limit curve (FLC) is applied to determine critical necking in this contribution. In the next step the stress-based model is investigated, including the forming limit stress curves (FLSC), which describe the critical necking depending on the critical principal stresses σ_1 and σ_2.

3.1 Strain-Based Damage Models and Results

The forming limit curve describes the forming capacity until the beginning of the instability by necking of a sheet metal at different strain states occurs (Fig. 2a). The

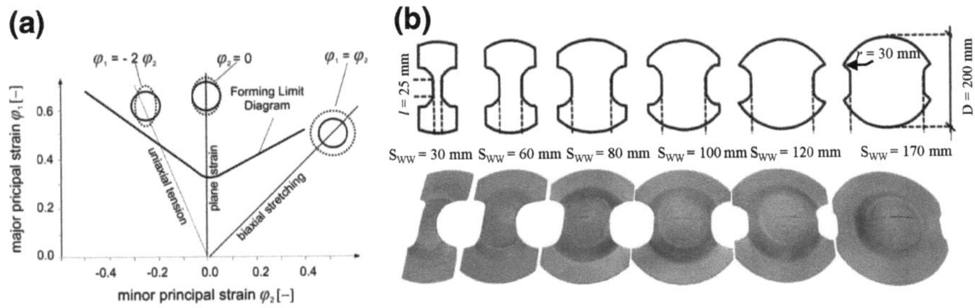

Fig. 2 **a** Schematic representation of an FLC. **b** FLC test specimens with different geometries

experimental investigation of the forming limit curve has been standardized since 2004 according to DIN EN ISO 12004 [8].

In addition to the specimens geometry, test equipment and conditions, evaluation method, documentation and test report are defined in the ISO 12004. Due to the fact that the evaluation method is one of the decisive factor for the quality of the result, it is discussed more detailed below. In the following, only methods determined using an optical measurement system will be considered. In ISO 12004-2, section line method described. Here, at least a minimum of five different FLC specimens geometries (see Fig. 2b) with different web widths (S_{WW}) are deformed until failure through cracking occurs. For the section line method (Sect ISO) five parallels lines are fitted through the middle of the test specimens and the major and minor principal strains are determined along the line. By fitting a parable in the major and minor principal strains at the vertex of the parabola the value pairs for the major and minor principal strains can be calculated for each test specimens, from which the mean value is formed.

In addition to the section line method, the strain-rate-dependent method (Line fit) exists for determination of the instability by necking. This method is currently not considered within the framework of ISO 12004 and has been published by Marron et al. proposed in 1997 [9]. The method is based on the incompressible behavior of metallic materials. On the basis of the measured geometric change, the major and minor principal strains as well as the degree of deformation in sheet thickness direction are calculated. Subsequently, the strain rate in the sheet thickness direction is derived. The range of instability is defined by a drastic increase in the rate of expansion by constriction. The point for the FLC is determined by two line fixtures, the stable degree fit in the range of the "quasi-constant" low strain rate, and the unstable degree fit in the range of the increasing strain rate. The point of intersection of these two line fits is then set in relation to the major and minor principal strains and is used as a pair of values for the FLC. In addition to the evaluation method, the sheet thickness, the strain reference size of the optical measuring grid, the strain path, the mechanical characteristics, the tool geometry, the temperature and the strain rate influence the result of the determination of the forming limit curve [10].

The experimental results of the FLC are shown in Fig. 3. The forming limit curves for a strain gauge length of a = 1 mm for the different evaluation methods

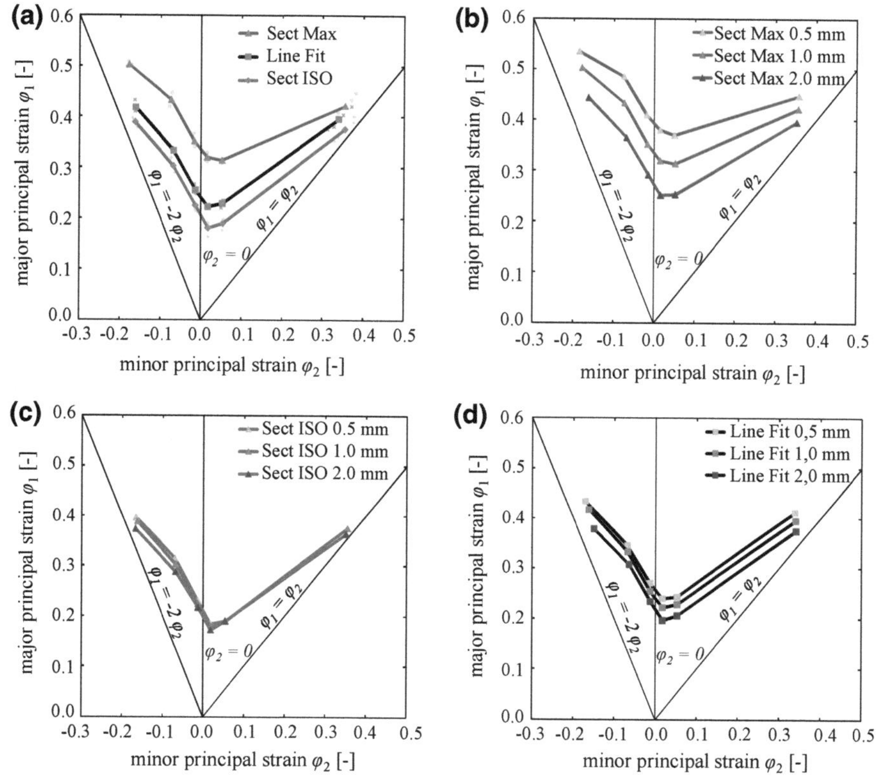

Fig. 3 Experimental results. **a** FLC comparison. **b** Sensitivity analysis based on SectMax. **c** Sensitivity analysis based on Sect ISO. **d** Sensitivity analysis based on line fit

are shown in Fig. 3a. The values determined by the section line method are lower than the strain-rate-dependent method with the similar boundary condition in evaluation. Furthermore, the maximum experimental investigated value pairs are shown as SectMax in Fig. 3a, which represents the maximum major and minor strain at crack. This FLC is calculated according to the section line method, whereby always the highest values of major and minor strain are chosen. In next step, the sensitivity analysis according to the different strain gauge lengths is carried out experimentally. The SectMax shows differences for the variation of the strain gauge length for a = 0.5, 1.0 and 2.0 mm. Particularly if local large effects, such as a constriction occur, a finer grid size results in a higher local resolution which then leads to higher quality of evaluated results. In Fig. 3c the sensitivity analysis for results evaluated by the section line method are shown. Here, a lower influence of the variation of the strain gauge length on the evaluation of the experimental results is shown. The variation of the strain gauge length has an influence on the evaluation of the experimental results, if the strain-rate-dependent method (Line Fit) is depicted (see Fig. 3d). In following studies the FLC determined by the ISO 12004 is used, due to the fact that lowest influence of the strain gauge length exists on evaluation of the experimental data, than by applying the strain-rate-dependent method.

3.2 Stress-Based Damage Models and Results

The validity of FLC is restricted to forming processes with a linear strain path. For this reason, the failure model based on FLC is of limited use for complex forming processes such as deep drawing, flanging, forming with stress superimpositions as well as in the case of active media-based forming and, partially, deep-drawing processes with first draw and initial high deformation or embossing. Opposite to this, the damage model based on FLSC (Forming Limit Stress Curve) is robust against changes in the strain path, because by its determination the hardening behavior of the material is taken into account [11, 12]. The basis for determining the FLSC is also the Nakajima test and the experimental determination of local necking. Afterwards, the FE-simulation of the Nakajima test must be carried out. Finally, in the postprocessing of FEA, the limiting principal stresses σ_1 und σ_2 corresponding to major and minor strains of the FLC are determined [11].

By numerical analysis of Nakjima test the presented material model is applied for calculation of three different linear strain paths, which represent a wide range of deformation states during the deep drawing process. For this purpose, the specimens with a web width of S_{WW} = 30 mm, S_{WW} = 100 mm and S_{WW} = 170 mm are investigated numerically using the FE software Abaqus CAE 6.13-1. In Fig. 4a the 3-D FE model of the specimen with S_{WW} = 30 mm is shown. The tool system components, consisting of die, blank holder and punch, are modelled as rigid bodies. Due to the tribological system used in the experiments the contact between punch and blank is assumed as friction-free. The contact between die, blank holder and blank is modelled with a friction coefficient of μ = 0.12. The comparison between numerical estimated and optical measured strain distribution is shown in Fig. 4b for a specimen with S_{WW} = 30 mm. By meshing the FE model the strain

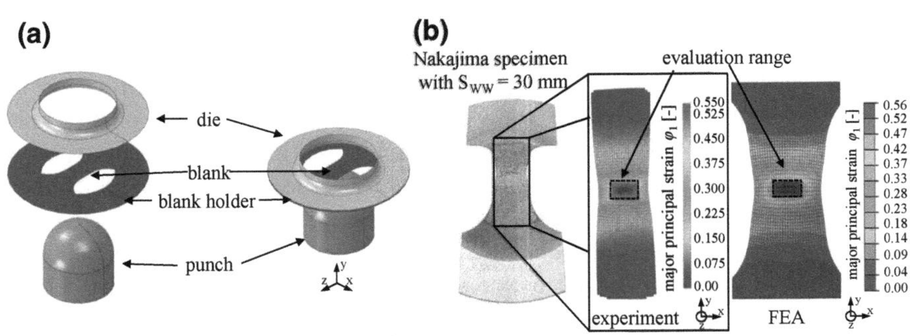

Fig. 4 HCT 600. **a** FE model for SWW = 30 mm. **b** Comparison of experimental and numerical results

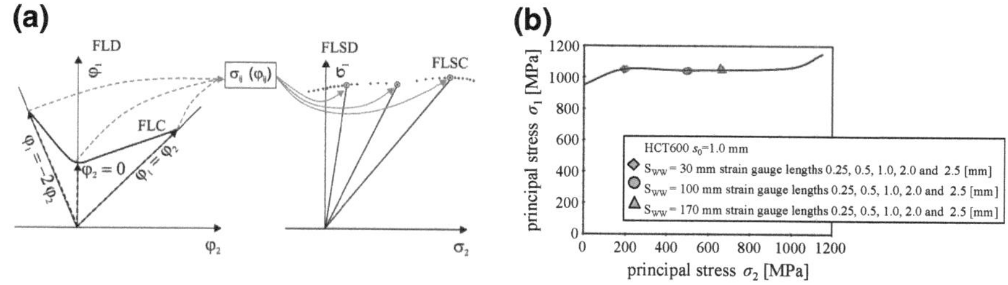

Fig. 5 **a** Transformation approach FLC to FLSC. **b** Sensitivity analysis

gauge length of a = 1.0 mm used by optical measurement is taken into account. Therefore, the FE model is discretized by the shell element of the length a = 1.0 mm. The comparison occurs in respect to the same forming step. For this purpose, the same pole height is taken into account. Here, an accurate agreement between numerical and experimental results is achieved.

In the next step, the calculated forming limit curve (FLC) is transformed into the forming limit stress curve (FLSC) The principal stresses σ_1 and σ_2 corresponding to major and minor strains of the FLC are shown schematically in FLSD (see Fig. 5a). Furthermore, the sensitivity analysis regarding the different strain gauge lengths is carried out numerically using the same FE model. Here, the strain gauge lengths of a = 0.25, 0.5, 1.0, 2.0 and 2.5 mm are considered by meshing the blank with shell elements. The numerical results show that different shell elements with corresponding element length dimension have no influence on FLSD (see Fig. 5b).

4 Deep Drawing with Additional Force Transmission

At the Institute of Forming Technology and Machines an innovative technology was developed for deep drawing and stretch-bending by using a counter punch for additional mechanical force transmission (Fig. 6a). With this technological approach the additional mechanical force is centralized in the bottom part area and radius crossover to the side wall during the forming process (Fig. 6b). As opposed to conventional deep drawing, the high stressed part areas, located at the punch radius and the transfer zone to the part wall where the maximal stretching of material occurs are reloaded. Thus, the initially high stressed area of the part is superimposed with pressure.

In previous work, the effect of additional force transmission by deep drawing was analyzed experimentally and numerically on rectangle cup profiles [2, 6]. Here, the reduction of the spring-back in the flange area and an improvement of the part bottom shaping were verified [2, 6]. Furthermore, a higher process reliability and higher process window is realized due to the activation of the additional force transmission in the bottom area of the drawn part. Here, the optimum blank holder force without force transmission $F_{NH,opt,o}$ was 300 kN and with force transmission

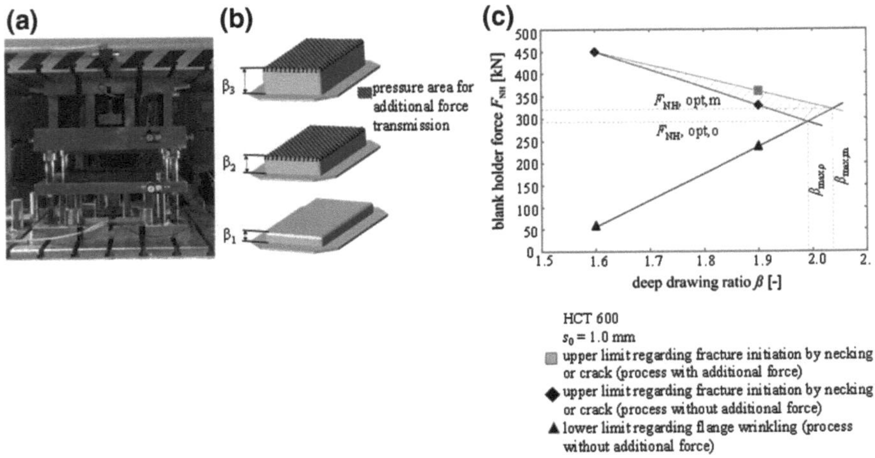

Fig. 6 Experimental setup for the deep drawing process with and without additional force transmission. **a** A photo of experimental installation. **b** Schematic process description. **c** Process area for HCT 600 s_0 = 1.0 mm

$F_{NH,opt,m}$ 330 kN, which are experimental determined related to the deep draw ratio of 1.9. By using the additional force transmission into the bottom of the cup the fracture initiation by necking or crack takes place at higher blank holder forces. Due to the fact that the transmission of the blank holder force into the flange area of the part is increased a higher limiting deep draw ratio is possible (see Fig. 6c).

5 FEA of Deep Drawing Process

Within the numerical investigations the FEA of the conventional process and the process with additional force transmission, is carried out for rectangle cup profile. Here, the HCT 600 is analyzed numerically regarding the blank holder forces of 300 and 330 kN. For presentation of the major and minor principal strains as well as the stress distribution, the area of bottom radius and wall at the corner of the rectangle cup profile was defined. In this area, failure occurs due to necking or crack during the forming process. First, the strain-based approach is applied. Here, the FLC according to ISO 12004 and strain-rate-dependent method are used and compared with numerical estimated strain distribution of conventional process (red mark) and process with additional force transmission (green mark) for both blank holder forces (see Fig. 7).

In Fig. 7 is shown, that for the HCT 600 the FLC according to ISO underestimates the material forming capacity for blank holder force 300 kN (Fig. 7a) and the strain-rate-dependent method overestimates the material forming capacity for blank holder force 330 kN (Fig. 7b). Furthermore, the same strain distributions by both processes and blank holder forces can be observed.

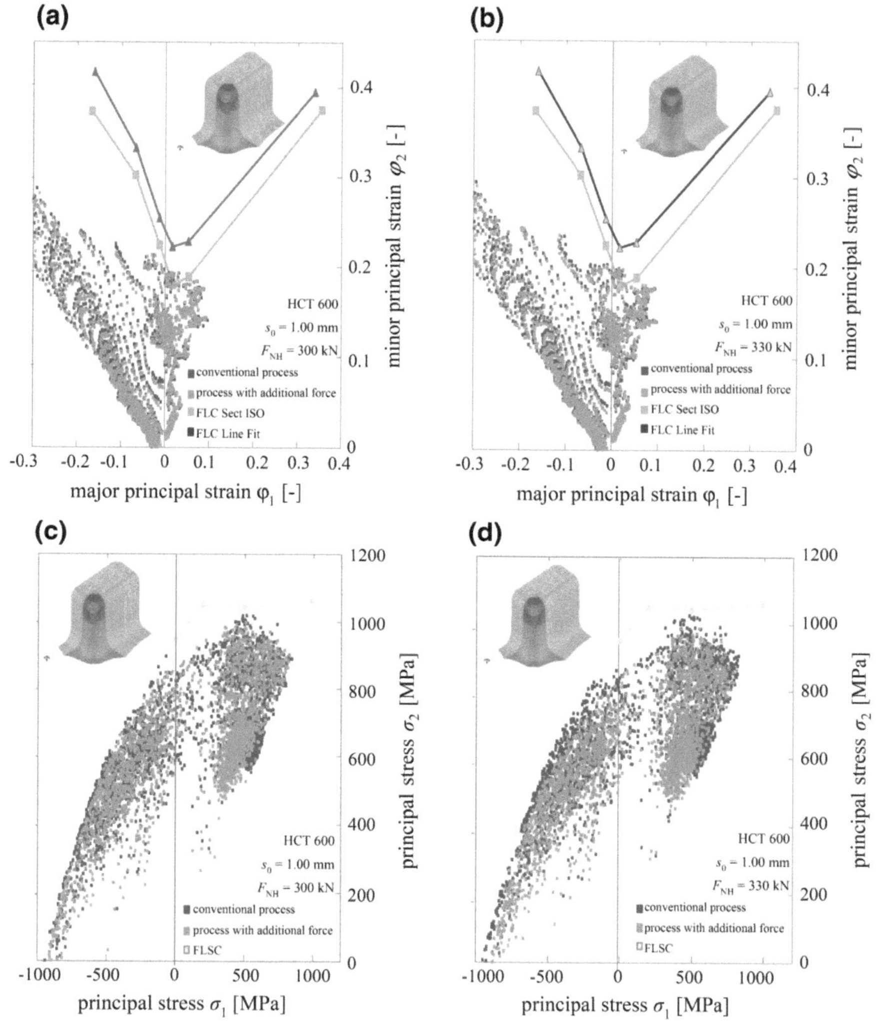

Fig. 7 Numerical damage study for HCT 600 $s_0 = 1.0$ mm with strain-based approach. **a** Blank holder force 300 kN. **b** Blank holder force 330 kN and stress-based approach. **c** Blank holder force 300 kN. **d** Blank holder force 330 kN

In the next step, the FLSC, generated from numerical results of Nakjima test (see Fig. 5b), is applied by FEA of conventional processes and process with additional force transmission for both blank holder force $F_{NH} = 300$ kN (Fig. 7c) and $F_{NH} = 330$ kN (Fig. 7d). Comparing to the conventional process a lower stress distribution can be observed for the process with additional force transmission as a result of the pressure superposition into the part bottom. Furthermore, the results show that the FLSC can correctly predict the material forming capacity for the applied blank holder forces regarding the conventional deep drawing for the

HCT 600. In addition, due to the consideration of the stress distribution, the lower tensile stresses can be demonstrated, which result from the pressure superposition. This effect is one of the reasons for the larger process window.

6 Summary and Outlook

In order to expend the process limits of existing sheet metal forming processes a new tool system was developed at IFUM, which allows the activation of an additional force transmission in the bottom area of the drawn rectangle cup part with a counter punch. Thus, a higher process reliability and a higher process window is realized. Using the Nakajima test the strain-based and, derived therefrom, the stress-based damage criterion are created for steel sheet material HCT 600 X in thickness of $s_0 = 1.0$ mm. Both damage modelling approaches are applied for FEA of conventional deep drawing and of the process with additional force transmission using FE software Abaqus. Here, the HCT 600 is analyzed numerically for blank holder forces of 300 and 330 kN. Using the strain-based damage criterion the discrepancies regarding the material forming capacity of numerical and experimental results are determined. Furthermore the FLSC as a stress-based criterion shows more suitability for an accurate prediction of material forming capacity for both processes and blank holder forces. Here, a sufficient agreement between the FE results and the experimental results is achieved.

In a next step, the presented forming process technology is to be further developed at IFUM. Hereby, other design concepts of the tool system with mechanical addition of the process forces have to be created.

Acknowledgements This paper is based on investigations of the project (BE169/139-1): "Deep drawing with additional force transmission", which is kindly supported by the German Research Foundation (DFG). The authors thank the DFG for project foundation.

References

1. Doege E, Behrens BA. Handbuch der Umformtechnik, Springer ISBN:978-3-642-04248-5
2. Behrens BA, Bouguecha A, Bonk C, Grbic N, Vucetic M (2017) Validation of the FEA of a deep drawing process with additional force transmission, 20th international ESAFORM conference on material forming, Dublin 2017 (accepted)
3. von Mises R (1913) Mechanik der festen Körper im plastisch deformablen Zustand, Göttin. Nachr. Math. Phys. 1:582–592
4. Hill R. A theory of the yielding and plastic flow of anisotropic metals. In: Proceedings of the royal society of London, Series A, vol 193, pp 281–197
5. Hallfeldt T, Hotz W, Leppin C, Keller S, Friebe H, Till ET, Müller R, Vučetić M, Vegter H (2014) Sheet bulge testing, In: Comprehensive Materials Processing, vol 1, pp 85–93, Elsevier Ltd., ISBN:978-008096533-8

6. Behrens BA, Bonk C, Grbic N, Vucetic M (2016) Numerical analysis of a deep drawing process with additional force transmission for an extension of the process limits. In: 4th international conference recent trends in structural materials COMAT2016, Pilsen 2016 (accepted)
7. Behrens BA, Bouguecha A, Vucetic M, Peshekhodov I (2012) Characterisation of the quasi-static flow and fracture behaviour of dual-phase steel sheets in a wide range of plane stress states. Arch Civ Mech Eng 12(4):397–406
8. DIN EN ISO 12004-2: Metallische Werkstoffe – Bleche und Bänder – Bestimmung der Grenzformänderungskurve – Teil 2: Bestimmung von Grenzformänderungskurven im Labor (ISO 12004-2: 2008), Beuth Verlag, 2009
9. Col A (2002) FLC's: past, present and future, IDDRG conference 2002, Nagoya, Japan. pp S107–S125
10. Banabic D (2004) Sheet metal forming processes, Springer, Heidelberg, doi:10.1016/j.ijplas.2004.04.003
11. Stoughton TB, Zhu X (1999) A general forming limit criterion for sheet metal forming. Int J Mech Sci, Elsevier, (pii:S0020-7403(98)00113-1)
12. Dic RE, Yoon JW, Stoughton TB (2015) Path-independent forming limit models for multi-stage forming processes. In: Thematic issue, Springer, Berlin, doi:10.1007/s12289-015-1220-4

Biologically Inspired Optimization Algorithms for Flexible Process Planning

Milica Petrović and Zoran Miljković

Abstract Flexible process planning belongs to one of the most essential functions of the modern manufacturing system. The aim of this function is to define detailed methods for manufacturing of a part in an economic and competitive manner starting from the initial phase (drawing of the target part) up to the final one (the desired shape of the target part). A variety of alternative manufacturing resources (machine tools, cutting tools, tool access directions, etc.) causes flexible process planning problem to be strongly NP-hard (non deterministic polynomial) in terms of combinatorial optimization. This paper presents a comparative analysis of biologically inspired optimization algorithms which are used to solve this problem. Four different optimization algorithms, namely genetic algorithms (GA), simulated annealing (SA), chaotic particle swarm optimization algorithm (cPSO), and ant lion optimization algorithm (ALO) are proposed and implemented in Matlab environment. Optimal process plans are obtained by multi-objective optimization of production time and production cost. The experimental verification is carried out by using real-world examples. The experimental results indicate that all aforementioned algorithms can be successfully used for optimization of flexible process plans, while the best performance shows ALO algorithm.

Keywords Flexible process planning · Genetic algorithms · Simulated annealing · Chaotic particle swarm optimization algorithm · Ant lion optimization algorithm

---

M. Petrović (✉) · Z. Miljković
Faculty of Mechanical Engineering, Production Engineering Department,
University of Belgrade, Kraljice Marije 16, 11120 Belgrade 35, Serbia
e-mail: mmpetrovic@mas.bg.ac.rs

Z. Miljković
e-mail: zmiljkovic@mas.bg.ac.rs

V. Majstorovic and Z. Jakovljevic (eds.), *Proceedings of 5th International Conference on Advanced Manufacturing Engineering and Technologies*,
Lecture Notes in Mechanical Engineering, DOI 10.1007/978-3-319-56430-2_31

1 Introduction

Computer Aided Process Planning (CAPP) belongs to one of the most important manufacturing functions. It can be defined as interface between computer aided design (CAD) and computer aided manufacturing (CAM). In other words, this manufacturing function establishes resources and procedures to be performed in order to manufacturing a part economically and concurrently. Two main activities in CAPP system include: (1) feature recognition and (2) selection and sequencing of machine operation. This paper focuses on the selection and sequencing of machine operation. According to fact that parts have multi alternative process plans, optimization of process plans belongs to NP hard (*non deterministic polynomial optimization*) problems.

Numerous efforts have been made in order to optimize flexible process planning problem. In recent years, biologically inspired optimization algorithms have been used as primary techniques for obtaining the optimal solutions the problem. Some of the most popular algorithms are: genetic algorithms (GA), genetic programming (GP), simulated annealing (SA), tabu search (TS), ant colony optimization (ACO), particle swarm optimization (PSO), or hybrid algorithms. Li et al. [4] proposed GP-based approach to optimize flexible process planning with minimum total processing time as criteria. Network representation was adopted to describe flexibility of process plans and efficient genetic representations and operator schemes were also considered. Using the same optimization objective and representation, Shao et al. [12] presented a modified GA-based approach for generating optimal and near optimal process plans. Nallakumarasamy et al. [8] used an efficient SA technique to obtain feasible optimal sequence with a minimal computational time based on the precedence cost matrix and reward-penalty matrix. Liu et al. [5] proposed ACO algorithm in order to obtain optimal process plans and cut down the total cost for machining process. Guo et al. [1] operation sequencing as a combinatorial optimization problem, and employed evolutionary particle swarm optimization (PSO) algorithm to solve it effectively. Lv and Qiao [6] proposed new approach called cross-entropy (CE) to optimize flexible process planning. They used AND/OR network to represented flexibility of process planning and established mathematical model for minimization of total processing time and total cost. Hybrid graph and genetic algorithm approach was applied in [2] to solve process planning problem for prismatic parts. Hybrid GA-SA algorithm used to solve flexible process planning problem with the objective of minimizing the production time was presented in [9].

Although these algorithms are able to find optimal/near optimal solutions of the problem, the so-called No Free Lunch theorem [13] allows researchers to propose new algorithms. This paper presents four algorithms, namely, genetic algorithms (GA), simulated annealing (SA), chaotic particle swarm optimization algorithm (cPSO), and ant lion optimization (ALO) algorithm, which are used to solve process planning problem with various flexibilities.

The structure of this paper consists of the following sections. In the Sect. 2 we introduce a flexible process planning problem and describe its representation. Mathematical model of the problem with two objective functions is formulated in Sect. 3. Section 4 briefly outlines four biologically inspired optimization algorithms. Section 5 shows comparative results and Sect. 6 gives concluding remarks. Finally, acknowledgement and references are stated at the end of the paper.

2 Flexibilities and Representation of the Process Planning Problem

In this paper, the following five types of flexibility for process planning optimization are considered: machine flexibility, tool flexibility, TAD flexibility, process flexibility, and sequencing flexibility. Machine flexibility relates to the possibility of performing one operation on different alternative machines, with various processing times and costs. Tool flexibility means that the same operation can be performed with alternative cutting tools. TAD flexibility assumes the possibility of performing the same operation with alternative TADs. Process flexibility refers to the possibility of producing the same part in different ways with alternative operations or sequences of operations. Sequencing flexibility implies the possibility to interchange the ordering of required manufacturing operations.

In order to corroborate these types of flexibility, three real sample parts are given in Fig. 1; part 1 consists of 5 manufacturing features, part 2 consists of 6 manufacturing features, and part 3 consists of 6 manufacturing features. The technical specification of parts including precision requirement, alternative operations, alternative machines, alternative tools, alternative TADs and corresponding processing times are given in details in [11]. Also, flexible process plan networks for sample part 1, part 2 and part 3 can be found in [11].

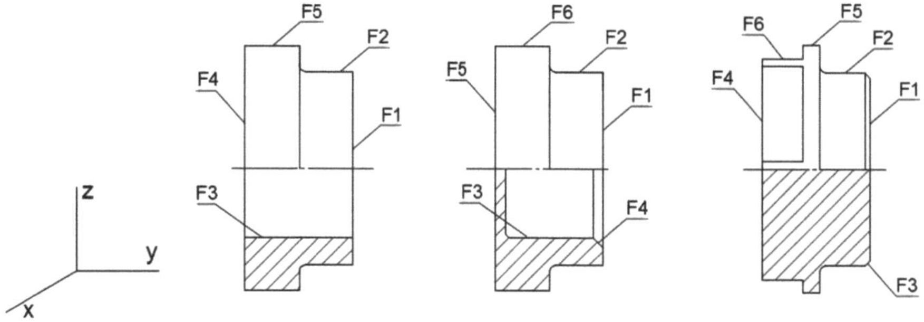

Fig. 1 The three sample parts (part 1, part 2, and part 3, respectively) with features

3 Mathematical Modeling and Optimization Objectives

Research in this paper focuses on total production time and total production cost as objectives of the optimization problem. The production time consists of (i) processing time (TW), (ii) transportation time (TT), (iii) total tool change time (TCT), and (iv) total setup change time (SCT). The detailed description of production time components is given in [10]. The total weighted production time (TWPT) is defined as follows:

$$TWPT = w_{11} \cdot TW + w_{12} \cdot TT + w_{13} \cdot TCT + w_{14} \cdot SCT \quad (1)$$

where w_{11}, w_{12}, w_{13}, and w_{14} are the weights.

The objective of production time minimization can be defined as:

$$f_1 = \max\left(\frac{1}{TWPT}\right) \quad (2)$$

The production cost is another criterion commonly used to optimize flexible process plans. The total production cost in this research consists of five components (i) machine cost (MC), (ii) tool cost (TC), (iii) machine change cost (MCC), (iv) set-up change cost (SCC), and (v) tool change cost (TCC). The details for each cost factor are described in [10]. All five costs are included in the total weighted cost (*TWC*) according to the following Eq. 3:

$$TWC = w_{21} \cdot MC + w_{22} \cdot TC + w_{23} \cdot MCC + w_{24} \cdot TCC + w_{25} \cdot SCC \quad (3)$$

where w_{21}, w_{22}, w_{23}, w_{24}, and w_{25} are the weights.

The objective function that defines the alternative process plans with the minimum production cost is given as follows:

$$f_2 = \max\left(\frac{1}{TWC}\right) \quad (4)$$

4 Biologically Inspired Optimization Algorithms

In this study, four different bioinspired optimization algorithms are employed. This section briefly introduces each method.

Algorithm 1 Pseudo code of GA algorithm	
1:	**Initialize** the parameters of the GA (size of population, probability of crossover, probability of mutation, maximum number of generations)
2:	**Initialize** randomly the population of individuals

(continued)

(continued)

Algorithm 1 Pseudo code of GA algorithm
Evaluate the fitness of the individuals (Eqs. 2 and 4)
Repeat:
Select the best individuals
Apply crossover and mutation operators
Evaluate the fitness of the individuals (Eqs. 2 and 4)
Until the maximum of generation is not met
Output the optimal process plan

Algorithm 2 Pseudo code of SA algorithm
Initialize the parameters of the SA (initial temperature T_0, the cooling factor T_c and the final temperature T_{lowest}); $T = T_0$
Initialize randomly the population of individuals
Take good individuals and call it current sequence
Repeat:
Make a change in the current sequence and call it temporary sequence
Evaluate the fitness's of current and temporary sequence
set $\Delta = f(temporary\ sequence) - f(current\ sequence)$
if $\Delta < 0$
set temporary sequence to be current sequence
else
randomly generate number r from [0, 1]
if $r < e^{\Delta/T}$
set temporary sequence be current sequence
else
set temporary sequence be unchanged
end if
end if
$T = T - T_c$
Until $T > T_{lowest}$
Output the optimal process plan

4.1 Genetic Algorithms

The genetic algorithms (GA) belong to the most widely used optimization methods. The main idea of the method is to apply GA operators such as selection, crossover and mutation in order to modify individuals in population over generations and to obtain optimal solution. Each individual in population represents one possible process plan. Encoding and decoding method for each process plan is given in [9].

According to [9], pseudo code of the GA approach for flexible process planning problem is described by Algorithm 1.

4.2 Simulated Annealing

Simulated annealing (SA) is a stochastic optimization algorithm which is widely applied to combinatorial optimization problems. This algorithm is inspired by the natural phenomena of annealing of solids; it implies the process of heating a material to the melting point and then slowly cooling until a thermal equilibrium is reached. According to [9], pseudo code of the SA approach for flexible process planning problem is described by Algorithm 2.

4.3 Chaotic Particle Swarm Optimization Algorithm

Inspiration for the development of particle swarm optimization (PSO) algorithm comes from the social behaviour of organisms in swarm. In order to search for food, each member of swarm (aka "particle") determines its velocity based on their personal experience as well as information gained through interaction with other members of the swarm. The equations of the traditional PSO algorithm are originally proposed by Kennedy and Eberhart in [3]. One of the major drawbacks of the traditional PSO is its premature convergence, especially while handling problems with more local optima. One of the ways to overcome this problem is to hybridize the PSO algorithm with chaos. In this study, we employ ten non-invertible, one-dimensional chaotic maps (see [10]) and replace the values of *rand*() and *Rand*() with chaotic number generators *chaos*() and *Chaos*(), respectively. Pseudo code of the cPSO approach is given by Algorithm 3.

4.4 Ant Lion Optimization Algorithm

Ant Lion Optimization (ALO) algorithm, originally proposed by Mirjalili [7], belongs to novel nature-inspired algorithm. The main inspiration of the ALO algorithm comes from the hunting mechanism of ant lions in nature. During the process of hunting, antlion makes funnel pits in soft sand and then waits patiently at the bottom of the pit. Mathematical modeling of the behavior of antlions and ants in domain of process planning problem is given in [9] as well as encoding/decoding scheme for process plans. Pseudo code of the ALO algorithm is given by Algorithm 4.

Algorithm 3 Pseudo code of cPSO algorithm

1:	**Initialize** the parameters of the cPSO for the process plans optimization (swarm size, maximum number of generation, inertia weight W, acceleration constants C_1 and C_2)
2:	**Initialize** a swarm of particles with velocities and positions $V_{id}^{t+1} = W \cdot V_{id}^{t} + C_1 \cdot chaos() \cdot (P_{ld}^{t} - X_{id}^{t}) + C_2 \cdot Chaos() \cdot (P_{gd}^{t} - X_{id}^{t})$ $X_{id}^{t+1} = X_{id}^{t} + V_{id}^{t+1}$
3:	**Evaluate** the fitness of each particle by using Eqs. 2 and 4
4:	find the global best position P_{gd} and local best position P_{ld}
5:	**Repeat:**
6:	generate next swarm by updating the velocities V_{id} and positions X_{id} of the particles
7:	round off the real number of positions for machine, tool, and TAD to the nearest integer number from machine, tool and TAD sets
8:	evaluate swarm
9:	compute each particle's fitness by using Eqs. 2 and 4
10:	find new global best position and local best position
11:	update the global best position of the swarm and the local best position of each particle
12:	**Until** the maximum of generation for process planning is not met
13:	**Output** the optimal process plan

Algorithm 4 Pseudo code of ALO algorithm

1:	**Initialize** the parameters of the ALO for the process plans optimization (maximum number of iterations, number of ants, number of antlions)
2:	**Initialize** a population of ant's and antlion's positions randomly
3:	**Evaluate** the fitness of each ant and antlion (Eqs. 2 and 4)
4:	find the elite antlion
5:	**Repeat:**
6:	**for** every ant
7:	select an antlion using roulette wheel selection
	create a random walk around antlion
	update position of ant
	end for
9:	compute each ant's fitness by using Eqs. 2 and 4
10:	replace an antlion with its corresponding ant it if becomes fitter
11:	update the elite antlions
12:	**Until** the maximum of iterations is not met
13:	**Output** the optimal process plan

5 Experimental Results

The proposed algorithms are implemented in Matlab environment running on a desktop computer with a 3.10 GHz processor (2 GB RAM). After some preliminary experiments, parameters of GA are set as follows: the size of population is 80, the probability of crossover is 0.60, the probability of mutation is 0.10, and the maximum number of generations is 200. The parameters of the SA algorithm are set as follows: the initial temperature T_0 is set at 1, the final temperature T_{lowest} is set at 0, and the cooling factor T_c is set at 0.00058. Appropriate parameters of the cPSO algorithm are set as follow: the size of population is 80, the maximum generations 200, the inertia weight W is set starting with 1.2 and is linearly decreased to 0.4. Acceleration constants C_1 and C_2 are set to 2.0. Parameters of ALO algorithm are: the size of population of ants is 80, the size of population of antlions is 80, and the maximum number of generations is 200. Three sample parts given in Fig. 1 are used to obtain optimal process plans. The available machining resources consist of 8 machine tools and 12 cutting tools. The transportation time between the machines is given in Table 1 and the information about machines, tools and their costs are given in Table 2. The aim of flexible process planning optimization is to maximize the objective function f_1, where w_{11}–w_{14} in Eq. 1 are set as 1, and f_2, where w_{21}–w_{25} in Eq. 3 are set as 1.

Tables 3, 4, and 5 show the best alternative optimal process plans for parts 1, 2 and 3 obtained under objective f_1. Tables 6, 7, and 8 show the best alternative optimal process plans obtained under objective f_2 for parts 1, 2 and 3, respectively. All results presented in Tables 3, 4, 5, 6, 7, and 8 are obtained using ALO algorithm.

The performance of the ALO algorithm was compared with GA, SA, cPSO for both conditions and the best result, average best result, and improved rate are reported in the Table 9 for comparison. The best result is the best one of all the best results in the 10 runs, average best result is the average of the 10 best results.

Comparative results for three sample parts obtained by cPSO and ALO algorithms under fitness function f_2 are presented in Table 10.

As it can be seen from Table 10, ALO algorithm finds better alternative process plans for part1 and part 3. Also, the best alternative process plan for part 2, with fitness function 0.0015243902 and total production cost 656, finds ALO algorithm.

Table 1 The transportation time between the machines

Machine	1	2	3	4	5	6	7	8
1	0	4	8	10	12	5	6	14
2	4	0	3	7	11	5	4	6
3	8	3	0	5	7	9	8	4
4	10	7	5	0	4	14	12	6
5	12	11	7	4	0	18	12	10
6	5	5	9	14	18	0	6	8
7	6	4	8	12	12	6	0	3
8	14	6	4	6	10	8	3	0

Table 2 Available machining resources and costs in a workshop environment

Machines		
No.	Type	Cost (MCI)
M1	CNC lathe	20
M2	Universal lathe	10
M3	Production lathe	20
M4	Universal milling machine	30
M5	CNC three-axis vertical milling machine	80
M6	CNC three-axis vertical milling machine	50
M7	Drilling machine	10
M8	Coordinate drilling machine	15
Tools		
No.	Type	Cost (TCI)
T1	Turning tool 1	12
T2	Turning tool 2	12
T3	Turning tool 3	12
T4	Turning tool 4	12
T5	Turning tool 5	15
T6	Drill	6
T7	Reaming tool 1	15
T8	Reaming tool 2	20
T9	Mill 1	20
T10	Mill 2	12
T11	Mill 3	20
T12	Mill 4	15
Cost and time indices		
MCCI = 80; SCI = 60; TCCI = 10; TCTI = 6; SCTI = 5		

Table 3 The optimal process plans for sample part 1 (objective f_1)

Operation	1	2	3	4	7	9
Machine	1	1	7	7	1	1
Tool	1	4	6	7	1	1
TAD	+z	+z	+z	+z	−z	−z

$TW = 16.4$; $TT = 12$; $TCT = 24$; $SCT = 10$; $TWTP = 62.4$; $f_1 = 0.0160$

Table 4 The optimal process plans for sample part 2 (objective f_1)

Operation	1	2	5	6	7	8	10
Machine	1	6	6	6	1	1	1
Tool	1	10	6	8	1	1	1
TAD	+z	+z	+z	+z	+z	−z	−z

$TW = 25.8$; $TT = 10$; $TCT = 24$; $SCT = 15$; $TWTP = 74.8$; $f_1 = 0.0134$

Table 5 The optimal process plans for sample part 3 (objective f_1)

Operation	8	9	10	11	13	14
Machine	6	1	6	6	1	6
Tool	10	4	9	9	4	12
TAD	−z	−z	−z	−y	+z	+z

$TW = 14.1$; $TT = 20$; $TCT = 24$; $SCT = 25$; $TWTP = 83.1$; $f_1 = 0.0120$

Table 6 The optimal process plans for sample part 1 (objective f_2)

Operation	1	2	3	4	7	9
Machine	2	2	7	7	2	2
Tool	1	4	6	7	1	1
TAD	+z	+z	+z	+z	−z	−z

$MC = 60$; $TC = 69$; $MCC = 160$; $TCC = 40$; $SCC = 120$; $TWC = 449$; $f_2 = 0.00222717$

Table 7 The optimal process plans for sample part 2 (objective f_2)

Operation	1	2	5	6	7	8	10
Machine	2	6	6	6	2	2	2
Tool	2	10	6	8	1	1	1
TAD	+z	+z	+z	+z	+z	−z	−z

$MC = 190$; $TC = 86$; $MCC = 160$; $TCC = 40$; $SCC = 180$; $TWC = 656$; $f_2 = 0{,}00152439$

Table 8 The optimal process plans for sample part 3 (objective f_2)

Operation	1	2	3	4	6	7
Machine	2	6	2	2	6	6
Tool	1	10	1	1	9	11
TAD	+z	+z	+z	−z	−z	+x

$MC = 180$; $TC = 88$; $MCC = 240$; $TCC = 40$; $SCC = 300$; $TWC = 848$; $f_2 = 0.00117924528$

Table 9 The results of flexible process planning for part 2

Objective	Method	Best result	Average best result	IR_{max} (%)	IR_{avg} (%)
f_1	GA	0.0132	0.0130	6.7	8.9
	SA	0.0125	0.0122		
	cPSO	0.0134	0.0133		
	ALO	0.0134	0.0134		
f_2	GA	0.0014	0.0013	21.6	33.3
	SA	0.0012	0.0010		
	cPSO	0.0015[a]	0.0014		
	ALO	0.0015[b]	0.0015		

[a] $TWC = 666$
[b] $TWC = 656$

Table 10 The results of flexible process planning for three sample parts—a comparative review of the results obtained by cPSO and ALO algorithm

Part	Alternative process plans	Algorithm/fitness	
		cPSO f_2 (TWC)	ALO f_2 (TWC)
1	1	0.00222717 (449)	0.00222717149 (449)
	2	0.002178649 (459)	**0.0021786492 (459)**
	3	0.002132196 (469)	0.0021786492(459)
2	1	0.00150150150 (666)	**0.0015243902 (656)**
	2	0.0014577259 (686)	0.00150150150 (666)
	3	0.0014367816 (696)	0.0014577259 (686)
3	1	0.001179245 (848)	0.001179245 (848)
	2	0.001165501 (858)	**0.00117924528 (848)**
	3	0.001138952 (878)	0.001165501 (858)

Note Bold values indicate the best results

6 Conclusion

In this paper, four biologically inspired optimization algorithms, namely, genetic algorithms (GA), simulated annealing (SA), chaotic particle swarm optimization algorithm (cPSO), and Ant Lion Optimization (ALO) algorithm, are proposed to optimize combinatorial NP-hard flexible process planning problem. The network representation method is adopted to describe process flexibility, sequencing flexibility, machine flexibility, tool flexibility, and TAD flexibility. Mathematical modeling of the objective functions i.e. production time and production cost is also presented. All the algorithms are implemented in Matlab environment. The performance of the presented ALO algorithm are verified and evaluated in comparison with the results obtained with GA, SA, and cPSO standalone algorithms. Experimental results indicate that the proposed ALO algorithm performs better in comparison with other bio-inspired optimization algorithms.

Acknowledgements This paper is part of the project "An innovative, ecologically based approach to the implementation of intelligent manufacturing systems for the production of sheet metal parts". The research in this paper was supported by the Ministry of Education, Science and Technological Development of the Serbian Government, Grant TR-35004 (2011–2017).

References

1. Guo YW, Mileham AR, Owen GW, Li WD (2006) Operation sequencing optimization using a particle swarm optimization approach. Proc Inst Mech Eng Part B J Eng Manuf 220 (12):1945–1958
2. Huang W, Hu Y, Cai L (2012) An effective hybrid graph and genetic algorithm approach to process planning optimization for prismatic parts. Int J Adv Manuf Technol 62(9–12):1219–1232

3. Kennedy J, Eberhart RC (1995) Particle swarm optimization. In: Proceedings of the IEEE international conference on neural network, vol 4. Perth, pp 1942–1948
4. Li XY, Shao XY, Gao L (2008) Optimization of flexible process planning by genetic programming. Int J Adv Manuf Technol 38(1–2):143–153
5. Liu XJ, Yi H, Ni ZH (2013) Application of ant colony optimization algorithm in process planning optimization. J Intell Manuf 24(1):1–3
6. Lv S, Qiao L (2013) A cross-entropy-based approach for the optimization of flexible process planning. Int J Adv Manuf Technol 68(9–12):2099–2110
7. Mirjalili S (2015) The ant lion optimizer. Adv Eng Softw 83:80–98
8. Nallakumarasamy G, Srinivasan PS, Raja KV, Malayalamurthi R (2011) Optimization of operation sequencing in CAPP using simulated annealing technique (SAT). Int J Adv Manuf Technol 54(5–8):721–728
9. Petrović M (2016) Design of intelligent manufacturing systems by using artificial intelligence. University of Belgrade—Faculty of Mechanical Engineering (Doctoral Dissertation), Serbian, pp 1–319
10. Petrović M, Mitić M, Vuković N, Miljković Z (2016) Chaotic particle swarm optimization algorithm for flexible process planning. Int J Adv Manuf Technol 85(9–12):2535–2555
11. Petrović M, Vuković N, Mitić M, Miljković Z (2016) Integration of process planning and scheduling using chaotic particle swarm optimization algorithm. Expert Syst Appl 64:569–588
12. Shao X, Li X, Gao L, Zhang C (2009) Integration of process planning and scheduling—a modified genetic algorithm-based approach. Comput Oper Res 36(6):2082–2096
13. Wolpert DH, Macready WG (1997) No free lunch theorems for optimization. IEEE Trans Evol Comput 1(1):67–82

Modelling and Analysis of 3-Axis Reconfigurable Hybrid Kinematics Mechanism with Translatory Actuated Joints

Goran Vasilic, Sasa Zivanovic and Branko Kokotovic

Abstract Modelling and analysis of a 3-axis reconfigurable hybrid kinematic mechanism is shown in this paper. Generalized model for solving the inverse and direct kinematic problem is presented. Generalized equations show the solution of kinematic problems of hybrid mechanism applied for any configuration of reconfigurable machine. 3-axis hybrid mechanism consists of a 2-axis parallel mechanism and added serial translatory axis. Chosen characteristic mechanism configurations are shown as the realization examples. Workspace is determined for chosen configurations and possible singular mechanism positions are analysed.

Keywords 3-axis reconfigurable hybrid kinematic mechanism · Inverse and direct kinematic problem · Workspace · Singularities

1 Introduction

Parallel kinematics machines (PKM) compared with serial structured machine tools have many advantages. PKM have higher rigidity, higher velocity and better mobility than serial structural machines and the loads are transferred to all operating axes with errors being averaged according to the number of axes, etc. Besides these characteristics which present the advantages of PKM, there are disadvantages like the workspace of irregular shape and small volume, complicated solving of

G. Vasilic (✉) · S. Zivanovic · B. Kokotovic
Faculty of Mechanical Engineering, Production Engineering Department,
University of Belgrade, Kraljice Marije 16, 11000 Belgrade, Serbia
e-mail: goran.v.vasilic@gmail.com

S. Zivanovic
e-mail: szivanovic@mas.bg.ac.rs

B. Kokotovic
e-mail: bkokotovic@mas.bg.ac.rs

V. Majstorovic and Z. Jakovljevic (eds.), *Proceedings of 5th International Conference on Advanced Manufacturing Engineering and Technologies*,
Lecture Notes in Mechanical Engineering, DOI 10.1007/978-3-319-56430-2_32

kinematic problems, existence of singularities, etc. In order to realize machine tools with the advantages, parallel and serial kinematics appeared as hybrid solutions of these machines by combining parallel and serial mechanisms.

Many researchers have contributed to the study of the hybrid type milling machines [1–10]. The problem solving of hybrid mechanisms is the subject of many researches. Within the scientific papers [1], hybrid mechanism with six degrees of freedom based on Stewart's platform [4] was considered and it contains two parallel mechanisms with serial connection. Hybrid mechanisms containing one parallel and one serial mechanism were also considered. In the paper [5], the combination of Hexapod parallel mechanism with six degrees of freedom was analysed where the rotating table with two degrees of freedom was added. As the result total of eight degrees of mobility of hybrid mechanism was achieved, five degrees were used for programming and three for optimal trajectory planning. The subject of paper [6] is also the considerations of combination of serial and parallel mechanisms. Serial mechanism with four degrees of freedom with wide workspace was used for the hybrid mechanism. Parallel, high precision, mechanism with three degrees of freedom was added to the serial mechanism.

In this paper we consider the upgrade of planar, 2-axis reconfigurable parallel mechanism MOMA [11] to 3-axis machine with hybrid kinematics by adding serial translatory axis. Parallel mechanism MOMA (Serbian acronym for Modular Machine Tool) is built according to Trijoint machines [7], two parallelogram mechanism [8], machine tool for triaxial machining of work pieces [9] and redundantly Actuated 3DOF planar parallel manipulator [10]. Parallel mechanism MOMA [11] is reconfigurable mechanism, so its configuration can be quickly changed [12]. According to the building program of parallel mechanism MOMA [13], reconfiguration of the software and hardware system of machine tool is being done. Five basic types of mechanism are defined by the building program (Fig. 1), and reconfiguration according to the current needs can be achieved without further costs [14].

Machine tool with hybrid kinematics MOMA3_HK (Fig. 2) is going to be considered in this paper. Respecting that hybrid mechanism contains reconfigurable parallel mechanism, machine MOMA3_HK is reconfigurable itself. The expected results of the analyses of the hybrid mechanism MOMA3_HK present setting the equations which can be used for configuring of the control system. Because of the reconfigurability of the mechanism, essential requirement for equations is that they should be generalized in order to be applicable for any possible configuration. Equations are also used for the analyses of reconfigurable hybrid mechanism, primarily for determining the workspace of the mechanism and then for forming the Jacobian matrix from which the singular configurations of mechanism are determined and which is utilized for optimization of some geometric dimensions of mechanism [15].

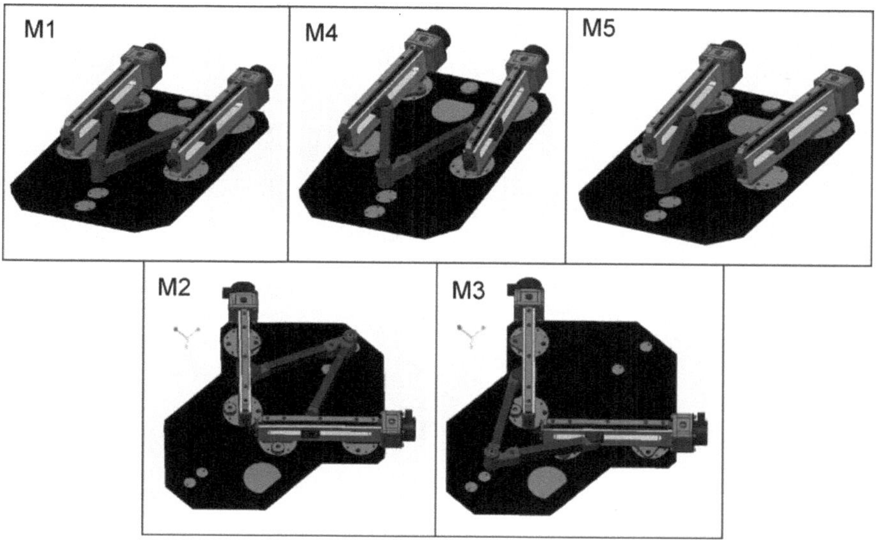

Fig. 1 CAD models of the basic types M1–M5 reconfigurable 2-axis parallel mechanism

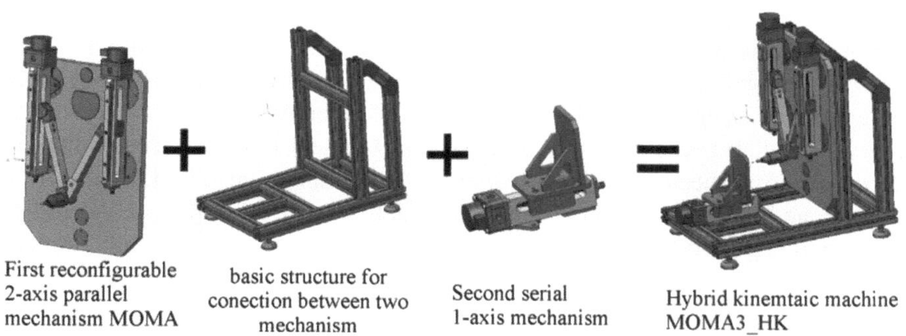

Fig. 2 Hybrid kinematic machine MOMA3_HK

2 Description of 3-Axis Hybrid Kinematics Mechanism

CAD model of machine with hybrid kinematics MOMA3_HK shown in Fig. 3 presents 3-axis horizontal milling machine with the tool moving in the direction of two axes (X, Y) and the workpiece moving along the third axis (Z). Parallel mechanism MOMA consists of two driving axes DA1 and DA2 and the sliders S_1 and S_2 which can be moved along its guides. Links of parallel mechanism L_1 and L_2 are connected to their sliders using revolute joints. Links are also connected to each other by the revolute joint which presents the platform of parallel mechanism P, which carries the main spindle. Moving the working table, with workpiece fixed on, is achieved by using additional translatory operating axis DA3.

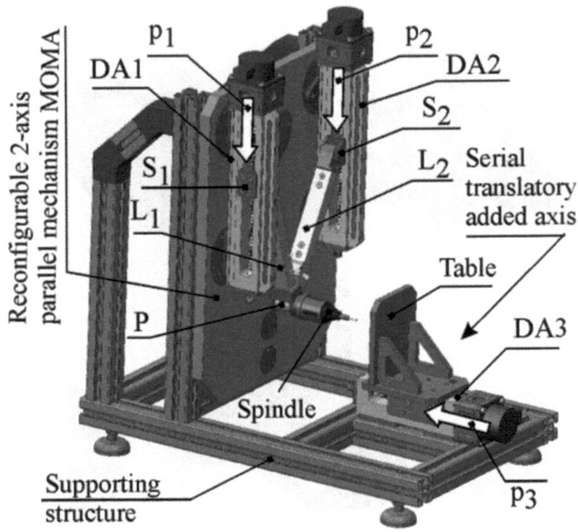

Fig. 3 CAD model of the hybrid mechanism MOMA3_HK

Driving axes DAi are used to control the internal coordinates of the hybrid mechanism pi, i = 1, 2, 3, also shown in Fig. 3, and by them required position of the top of the tool relative to the workpiece is achieved. The tool position relative to the workpiece presents the external coordinates of mechanism MOMA3_HK achieved either by manual programming of numeric driven machine tool or by programming using CAD/CAM system. Links of different length can be used in the case of parallel mechanism and driving axes DA1 and DA2 can change the orientation during mechanism reconfiguration. Reconfiguration of MOMA3_HK is being done by the orientation change of operating axes DA1 and DA2. According to the building program [13], links can have lengths l = 180 mm, l = 195 mm and l = 250 mm. Orientation of the operating axis DAi is defined by the angle α_i which presents the angle between the X axis and the direction of operating axis, while the position of each operating axis is defined relative to the respecting referent point Ri, i = 1, 2, 3, which is fixed for particular driving axis. By the building program of parallel mechanism MOMA [13] five types of mechanism are defined and we'll consider configurations M1 (Fig. 4a, b), configurations M4 (Fig. 4c, d) and configurations M5 (Fig. 4e, f), and all their subvariants which can be the result of reconfiguring. For each of the configurations in Fig. 4 orientation angle of operating axis α is determined by auxiliary angle γ_i, and it is $\alpha_i = -\pi/2 + \gamma_i$ where angle γ_i can have the value $\gamma_i = +5°$, $\gamma_i = -5°$ and $\gamma_i = 0°$.

3 Geometric Model of Hybrid Mechanism MOMA3_HK and Solving the Kinematic Problem

Geometric model of hybrid mechanism MOMA3_HK is shown in Fig. 5. For the further analyses of the mechanism coordinate systems are adopted: {W} in connection with workpiece and stationary coordinate system $\{B_1\}$. The programming

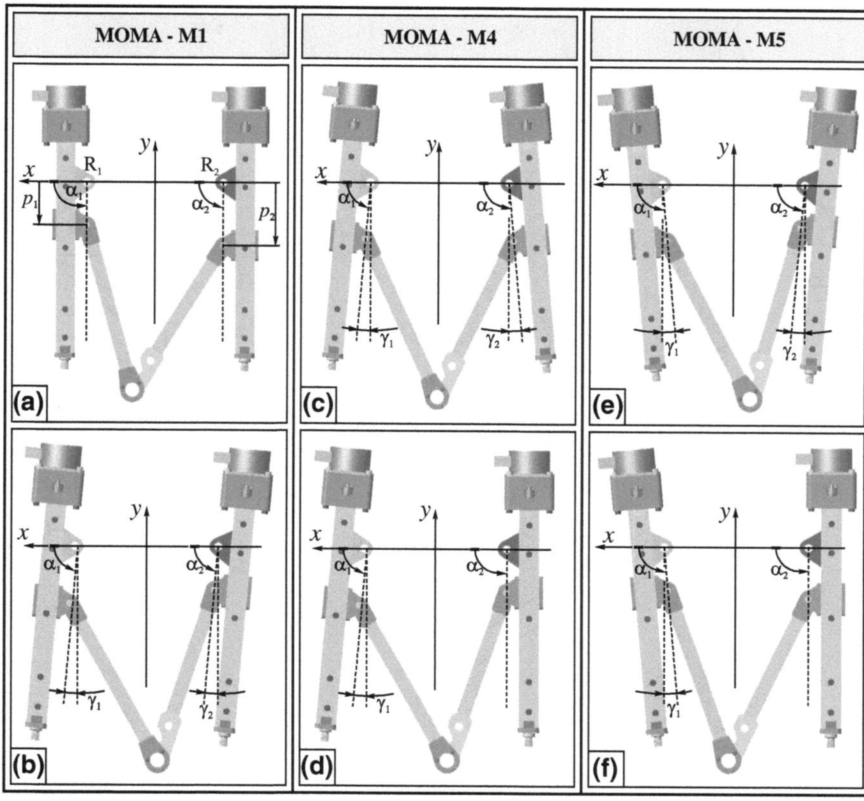

Fig. 4 Configurations of reconfigurable 2-axis parallel mechanism with their basic parameters

of machine tool is being done in coordinate system {W} so the vector which determines the position of the top of the tool in the coordinate system of the workpiece is known and its coordinates are $^{OW}\boldsymbol{p}_T = [X_{Tw}\ Y_{Tw}\ Z_{Tw}]^T$ and these coordinates are the internal coordinates of the hybrid mechanism at the same time. The coordinates of the vector of the top of the tool in coordinate system {B_1} are defined by vector $^{OB1}\boldsymbol{p}_T = [X_{T,B1}\ Y_{T,B1}\ Z_{T,B1}]^T$ and they change with the shift of internal coordinates p_1, p_2 and p_3. Finding the subordinate $p_i = f(^{OW}\boldsymbol{p}_T)$, or the relation between internal and external coordinates presents the solution of inverse kinematic problem (IKP). Contrary to that, finding the relation between external and internal coordinates $^{OW}\boldsymbol{p}_T = f(p_i)$ presents the solution of direct kinematic problem (DKP).

Vectors and values necessary for solving kinematic problems are shown in Fig. 5:

- $^{B1}\boldsymbol{p}_{OW}$—vector which defines the position of the coordinate system of workpiece {W} in coordinate system {B_1}. After setting the preparation workpiece on working table this vector becomes known;

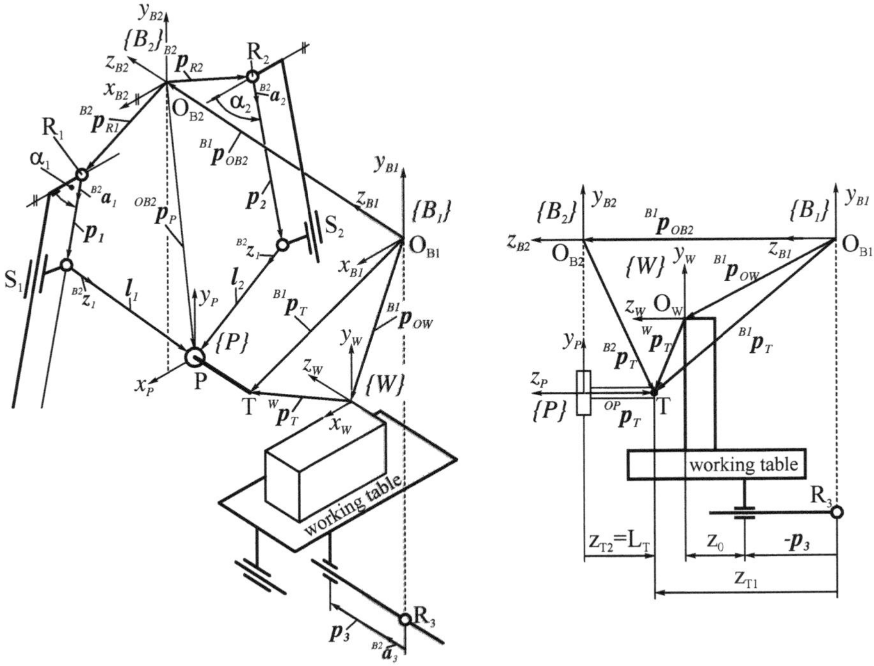

Fig. 5 Geometric model of MOMA3_HK hybrid mechanism

- ${}^{B1}\boldsymbol{p}_{OB2}$—vector which defines the position of coordinate system $\{B_2\}$ in relative to the coordinate system $\{B_1\}$. The movement of parallel mechanism is observed in plane $Ox_{B2}y_{B2}$. Vector is defined by the geometry of the hybrid mechanism so it is known;
- ${}^{B2}\boldsymbol{p}_P$—vector which defines the position of the platform in coordinate system $\{B_2\}$. This vector depends on the internal coordinates p_1 and p_2;
- ${}^{Bi}\boldsymbol{p}_T$—vector which determines the position of the top of the tool in relation to the coordinate systems $\{B_1\}$ and $\{B_2\}$;
- $\boldsymbol{p}_i = {}^{B2}\boldsymbol{a}_i \cdot p_i$—vectors of the internal coordinates determined by the unit vectors of direction ${}^{B2}\boldsymbol{a}_i$ and scalar values of internal coordinates;
- $\boldsymbol{l}_i = {}^{B2}\boldsymbol{z}_i \cdot l_i$—vectors determined by the length of the links and the directions of unit vectors ${}^{B2}\boldsymbol{z}_i$;
- z_0—Value determined after setting the workpiece.

According to the shown geometric model, the following vector equations can be obtained:

$$ {}^{B_1}\boldsymbol{p}_T = {}^{B_1}\boldsymbol{p}_{OW} + {}^{W}\boldsymbol{p}_T = {}^{B_1}\boldsymbol{p}_{OB_2} + {}^{B_2}\boldsymbol{p}_{OP} + {}^{OP}\boldsymbol{p}_T \quad (1) $$

$$ {}^{B_2}\boldsymbol{p}_T = {}^{B_2}\boldsymbol{p}_{OP} + {}^{P}\boldsymbol{p}_T \tag{2}$$

The following equations can also be obtained:

$$ {}^{B_2}\boldsymbol{p}_{OP} = {}^{B_2}\boldsymbol{p}_{Ri} + \boldsymbol{p}_i + \boldsymbol{l}_i = {}^{B_2}\boldsymbol{p}_{Ri} + {}^{B_2}\boldsymbol{a}_i \cdot p_i + {}^{B_2}\boldsymbol{z}_i \cdot l_i \tag{3}$$

or in alternative notation:

$$ {}^{B_2}\boldsymbol{z}_i \cdot l_i = {}^{B_2}\boldsymbol{p}_{OP} - {}^{B_2}\boldsymbol{p}_{Ri} - {}^{B_2}\boldsymbol{a}_i \cdot p_i = k_i \cdot {}^{B_2}\boldsymbol{w}_i - {}^{B_2}\boldsymbol{a}_i \cdot p_i \tag{4}$$

By squaring and fixing the Eq. (4) we can get the general implicit equation of parallel mechanism which connects internal coordinates p_1 and p_2 with external coordinates x_{Tw} and y_{Tw}:

$$ p_i^2 - 2p_i\left({}^{B_2}\boldsymbol{a}_i \cdot k_i^{B_2}\boldsymbol{w}_i\right) + \left(k_i \cdot {}^{B_2}\boldsymbol{w}_i\right)^2 - l_i^2 = 0 \tag{5}$$

Implicit equation which connects internal coordinate p_3 with external coordinates z_{Tw} can be obtained using Eq. (1) and by projecting on z axis:

$$ -p_3 + z_0 + Z_{Tw} - z_{T_1} = 0 \tag{6}$$

The Eqs. (5) and (6) are the basis for further solving of kinematic problems.

3.1 The Solution of Inverse Kinematic Problem

Considering the Eq. (2) and coordinates of vector ${}^{P}\boldsymbol{p}_T = [00L_T]^T$ and the fact that vector ${}^{B1}\boldsymbol{p}_{OB2} =$ const. the coordinates of the top of the tool and the platform of the parallel mechanism in coordinate system $\{B_2\}$ can be obtained based on the following matrix equation:

$$ \begin{bmatrix} x_P \\ y_P \end{bmatrix} = \begin{bmatrix} x_T \\ y_T \end{bmatrix} = \begin{bmatrix} {}^{B2}x_{OW} \\ {}^{B2}y_{OW} \end{bmatrix} + \begin{bmatrix} X_{Tw} \\ Y_{Tw} \end{bmatrix} \tag{7}$$

By substituting index i = 1, 2 two equations can be obtained with only one internal coordinate of two external coordinates.

$$ \begin{aligned} p_1^2 &+ 2p_1\left[a_{x1}(x_P - x_{R1}) + a_{y1}(y_P - y_{R1})\right] \\ &+ (x_P - x_{R1})^2 + (y_P - y_{R1})^2 - l_1^2 = 0 \end{aligned} \tag{8}$$

$$ \begin{aligned} p_2^2 &+ 2p_2\left[a_{x2}(x_P - x_{R2}) + a_{y2}(y_P - y_{R2})\right] \\ &+ (x_P - x_{R2})^2 + (y_P - y_{R2})^2 - l_2^2 = 0 \end{aligned} \tag{9}$$

After the introduction of substitutions for Eqs. (7) and (8):

$$\begin{aligned} B_i &= a_{xi}(x_P - x_{Ri}) + a_{yi}(y_P - y_{Ri}) \\ C_i &= (x_P - x_{Ri})^2 + (y_P - y_{Ri})^2 - l_i^2; \mathrm{i} = 1,2 \end{aligned} \tag{10}$$

and solving squared equations we can obtain the equations which give dependence to internal coordinates p_1 and p_2 from the platform coordinates x_p and y_p and they are:

$$p_i = B_i \pm \sqrt{B_i^2 - C_i} \tag{11}$$

As the coordinate of the top of the tool z_{T1} in coordinate system $\{\mathrm{B}_1\}$ is constant and value z_0 doesn't change after setting the workpiece, using the Eqs. (6), the dependence of internal coordinate p_3 from external coordinate z_{Tw} can be obtained:

$$p_3 = z_0 + z_{Tw} - z_{T1} = z_{cor} + z_{Tw} \tag{12}$$

Compensation of the tool length as well as the position of the workpiece in the direction of the z axis on the working table can be done by value z_{cor}. Equations (11) and (12) describe dependence of internal coordinates p_1, p_2 and p_3 from external coordinates x_{Tw}, y_{Tw} and z_{Tw} and they present solution of inverse kinematic problem of hybrid mechanism MOMA3_HK.

3.2 The Solution of Direct Kinematic Problem

By solving the square Eqs. (8) and (9) by coordinates x_P and y_P and expressing z_{Tw} from the Eq. (12), we can obtain:

$$y_P = \frac{-v_{10} - \sqrt{v_{10}^2 - 4v_9 v_{11}}}{2v_9} \tag{13}$$

$$x_P = v_7 - y_P v_8 \tag{14}$$

$$z_{Tw} = p_3 - z_{cor} \tag{15}$$

In the Eqs. (11) to (13) the following substitutions were used:

$$
\begin{aligned}
v_1 &= 2(p_1 a_{x1} + x_{R1});\ v_2 = 2(p_1 a_{y1} + y_{R1}) \\
v_3 &= p_1^2 + 2p_1\left(a_{x1}x_{R1} + a_{y1}y_{R1}\right) - l_1^2 + x_{R1}^2 + y_{R1}^2 \\
v_4 &= 2(p_2 a_{x2} + x_{R2});\ v_5 = 2(p_2 a_{y2} + y_{R2}) \\
v_6 &= p_2^2 + 2p_2\left(a_{x2}x_{R2} + a_{y2}y_{R2}\right) - l_2^2 + x_{R2}^2 + y_{R2}^2 \\
v_7 &= (v_6 - v_3)/(v_4 - v_1);\ v_8 = (v_2 - v_5)/(v_4 - v_1);\ v_9 = 1 + v_8^2 \\
v_{10} &= 2v_7v_8 - v_1v_8 - v_2 : v_{11} = v_7^2 - v_1v_7 + v_3
\end{aligned} \tag{16}
$$

Generalized Eqs. (13–15) present the solution of direct kinematic problem and they are valid for hybrid mechanism MOMA3_HK with any configuration of parallel mechanism shown in Fig. 4.

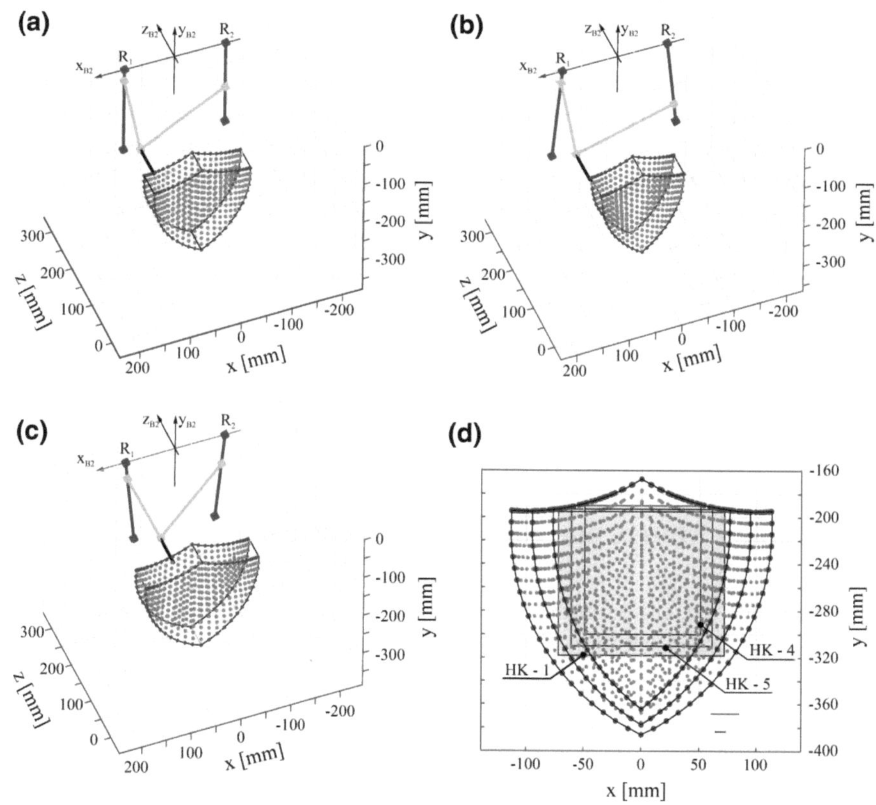

Fig. 6 Workspaces of the hybrid mechanism MOMA3_HK: (**a**) MOMA_3HK-1; (**b**) MOMA_3HK-4; (**c**) MOMA_3HK-5; (**d**) real workspaces of three configurations of hybrid mechanism MOMA3_HK

4 Workspace of Hybrid Mechanism MOMA3_HK

For determining the workspace of hybrid mechanism MOMA3_HK equations which present the solution of DKP were used. Values of internal coordinates are put into Eqs. (13–15) starting from values $p_i = 0$ to the maximum values which are allowed by the length of the guides. For thus defined values the values of external coordinates can be obtained. In this way we can determine theoretically attainable workspace of mechanism which is different from the real workspace.

In Fig. 6 we can see the workspaces of the hybrid mechanism with different configurations of parallel mechanism MOMA. For all three types of parallel mechanism links with the length $l_1 = l_2 = 195$ mm were used and guides with the stroke length Lv = 200 mm. The coordinates of the referent points are $R_1 = (100\ 0\ 300)$, $R_2 = (-100\ 0\ 300)$ and $R_3 = (0\ -5150)$ in coordinate system $\{B_1\}$. The angles of guide orientation for different configurations of parallel mechanism are: (i) M1: $\alpha_1 = -\pi/2$ i $\alpha_2 = -\pi/2$, (Fig. 6a); (ii) M4: $\alpha_1 = -\pi/2 + 5°$, $\alpha_2 = -\pi/2 - 5°$, (Fig. 6b); (iii) M5: $\alpha_1 = -\pi/2 - 5°$, $\alpha_2 = -\pi/2 + 5°$, (Fig. 6c). These figures show the maximum attainable theoretical workspace. Some areas of these workspaces can't be reached because of physical limits. Such areas relate to the points of the workspace which are placed on the left side of the guide of the operating axis DA1 and on the right side of the guide of the operating axis DA2. Figure 6 shows the bounded parts of workspaces for all three considered configurations of hybrid mechanism which are used for the machining. These bounded parts of the workspace have regular geometric shape and every point in this space is reliably attainable regardless of the construction and geometry of parallel mechanism. Dimensions of the workspaces in the directions of X and Y axis for considered configurations of hybrid mechanism are: (i) b × h = 125 × 100 [mm] for MOMA3_HK-1; (ii) b × h = 120 × 90 [mm] for MOMA3_HK-4; (iii) b × h = 130 × 125 [mm] for MOMA3_HK-5.

5 Jacobian Matrix and Singularity Analysis

If we mark the Eq. (8) as implicit equation f_1, Eq. (9) as implicit equation f_2 and Eq. (6) as implicit equation f_2 then the Jacobian matrix of hybrid mechanism is:

$$J = J_P^{-1} \cdot J_x = \begin{bmatrix} \frac{\partial f_1}{\partial p_1} & \frac{\partial f_1}{\partial p_2} & \frac{\partial f_1}{\partial p_3} \\ \frac{\partial f_2}{\partial p_1} & \frac{\partial f_2}{\partial p_2} & \frac{\partial f_2}{\partial p_3} \\ \frac{\partial f_3}{\partial p_1} & \frac{\partial f_3}{\partial p_2} & \frac{\partial f_3}{\partial p_3} \end{bmatrix}^{-1} \cdot \begin{bmatrix} \frac{\partial f_1}{\partial x_{Tw}} & \frac{\partial f_1}{\partial y_{Tw}} & \frac{\partial f_1}{\partial z_{Tw}} \\ \frac{\partial f_2}{\partial x_{Tw}} & \frac{\partial f_2}{\partial y_{Tw}} & \frac{\partial f_2}{\partial z_{Tw}} \\ \frac{\partial f_3}{\partial x_{Tw}} & \frac{\partial f_3}{\partial y_{Tw}} & \frac{\partial f_3}{\partial z_{Tw}} \end{bmatrix} \tag{17}$$

Partial derivatives of implicit functions f_i by internal and external coordinates are given in Eqs. (18) to (22).

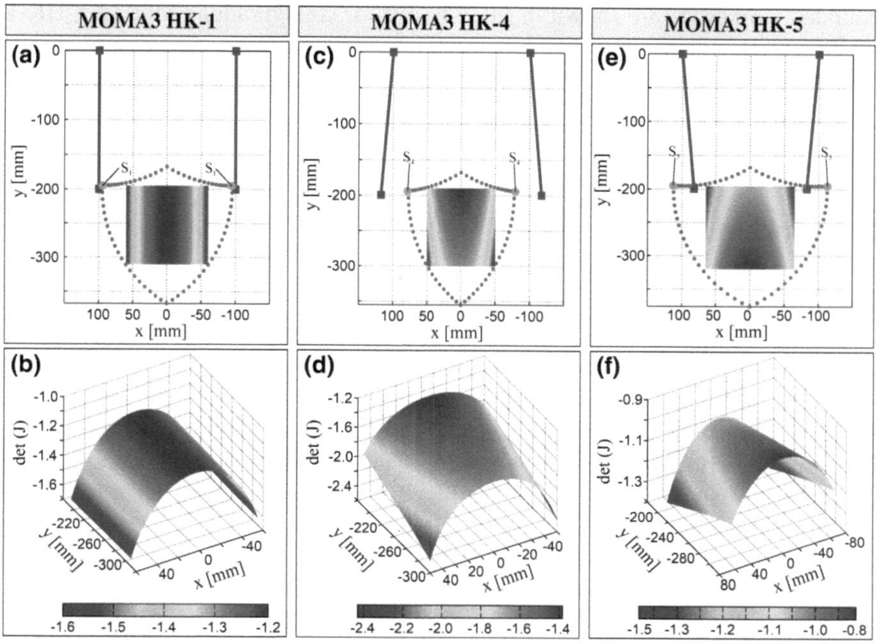

Fig. 7 The distribution of det(J) for different configurations of hybrid mechanism MOMA3_HK

$$\frac{\partial f_i}{\partial p_i} = 2p_i\left[a_{xi}(x_{Tw} - x_{Ri}) + a_{yi}(y_{Tw} - y_{ri})\right]; \quad i = 1, 2 \tag{18}$$

$$\frac{\partial f_i}{\partial p_j} = 0; \quad i = 1, 2, 3; \; j = 1, 2, 3; \; i \neq j \tag{19}$$

$$\frac{\partial f_3}{\partial p_3} = -1 \tag{20}$$

$$\frac{\partial f_i}{\partial x_{Tw}} = 2(x_{Tw} - x_{Ri}) - 2p_i a_{xi}; \; \frac{\partial f_i}{\partial y_{Tw}} = 2(y_{Tw} - y_{Ri}) - 2p_i a_{yi}; \quad i = 1, 2 \tag{21}$$

$$\frac{\partial f_3}{\partial x_{Tw}} = \frac{\partial f_3}{\partial y_{Tw}} = 0; \quad \frac{\partial f_3}{\partial z_{Tw}} = 1 \tag{22}$$

By the analysis of the matrix from the Eq. (17) and calculating the values of det(J) for each point of theoretic workspace of hybrid mechanism MOMA3_HK singularities can be obtained. For three considered configurations of hybrid mechanism singularities are placed on the margins of theoretic workspace and they are marked as S1, S4 and S5 in Fig. 7a, c, e. The points where singularities appear are placed outside the bounded workspace which is used for the work of mechanism, as shown in Fig. 7a, c, e. Distribution of the values of det(J) for the points inside the bounded workspace are shown in Fig. 7b, d, f for configurations MOMA3_HK-1, MOMA3_HK-4 i MOMA3_H-5 respectively, and we can say for sure that in boundaries of workspace which is used for the work of mechanism there are no singularities.

6 Conclusion

The basic aim of this paper was to establish the mathematical model for a machine from the family of hybrid machines based on reconfigurable 2-axis parallel mechanism. This paper also considers 3-axis machine tool with hybrid kinematics MOMA3_HK. For this purpose the equations which present the solution of inverse and direct kinematic problem are given. By the equations of IKP and DKP we can absolutely describe the work of hybrid mechanism, and because they are given in general form they are valid for any configuration of parallel mechanism within the hybrid mechanism.

The equations of IKP and DKP were primarily used to analyse the workspace of hybrid mechanism and then to determine Jacobian matrix in order to analyse the singularities of the mechanism. As all three configurations have similar shape and the size of workspace, given equations are further used for determining the optimal configuration of mechanism according to some given criteria. Further research plan is to carry out the analysis of hybrid mechanism with more than 3 axes and for different material processing methods like: 4-axis and 5-axis milling, wire cutting EDM, production of parts by additive technologies, etc.

References

1. Romdhane L (1995) Design and analysis of a hybrid serial-parallel manipulator. Mech Mach Theory 34:1037–1055
2. Tanev T (2000) Kinematics of a hybrid (parallel-serial) robot manipulator. Mech Mach Theor 34:1183–1196
3. Zheng X, Bin Z, Luo G (2004) Kinematic analysis of a hybrid serial-parallel manipulator. Int J Adv Manuf Tech 23:925–930
4. Stewart D (1965) A platform with six degree of freedom. Proc Inst Mech Eng London 180 (1):371–386
5. Harib K, Sharif A, Hammami A (2007) Hexapod-based machine tool with hybrid structure: kinematic analysis and trajectory planning. Int J Mach Tools Manuf 47:1426–1432
6. Choi H, Han C, Lee K, Lee S (2005) Development of hybrid robot for construction works with pneumatic actuator. Autom Constr 14:452–459
7. Šika Z, Hamrle V, Valášek M, Beneš P (2012) Calibrability as additional design criterion of parallel kinematic machines. Mech Mach Theory 50:48–63
8. Jinson W, Tiemin L, Liping W (2007) Dynamic analysis of the 2-DOF planar parallel manipulator of a heavy duty hybrid machine tool. Int J Adv Manuf Technol 34:413–420
9. Hanrath G, Stengele B (2001) Machine tool for triaxial machining of workpieces. US Patent 6,328,510 Dec. 2001

10. Jun W, Jinsong W, Liping W (2008) Optimal kinematic design and application of a redundantly actuated 3DOF planar parallel manipulator. J Mech Des 130(5). doi:10.1115/1.2890118
11. Vasilic G, Zivanovic S (2016) Modeling and analysis of 2-axis reconfigurable parallel mechanism MOMA with translatory actuated joints. TECHNICS special edition. In: Magazine of the Society of Engineers and Technicians of Serbia, Mechanical Engineering, vol LXXI, pp 59–66. doi:10.5937/tehnika1601057V
12. Mehrabi MG, Ulsoy AG, Koren Y (2000) Reconfigurable manufacturing systems and their enabling technologies. Int J Manuf Technol Manage 1(1):114–131
13. Zivanovic S, Vasilic G (2014) Variants of configuring the 2-axis reconfigurable parallel mechanism—MOMA. In Proceedings of 2nd International Scientific Conference on Mechanical Engineering Technologies and Applications COMETa 2014 (in Serbian). University of East Sarajevo, Faculty of Mechanical Engineering, Jahorina, B&H, Republic of Srpska, pp 33–40, Dec 2014
14. Landers G, Mid K, Koren Y (2001) Reconfigurable machine tools. CIRP Ann Manuf Technol 50(1):269–274
15. Vasilic G, Zivanovic S, Kokotovic B, Glavonjic M (2014) Optimization of the length of legs of the 2-DOF reconfigurable parallel mechanism—MOMA. In Proceedings of the 39th JUPITER Conference, 35th Symposium "NC-ROBOTS-FMS" (in Serbian). Faculty of Mechanical Engineering, Belgrade, Serbia, Oct 2014, pp 3.28–3.35

Influence of Ultrasonic Assistance on Delamination During Machining of CFRP Composite

Marcel Kuruc, Martin Necpal, Tomáš Vopát, Vladimír Šimna and Jozef Peterka

Abstract Many nowadays researches are focused on decreasing the weight of components and increasing their strength, or obtaining unique properties. One of the results of the researches are composite materials. However, different material properties of matrix and reinforcing material cause certain problems during conventional machining, such as rapid tool wear and delamination of composite. Therefore, other researches start focusing on machinability of this kind of materials. Scientists have discovered several ways of how to satisfactorily machine composite materials. Seco Tools Company starts developing a new design of the tools. Researchers of the Stanford University have devised a diamond-like carbon (DLC) coating available to effectively cut matrix as well as reinforce fibers. DMG Mori Company has implemented ultrasonic assistance into machining process and reached enhanced machinability. This article compares conventional milling and ultrasonic assisted milling of the carbon fiber reinforced polymer (CFRP) composite material by a special designed cutting tool. The results could be applied to determine when ultrasonic assistance is advantageous and what advantages it brings about.

Keywords Ultrasonic assisted machining · CFRP · Composite · Delamination

M. Kuruc (✉) · M. Necpal · T. Vopát · V. Šimna · J. Peterka
Institute of Production Technologies, Faculty of Materials Science and Technology, Slovak University of Technology, Bottova 25, 917 24 Trnava, Slovakia
e-mail: marcel.kuruc@stuba.sk

M. Necpal
e-mail: martin.necpal@stuba.sk

T. Vopát
e-mail: tomas.vopat@stuba.sk

V. Šimna
e-mail: vladimir.simna@stuba.sk

J. Peterka
e-mail: jozef.peterka@stuba.sk

V. Majstorovic and Z. Jakovljevic (eds.), *Proceedings of 5th International Conference on Advanced Manufacturing Engineering and Technologies*, Lecture Notes in Mechanical Engineering, DOI 10.1007/978-3-319-56430-2_33

1 Introduction

Different mechanical properties of each compound in composite material cause challenges in its machining. Recommended machining parameters for hard and strong materials such as reinforcing fibre are different from the machining parameters for soft and tough materials such as matrix. That means, the machining parameters proper for one compound may be improper for others compounds. Improper machining parameters could cause melting of the matrix, tearing out the fibers, edge-chipping of the composite, rapid tool wear, formation of build-up-edges (BUE), etc. Different attitudes have been suggested to solve the problems, one of them being enchanting of the cutting tool geometry and design. Newly developed cutting tools for machining composite materials are typical by unusual shape. They could have additional notches on the teeth, and/or arrow shaped double screw, or pineapple-like shape. Another attitude is focused on advanced coatings, such as diamond or DLC (diamond like carbon) coated cutting tools. Another way to improve machining of composite materials is ultrasonic assistance. Ultrasound avoids creation of BUE, reduces cutting force and process heat. Resultant cutting parameters are still affected by the type of matrix and the type of reinforcing material. Also, different cutting parameters will be set for roughing and finishing [1–9].

2 Machining Method

A machine tool for rotary ultrasonic machining (RUM) was applied in this experiment. RUM usually utilize ultrasonic tool with diamond particles and undefined geometry. This tool rotates around the vertical axis and oscillates by ultrasonic frequency in vertical direction. The coolant enhances this machining process, and therefore it is fed on the tool-workpiece interface. Accordingly, very hard and brittle materials (such as ceramics) can be machined. Advantages of this process include decrease of the cutting force, reduction of heat generation, no chemical affection of workpiece, increase of tool life, improvement of machined surface, etc. When a tool with defined geometry (e.g. conventional milling cutter) is applied, the process is called ultrasonic assisted machining (UAM). RUM has a lot of advantages, such as decrease of cutting force, reduction of heat generation, elimination of BUE formation, superior machined surface, etc. Generally, RUM is

Fig. 1 Milling cutter for composite materials [16]

proper for the hard and brittle materials (such as optic glass), while UAM is proper for the soft and tough materials (such as metals and polymers). RUM and UAM execute similar movements, and therefore they could be usually performed on the same machine tool.

In this experiment, rotary ultrasonic milling machine DMG ULTRASONIC 20 linear was used. This machine tool is able to operate continuously in five axes and it can operate as a conventional milling machine, high speed cutting machine, ultrasonic assisted milling machine and rotary ultrasonic milling machine. It is therefore able to machine almost every material [10–12].

As a cutting tool, a milling cutter for composite materials Fiber Line Honeycomb Hexacut—Fräse (Hexacut—End mill) with internal labeling 068HOA050 manufactured by Hufschmied Zerpanungs system was used. This tool has the cutting diameter of 5 mm, cutting length of 18 mm, shank diameter of 6 mm, overall length of 60 mm and 8 flutes (Fig. 1). A 3D model of the workpiece was created by CAD software Autodesk Power Shape Ultimate 2017 and the NC program for machining (for control system Siemens Sinumerik 840D Solution line) was generated by CAM software Autodesk Power Mill Ultimate 2017 [13–16].

3 Description of the Experiment

As machined materials, a carbon fibre reinforced polymer (CFRP) was used. CFRP material was provided by the Czech company Sanax (type S). The specimen was of a plate shape with the dimensions of 50 × 180 × 1.4 mm. The following cutting conditions were set up: harmonic frequency of the tool 23,500 Hz, cutting speed 100 m/min (spindle speed 6366 rpm), feed per tooth 0.015 mm (feed rate 764 mm/min) and depth of cut 0.7 mm. During ultrasonic machining, it is recommended to use harmonic frequency of the tool. Other cutting parameters were adjusted according to the SECO catalogue [17].

Several notches were performed into CFRP. Two notches were parallel with the fibres, two notches were perpendicular to the fibres. They were placed on the biggest surface of the rod. On the edges of CFRP, four notches to the previous ones

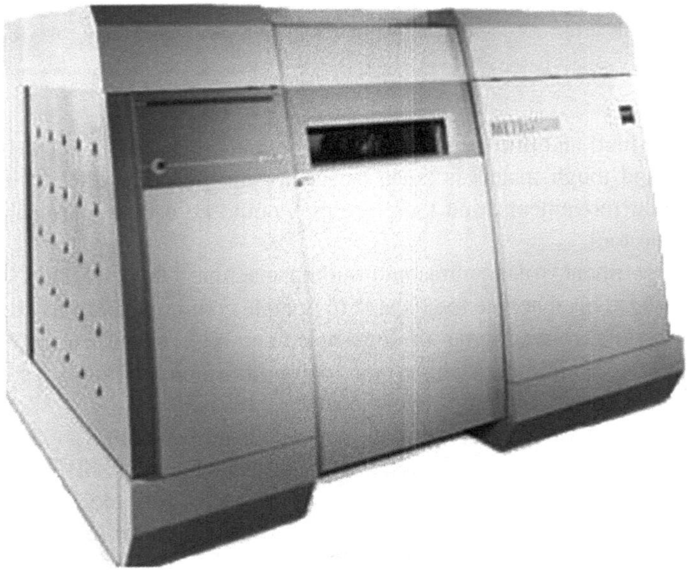

Fig. 2 Scanning device and prepared specimen [18]

were placed. Half of the notches were machined without ultrasonic assistance and the rest of them with the assistance of ultrasound. The notches were 0.7 mm deep. After machining, the specimens were digitized by Zeiss Metrotom 1500 computer tomography. The device is shown in Fig. 2.

Digitized models were compared with the original CAD model. Delamination was evaluated as the highest width of each notch (original width was 5 mm). Besides delamination, the machine load (in Z direction) and surface roughness were also evaluated (they are often connected with delamination itself). Machine load was obtained directly by the machine tool. Surface roughness was measured by ZEISS Surfcom 5000, which is shown in Fig. 3.

4 Results of the Experiment

Several factors of the delamination were obtained and evaluated, such as delamination width, machine tool load and surface roughness of CFRP. They were achieved for each machined notch (blind ones). Labelling of the notches is shown in Fig. 4. An extra notch (but continuous)was produced for easier determination of position. The green ones (letter "a") were machined without ultrasound. The blue ones (letter "b") were machined with ultrasonic assistance. Direction of the fibres is parallel with the grooves numbered 1 and 4.

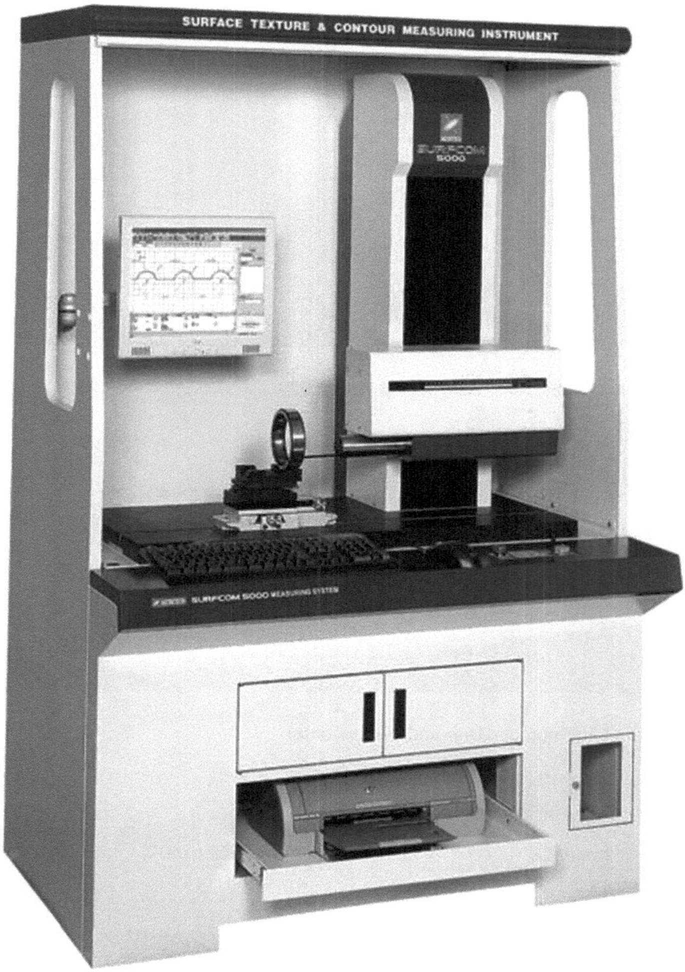

Fig. 3 Surface roughness measuring device [19]

Results of the measurements are summarised in Table 1. Recorded are only average values (each measurement was repeated three times). At the bottom of the Table, summarised are the average values of all previous experiments. In the Table, machine tool load in Z direction is labelled as *Load Z*, and it is the percentage of the maximum available load of the machine tool. Width of the delamination is labelled as *Df* and it is the percentage difference between the recommended width of the notch and the real width of the notch. Its average (*Ra*) and maximum (*Rz*) values were chosen as roughness parameters.

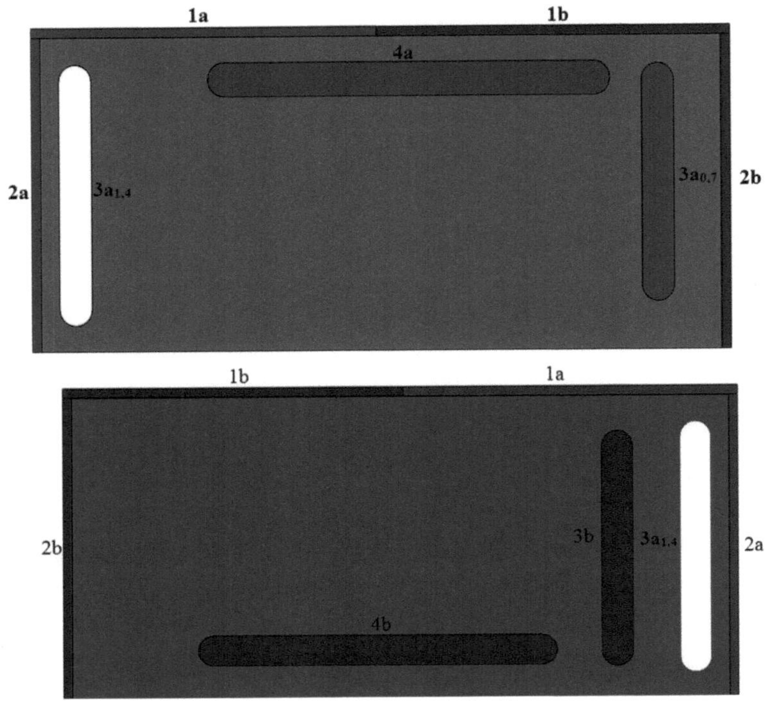

Fig. 4 Labelling of the notches (*top* and *bottom* side)

Table 1 Results of the measurements

CFRP	Load Z (%)	Df (%)	Ra (μm)	Rz (μm)
1a	15	3.8	1.25	7.43
1b	5	1.9	1.15	7.34
2a	10	29.4	1.09	7.83
2b	5	38.1	0.92	6.92
3a	15	24.6	1.96	17.08
3b	5	17	2.15	18.60
4a	10	21.8	1.46	8.71
4b	5	14.2	1.45	9.42
Avg. a	12.5	19.9	1.44	10.26
Avg. b	5	17.8	1.42	10.57

5 Conclusion

The obtained results allow us to conclude that ultrasonic assistance has a great influence on the reduction of machine tool load (decrease by approx. 60%). The surface roughness parameters were not significantly affected. Width of the delamination was also very similar (reduction by approx. 11%). Relatively low ultrasonic influence on delamination and surface roughness parameters could be caused by: slightly worn tool (beginning without ultrasound); relatively low amplitude (tool was not directly manufactured for ultrasonic assistance—absence of concentrator); and purpose of the tool was the machining of composites (therefore not enough space remained for additional improvement). We also observed much lower delamination on the edge surface parallel to the fibres direction of the CFRP plate. To sum up, ultrasonic assistance is proper especially in terms of decreasing the machine tool load (increasing material removal rate, i.e. increasing productivity). If a tool is designed for ultrasonic assistance, its improvement will be even greater. Further research will be focused on the determination of influence of ultrasonic assistance on different composite materials, such as carbon fibre reinforced polymers of different types, glass fibre reinforced polymer, metal matrix composite, or ceramic matrix composite.

Acknowledgements This contribution is a part of the GA VEGA project of the Ministry of Education, Science, Research and Sport of the Slovak Republic, No. 1/0477/14 "Research of influence of selected characteristics of machining process on achieved quality of machined surface and problem free assembly using high technologies".

References

1. Lee DG et al (2017) Adhesion characteristics of fiber-exposed glass composites. Compos Struct 165:9–14
2. Ning F, Wang H, Cong W, Fernando PKS (2017) A mechanistic ultrasonic vibration amplitude model during rotary ultrasonic machining of CFRP. Ultrasonics 76:44–51
3. Su Z, Jia Z, Niu B, Bi G (2017) Size effect of depth of cut on chip formation mechanism in machining of CFRP. Compos Struct 164:316–327
4. Khan MA, Kumar AS (2011) Machinability of glass fibre reinforced plastic (GFRP) composite using alumina-based ceramic cutting tools. J Manuf Process 13(1):67–73
5. Sonkar V, Abhishek K, Datta S, Mahapatra SS (2014) Multi-objective optimization in drilling of GFRP composites: a degree of similarity approach. Procedia Mater Sci 6:538–543
6. Sreenivasaulu R (2013) Optimization of surface roughness and delamination damage of GFRP composite material in end milling using Taguchi design method and artificial neural network. Procedia Eng 64:785–794
7. Akbari J et al (2013) Applying ultrasonic vibration to decrease drilling-induced delamination in GFRP laminates. Procedia CIRP 6:577–582
8. Palanikumar K (2011) Experimental investigation and optimisation in drilling of GFRP composites. Measurement 44(10):2138–2148
9. Bosco MAJ, Palanikumar K, Prasad BD, Velayudham A (2013) Influence of machining parameters on delamination in drilling of GFRP-armour steel sandwich composites. Procedia Eng 51:758–763
10. Kuruc M, Vopát T, Peterka J (2015) Surface roughness of poly-crystalline cubic boron nitride after rotary ultrasonic machining. Procedia Eng 100:877–884
11. Kuruc M, Zvončan M, Peterka J (2014) Investigation of ultrasonic assisted milling of aluminum alloy AlMg4.5Mn. Procedia Eng 69:1048–1053

12. Kuruc M (2015) Machine tool loads in rotary ultrasonic machining of alumina, CBN and synthetic diamond. In: Proceedings of the 26th DAAAM International symposium, pp 519–523. ISSN 1726-9679. ISBN 978-3-902734-07-5
13. Delcam—Advanced manufacturing Solutions. PowerMILL. Available on the Internet: http://www.powermill.com/
14. Zvončan M, Kováč M, Beňo M (2012) Machine tool's DMG ultrasonic 20 linear simulation in Powermill CAM software. In: CA systems in production planning, vol 12, no 1, pp 90–93. ISSN 1335-3799
15. Siemens. Sinumerik 840D. Available on the Internet: http://w3.siemens.com/mcms/mc-systems/en/automation-systems/cnc-sinumerik/Pages/cnc-systems.aspx
16. Hufschmied: Catalogue Plastic-Machining. Available on the Internet: http://www.hufschmied.net/tl_files/dokumente/HUF-Kuststoff-Low-080415.pdf
17. SECO Tools. Monolithic Carbide Mills. Available on the Internet: https://www.secotools.com/CorpWeb/Czech%20Republic/katalogy/2015/CZ_Catalog_Solid%20end%20mills_2015_Inlay_LR.pdf
18. Zeiss. Metrotom 1500. Available on the Internet: http://www.zeiss.com/industrial-metrology/en_de/products/systems/computedtomography/metrotom-1500.html
19. Zeiss. SURFCOM 5000. Available on the Internet: http://www.zeiss.com/industrial-metrology/en_de/products/systems/form-andsurface/surface-and-contour/surfcom-5000.html

Microfluidic Chip Fabrication for Application in Low-Cost DIY MicroPIV

Jelena Pejović Simeunović, Ivana Gadjanski, Željko Janićijević, Milica M. Janković, Marko M. Barjaktarović, Novica Z. Janković and Đorđe S. Čantrak

Abstract The article presents the process of fabrication of the microfluidic chip to be used with the do-it-yourself (DIY) micro-PIV system previously made and compared to the classic PIV setup. This pilot study is an example of research being conducted in the Scientific fab lab (fabrication laboratory), founded at the Faculty of Mechanical Engineering, University of Belgrade. Fab labs and DIY principle are becoming more and more accepted by the scientific community and this article aims to contribute to such trend.

Keywords Microfluidics · MicroPIV · Do-it-yourself · Fab lab · Microfluidic chip

---

J.P. Simeunović
Central Institute of Technology (CEITEC), Brno University of Technology, Brno, Czech Republic
e-mail: jelena.pejovic@ceitec.vutbr.cz

I. Gadjanski (✉)
Innovation Center, Faculty of Mechanical Engineering, University of Belgrade, Belgrade, Serbia
e-mail: igadjanski@gmail.com

Ž. Janićijević · M.M. Janković · M.M. Barjaktarović
School of Electrical Engineering, University of Belgrade, Belgrade, Serbia
e-mail: zeljkoj@etf.bg.ac.rs

M.M. Janković
e-mail: piperski@etf.rs

M.M. Barjaktarović
e-mail: mbarjaktarovic@etf.rs

N.Z. Janković · Đ.S. Čantrak
Faculty of Mechanical Engineering, University of Belgrade, Belgrade, Serbia
e-mail: njankovic@mas.bg.ac.rs

Đ.S. Čantrak
e-mail: djcantrak@mas.bg.ac.rs

V. Majstorovic and Z. Jakovljevic (eds.), *Proceedings of 5th International Conference on Advanced Manufacturing Engineering and Technologies*, Lecture Notes in Mechanical Engineering, DOI 10.1007/978-3-319-56430-2_34

1 Introduction

Fab labs (fabrication laboratories) and the tools they offer (CAD, rapid prototyping, 3D printing, electronics) are becoming an important component in universities and research centers, not only relating to the mechanical and electrical engineering courses and research but also concerning bioengineering, chemical, biological and even medical topics [1]. One of the most interesting techniques for the biomedical field is microfluidics which can be used for a number of biomedical experiments with potential clinical applications [2].

However, some of the main problems with microfabrication i.e. fabrication of microfluidic chips are the need for clean room facilities and rather costly and lengthy process of mask fabrication, required for the mask-based photolithography. This is why there is a number of "maskless" rapid prototyping approaches being currently developed, such as inkjet- or 3D-printing, laser micromachining, micro-electrical discharge machining [3] and/or out-of-clean room rapid mask prototyping alternatives such as the one described by Maisonneuve et al. [4]. Fablabs will become even more influential for the microfabrication once such approaches get more efficient.

Another problem concerning wider application of microfluidics is the inability to use traditional flow diagnostic tools, due to the small length scales of the flow passages. This was the reason micron-resolution particle image velocimetry (micro-PIV) system has been developed to measure flow fields in micron-scale fluidic devices [5]. However, microPIV equipment is also expensive which presents another problem for the wider use.

Fablabs are strong gates for introduction of the (low-cost) do-it-yourself science approach that is receiving more and more acceptance even in the notoriously conservative scientific community. This is particularly the case for the do-it-yourself biology (DIYbio) which fosters open access to resources permitting modern molecular biology, synthetic biology, biomedical engineering, most importantly as a source of cheaper and simpler solutions for environmental monitoring, personal diagnostic and the use of biomaterials [6].

As part of the multidisciplinary research we conduct within the Scientific fab lab at the Faculty of Mechanical Engineering, University of Belgrade, which we founded and are expanding [7–10], we started developing the low-cost do-it-yourself version of the microPIV setup [11] and intend to develop low-cost microfabrication methods and implement both for more DIYbio research, focusing particularly on the cell biology experiments.

However, for the current pilot-study for the use of DIY microPIV in microfludics, we decided to use the current state-of-the-art microfabrication method, since we wished to have the standard chip to test it for the use in all the PIV variants: classic microPIV and low-cost DIY microPIV [11]. In further research, we will focus on alternative low-cost DIY methods of microfabrication described above.

2 Fabrication of a Chip

Rapid prototyping of the chip starts with creating a design for a device in a computer-aided design program. A high-resolution printing is used to print the design on a chrome transparency mask. This transparency serves as a photomask in photolithography to produce a positive relief of the photoresist on silica wafer.

Photolithography was used to fabricate mold with microchannels. Next steps were performed to fabricate mold with channels (Fig. 1).

- A blank 4¨ Silicon wafer (Si(100) wafer, N-type, 500 nm of SiO_2 layer) was rinsed with isopropanol, DI water, following with drying with N_2 gun. To increase adhesion of the photoresist on the silicon, wafer surface was treated with Hexamethlyldisilazane (HDMS) for 5 min
- AZ 1518 positive photoresist was spun (2-step spinning parameters: step one-500 rpm, 500 rpm/sec, 5 s. Step two-4000 rpm, 500 rpm/sec, 45 s) to create uniform film ~2 μm thick. Wafer was pre-baked on a hot plate (200 °C for 10 min)

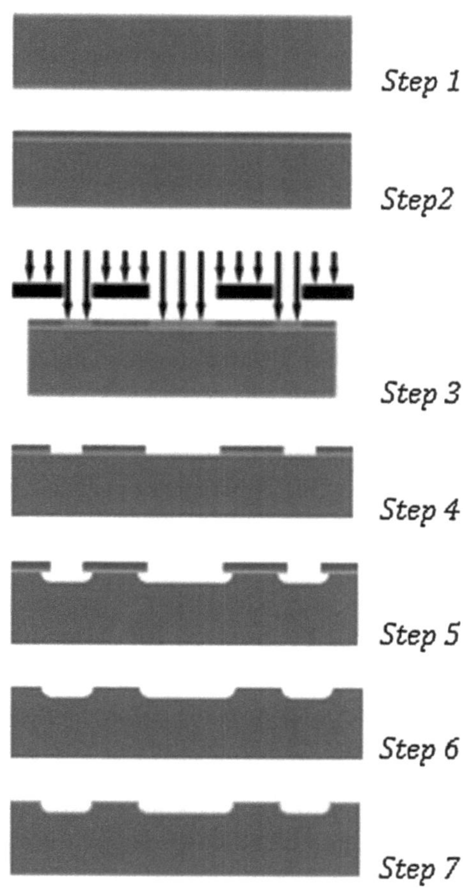

Fig. 1 Steps for fabrication of the silicon mold

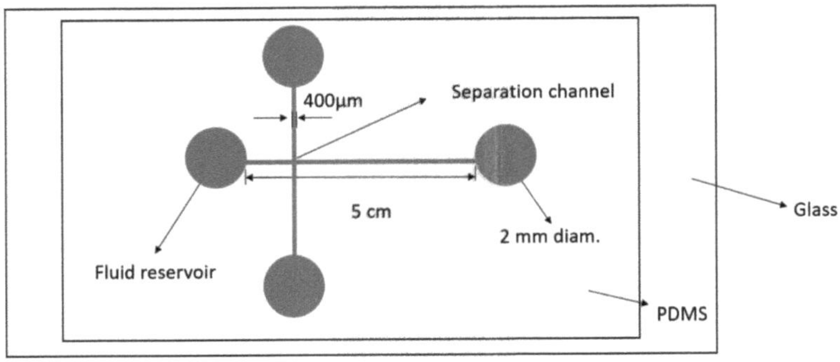

Fig. 2 Cross-junction on the chip

- Wafer was exposed with 2500 mJ/cm^2 at 405 nm through a photomask. Post/bake was done on a hot plate at 100 °C for 1 min. Wafer was developed in AZ 327 mif for 40 s, rinsed in isopropanol and dried with N_2 gun
- Etching of silicon dioxide, to remove areas of silicon dioxide unprotected by the photoresist, was done using buffered oxide etch (BOE) solution (6 parts of 40% NH_4F and 1 part of 49% HF) for 20 s
- XeF_2 was used for dry isotropic etching of silicon
- Rinsing with isopropanol was done to remove the photoresist. Polishing of the wafer surface with microchannels was done using mixture of NHO_3 and BOE (100:1)
- In the last step wafer mold was cleaned using piranha solution for a few seconds

Silica mold with two same sized channels 400 µm wide, 100 µm high and 5 cm long was made in described way and used for rapid molding of polydimethylsiloxane (PDMS). Simple cross-junction was created as a separation channel (Fig. 2).

A mixture of PDMS, Dow Corning Sylgard 184 base prepolymer and a curing agent at a ratio of 10:1 (w/w) was completely degassed, poured onto the silica mold, cured for 2 h at 65 °C, and finally peeled off. After air-drying of the channels, PDMS is punched to make holes (2 mm diam.) on the fluid reservoirs. The patterned side of PDMS was treated with oxygen plasma (200 W, 15 min) and bonded permanently with a plasma-treated glass substrate to form a closed fluidic system.

3 MicroPIV System for Microfluidic Purposes

A micron-resolution particle image velocimetry (microPIV) systems are broadly used for accurate measurements of velocity profiles in microchannels [5]. MicroPIV system includes specific camera (with camera link interface), laser with synchronizer and microscope. The main reason why those systems are not widely used is their high price. Low-cost solution for conventional PIV measurement based on cheap continuous-wave laser and fast high-resolution camera, has already been published by the authors [11]. Similar principle can be utilized for the microPIV setup. LED lamp could illuminate continuously the measuring area, so there is no need for the synchronizer as it is the case in classic microPIV systems, Fig. 3. Image acquisition system controls the camera parameters, like shutter speed, gain and acquisition frame rate.

Figure 4 shows a pilot experiment performed by commercial microPIV system (TSI, USA) and do-it-yourself (DIY) system based on low-cost industrial camera Basler acA2500-14 gm (Basler, Arhensburg, Germany). The sensor is CMOS type with 1/2.5 in. size, high SNR (about 40 dB) and low dark current. Although camera has a rolling shutter sensor, smearing is not introduced due to the low fluid speed.

Vision acquisition and processing by TSI system was performed by Insight software (TSI, USA). In the DIY system, vision acquisition is accomplished using LabVIEW 15.0 programming environment and add-on module NI Vision Acquisition Software (National Instruments, Texas, Austin), frame rate was set to 10 fps. Streaming images to disk do not reduce fps since BMP file format is used and disk is SSD type. After acquisition images are processed offline using OpenPIV library [12]. To calculate velocities profile over time, pairs of two successive images are analysed. Images are divided in several interrogation areas. For each interrogation area in the first image, best matching area is found in the second image. Displacement vector is equal to the difference between centroids of interrogation areas found in previous step, while the velocity of the fluid is displacement divided by the acquisition time interval. An example of the velocity profile in microfluidic chip acquired by the DIY system is presented in Fig. 5. The interrogation window size was set to 128 × 128 pixels and horizontal/vertical overlapping was set to 64 pixels. The velocity value is estimated 107 μm/s.

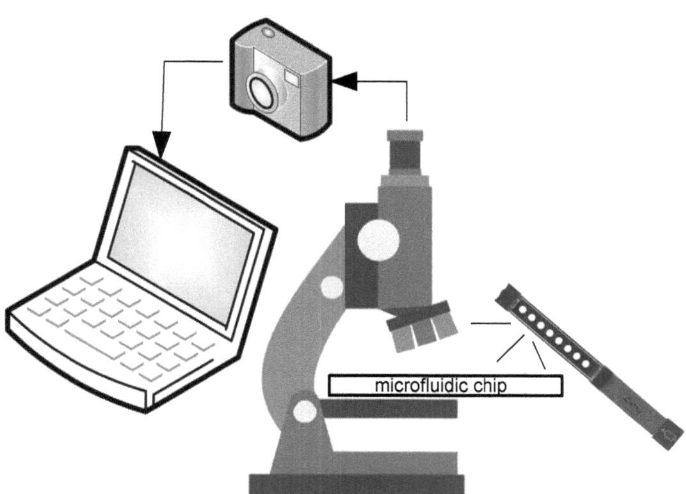

Fig. 3 Scheme of low-cost DIY microPIV system

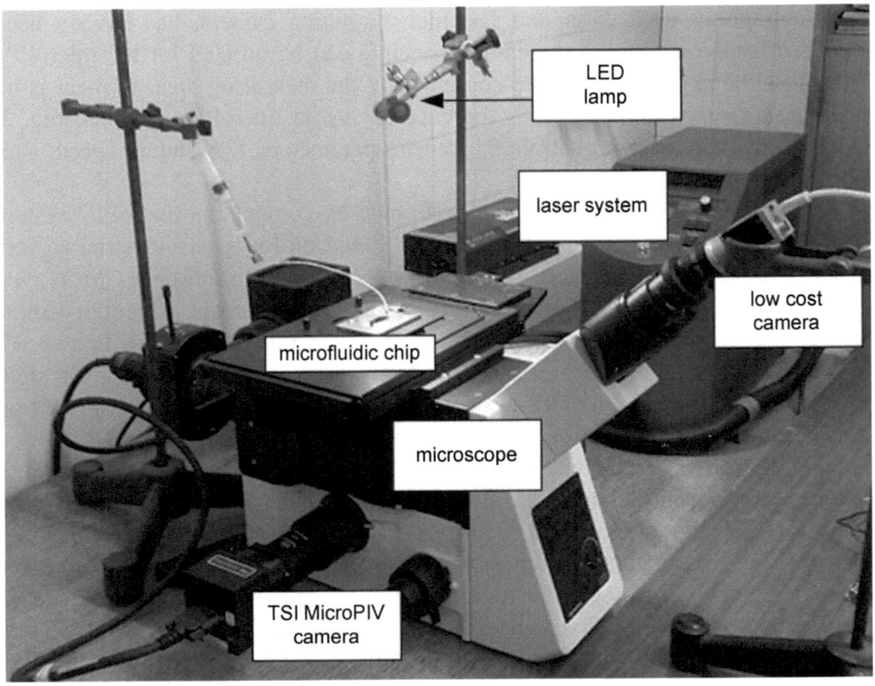

Fig. 4 MicroPIV system setup

Fig. 5 An example of the velocity profile

4 Properties, Possible Modifications and Potential Applications of the Microfluidic System

Most microfluidic channels are characterized by the laminar creeping flow with the flattened unidirectional velocity profiles for high aspect ratios of channel width/channel height. Our geometry provides the possibility to obtain different velocity profiles in the junction region of the microfluidic system. This result can be attained by varying the number of inlets and outlets and using clever flow rate adjustments.

For example, if one inlet and three outlets are used, a simple model of a bifurcating blood vessel is obtained (see Fig. 6). On the other hand, if two pairs of inlets and outlets corresponding to perpendicular channels are utilized, a defined interface can be achieved across which two flowing phases can collide and chemically interact with diffusion limited mixing (see Fig. 7). Shape and position of such interface can be modulated by the flow rate and composition of the flowing media.

Distributions of velocity magnitude in Figs. 6 and 7 are of qualitative nature and serve mainly to illustrate the concepts.

The bottom surface of the channel can be modified to increase the range of possible applications of the system. The surface in the central region can be patterned to increase the roughness and provide better mixing of the fluids. Dedicated applications may require electrical excitation to be performed through specific microelectrode geometry. Electric fields could contribute to cell separation and focusing, or even modulate physicochemical interactions of the flowing phases.

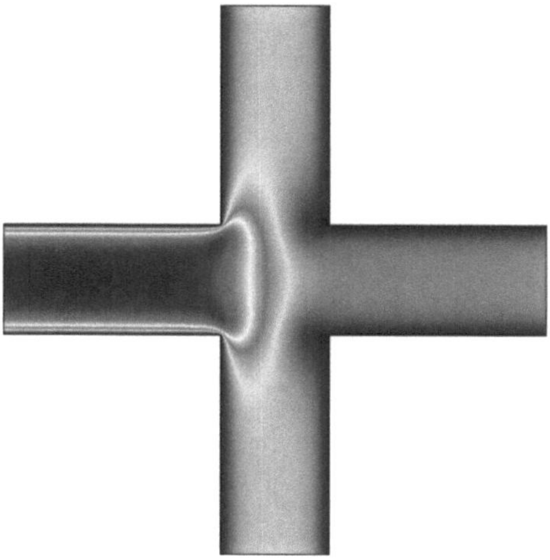

Fig. 6 Velocity magnitude distribution in the cross-junction region with one inlet and three outlets

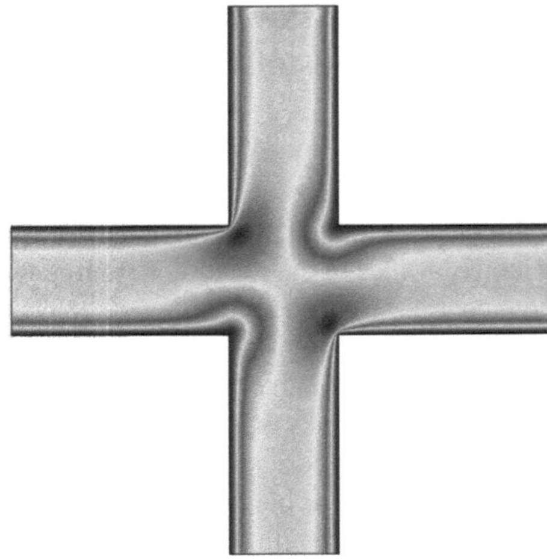

Fig. 7 Example of the velocity magnitude distribution in the cross-junction region with two inlets and two outlets

5 Conclusion

The current pilot-study describes the process of fabrication of the microfludic chip which is to be used with our DIY microPIV system. We also explain how the low-cost DIY microPIV achieves similar results as the classic microPIV. We are currently testing the fabricated microfluidic chip in the DIY microPIVsystem and simultaneously developing maskless microfabrication method that will bring significant decrease in the cost of chip fabrication and will be used in the Scientific fab lab at the Faculty of Mechanical Engineering, University of Belgrade.

Acknowledgements This work was supported by the Ministry of Education, Science and Technological Development Republic of Serbia—Projects No.TR 35046, ON174028, III41007, OS 175016.

References

1. Carlo F, Enrique C (2016) Making ideas at scientific fabrication laboratories. Phys Educ 51 (6):065016
2. Chiu DT et al (2017) Small but perfectly formed? Successes, challenges, and opportunities for microfluidics in the chemical and biological sciences. Chem 2(2): 201–223
3. Leester-Schädel M et al (2016) Fabrication of microfluidic devices in microsystems for pharmatechnology. Springer, pp 23–57
4. Maisonneuve B et al (2016) Rapid mask prototyping for microfluidics. Biomicrofluidics 10 (2):024103
5. Santiago JG et al (1998) A particle image velocimetry system for microfluidics. Exp Fluids 25 (4):316–319
6. Landrain T et al (2013) Do-it-yourself biology: challenges and promises for an open science and technology movement. Syst Synth Biol 7(3):115–126

7. Gađanski II, Čantrak ĐS (2016) Kickstarting the fab lab ecosystem in Serbia: SciFabLab and FABelgrade conference. In: Proceedings of international conference multidisciplinary engineering design optimization (MEDO), IEEE
8. Gađanski ĐČ, Matijević M, Prodanović R (2015) Stimulating innovations from university through the use of digital fabrication—case study of the SciFabLab at Faculty of Mechanical Engineering, University of Belgrade. In: International conference on WBCInno2015, Novi Sad, Serbia
9. Janković NZ et al (2016) Application of 3D printing in M. Sc. studies—axial turbocompressors. In: International conference multidisciplinary engineering design optimization (MEDO), IEEE
10. Čantrak ĐS et al (2016) Centrifugal pumps, impellers design and digital fabrication. In: International conference multidisciplinary engineering design optimization (MEDO), IEEE
11. Janković NZ, Janković BMC, Milica M, Čantrak DS (2016) First steps in new affordable PIV measurements. In: 24th Telecommunications Forum (TELFOR), Belgrade: IEEE
12. OpenPIV. OpenPIV software, open-source library. Available from: http://www.openpiv.net/

R&D in a Fab Lab: Examples of Paste Extrusion Method

Borko Jovanović, Ivana Gadjanski, Jela Burazer, Luka Nikolić, Nina Babić and Milan Lečić

Abstract The authors report results of modification performed on the standard, open-source Cartesian style 3D printer and customized hardware (Printrbot Simple Metal) in order to perform paste extrusion with three different materials (hydroxyapatite, clay and chocolate). These results are shown as examples of research and development performed in an informal institution—a fab lab, in connection with formal research institutions. The authors aimed to show how such collaboration of various players in the research community, coupled with multidisciplinary background of participants, produces new and innovative aspects, which is one of the benefits of fab labs.

Keywords Fab lab · Paste extruder · 3D printing · Hardware · Firmware · Software

B. Jovanović · I. Gadjanski (✉) · L. Nikolić · N. Babić
Polyhedra Fab Lab, Belgrade, Serbia
e-mail: igadjanski@gmail.com

B. Jovanović
e-mail: borko.jovanovic@polyhedra.com

N. Babić
e-mail: ninocccka@gmail.com

J. Burazer · M. Lečić
Faculty of Mechanical Engineering, University of Belgrade, Belgrade, Serbia
e-mail: jburazer@gmail.com

M. Lečić
e-mail: mlecic@mas.bg.ac.rs

I. Gadjanski
Innovation Center, Faculty of Mechanical Engineering, University of Belgrade, Kraljice Marije 16, Belgrade 11000, Serbia

V. Majstorovic and Z. Jakovljevic (eds.), *Proceedings of 5th International Conference on Advanced Manufacturing Engineering and Technologies*, Lecture Notes in Mechanical Engineering, DOI 10.1007/978-3-319-56430-2_35

1 Introduction

Fab lab is a workshop equipped with tools for digital fabrication providing widespread access to modern means for invention [1]. Fab labs began as an outreach project from MIT's Center for Bits and Atoms (CBA), and became a collaborative and global network sharing knowledge and expertise world-wide, including Serbia [2–4].

Polyhedra was established in February 2015 as a mixture of fab lab and makerspace, and has been building up both the tools (fab lab component) and the community (makerspace component) ever since. The list of tools includes various 3D printers, CNC mills, laser cutters, and electronics, but also specific custom-made devices and high-temperature kilns.

With over a hundred visitors per month and dozen active members on a daily basis, Polyhedra serves as:

- a community hub providing meeting place for its members
- an informal educational facility for digital fabrication
- a prototyping center providing jobs to its members
- a research facility for the new digital fabrication tools and materials

Through various projects over the last two years, Polyhedra has established close relationship with all relevant organizations in Serbia, including but not limited to: Fab Initiative (http://www.fablab.rs), Petnica Science Center, University of Belgrade (Faculty of Mechanical Engineering, Faculty of Electrical Engineering, Faculty of Architecture, and Faculty of Philosophy), University of Arts in Belgrade (Faculty of Applied Arts), Vinca Institute of Nuclear Science, etc.

Some of the notable research and development projects done in cooperation with the above mentioned organizations include Paste Extrusion, an additive manufacturing process similar to 3D printing. Examples are Paste Extrusion with Clay, Paste Extrusion with Tempered Chocolate and Paste Extrusion with Modified Hydroxy-Apatite.

1.1 3D Printer Modifications

In order to use Paste Extrusion, a custom-made machine fitted with extrusion block was needed. For this purpose, we used a standard, open-source Cartesian style 3D printerand customized hardware (Printrbot Simple Metal), firmware (Marlin) [5] and printing software (Slic3r) [6].

Hardware modifications: Filament extruder, direct drive stepper motor, hot end, inductor probe, and fan were removed from the printer. Paste Extruder was constructed with aluminum heat block designed to fit 60 ml syringe with "luer lock" needles, plunger using a previously removed stepper motor, ¼" Acme push rod and nut, two 3D printed gears (small—13 teeth and large—43 teeth), temperature sensor

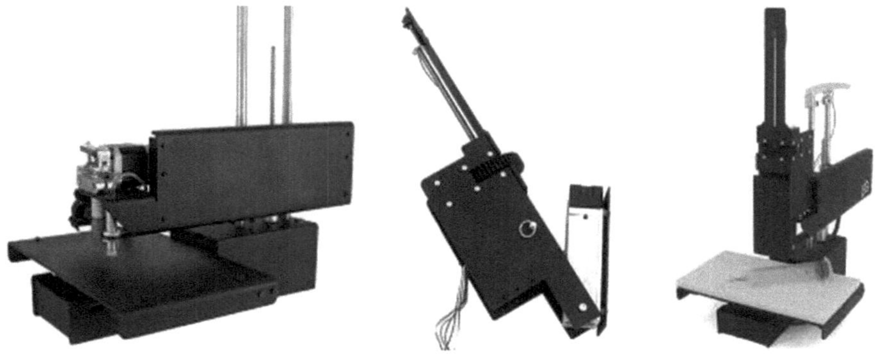

Fig. 1 Printrbot simple, custom extruder and modified simple

placed within the heater block, and two end stops for min and max limits. Paste Extruder was attached to the printer using 4 screws. All components are attached to appropriate ports on the Printrboard rev. F6 board (Fig. 1).

Firmware modifications: There were several modifications to Marlin 3D Printer Firmware, which is a most widely used firmware for 3D printers and CNC machines. Marlin has a GPL license, so all modifications were posted back to the public and can be found at: https://github.com/mickbalaban/Marlin/tree/printrbot-food/Marlin. We also updated EPROM with new custom settings for the extruder.

- M92 Z8467.69 (to accommodate 13:43 gear ratio)

Software modifications: For model preparation and printing control, we used Repetier-Host v1.6.2 [7], open-source software with support for Marlin firmware and its derivatives. For slicing (G-code generations), we modified Slic3r to fit the 60 ml syringe:

- Filament Settings: Filament: Diameter: 27 mm
- Filament Settings: Colling: Enable: False
- Printer Settings: Custom G-code: Start G-code: G92 X0 Y0 Z0 E0
- Printer Settings: Extruder 1: Nozzle Diameter: xx (needle size)
- Printer Settings: Extruder 1: Retraction: Length: 0 (disable).

2 Paste Extrusion with Clay

Luka Nikolić undertook a project to create 3D printable water filters, one of the major components being built from clay. In order to successfully 3D print clay, we had to develop a special recipe to assure consistency through the Paste Extruder as well as print stability for layer building. Ingredients include:

- Clay—Terracotta
- Organic material—Sawdust, wheat flour
- Water—Tap water

As project assumed remote location and limited tools, no exact measurements are recorded, but rather the consistency and feel of the printing material—sticky enough not to ooze out of the mixing bowl, soft enough to be able to transfer it to the syringe (Fig. 2).

Quality of print and finishing was not of concern, so several shapes were made, then fired in the kiln with the following curve:

- Up to 500 °C in 3 h, stabilize at 500 °C for 15 min
- Up to 1200 °C in additional 3 h, Stabilize at 1200 °C for 15 min
- Cooldown and test (Fig. 3).

Fig. 2 Testing initial recipe

Fig. 3 Final prototypes

3 Paste Extrusion with Chocolate

Chocolate printing was a popular topic in our community, so we had an easy time selecting the right partner—Nina Babić, creator of Gospodja Mirkov fine chocolates (http://gospodjamirkov.tumblr.com/).

Cocoa butter is the fat in the cacao bean that gives chocolate its unique mouth-feel and stable properties. Once you melt a chocolate bar, the fatty acid crystals separate. The trick is in "tempering" melted chocolate to entice the disparate fatty acid crystals of cocoa butter back into one stable form. This effectively means that the chocolate has to be kept at 21 °C (varies slightly depending on the type of chocolate) and allowed to cool while forming a shape.

To assure constant temperature of the main batch, we used a water bath with sensor to control the temperature and keep the chocolate at 21 °C (Fig. 4).

Once the chocolate is poured in the syringe, the syringe is placed in the heater block where temperature remains at 21 °C until disposal. The temperature of the heater block is controlled though the Slic3r software with the following settings:

- Filament Settings: Filament: Temperature: Layer: 21

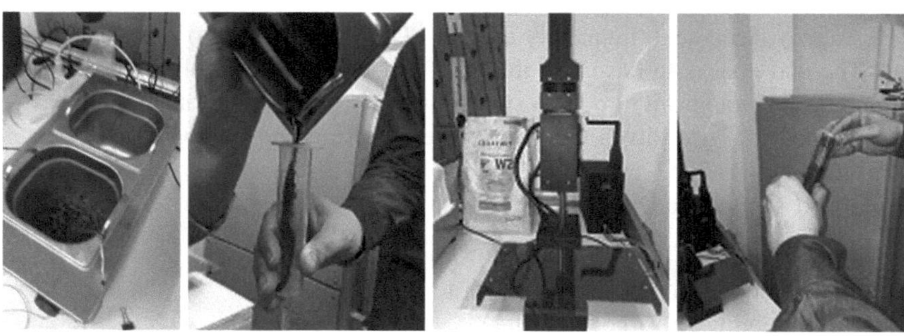

Fig. 4 Tempering and loading chocolate

Fig. 5 Chocolate airplane

Also, since we were using a needle with the 1 mm tip, the following settings were modified:

- Printer Settings: Extruder 1: Nozzle Diameter: 1 mm
- Print Settings: Layers and perimeters: Layer Height: 1 mm (Fig. 5).

4 Paste Extrusion with Modified Hydroxy-Apatite

Team of experts has developed a special formula for hydroxy-apatite (HA) as a material for bone tissue engineering [8].

Initially, it was meant to be manufactured in cubes of various sizes and implanted in patients after removing enough bone to fit the shape. Generating 3D model of missing bone and 3D printing an exact fit would not only reduce the amount of healthy bone being taken out to fit the pre-shaped scaffold, but will benefit from ability to use multiple types/densities of HA within the same scaffold (Fig. 6).

However, HA is a powder and not suitable for extrusion printing, so we set out to further modify it by adding very small amounts of bio-degradable materials such as PLA (poly-lactic acid, frequently used in FDM 3D printing) to serve as powder glue. The end goal is to have less than 0.1% of biodegradable plastics within the HA and be able to push it through a syringe without losing any of its properties. We have reached the viscosity needed for paste extrusion and compared all relevant quantifiers.

The next steps include:

- multiple extruders
- automatic conversion of scanned data to a 3D model
- mirroring abilities for modeling
- custom control software.

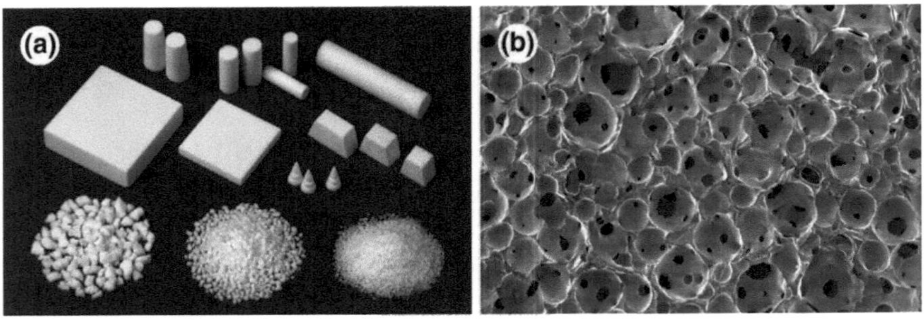

Fig. 6 Interconnected porous hydroxy-apatite [8]

5 Discussion

Paste Extrusion is a complex process requiring specific hardware, firmware and software components, as well as in-depth knowledge of the materials used. For example, in order to use paste extrusion with modified hidroxyapatite, the team working on the project must have experts on mechanical and electrical properties of the machine, electronics and firmware, software development, chemistry, biology, etc.

It is very hard to gather such a versatile team in any of the formal institutions and this is where fab labs can help research institutions and bridge this gap though its multi-disciplinary community [2]. This can either be done through formal agreements between legal entities or by simply connecting community members with relevant institutions.

Also, fab labs use tools of digital fabrication often missing from the research laboratories and other institutions, such as 3D printers, CNC mills, laser cutters and electronics—all very useful in the prototyping phase.

Acknowledgements JB and ML were supported by the Ministry of Education, Science and Technological Development Republic of Serbia—Project No.TR 35046, while IG was supported by Projects ON174028 and III41007.

References

1. Gadjanski I (2015) Fabrication laboratories—fab labs—tools for sustainable development
2. Gadjanski Č, Matijević M, Prodanović R (2015) Stimulating innovations from university through the use of digital fabrication—case study of the SciFabLab at Faculty of Mechanical Engineering, University of Belgrade. In: proceedings of international conference on WBCInno, Novi Sad, Serbia
3. Gadjanski I et al (2016) Formation of Fab lab Petnica. In: International conference on multidisciplinary engineering design optimization (MEDO), IEEE
4. Gađanski II, Čantrak ĐS (2016) Kickstarting the fab lab ecosystem in Serbia: SciFabLab and FABelgrade conference. In: International conference multidisciplinary engineering design optimization (MEDO), IEEE
5. marlinfw.org (2017) Available from: http://marlinfw.org/
6. slic3r.org (2017) Available from: http://slic3r.org/
7. repetier.com (2017) Available from: https://www.repetier.com/
8. Jokanović V et al (2016) Extraordinary biological properties of a new calcium hydroxyapatite/poly (lactide-co-glycolide)-based scaffold confirmed by in vivo investigation. Biomed Eng/Biomedizinische Technik

Manufacturing Technology of Aircraft and Wind Turbine Blades Models, Plugs and Moulds

Zorana Trivković, Jelena Svorcan, Ognjen Peković and Toni Ivanov

Abstract Rapid prototyping/manufacturing technology, as one of the fastest growing technologies, made its way into aeronautical industry long time ago. Beside huge cost and time savings, its main advantage lies in designers' freedom to create new and innovative parts and systems. Today, there are many different rapid prototyping technologies available. This paper presents one possible manufacturing technology for producing accurate parts directly from computer-aided-design (CAD) models using 3-axis numerically controlled machine (in the form of subtractive milling). This process allows optimization of the design since changes can be easily incorporated into the three-dimensional (3D) model. Technology is applied on small aircraft and wind turbine blade models, and also on wind turbine blade mould, which can be used for both experimental and teaching purposes. Main advantages of the presented processes include: standardization, increased quality and accuracy, repeatability, cost and time reduction.

Keywords Rapid prototyping · CAD/CAM technologies · Aircraft · Wind turbine blades

Z. Trivković · J. Svorcan (✉) · O. Peković · T. Ivanov
Department of Aeronautics, Faculty of Mechanical Engineering,
University of Belgrade, Belgrade, Serbia
e-mail: jsvorcan@mas.bg.ac.rs

Z. Trivković
e-mail: zposteljnik@mas.bg.ac.rs

O. Peković
e-mail: opekovic@mas.bg.ac.rs

T. Ivanov
e-mail: tivanov@mas.bg.ac.rs

V. Majstorovic and Z. Jakovljevic (eds.), *Proceedings of 5th International Conference on Advanced Manufacturing Engineering and Technologies*,
Lecture Notes in Mechanical Engineering, DOI 10.1007/978-3-319-56430-2_36

1 Introduction

Aircraft/aeronautical structures are nonstandard parts due to their complex geometry, high cost, use of contemporary materials, dynamic loads, etc. and are therefore usually produced in small numbers. They also most often require employment of advanced and innovative manufacturing technologies to ensure necessary precision, automation, time efficiency, cost reduction etc.

In comparison to fast developing layered manufacturing (of limited accuracy) subtractive prototyping (as a conventional machining process) gives the user the opportunity to design, prototype, and manufacture end-use parts and/or products. The use of well known techniques is an appropriate choice for parts used for both small and large volume production runs, to obtain specific finishes, or to obtain specific mechanical properties. Applications are practically limitless: from initial prototypes, master plugs and pilot production models to testing models and final products, referring to vehicle structures (cars, boats and aircrafts), industrial parts or even art objects.

Many opportunities for innovative designs still exist with endless choices in aerodynamic shapes, material selection and composite layout. Close coordination between these research fields together with adequate manufacturing techniques must exist to successfully complete aircraft part or blade design.

2 Current Concepts and a Connection to Fab Lab

With the use of composite materials, contemporary manufacturing techniques and tooling almost any geometry is made possible. This fact is particularly interesting to academic researchers since products of their detailed computational analyses and optimization processes can now be quickly and relatively cheaply produced and tested. Academic community is no longer bound solely by the economical aspects and policies of great, leading companies but can instead, on its own, realize smaller models in relatively modest and humbly equipped laboratories or workshops. This idea is now so widespread that it resulted in several international collaborations in the form of Fab Lab academies, educational projects etc.

Although the Laboratory of the Department of Aeronautics at the Faculty of Mechanical Engineering in Belgrade is not a formal Fab Lab, it offers such possibilities through the use of CAD/CAE/CAM tools and a CNC router. This paper gives a short review of the models and moulds produced at the laboratory.

A single source of engineering data (in the form of 3D CAD model closely related to CAM technology) increases accuracy and fidelity of the final product. It enables achieving stringent aerospace quality at industrial cost. Here, CATIA v5 modeling is used for design of geometrical surfaces as well as for g-code generation (NC programming is quite simplified by the automatic tool path fabrication by a commercial software package). Presented technology is particularly applicable to custom designs made for specific working regimes or locations.

There are two possible approaches to advanced, computer controlled manufacturing. Although additive manufacturing is in tremendous focus nowadays, and is considered by some the primary future production direction, it is still not cost efficient enough to be used on larger parts nor in university laboratories. Also, many authors emphasize that the choice of manufacturing technology depends on the desired geometry and used materials. While additive manufacturing is more suitable to lattice models, subtractive manufacturing is better applied to compact models. Also, while structural quality and reliability of printed or melted models might still be questionable, it is well known that subtractive manufacturing models can be made of any material that can be milled on CNC machines with satisfactory outcome [1, 2]. Therefore, in this research, CAD models are uploaded to a CNC machine (where the only inputs are CAD model and material and machine properties).

3 Process Description

Design process begins with a 3D geometry definition followed by an initial production part modeling. Depending on the part in question, it might be necessary to scale and/or divide the initial geometry into several distinct segments that will latter on be merged or glued together. After making a choice of manufacturing parameters and tools, it is possible to review the manufacturing process simulation and optimize the tool path or production time or define the starting geometry that will result in the least excess material.

All presented processes were realized on a 3-axis CNC router with the working envelope of 3.2 m × 2.2 m × 0.6 m. On this machine, it is possible to process a number of (not so hard) materials: plywood, wood, balsa, different foam and core materials (styrofoam, block material), aluminum.

Several different machining processes were conducted, starting from rough to smooth finish ensuring the satisfactory quality of the final surfaces and requiring the least additional work.

By producing initial models and prototypes, all previously unperceived mistakes can be easily and quickly corrected, thus enabling a significant reduction in production time and overall cost.

4 Possible Applications and Produced Parts

4.1 Amphibian Aircraft Model

The model of the light amphibian aircraft is characterized by extremely complex geometry necessary to enable top aerodynamic performances whilst landing on water surfaces, Fig. 1.

Lifting surfaces (such as wing and tail surfaces) are of high aspect ratio and had to be produced separately and then glued together, Fig. 2. Generated light aircraft model shown in Fig. 3 can be used in a number of experimental investigations: aerodynamic, structural, hydrostatic or hydrodynamic.

4.2 Composite Blade Master Plug

Wind turbine blade is one of the fundamental parts of a wind turbine. It is also distinguished by highly curved and slender geometry. Furthermore, since blades are almost always composite nowadays, necessary tools must be specifically prepared. In order to produce composite blade mould it is first necessary to produce a starting master plug (model). If produced accurately and adequately, a single plug can be

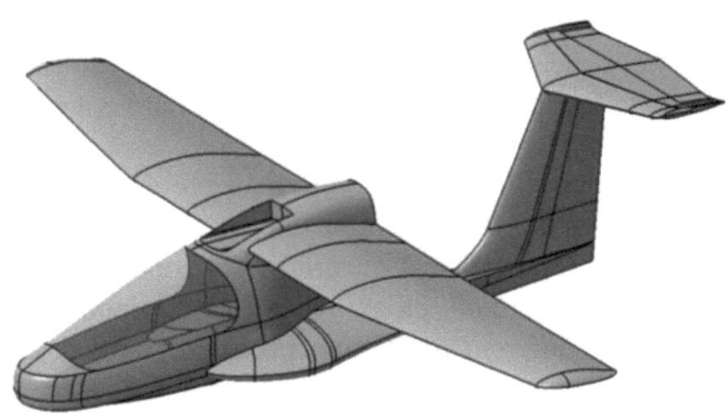

Fig. 1 CAD model of the amphibian aircraft

Fig. 2 Simulation of the rough machining of the amphibian wing suction side

Fig. 3 Produced model of the light amphibian aircraft

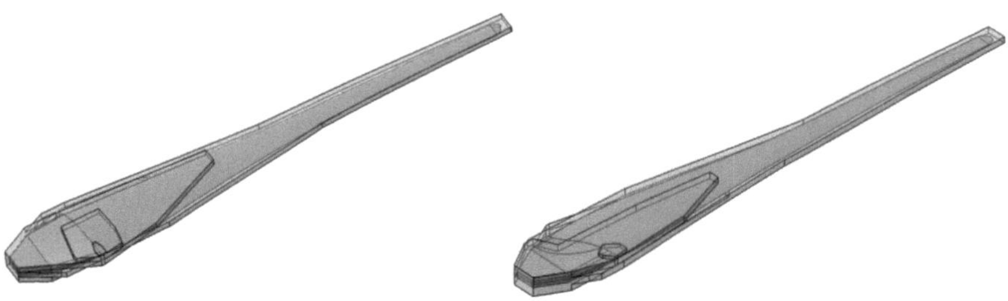

Fig. 4 CAD model of the wind turbine blade master plug (pressure and suction sides)

used to produce a number of blade moulds without the need for major repairs or loss of accuracy. The resulting aerodynamic, low weight and reliable blades are then highly competitive and much cheaper.

Although this approach (CNC milling directly from CAD model) to plug production is common in developed countries [3], it is relatively new in Serbia. Its benefits are numerous, with the most noticeable being: significant reduction in production cost, improved quality and standardization. Presented wind turbine blade model, somewhat over 3 m in length, was manufactured in less than 25 operational hours, Fig. 4.

Pressure and suction sides were sequentially machined. Special attention was given to trailing edge finish, Fig. 5.

4.3 *Wind Turbine Production Mould*

As with wind turbine blades, special care must be paid to achieve high mould productivity. Some of the common requirements are that moulds be made compact

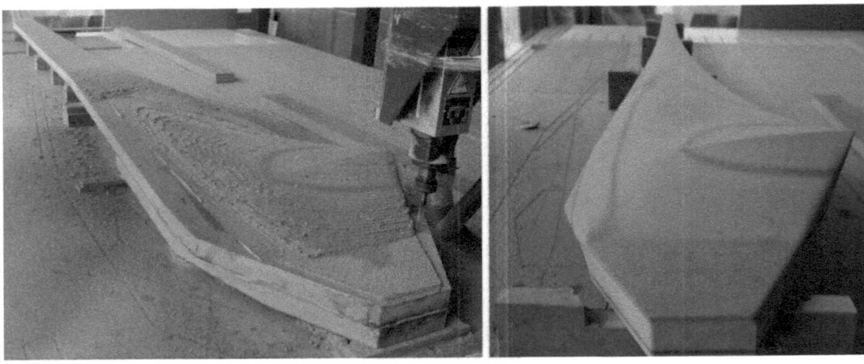

Fig. 5 Rough machining of the blade pressure side and finished wind turbine blade model

and maneuverable (easily accessed, opened and closed) to provide efficient ply arrangement. They must be accurately produced to ensure required aerodynamic performance and short blade assembly time. They should be of sufficient structural quality and integrity and require the least possible maintenance time, as well as high production speed. By producing moulds on NC machines many production mistakes and errors can be avoided [4].

This final example refers to a conventional, two-part, wooden mould, Fig. 6. Its average dimensions are 3.35 m × 0.7 m × 0.3 m. It is intended for wind turbines blades of 10 kW rated power.

The mould was produced in approximately 60 working hours. Again, leading and trailing edges were machined with special care, Fig. 7.

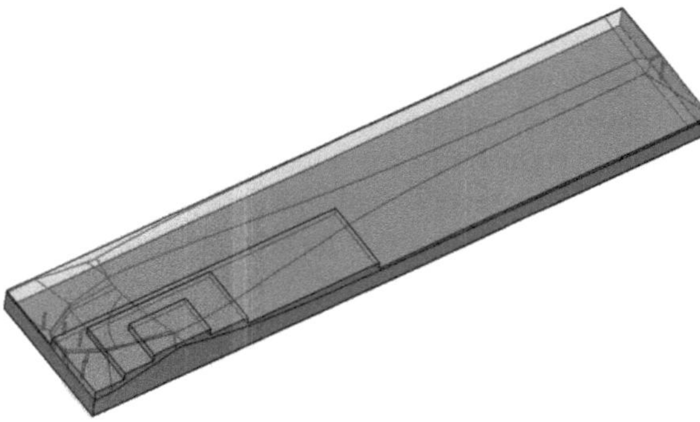

Fig. 6 CAD model of the wind turbine blade two-part mould

Fig. 7 Rough machining of the lower part of the mould and finished wind turbine blade mould

5 Conclusions

Presented manufacturing technologies have many advantages that could eventually lead to a significant increase in production capacity. Produced models are light and compact. Production processes are fast and material consumption is optimized. Greater geometrical accuracy and precision is achieved through the use of computer controlled machine resulting in high quality aircraft and wind turbine blades models and moulds. Both the initial investment and is the workload are significantly reduced. Less finishing work is needed.

Additional fields of application are: quick generation of models of increased quality and precision needed for further testing, achievement of high quality moulds, appropriate for various subsequent manufacturing processes, as well as fast and economic accomplishment of geometrically different models, necessary in academic research and education.

Acknowledgements The research work is funded by the Ministry of Science, Education and Technological Development of the Republic of Serbia through Technological Development Project no. 35035.

References

1. Pham DT, Gault RS (1998) A comparison of rapid prototyping technologies. Int J Mach Tools Manuf 38:1257–1287
2. Yan Y, Li S, Zhang R et al (2009) Rapid prototyping and manufacturing technology: principle, representative technics, applications and development trends. Tsinghua Sci Technol 14(S1): 1–12
3. Ding Y, Lan H, Hong J et al (2004) An integrated manufacturing system for rapid tooling based on rapid prototyping. Rob Comput-Integr Manuf 20:281–288
4. Stege J, Stiesdal H (2012) Direct production of a negative mold for producing wind turbine blades. DK Patent 20120181724, 19 July 2012